BIOETHICS IN A EUROPEAN PERSPECTIVE

INTERNATIONAL LIBRARY OF ETHICS, LAW, AND THE NEW MEDICINE

VOLUME 8

BIOETHICS IN A EUROPEAN PERSPECTIVE

Edited by

HENK TEN HAVE

and

BERT GORDIJN

Department of Ethics, Philosophy and History of Medicine,
University Medical Centre Nijmegen, Nijmegen, The Netherlands

SPRINGER-SCIENCE+BUSINESS MEDIA, B.V.

A C.I.P. Catalogue record for this book is available from the Library of Congress.

ISBN 978-90-481-5872-0 ISBN 978-94-015-9706-7 (eBook)
DOI 10.1007/978-94-015-9706-7

Printed on acid-free paper

TABLE OF CONTENTS

PREFACE

Writing, completing and producing a book often is a long-term project. The first ideas to develop this book were discussed at a time (1991) when the Department of Ethics, Philosophy and History of Medicine was established at the Faculty of Medical Sciences of the University of Nijmegen, the Netherlands. Like in other universities, the department has emerged because three separate units have been combined. Willing to explore the potentials of all three disciplines, plans have been made to concentrate on ethics in health care but with a major emphasis on the philosophical background and historical development of moral notions and ethical problems. Staff and faculty of the department collaborated in this effort to create and elaborate such a broader approach to bioethics. This textbook is the result of this ongoing effort.

In 1992, the Department initiated an annual European Bioethics Seminar. This five days intensive programme aimed at providing an encompassing introduction into the major approaches and topics of bioethics. However, the focus also was on typically European issues and perspectives in present-day bioethics. Over the years, many colleagues have contributed as teachers in these seminars, lecturing on important topics, and as moderators in small group sessions, discussing clinical cases and analysing practical problems. Approximately two hundred students have participated, from different countries. Teachers as well as students have benefitted from the exchange of cultural experiences and moral views. Reading materials have been developed in these seminars, and these have been perfected during years of educational use. These materials finally have been brought together in this book. The annual seminar, initiated through the organisational talents of Jos Welie, have since 2000 been incorporated into a more extensive educational programme, the European Master in Bioethics. This programme offers 16 courses on all major subjects of bioethics. It is a cooperative endeavour of the universities of Nijmegen, Leuven, Padova, Madrid and Basel.

I am of course indebted to many colleagues who have contributed to the annual seminars and are now cooperating in the Masters programme. Most of them have graciously contributed to this textbook. I would also like to acknowledge the assistance of several colleagues who provided information about bioethics resources in their countries, in particular Bela Blasszauer (Hungary), Pierre Boite (France), Eugenijus Gefenas (Lithuania), Jozef Glasa (Slovak Republic), Sefik Gorkey (Turkey), Lennart Nordenfelt (Sweden), Renzo Pegoraro (Italy), Zbigniew Szawarski (Poland), Pavel Tischenko

(Russia), Lora Vidic (Slovenia), Joachim Widder (Austria), Henrik Wulff (Denmark). The finishing touch of the materials included in this book was possible because the Center for Health Policy and Ethics, Creighton University, Omaha (USA) awarded me a senior visiting fellowship. The academic atmosphere and tranquillity, as well as the superb infrastructure of one of the most beautiful ethics buildings in the world, provided the opportunity to complete the whole manuscript. I am particularly grateful to Ruth Purtilo, Director of the Center for Health Policy and Ethics, for hospitality, encouragement and friendship. Writing of a book is impossible without adequate secretarial and administrative support within the enterprise as whole. I want to thank Valesca Hulsman for the editing and revision of the language, and in particular Marian Poulissen for her unstinting supply of assistance throughout this and numerous other pieces of work.

Nijmegen
July 2001

Henk ten Have

CONTRIBUTORS

BELA BLASSZAUER

Lawyer and bioethicist. He worked as a teacher of medical ethics at the Medical University of Pecs, Pecs, Hungary. He is President of the Central and East European Association for Bioethics.

RUTH CHADWICK

Philosopher, and Professor of Bioethics. Institute for Environment, Philosophy and Public Policy, Fumess College, Lancaster University, Lancaster, UK

WIM DEKKERS

Physician and Philosopher, Department of Ethics, Philosophy and History of Medicine, University Medical Centre, Nijmegen, the Netherlands

WILLIAM ELLOS

Bioethicist, Clinical Professor, Department of Medicine, University of Texas Health Science Center, San Antonio, Texas, U.S.A.

MARTYN EVANS

Philosopher, and Director of Centre for Philosophy and Health Care, School of Health Science, University of Wales Swansea, Swansea, Wales, United Kingdom

EUGENIJUS GEFENAS

Psychiatrist and Philosopher, and Associate Professor, Medical Faculty, Institute of Philosophy and Sociology, University of Vilnius, Vilnius, Lithuania

BERT GORDIJN

Philosopher and Clinical Ethicist, Department of Ethics, Philosophy and History of Medicine, University Medical Centre, Nijmegen, the Netherlands

DIEGO GRACIA
Psychiatrist and Philosopher, and Professor and Director, Department of Public Health and History of Science, Complutense University, Madrid, Spain; he is also Director of the Institute of Bioethics, Health Sciences Foundation, Madrid, Spain.

HENK TEN HAVE
Physician and Philosopher, and Professor and Head of Department, Department of Ethics, Philosophy and History of Medicine, University Medical Center, Nijmegen, the Netherlands

FRANZ JOSEF ILLHARDT
Theologian and Bioethicist, Executive Manager of the Research Ethics Committee in the Centre of Ethics and Law, University Hospital Freiburg, Freiburg, Germany

RIEN JANSSENS
Moral theologian, Department of Ethics, Philosophy and History of Medicine, University Medical Center, Nijmegen, the Netherlands

RITA KIELSTEIN
Nephrologist, Department of Renal Diseases, Medical Faculty, Otto von Guericke University, Magdeburg, Germany

CHRISTOPHER MELLEY
Philosopher, University of Maryland, European Division, Heidelberg, Germany

ROBERTO MORDACCI
Philosopher, and Associate Professor of Philosophical Anthropology and Bioethics, University Vita-Salute San Raffaele, Milan, Italy

LINDA SCHEIRTON
Assistant Dean for Academic Affairs, School of Pharmacy and Allied Health Professions; Faculty Associate, Centre for Health Policy and Ethics; Associate Professor of Occupational Therapy; Associate Clinical Professor, School of Dentistry, Department of Periodontology, Creighton University, Omaha, Nebraska, U.S.A.

PAUL SCHOTSMANS

Moral theologian, and Professor and Director of Department, Centre of Biomedical Ethics and Law, School of Medicine, Catholic University of Louvain, Louvain, Belgium. He is President of the European Association of Centres of Medical Ethics.

ZBIGNIEW SZAWARSKI

Philosopher, Professor of Philosophy, Department of Philosophy, Warsaw University, Warszawa, Poland

MARCEL VERWEIJ

Philosopher, Centre for Bioethics and Health Law, University of Utrecht, Utrecht, the Netherlands

HUB ZWART

Philosopher and Psychologist, and Professor and Head of Department, Department of Philosophy and History, Faculty of Science, Mathematics and Informatics, Nijmegen, the Netherlands

HENK A.M.J. TEN HAVE

INTRODUCTION

Bioethics and European Traditions

1. THE EMERGENCE OF BIOETHICS

The evolution of bioethics into an autonomous discipline is an unprecedented story of success. Not only the moral problems of medicine and health care have burgeoned over the last three decades, but also the preferred methods and concepts to scientifically approach these problems have been moulded into a separate discipline. Two developments are usually regarded as determinative for the rise of bioethics. First, the enormous advances in biotechnology, molecular biochemistry and pharmacology have led to drastic changes in medical knowledge and practice. Many people are not aware how relatively novel many benefits of present-day medicine actually are. Medical students, for example, are surprised to find out that it was only from 1960 onwards that the major medical journals began to describe experiences with fibre-glass endoscopy, coronary contrast radiology, artificial lens implantation, cardioresuscitation technique, and oral contraceptives. New and more effective diagnostic and therapeutic interventions have increasingly called into question the usual goals of medicine. The second development that has transformed the traditional notion of medical ethics is the changing socio-cultural context of medical practice. Not only a plurality of values has emerged, but also a non-religiously, secularly-grounded normative view of human life has become more influential. This view emphasizes personal autonomy and each patient's right to make his or her own health care decisions.

Whatever the precise determinants, traditional medical ethics has evolved rapidly into bioethics. 'Medical ethics' used to refer to the deontology of the medical profession. In this perspective, it is a system of moral rules, rules of etiquette and rules for professional conduct. This system is immediately intelligible to the medical practitioner, since it emerges from the internal morality of medicine, those values, norms and rules intrinsic to the actual practice of health care. This system is also endorsed with the entrance of new

H.A.M.J. ten Have and B. Gordijn (eds.), Bioethics in a European Perspective, 1-11.

practitioners into the medical profession. Being a physician implies the acceptance of particular moral views. Medicine is not considered a merely technical enterprise that can be morally evaluated from some exogenous standpoint. Rather, the professional practice of medicine always presumes and implies a moral perspective or point of view; good clinical practice is determined by the shared rules and standard procedures of the profession.

The emergence of modern bioethics and the concomitant atrophy of medical ethics in the traditional sense, is visible in at least three ways. It has first of all, produced a new expert, the 'bioethicist' or the health care ethicist. This professional has a specific body of knowledge and particular cognitive skills; he or she publishes in specialized journals, participates in new societies, and teaches in special centres, institutes, and departments. Second, it has produced an ever growing public interest in moral issues in health care; it has also promoted a lively public debate on all sorts of problems, cases, dilemmas that arise from new developments in medical science and medical technology. Whenever issues of reproduction, education, relationships, sexuality, suffering, handicaps lead to public concerns, ethicists are public spokespersons to analyze, explain and resolve these matters. But also the public itself are increasingly engaged in re-thinking the bioethical dimensions of present-day health care. Third, it has led to certain institutional changes: bioethics laws, regulations, statutes, review boards, codes of conduct, practice guidelines. Most prominent has been the establishment of ethics committees; these are institutional platforms for moral debate, involving more professionals than only medical ones, and sometimes also lay-persons. Such committees at least transform the formerly private character of moral deliberation from the context of the doctor-patient relationship into an inter-professional and inter-personal debate over moral matters, with the intention to reach consensus. 'Bioethics' has come to be regarded as the major public vehicle to address, explicate and give meaning to a broad range of problems generated by science and technology.

The above phenomena are obviously not specifically linked to the European or North-American context. They point to developments and characteristics of Western culture in general. In his recent study of the birth of bioethics, Jonsen (1998) interprets bioethics as a quintessential American phenomenon. Although much can be said about peculiar features of present-day ethical debate in relation to American cultures and traditions, this view neglects the more fundamental background of bioethics. At the same time, it would be a mistake not to recognize the particular cultural articulation of bioethics. The dominant conception of present-day bioethics has evolved from a North-American context. Inasmuch as it makes sense to reflect on the specific cultural dimensions of bioethics, it is also worthwhile to examine the question: What is typically European about bioethics?

2. EUROPEAN TRADITIONS

The concept of Europe has a long and diversified tradition. It is first of all a geographical term, first used in classical Greek civilisation around 700 BC. 'Europe' is the name of a Phoenician princess, abducted and raped by the father of the gods, Zeus, and taken to Crete. Perhaps with this mythical story the Greeks wanted to recognize their many cultural debts to Asia. At the same time, the Greeks wanted to explain how their civilisation had been shaped and why it was different. Europe then is not merely a distinct geographical entity, but rather a political and cultural concept. In historical perspective, Europe has become an idea. As a concept, an idea, Europe is difficult to define. Cultural studies aiming at elucidating the idea of Europe are necessarily broad (Rietbergen, 1998). In some cases, they simply identify core components of the idea, such as freedom, Christianity and civilisation (den Boer, 1997). These studies illustrate that the concept of Europe has shifting contours and varying characteristics in a long historical tradition. Nevertheless, the concept refers to an area with a relative unity because of similar ways of life and thinking. Although there are a number of traditions, together they constitute a coherent culture, a specific cultural sphere.

> 'Europe' is manifested outwardly as a relative unity. It partly legitimizes itself by pointing to certain economic and political choices and achievements that are said to imply moral choices as well. More importantly, it tries to defend certain values, the results of a rich cultural tradition (Rietbergen, 1998, p. 461).

This cultural sphere has been strongly influenced by the development of Christianity, by political changes as the French Revolution, by philosophical ideas about humanism and Enlightenment, by scientific and technological progress. Various catalogues of 'European' values have been proposed: freedom, tolerance, equal opportunity, social justice, human dignity, solidarity.

> Indeed, there seems to be a consensus in Europe that looks at man in society under a threefold aspect: each man is unique, each person has to make his own choices for good and evil but, first and foremost, being human means taking responsibility for others, that means, protection of others to preserve the quality of society at large (Rietbergen, 1998, p. 465).

However, even if it is possible to identify a common set of values, a continuous effort will be required in order to critically assess the actual meaning of these values as articulated and codified in the past, and to evaluate and rephrase the underlying traditions. The challenge is to constantly reaffirm the values which are deemed crucial for a European perspective.

3. BIOETHICS AND EUROPE

The search for specifically European perspectives in bioethics has started not long ago (Welie and ten Have, 1992). A series of developments have motivated the attempts to articulate these perspectives. At a political and economical level, the need to articulate the specific European identity *vis à vis* North America, Russia and Asian countries, intensified after 1989 when the political changes in Central and Eastern European countries started a period of transition and transformations all over the continent. Now that ideological barriers have fallen, new nations have come into existence, and the map of Europe has been changed. In a short timespan, the Council of Europe has almost doubled in size, with its membership growing from 23 states in 1989 to 40 in 1996. Along with the outbreak of nationalism, there were many efforts to reassert the common cultural values and to stress human rights and fundamental freedoms. Since 1989, the European Community, for example, has started intensive cooperation programmes with other countries in Europe, not only to support transition in economic and scientific terms, but also to enhance the protection of human rights and democracy, as well as to endorse particular values.

A significant development has been the gradual expansion of the European Union (EU). The EU will further expand in the near future and at the same time will move towards greater internal unity. The members of the European Union no longer exclusively focus on economical issues, aiming at harmonisation of their respective national legislations to achieve the common market. Since the Maastricht Treaty on the European Union (1992) public health has become an object of EU policy. It is explicitly stated that the EU will contribute to a high level of health protection, by promoting cooperation between member states, supporting and complementing their efforts where necessary but stopping short of harmonisation. This statement has consequences for example in the area of research. Whereas previous research programmes were primarily intended to strengthen the competitive position of the European industry, the Treaty calls for research that meets the public health needs of the European citizen. Since 1991 the European Union has also funded research in the area of biomedical ethics, creating networks of researchers in bioethics. In the period between 1994-1998, an amount of 27 million ECU was spent to fund research projects in bioethical matters.

Bioethics also moved into the foreground within Europe as an unavoidable consequence of the introduction of the single European Market in 1993 (Gillon, 1993). It has become clear that the increased economic and political unity has opened up new possibilities, some of which cannot simply be left to be dealt with by the separate member states. Pharmaceutical companies and research institutions nowadays have established themselves in many countries and if some project meets ethical concern in one state, it can simply be moved to another. Health insurance companies are operating as multinationals; the rules and regulations for health benefits are increasingly harmonized. With the intensive travel of people and emigration to other European countries due to increased freedom of employment, new questions have arisen, for example about the ownership of donor organs. Are organs donated by a Dutch national the property of Dutch people only? Or do they have at any rate more of a right to the organs? Should it be possible to do research on human embryos in one country when it is forbidden across the border? Should a woman over 50 years of age who does not qualify for IVF at home, be able to have it elsewhere in Europe? The question arises what concerning these ethical matters should be left to individual member states, and what should be dealt with at the supranational European level (Riis, 1993).

As a result of the above developments many ethical bodies have come into existence at a supranational level. In 1985, the Committee of Ministers of the Council of Europe decided that bioethical issues should be dealt with by a single specialized committee (Quintana, 1993). This committee obtained a permanent status in 1992 as the Steering Committee on Bioethics (Comité Directeur de Bioéthique, or CDBI). The purpose of the documents of the Committee is to reaffirm the major principles and values which must guide any regulation on bioethics and also indicate which limits must at all costs be respected. Since 1990, the Council of Europe has issued several important documents in the area of bioethics, for example on medical research on human beings, prenatal diagnosis, genetic screening, autopsy rules, DNA analysis for forensic purposes, and biotechnology (Rogers and Durand de Bousingen, 1995).

In 1991, the Commission of the European Union set up a Group of Advisers on the Ethical Implications of Biotechnology (GAEIB). This group of experts from various member states should help to identify and define ethical issues raised by biotechnology. It should also appraise the ethical aspects of the Union's activities in the field of biotechnology and their potential impact on society and the individual. It should finally advise the Commission on ethical aspects of biotechnology with a view to improving public understanding and acceptance of it. Recently, the GAEIB issued an Opinion on the ethical consequences of cloning (Advisers, 1997). In 1999, the name of the group was changed into 'European Group on Ethics in Science and New Technologies'.

A landmark in this process of articulating European perspectives in bioethics is the Convention on Human Rights and Biomedicine, developed by the Council of Europe and signed in 1997 by 21 member states in Oviedo, Spain (Dommel and Alexander, 1997). The objective of the Convention is to establish a number of general principles which will protect human rights in the changing context of medical practice (Council of Europe, 1996).

The Convention identifies basic principles, necessary for the application of medicine and the life sciences. It sets out to protect the dignity and identity of all human beings: "The interests and welfare of the human being shall prevail over the sole interest of society or science" (art. 2). It also requests that appropriate measures be taken to provide equitable access to health care of appropriate quality. The principle of respecting the free and informed consent of the person is clearly stressed. The issue of interventions on persons unable to provide an informed consent has been controversial for a long time. It has now been stated that interventions may be carried out on persons with impaired decision-making capacities, but only for their benefit, and only if minimal risks and minimal burdens are imposed. Consensus apparently exists over a broad range of issues. Privacy and free access to information are defined as rights. Discrimination against a person on the basis of his or her genetic heritage is prohibited. Sex selection in medically assisted procreation is prohibited (except to avoid serious hereditary, sex-related diseases). Financial gain from using the human body and its parts is explicitly prohibited.

The more controversial issues in bioethics are not addressed in this Convention, however. An exception is research on embryos in vitro. Article 18 states that "The creation of human embryos for research purposes is prohibited". Specific problems can be elaborated in special protocols, supplementing the Convention. Recently, nineteen countries signed a protocol to forbid any attempt to clone human beings (News, 1998).

The European Convention on Human Rights and Biomedicine is a well-intended and carefully prepared document that may stand as a landmark in the evolution of bioethics in Europe. It builds on the earlier foundations of the Universal Declaration of Human Rights as well as on the European Treaty for the Protection of Human Rights and Fundamental Freedoms. It identifies basic moral principles and moral procedures. For ethicists from North-western Europe the Convention could be disappointing, due to the general character of its formulations. But given the status of bioethics throughout Europe, in particular the wide variety of theories and practices, the lack of bioethics education in many countries, the embryonic state of many procedures and committees in hospitals, as well as the poor quality of public debate, this Convention could lead to more concerted approaches to enhance the sophistication of bioethics in all European nations.

At the same time, the Convention raises questions and problems (ten Have and Sass, 1998). Because its focus is on building a general frame of reference for public policy and international cooperation, it must be regarded primarily as a legal document, and therefore a milestone in health care law, rather than bioethics. However, it is undeniable that the text is clearly informed by ethical principles (de Wachter, 1997). A more fundamental question is whether it really is desirable that so many different ethical approaches are brought together in very general statements, suggesting some kind of European consensus on health care issues. The validity and truth of moral views cannot be decided by general agreement. Even if all European nations would agree on some set of ethical principles, these principles could still be all morally wrong.

Codification of moral principles also implies a specific view on ethics as a product of reflection. Ethics, and philosophy in general, is usually valued as a process of reflection, leading to various and specific products over time. What is most important is the continuous, critical thinking; the specific results of thought are themselves perpetuous occasions to re-start thinking and to develop new ideas and systems of thought. Although the products of philosophical thought (mainly in the form of books and other publications) catch our attention, philosophy itself is not the reproduction of such products, but the process of thinking and rethinking the notions, views and ideas that are precipitated in the works and products of other philosophers.

4. EUROPEAN APPROACHES TO BIOETHICS

If this interpretation of philosophy makes sense, then we should again ask the question whether there are specific European approaches to bioethics, not as specific products, but rather as typical processes. Also here, it is problematic to identify typically European perspectives. Continental philosophy is typified as an amazing variety of philosophical theories and methods without any major and dominating school. The same is true for ethics. Here, many schools and approaches flourish in practice and literature: applied ethics, hermeneutical ethics, casuistry, clinical ethics, narrative ethics, care ethics. But this situation seems not too different from North America. There, also a variety of approaches exists. Although there are many similarities, there is, nonetheless, truth in the proposition that American bioethics has predominantly been developed around a common methodological structure and a particular set of ethical principles. In European approaches the spectre of conceptions in medical ethics seems broader than in the U.S. On the one hand, in many European countries, ethics is very much under the influence of philosophical and theological traditions, and multifaceted in philosophical substance, - not dominated by analytical philosophy. On the other hand, only in a very few countries, such as the United Kingdom, the Netherlands and the Nordic countries, medical ethics in particular is the specialized enterprise of a new profession; most often it is the recognized business of medical practitioners or lawyers, who therefore dominate public debate. This is, presumably, also one of the reasons why the term 'bioethics' is not as frequently used as 'medical ethics' or 'health care ethics'. Sometimes, there even is reluctance to use the term 'bioethics' at all. One reason is that 'bioethics' often is identified not with a discipline of moral philosophy or moral theology, but with a specific approach of moral problems. Sometimes, such identification leads to negative responses, accusing 'bioethicists' of facilitating medical technologies and attempting to soften moral resistance against innovations. It is also argued that bioethics, as a typically American approach, is onesidedly emphasizing individual autonomy, implying an underdevelopment of beneficence and justice which are concepts more characteristic of European approaches (Holm, 1995).

When we try to identify what is typical of European approaches to bioethics, we will notice that European literature in the area of bioethics tends to put more emphasis on (1) the historical background of ethical issues, (2) the sociocultural context, and (3) substantive normative viewpoints.

European authors tend to locate present-day moral problems within the context of historical developments. They emphasize that better understanding of the current problems will follow from a thorough analysis of their evolution. In their view, the present-day interest in medical ethics should not be regarded as completely new, creating a really new situation for humankind. Bioethics today should be regarded as the latest phase in a tradition of theoretical reflection upon medicine. Without, for example, a long phase of anthropological reflection within medicine itself, it would have been impossible to become aware of a new kind of subjectivity of the patient as well as the health care professional. Many scholarly studies on the doctor-patient relationship pointed out the basic human nature of this relationship (see, for example: Lain, 1969). They more or less disclosed the moral dimension of medicine for public reflection, because they showed that medicine itself is a normative science of life. They also introduced a secular, 'humanistic' discourse to clarify and debate the moral dimensions of health care.

Secondly, European authors tend to emphasize the social and cultural context of many ethical debates. They are focusing attention on the structure and organisation of the health care system, as well as the network of social values in which the moral problems are presented. They criticize the individualistic focus of dominant bioethical discourses and the relative negligence of community values, interpersonal relationships and solidarity. Individualist ethics in their view should be complemented with social ethics. In his analysis of the differences between the European and American tradition, Wulff (1998) criticizes the latter because of its emphasis on personal autonomy and individual rights. When rights are given priority, primary ethical concerns have been neglected; a medical ethics that is discussing the respective rights of patients and doctors, has already missed the point because discussing these rights implies that the basic qualities of the doctor-patient relationship such as mutual trust and the doctor's concern for his patient as a fellow human being, have been lost. Relations analyzed in terms of rights have lost core qualities like cooperation, friendship and kindness. Instead of dealing with medical ethical issues in a quasi-legalistic manner, Wulff wants the moral debate to focus on ideas of reciprocity, mutual obligations, social justice, relationships between individuals:

> Human beings ... depend both on each other and on their environment, and a realistic appraisal of the present predicament of mankind may yet force us to emphasize more strongly the acceptance of mutual obligations for the common good (Wulff, 1998, p. 71)

The sociocultural context is also considered important for both the perception and management of moral issues in medicine. Certain issues do not seem to be located on the bioethical agenda. From an individualistic perspective, focusing on individual choice and respect for autonomy, issues such as homelessness, drug addiction or insurability do not seem to raise basic moral questions. However, from a social perspective, they call into question fundamental assumptions concerning the moral quality of communities and social responsibility.

Third, European approaches are at least as pluralistic as North American approaches. However, it seems that they are less directed at developing a procedural ethics as the privileged solution to moral controversies. One dimension is that there exists a rather broad consensus regarding fundamental principles of 'European bioethics': respect for human dignity, protection of individual integrity, public responsibility regarding the application of biomedical sciences, prohibition of all commercial agreements concerning the human body and its organs (Byk, 1993), but also universal availability and equal accessibility of health care for every citizen (Holm, 1992). Such principles are not well-ordered within a coherent hierarchical framework and application rules.

Another dimension is that diversity of opinion in moral matters is not taken as a problem that must be resolved, but as a source of inspiration to further articulate the perennial tension between universality and particularity. Diversity of opinion and method does not mean discrediting bioethics; it is a stimulus to involve more and more participants in the debate about the moral dimensions of health care. The more heterogeneity, the more motive to attempt to define some grounds of common understanding.

Both dimensions encourage the affirmation of substantive moral positions.

5. THIS VOLUME

The contributions to this volume have developed from a common educational experience. Since 1992, many authors have been cooperating in the designing and teaching of an intensive bioethics course. This course, taught annually in Nijmegen, the Netherlands, offered a week-long introduction into the major issues of contemporary bioethics. It also aimed at explicitly developing European perspectives on bioethics. Over the years, many scholars from various professional, religious and national backgrounds have attended this course. Especially the diversity and plurality of viewpoints have been instructive for teachers as well as students. The interest in articulating European perspectives, in analyzing various theoretical arguments, and comparing different health care practices and professional experiences have stimulated the production of written materials.

The contents of this book follow the format of our teaching experiences. The authors are from different (although not exclusively) European countries. They have made two types of contributions: chapters and case analyses. The first are scholarly expositions of the state of the art in bioethical thinking on particular subjects. The second are examples of practical applications to specific problems and cases from health care. The book comprises five major parts, focusing respectively on the foundations and history of bioethics, on issues relating to the human person and his or her body, on the relationship of person and community, on moral issues regarding the beginning of human life, and finally on moral problems related to the end of human life. The closing section presents practical information about the discipline of bioethics, such as journals, textbooks and learned societies. All contributions together hope to provide reflective and critical expositions and interesting information concerning European perspectives in present-day bioethics.

REFERENCES

Advisers to the President of the European Commission on the ethical implications of biotechnology: 1997, Ethical aspects of cloning techniques. *Journal of Medical Ethics* 23: 349-352.

Boer, P. den: 1997, *Europa. De geschiedenis van een idee*. Prometheus, Amsterdam.

Byk, C.: 1993, The European Convention on bioethics. *Journal of Medical Ethics* 19: 13-16.

Council of Europe: 1996, *Convention for the protection of human rights and dignity of the human being with regard to the application of biology and medicine: Convention on human rights and biomedicine*. Directorate of Legal Affairs, Strasbourg.

Dommel, F.W. and Alexander, D.: 1997, The convention on human rights and biomedicine of the Council of Europe. *Kennedy Institute of Ethics Journal* 7(3): 259-276.

Gillon, R.: 1993, Biomedical ethics in Europe - a need for the POBS? *Journal of Medical Ethics* 19: 3-4.

Have, H.A.M.J. ten and Sass, H.-M.: 1998, Introduction: Consensus formation in health care ethics. In: Have, H.A.M.J. ten and Sass, H.-M. (eds.): *Consensus formation in health care ethics*. Kluwer Academic Publishers, Dordrecht, p. 1-14.

Holm, S.: 1992, A common ethics for a common market? *British Medical Journal* 304: 434-436.

Holm, S.: 1995, Not just autonomy - the principles of American biomedical ethics. *Journal of Medical Ethics* 21: 332-338.

Jonsen, A.: 1998, *The birth of bioethics*. Oxford University Press, New York/Oxford.

Lain Entralgo, P.: 1969, *El médico y el enfermo*. Ediciones Guadarrama, Madrid.

News: 1998, Europe moves to ban human cloning. *Bulletin of Medical Ethics*, nr. 134, p. 3-5.

Quintana, O.: 1993, International bioethics? The role of the Council of Europe. *Journal of Medical Ethics* 19: 5-6.

Rietbergen, P.: 1998, *Europe. A cultural history*. Routledge, London and New York.

Riis, P.: 1993, Medical ethics in the European Community. *Journal of Medical Ethics* 19: 7-12.

Rogers, A. and Durand de Bousingen, D.: 1995, *Bioethics in Europe*. Council of Europe Press, Strasbourg.

Wachter, M.A.M.: 1997, The European Convention on bioethics. *Hastings Center Report* 27(1): 13-23.

Welie, J.V.M. and Have, H.A.M.J. ten: 1992, Bioethics in a supranational European context. In: Lustig, B.A., Brody, B.A., Engelhardt, H.T. and McCullough, L.B. (eds.): *Bioethics Yearbook. Volume 2: Regional developments in bioethics: 1989-1991*. Kluwer Academic Publishers, Dordrecht, p. 97-126.

Wulff, H.R.: 1998, Contemporary trends in health care ethics. In: Have, H.A.M.J. ten and Sass, H.-M. (eds.): *Consensus formation in health care ethics*. Kluwer Academic Publishers, Dordrecht, p. 63-72.

PART 1

FOUNDATIONS AND HISTORY OF BIOETHICS

HENK A.M.J. TEN HAVE

FOUNDATIONS AND HISTORY OF BIOETHICS

Introductory Comments

In *Danubio* (1986), a cultural history of Central Europe, the Italian writer Claudio Magris explains how he started the research for the book. His idea was to follow the course of the Danube, one of the longest rivers in Europe, and then narrate the history and culture of all civilisations and peoples along the stream. However, this task turned out to be more difficult than expected. One of the puzzles was at the beginning: where exactly was the source of the river? When you follow the river upstream, somewhere somehow the first signs of water, however minimal they may be, indicate that you have found the river's source. This assumption, however, was too simple. For centuries two small towns in the Black Forest have both claimed to be the location where the Danube originates. Is the starting-point of what we call 'the Danube' the confluence of two brooklets or is one of these the original stream?

The story of Magris exemplifies that ambiguity is a significant dimension of our world. We used to have particular expectations and assumptions concerning the world around us, but upon further reflection, only few issues are immediately clear and self-evident. More often than not, reality requires interpretation.

Humankind has developed diverse strategies to deal with the ambiguity of the world. Through tradition, custom, and routine we try to reduce and manage the possible uncertainties and ambivalences that can confront us in daily life. Scientific research, legal arrangements, social agreements are other ways to master reality's ambiguity. Likewise, literature, poetry, music, and arts in general, provide possible interpretations of our world that can yield clarity and certainty. Last but not least, ethics is one of our cultural instruments to attenuate the above ambiguity. Unlike the arts, the approach of ethics is more rational, scientific, methodical. Unlike the sciences, the modus operandi of ethics is more evaluative, qualitative, engaged.

The history and theoretical models of bioethics are the central concern of the first part of this book.

H.A.M.J. ten Have and B. Gordijn (eds.), Bioethics in a European Perspective, 15-16.

Diego Gracia (Spain) examines the roots of bioethics as a present-day phenomenon. At first sight, bioethics does not have a long history, but ethical concerns have always been part of medical practice. Gracia explores the reasons for the current interest in bioethics, showing how in fact the ethical principles of beneficence, autonomy, and justice have a long tradition in the history of health care, law and politics. The convergence of the three principles created the arena in which new moral conflicts could arise and a new discipline could emerge.

In his contribution, Henk ten Have (the Netherlands) analyzes the various theoretical approaches to ethics. Currently, bioethics is dominated by the applied ethics approach or principlism. The principles discussed in Gracia's historical exposition have been elaborated into a coherent model that is extremely useful to clarify and resolve practical problems in the health care setting. Nonetheless, the dominance of this model is criticized and alternative models and approaches are now fashionable in bioethical discourse. Ten Have also explores the question whether there are typically European approaches or issues in bioethics. He finally suggests that bioethics should primarily be regarded as an interpretive discipline.

Two case analyses follow the above chapters. The objective of these analyses is to illustrate and elaborate some of the core notions discussed in the chapters within the context of practical health care. Usually, they will apply some of the theoretical ideas and argumentation to particular cases or specific problems. The first case, presented by Martyn Evans (United Kingdom), addresses the difficulties of respecting the autonomy of the patient. The question is discussed whether and how a person with high blood pressure can freely consent to long-term medication with potential side-effects. Evans scrutinizes the conditions for an autonomous choice: is it informed, free and intended? Evans shows that in practical circumstances these conditions are easily compromised, and therefore in need of continuous re-evaluation.

The second case calls attention to the health care professional. As a counterpart of patient's autonomy, it is often argued that it is inherent in the internal morality of health care that the health care provider has professional autonomy. But what does this autonomy imply? Linda Scheirton (U.S.A.) uses the example of advertising to examine two perspectives on the physician-patient relationship: it can be viewed as a commercial or as a professional interaction. Considered superficially, advertising is a way of providing information to patients and the general public. Upon further consideration, advertisements are easily misleading and biased because their primary intention is to promote business rather than to increase patient autonomy. Most European codes of ethics endorse a restrictive policy as regards advertisements by health care professionals.

DIEGO GRACIA

HISTORY OF MEDICAL ETHICS

1. THE WORD AND ITS MEANING

The term 'bioethics' was used for the first time by Potter in 1970 and has completed its first twenty-eight years of existence (Potter, 1970, 1971). This term is coming of age, which explains why it is beginning to have a sufficiently mature and systematic doctrine behind it.

Warren Reich has recently analyzed the history of the term 'bioethics' and the beginning of this movement (Reich, 1993, 1994, 1995b). His thesis is that bioethics had a 'bilocated birth'. One of the birthplaces was Madison, Wisconsin, where Van Rensselaer Potter, the person who coined the word, was living and working, and the other Washington, D.C., where André Hellegers used this word for the first time in an institutional way to designate the focused area of inquiry that became an academic field of learning and a movement regarding public policy and the life sciences. The different backgrounds of these two pioneers, biology in the case of Potter, and medicine in Hellegers, caused the content of this new discipline to be oriented differently. Hellegers understood bioethics as a new way of approaching and resolving the moral conflicts raised by the new medicine, whilst Potter's view was much more comprehensive and global, as the moral analysis of the equilibrium of life over the earth, and the present and future of life and quality of life. Reich has shown that Hellegers in fact proposed a global approach to bioethics, bringing his vision much closer to Potter's evolving view than what previously was acknowledged. In any case, immediately after its beginning, bioethics became little by little a medical discipline, specially focused on the ethical problems raised at the beginning and at the end of human life. Today, the dominant model is widely professionalised and medicalized, and Potter's approach remains clearly marginalized. This is the reason why bioethics is today used to designate the ethical problems raised by medicine and health care professions. This permits understanding why the discipline Bioethics is also called today 'Biomedical Ethics' (Beauchamp and Childress, 1979) and 'Medical Ethics'.

H.A.M.J. ten Have and B. Gordijn (eds.), Bioethics in a European Perspective, 17-50.

This means that bioethics has had in its short history at least two different meanings, and one wider than the other. The first and wide sense of bioethics understands it as a general approach analyzing all kinds of ethical problems. We are in a moment of the history of mankind in which life, the present and the future of life in general, and of human life in particular, is at stake. This is perhaps the most important characteristic of the ethical situation of our societies at the end of this second millennium. All other ethical problems are finally reduced in our ethical debates to the problem of the sustainability of life and the quality of life. If in the sixties the great ethical debate was always the confrontation between East and West, the liberalism of the First World and the socialism of the Second, today the great confrontation is between North and South, the unsustainable overdevelopment of the First World and the also unsustainable underdevelopment of the Third World. Therefore, the need for a real convergence around a third model, the so-called sustainable development, in order to preserve life, quality of life, now and in the future. Bioethics is, therefore, a new mentality, a new sensibility of respecting things, either natural or human. Human beings are ends in themselves, as Kant said, but natural things are also ends in some way, and not only means, because nature is the necessary context for humanity. Human beings are not angels but animals, and only in the interior of the ecological equilibrium of nature and life can they develop their own lives. This is the new mentality promoted by the bioethical movement. It is a new way of thinking and working, a new form of living, re-sacralizing reality and therefore promoting a new ethics of respect not only for human beings but also for life in general, and for nature. This was the original idea of Potter. Therefore, bioethics cannot be identified with or mistaken for some kind of professional ethics, like for example the ethics of the health care professionals, but instead it must be considered to be the general or civil ethics of our societies at the end of our century and millennium. There is not a particular chapter of ethics but a particular perspective or a new approach when analyzing all ethical problems. There are, therefore, many kinds of bioethics. There is an ecological bioethics, a biological bioethics, a medical bioethics, etc. A good convention could be using the substantive 'bioethics' always beside a determinative adjective, as 'global', 'ecological', 'medical', 'clinical', etc. In his last book, Potter proposes the title 'Global Bioethics' to designate the entire field of bioethics, distinguishing in its structure different parts, which he calls 'Medical Ethics', 'Ecological Ethics', etc. (Potter, 1988, 1990, 1995, 1998). In the following, I will restrict my analysis to the history of 'clinical or medical bioethics', focusing, therefore, the analysis on the historical development of bioethics as a health care discipline.

2. REASONS FOR THE BIRTH OF CLINICAL BIOETHICS AS A DISCIPLINE

The practice of medicine has always raised ethical issues and required that the doctor possess high moral standards. The uninterrupted series of deontological documents that sprinkle the history of Western medicine from the Hippocratic oath to today demonstrate this point. Without question, no other epoch in medicine has faced such moral issues as does ours today. Now more than ever, health professionals require adequate ethical training. This explains why publications on medical and clinical ethics have grown exponentially in recent years. Such growth has given rise to the elaboration of a large body of doctrines, which today are indispensable to the training of good doctors. There are three different reasons for this change.

2.1. The Emancipation of the Patient

The first reason has to do with the changes that have occurred in recent years in the physician-patient relationship. Classical medical ethics was based on the principle that disease not only altered the individual's physical equilibrium, it altered mental and moral equilibrium as well. Pain alters a man's capacity of judgement to such a point, as Aristotle has already said, that it prevents him from making wise decisions (Aristotle, 1960b, 1104b13-28; 1140b13-19). For this reason, the primary virtue (and almost the only one required) of the patient is obedience. The physician-patient relationship is by its very nature asymmetric and vertical since a physician's function is to command, and a patient's function is to obey. According to the classical stereotype, a good patient was always passive, neither asked questions nor protested, and established a relationship with the physician that was extremely similar to that of a boy and his father. Classical medical relationships were thus paternalistic. In the physician-patient relationship, the physician assumed the role of a father seeking what was best for the patient, without, however, taking the patient's own will and desires into account. This explains why for a long time in the history of medicine it was possible to find a vast output of ethical codes for physicians, but not for patients (Gracia, 1989a, p. 23-120). Ethical codes for patients only came into existence twenty-eight years ago. It was only at the beginning of the 1970s that the physician-patient relationship began to be interpreted as a process of relations between adults, each one autonomous and responsible. In this manner, the patient ceased to be treated like an infant, as was previously the norm in the classical model. Today it is said that the more medical relations are conducted as adults the more mature they will become. To achieve this goal, it was necessary to stop treating the patient as a child. The patients themselves who wanted this type of relation, demanded also that a body of rights be respected (which starting from 1972 gave rise to codes of patients' rights). The most important of these rights was that of informed consent. In the physician-patient relationship, the physician has the technical information and the patient has the capacity to consent or decide; neither can operate without the other and each is necessary. Every medical act is a negotiation and explanation process between the two parts, just as occurs in other aspects of human life, such as in marriage, in the family, in social and political life, etc. In all these areas of human life, relations went from vertical and compulsory to horizontal and participatory. Medicine was no exception to this trend.

This process made relations more mature, but also increased conflict within them. It is however a phenomenon that has occurred in all other areas of human relations. Such relations were made unstable, but at the same time they gained in maturity. Indeed, it is not even necessary that these relations be stable in order to be human or ethical. On the contrary, what very often occurs is that the human equilibrium is always unstable and requires a continuous process of adjustment (Gracia, 1989a, p. 121-198).

2.2. The Coming of New Biotechnologies

Another factor that caused a multitude of problems and ethical conflicts is the enormous progress of medical technology in the last forty years. Starting from the 1960s, the development of various substitution procedures for organic functions considered vital (dialysis and kidney transplant in the case of kidney function, artificial respiration, resuscitation techniques, defibrillation, parenteral nutrition, etc.) has allowed the medicalisation of the last phase of human life (unthinkable until a short time ago) and even a revision of the definition of death. The concept of brain death today allows people whose heart still beats and who, according to classical definition, are alive, to be diagnosed of death. On the other hand, the concentration of new, life-saving technologies in some hospital services (known since the 1970s as Intensive Care Units) raises a new group of ethical issues: who will and who will not be allowed this intensive care? When can the respirator be withdrawn? Which patients should be resuscitated and which should not? In all of these issues, should medical criteria alone be taken into account, or should the patients' desires be taken into account as well? What role in decision-making do the following actors have: the relatives, the insurance companies, and the state?

Even more extraordinary than technologies at the end of life are those that permit the manipulation of the beginning of life. These technologies include genetic engineering, artificial insemination, in vitro fertilisation, embryo transfer, prenatal diagnosis, and so on. What ethical principles should guide medical practice in these areas? How can we determine what is moral and what is immoral in a pluralist society such as ours in which it is not easy for people to agree on concepts of good and evil? The list of questions could easily be added to.

2.3. Medicine as Social Organisation

The third area of ethical issues regards equal access to health services and equitable distribution of limited and scarce economic resources. In our era, and for the first time in history, every citizen's access to health care has been promoted. It seems that the very idea of justice demands that all individuals satisfy necessities that are as fundamental as medical care. Well, what are medical needs? How can we differentiate in health matters between the necessary and the superfluous? Economics assure us that in the health field, supply creates its own demand and thus the consumption of health goods is practically unlimited. Is there a moral obligation to satisfy these growing needs in light of the principle of justice? How can rational limits be established? Given that in the realm of health the needs will always be greater than the available resources, what criteria should be used for the distribution of scarce resources?

The convergence of these three types of factors has meant that current medicine is completely distinct from that of any previous period. It can be confirmed without question that the physician-patient relationship has changed more in the past thirty years than in the past thirty centuries, that is, from the beginnings of Western medicine up to the 1960s. This has meant that medical ethics is now more important than it has been at any other time. A new discipline had to be created that no longer identifies with traditional professional deontology. For this reason, the previous name was discarded and the term bioethics appeared (Gracia, 1989b).

The discovery of these problems led in the mid seventies to the identification of three basic ethical principles, namely respect for persons, beneficence, and equity. This identification was made in the U.S.A. by the National Commission for the Protection of Human Subjects of Biomedical and Behavioural Sciences between 1974 and 1978. Its final report, known as the Belmont Report, established the system of three principles, which have been the common language of bioethics from its beginning until now (Jonsen, 1998, p. 103f). These principles were defined in the following terms:

> 1. Respect for Persons.
> An autonomous person is an individual capable of deliberation about personal goals and of acting under the direction of such deliberation. To respect autonomy is to give weight to autonomous persons' considered opinions and choices while refraining from obstructing their actions unless they are clearly detrimental to others. To show lack of respect for an autonomous agent is to repudiate that person's considered judgements, to deny an individual the freedom to act on those considered judgements, or to

withhold information necessary to make a considered judgement, when there are no compelling reasons to do so.

2. Beneficence.
Persons are treated in an ethical manner not only by respecting their decisions and protecting them from harm, but also by making efforts to secure their well-being. Such treatment falls under the principle of beneficence. The term 'beneficence' is often understood to cover acts of kindness or charity that go beyond strict obligation. In this document, beneficence is understood in a stronger sense, as an obligation. Two general rules have been formulated as complementary expressions of beneficent actions in this sense: (1) do not harm and (2) maximise possible benefits and minimise possible harms.

3. Justice.
Who ought to receive the benefits of research and bear its burdens? This is a question of justice, in the sense of 'fairness in distribution' or 'what is deserved'. An injustice occurs when some benefit to which a person is entitled is denied without a good reason or when some burden is imposed unduly. Another way of conceiving the principle of justice is that equals ought to be treated equally (Jonsen, Veatch, and Walters, 1998, p. 23f).

The Belmont Report not only defined the three ethical principles implied in modern biology and medicine, but also the most important ways to apply each one of them. The main application of the principle of respect for persons is informed consent, that of the principle of beneficence, the assessment of risks and benefits, and the practical consequence of the ethical principle of equity is the equitable selection of subjects. The Belmont Report describes them in the following terms.

1. Informed consent.
Respect for persons requires that subjects, to the degree that they are capable, be given the opportunity to choose what shall or shall not happen to them. This opportunity is provided when adequate standards for informed consent are satisfied.
While the importance of informed consent is unquestioned, controversy prevails over the nature and possibility of an informed consent. Nonetheless, there is widespread agreement that the consent process can be analyzed as containing three elements: information, comprehension and voluntariness.

2. Assessment of Risks and Benefits.
The assessment of risks and benefits requires a careful array of relevant data, including, in some cases, alternative ways of obtaining the benefits sought in the research. Thus, the assessment presents both an opportunity and a responsibility to gather systematic and comprehensive information about proposed research.

3. Selection of Subjects.
Justice is relevant to the selection of subjects of research at two levels: the social and the individual. Individual justice in the selection of subjects would

> require that researchers exhibit fairness: thus, they should not offer potentially beneficial research only to some patients who are in their favour or select only 'undesirable' persons for risky research. Social justice requires that distinction be drawn between classes of subjects that ought, and ought not, to participate in any particular kind of research, based on the ability of members of that class to bear burdens and on the appropriateness of placing further burdens on already burdened persons. Thus, it can be considered a matter of social justice that there is an order of preference in the selection of classes of subjects (e.g., adults before children) and that some classes of potential subjects (e.g., the institutionalised mentally infirm or prisoners) may be involved as research subjects, if at all, only on certain conditions (Jonsen, Veatch, and Walters, 1998, p. 25-27).

The three ethical principles identified and described by the Belmont Report have evolved in a very different manner, and they can be used to structure the historical evolution of medical ethics. There are other ways of analyzing the history of medical ethics (Reich, 1995a, vol. 3, p. 1439-1646; Jonsen, 1997a; 1997b; 1998; Jonsen, Veatch and Walters, 1998), more concerned on concrete events and particular data, but at the same time less comprehensive (Wear, Geyer-Kordesch and French, 1993; Dell'Oro and Viafora, 1996). Other times, the history of medical ethics has been more story than actual history (Rothman, 1991). The latter must necessarily overpass concrete data, looking for the interpretation of the global tendencies and the structural changes. Only very recently this kind of hermeneutic approach has become possible (Chambers, 1998). No doubt, the near future will introduce important novelties in the historical understanding of this complex phenomenon, which is bioethics.

My hypothesis in this chapter is that the analysis of the origin and evolution of the three bioethical principles can be a good way of understanding the paradigms and revolutions that have taken place in medical ethics throughout history. In the following, I will describe this history by studying the origin and evolution of each ethical principle, and in the conclusion, I would like to explain the reason why these three principles have converged today, defining a substantially new and conflicting situation. Bioethics is incomprehensible without the understanding of this historical evolution.

3. THE MEDICAL TRADITION AND THE ETHICAL PRINCIPLE OF NON-MALEFICENCE/BENEFICENCE: MEDICAL PATERNALISM

Since the beginnings of Western medicine, which is to say from the time of the writings which tradition has ascribed to the Greek physician Hippocrates of Cos, medical ethics has made use of a 'naturalistic' criterion to distinguish good from bad. This criterion, irrespective of whether it has involved what has been known since the start of this century as the 'naturalistic fallacy' (Moore, 1994, p. 89-110), has customarily identified good with the 'natural' order, while considering any departure from that order to be bad. Nature is the work of God, said the Christian theologians of the Middle Ages, and so the natural order is essentially good.

This explains why medieval culture revolved around the idea of 'order,' which embraced not only those things we customarily call natural but also men, society, and history. In the area of medicine, any disordered or unnatural use of the body or any of its organs was considered bad; and it was also felt that the physician-patient relationship, like other social and human relations, had to conform to a certain order.

This order was not univocal, since within it the physician was considered to be the subject agent and the sick the subject patient. The physician's duty was to 'do good' for the patient, and that of the patient was to accept this. The morality of the physician-patient relationship thus had to be a characteristic 'morality of beneficence.'

What the physician was attempting to achieve was an 'objective' good, *viz.* the restitution of the natural 'order,' for which reason he had to impose this order on the patient, even against the patient's own wishes. It is true that the patient might not consider what the physician was advocating to be good, but this was due to a 'subjective' error which, obviously, could not be expected to possess the same merits as the objective truth.

As a result, within the bounds of the physician-patient relationship the physician was not only a technical agent but also a moral one, while the sick person was a patient in need of both technical and ethical help. The one possessing knowledge of the natural order, in the case of disease, was the physician, who was both able and obliged to proceed on the basis of this knowledge, even in opposition to the patient's desires. It was the essence of 'paternalism,' a constant in all medical ethics of the natural 'order.'

Few literary documents show this as clearly as Plato's *Republic*, which has shaped Western political thought for more than a millennium. According to Plato, any well-constituted political society must consist of several types of people, as follows:

One type includes those within the city who dedicate themselves to the cultivation of the so-called servile or mechanical arts (agriculture, manufacturing, carpentry, blacksmithing, masonry, etc.) As a consequence of their work, Plato says, such people are deformed in body and ignoble of spirit. In them there is no possible health or morality. For this reason their political status cannot be that of free persons, but instead must be that of serfs or slaves. They are thus without political or civil liberties.

The opposite is true for other men who dedicate themselves to the cultivation of the liberal or scholarly arts (grammar, rhetoric, poetry, arithmetic, geometry, music, astronomy), upon whom Plato confers the estate of guardians. They must fulfil two functions within the city, that of defending the city from external threats (for which purpose they must be healthy and strong of body), and that of imposing order and peace upon internal disputes (something that cannot be accomplished except through a good moral education coupled with an exquisite sense of the four cardinal virtues: prudence, justice, fortitude, and temperance). If the artisans are considered to be of diseased and low moral condition, the guardians, in contrast, are considered healthy in body and soul. They can thus be free men and they can enjoy liberties.

From the best of the guardians come the governors, who Plato feels represent the category of perfect men. From this derives the fact that the rank of philosopher, together with mastery of the highest science, dialectics, is inherent to the Governor of the Republic.

Through dialectics the philosopher is able to differentiate the true from the false, the good from the bad, the just from the unjust, and to convey it, inasmuch as he is the monarch, to the community. In this manner the platonic governor 'imposes' values on the other members of the social body. He is an absolute and absolutist sovereign, the polar opposite of a democratic governor. Human beings, the inhabitants of the city, are not the prime holders of rights and political liberties, some of which they delegate to the sovereign; on the contrary, the governor by nature is the prime holder of these things, and the liberties enjoyed by the citizens are imposed upon them from above.

In concrete terms, the moral order seen by Plato is derived from the privileged view that the monarch has of the world of ideas, above all the idea of goodness. And the governor's function is none other than that of mediating between the world of ideas and the world of men. However strange it may appear, then, the moral order does not derive from free acceptance but from imposition. It is well known that in the Socratic tradition such imposition does not conflict with freedom, since whoever sees the good cannot fail to yearn for it. What is free is not in opposition to what is necessary. Compelling his subjects to comply with the imposed moral order, the platonic governor in fact promotes the freedom of each and every individual.

Such is the moral justification of political absolutism. And if the term 'monarch' or 'governor' is substituted for 'physician', and the term 'subject' for 'patient,' one arrives at a strictly faithful image of the traditional enlightened despotism of the physician. The physician has always been to the body what the monarch has been to the republic, an absolute and absolutist sovereign until the democratic revolutions of modern times, one perpetually oscillating between the paternalism of family relations and the despotism of slave relations. Aristotle described this oscillation between paternalism and despotism in the following terms:

> The rule of a master, although the slave by nature and the master by nature have in reality the same interests, is nevertheless exercised primarily with a view to the interest of the master, but accidentally considers the slave, since, if the slave perishes, the rule of the master perishes with him. On the other hand, the government of a wife and children and of a household, which we have called household management, is exercised in the first instance for the good of the governed or for the common good of both parties, but essentially for the good of the governed, as we see to be the case in medicine, gymnastic, and the arts in general, which are only accidentally concerned with the good of the artists themselves (Aristotle, 1960a, 1278b32-1279a1).

This mixture of paternalism and despotism permits understanding of the Hippocratic attitude towards the disclosure of information to patients. For instance, in the book entitled *Decorum* we can read the following:

> Perform all this calmly and adroitly, concealing most things from the patient while you are attending to him. Give necessary orders with cheerfulness and serenity, turning his attention away from what is being done to him; sometimes reprove sharply and emphatically, and sometimes comfort with solicitude and attention, revealing nothing of the patient's future or present condition. For many patients through this cause have taken a turn for the worse, I mean by the declaration I have mentioned of what is present, or by a forecast of what is to come (Hippocrates, 1981, p. 297-299).

In the Hippocratic tradition, this has been the classic formulation of the principle of non-maleficence/beneficence: my duty as a physician is to help, or at least to do no harm to patients instead of informing or respecting their decision. Traditionally the principle of non-maleficence/beneficence has been interpreted as absolute and independent from the feeling and opinion of the patient. This is the meaning of the famous sentence written in the Hippocratic book entitled *Epidemics*:

> As to diseases, make a habit of two things: to help, or at least to do no harm. The art has three factors, the disease, the patient, the physician. The physician is the servant of the art. The patient must co-operate with the physician in combating the disease (Hippocrates, 1972, p. 165).

The physician has a duty to fight against the disease. This duty is not primary and unconditioned, but the consequence of the legal order of nature. Ancient philosophy grounded the idea of duty on that of law and right. This was the consequence of its naturalism, the belief that there are natural laws and natural rights, and that these form the background of duties, at least of the so-called negative duties or precepts, directly derived from the legal order or nature. These duties may be compelled by force, because in any other case natural law would be submitted to continuous transgressions. Negative precepts were conceived as absolute and therefore compulsory for all. They have not only moral but also legal status. In fact, they are primarily laws, laws with universal validity, prior to and independent from the willingness of individuals. That is the reason why in order for them to be fulfilled coercion could be used.

In ancient ethics, *negative* precept was the name given to those duties we call today duties of non-maleficence. The idea of a natural order permitted an objective and universal definition of its content, equal for all and obtained with coercion, if necessary. This was the difference with other duties called *positive*, which did not have the form of prohibitions or mandates but of promotions or counsels. An example of the first is: "do not kill others!" and an example of the second: "love your neighbour!" This last is not in the field proper of precepts but of counsels, the old expression of defining what we think now as covered by the principle of beneficence. Traditionally non-maleficence was identified with the so-called negative duties, whilst beneficence with the positive. The first were also called duties of prohibition, and the second duties of promotion or virtue. Human beings were considered autonomous to manage the second, but not the first, which were established by God and nature throughout natural laws. This was the classical distinction between non-maleficence and beneficence. The old idea was that the first should be expressed legally, and the second by ethics, etiquette, and ascetics. Therefore, the opinion of old philosophy was that law is prior to ethics, rights to duties, and non-maleficence to beneficence. Duty is always the consequence of a right. If I have a right, then another person must have a duty. This is the idea about rights and duties proper of the natural law theory.

With these ideas in mind, we must return to the analysis of the Hippocratic text. It could be thought that the sentence: "to help, or at least to do no harm!" means the same as: "be beneficent, or at least be non-maleficent!" Physicians should be beneficent, but in every case they must not be maleficent. This interpretation seems correct, because it is coherent with our modern mentality, but it is completely strange to the Hippocratic mind. Generally, Hippocratic texts do not use both expressions, 'to help' and 'to do no harm', but only the positive one, 'to help' or 'to benefit' (*ophélein*). This is the main word in Hippocratic ethics. It does not make distinctions between to benefit and not to harm, because it identifies both, it considers that both are the same. The reason is that disease is always and necessarily a disorder, an alteration of the natural order, and therefore a transgression of the natural law. In the ancient culture, there was something illegal and immoral about disease. Consequently, it is always maleficent, and the duty of helping or curing proper of the physician is a precept of non-maleficence. In medical practice there was no distinction in ancient times between non-maleficence and beneficence. To benefit the patient, to restore his natural order, was a duty of non-maleficence. That is why the patient had no place in these kinds of decisions, he could not intervene or decide about his health, because the duties of non-maleficence were out of his capability.

The lack of distinction between the duties of non-maleficence and those of beneficence, has traditionally been used to define the specificity of professions as opposed to occupations. Occupations were all the social roles whose members were legally obliged to be non-maleficent, avoiding ignorance, incompetence or negligence, but with no special duties of beneficence; beneficence was for them only a sign of virtue and excellence. On the other hand, social roles such as priesthood, political government, and medicine, in which, due to the importance of the things they were taking care of, not only was non-maleficence morally and legally compelling but also beneficence; in other words, professions were these occupations in which the duties of beneficence were conceived exactly as duties of non-maleficence. This is the reason why professionals have been thought of traditionally as strongly differentiated from the common morality, and therefore with special norms and principles (Goldman, 1980). Professionals were thought of as binded with specific duties throughout public covenants or oaths. The Hippocratic *Oath* is not the only example, but it is without doubt the most famous, important and well known one.

Now medical paternalism can be defined in a new way, saying that it was the consequence of the lack of distinction between non-maleficence and beneficence in the care of patients. When beneficence is confused with non-maleficence, and all health care duties are thought of as maleficent or non-maleficent, then paternalism is unavoidable. Hippocratic writings generally use the verb *ophéllo* to define the moral duties of physicians, but they understand them as duties of non-maleficence, that is, duties which are objective, universal and absolute, equal for all, and required if necessary by coercion or force. That is why we have translated *ophélein* as non-maleficence/beneficence, as it appears in the text of *Epidemics* quoted before.

The physician has a moral duty of non-maleficence/beneficence that is absolute and must be realised also against the willingness of the patient. This is what has sometimes been called 'hard' paternalism, to be distinguished from 'soft' paternalism, which was most frequent after the 18th century. As Gerald Dworkin has written, "it is useful to distinguish between 'hard' and 'soft' paternalism. By soft paternalism, I mean ... that the person for whom we are acting paternalistically is in some way not competent ... By hard paternalism, I mean the view that paternalism is sometimes justified even if the action is fully voluntary" (Dworkin, 1983, p. 107).

The paternalism proper of the history of medicine has always been *hard*. The patient has been treated not only as corporally ill, but also as morally ill, or incompetent. Therefore, all through its history, medicine has been mainly paternalistic. For many centuries, from the Hippocratics to the 18th century, this hard paternalism was *complete*, and applied to all patients in all circumstances. As opposed to this, hard paternalism, from the 18th century until now, has been incomplete, because medicine has applied this hard paternalism only to grave and hopeless cases, like cancer patients. Perhaps it can be said that until the 18th century patients were treated as infants or completely incompetent boys and from the 18th century onwards as adolescents.

Some authors have studied the lack of information for cancer patients. All the studies in general reach the main conclusion, namely that there is a strong and general tendency to withhold from the patient the information that he has cancer. Almost ninety percent of the internists, surgeons, and generalists reported, assumed that their usual policy is not to tell the patient that he has cancer. In general, hospitals do not have written policies of informing every patient. And some physicians avoid even the slightest suggestion of neoplasm and quite specifically substitute another diagnosis. Almost everyone reported resorting to such falsification on at least a few occasions, most notably when the patient was in a far advanced state of illness at the time he was seen.

Paternalism is today generally understood as a negative moral attitude. That is the consequence of taking into account a new moral principle called autonomy. Only this way can beneficence be distinguished from non-maleficence, setting up moral life on new grounds. This is the second part of the story.

4. THE LEGAL TRADITION AND THE PRINCIPLE OF AUTONOMY: THE RIGHTS OF PATIENTS

The old and paternalistic intellectual universe did not undergo any substantial change until the modern world was well established. Indeed, if the Protestant Reformation sought and obtained something, it was the substitution of the idea of 'order' for that of 'autonomy,' and of the 'natural' order or order of freedom for the 'moral' order. From this arose the second major moral paradigm of Western history, whose origins are intertwined, with the progressive discovery of human rights from Locke's time to the present.

Human rights were at the beginning conceived according to the old categories proper of the natural law theory, and were defined as the fundamental rights every human being has by the only fact of being human, pertaining to the human species or enjoying human nature. This condition applies to every man with human rights, and generates in all other human beings the duty of their respect. The only difference with antiquity is that the authors of the seventeenth century understood the natural law theory not exactly as the legal order of nature but as the intrinsic order of reason. This was the thought of Descartes, Spinoza, Leibniz, and also of Locke. At the beginning this difference seemed not to be very important, but later, in the eighteenth century, it led to the conclusion that the primary moral concept of reason is not that of right but of duty, and that the true and primary law of reason is the law of duty, the moral law. This was the Kantian categorical imperative. Duty is prior to right. The only primary law is moral law, the categorical imperative. There are other laws, but they all derive from moral duties. The foundation of human rights is therefore not natural but moral, not the heteronomous laws of nature, but the autonomous law of reason. Autonomy is now the main concept, the first principle of morality. All the rest must derive from it.

The discovery of autonomy as a moral principle leads to redefining the content of the principles of non-maleficence, and beneficence. The new idea that begins to arise is that nobody can be beneficent with another without respecting his autonomy, and therefore his willingness. Beneficence is now inseparable from autonomy. Non-maleficence also, but in a different way. I must always be non-maleficent, but I must be beneficent *only* when the patient consents to my action, or agrees with it. Therefore, I must be nonmaleficent with others, but I am not obliged to be beneficent with them in the same way or to the same extent. The non-maleficence principle is more compelling than that of beneficence. That is why the duties derived of the first principle are generally called 'perfect', and that of the second 'imperfect'.

> The latter are those in which, though the act is obligatory, the particular occasions of performing it are left to our choice, as in the case of charity or beneficence, which we are indeed bound to practise but not toward any definite person, not at any prescribed time. In the more precise language of philosophic jurists, duties of perfect obligation are those duties in virtue of which a correlative *right* resides in some person or persons; duties of imperfect obligation are those moral obligations which do not give birth to any right (Mill, 1995a, p. 1160).

The ideas of good are different in human beings and proper of every one, and therefore the duties of beneficence are personal and not transferable. On the contrary, the content of the principle of non-maleficence should be the same for all, and therefore must be established using the rule of universalisation, impartiality, or symmetry. And because the duties of non-maleficence generate rights in other persons, as Mill said, they are not only moral but also legal, and take the form of positive laws, which can be implemented by force. These duties must be stabilised between all, throughout public consensus, and required equally to all members of society. Autonomy obliges us to respect the plurality of ideas of good, and therefore to respect the diversity of contents of the principle of beneficence: at the same time it leads us to define publicly, taking into account all actual and virtual human beings, the content of the principle of non-maleficence. John Stuart Mill wrote in his book *On liberty*:

> The object of this essay is to assert one very simple principle, as entitled to govern absolutely the dealings of society with the individual in the way of compulsion and control ... That principle is, that the sole end for which mankind are warranted, individually or collectively, in interfering with the liberty of action of any one of their number, is self-protection. That the only purpose for which power can be rightfully exercised over any member of a civilised community, against his will, is to prevent harm to others [Non-maleficence]. His own good [Beneficence], either physical or moral, is not a sufficient warrant (Mill, 1995b, p. 1176-7).

As this new way of thinking was taught, the old human relationships established in conformity with the medieval idea of hierarchic order came to seem excessively vertical, monarchic, and paternalistic. As an alternative to these relationships, others of a more horizontal, democratic, and symmetrical nature were proposed. The democratisation and generalisation of the basic autonomy of all human beings began in the 17th century, when the theory of civil and political rights was constructed. All human beings have the same basic human rights, which make them fundamentally equal. These rights are thought to belong to the individual under natural law as a consequence of his being human. Describing the state of nature, John Locke wrote:

> But though this [state of nature] be a status of liberty, yet it is not a state of license; though man in that state has an uncontrollable liberty to dispose of his person or possessions, yet he has not liberty to destroy himself, or so much as any creature in his possession, but where some noble use than its bare preservation calls for it. The state of nature has a law of nature to govern it, which obliges everyone; and reason, which is that law, teaches all mankind who will but consult it, that, being all equal and independent, no one ought to harm another in *his life, health, liberty and possessions* (Locke, 1823, p. 341).

The Universal Declaration of Human Rights of the United Nations of 1948 affirms, in the same way, to have "faith in fundamental human rights, in the dignity and worth of the human person, in the equal rights of men and women and of nations large and small".

The liberal theory introduced the civil and political rights in the 17th and 18th centuries, whilst in the 19th century democratic socialism corrected the liberal theory of human rights, adding the economic, social, and cultural rights. The former rights of the liberal theory were called negative human rights, since they preceded the formation of the State and could be demanded before the existence of any positive law. The latter were considered positive human rights, because they could be implemented only by the State, and therefore had no value other than that conferred on them by positive law.

Both types of rights are the consequence of the two faces of human freedom, the so-called *freedom from* external coercion, and the *freedom to* achieve the own goals in human life. The civil and political rights protect the first kind of freedom, and the economic, social and cultural rights promote the second. However much one may be 'free from' external coercion, one cannot live in society under adequate conditions if one does not have 'freedom to' work, have a family, raise children, etc., such freedoms being granted by economic, social, and cultural rights. For this reason, socialism began to consider the 'freedom from' as purely *formal* human rights vis-à-vis the 'freedom to', which were seen as *real* rights.

In the second half of the nineteenth century, democratic socialism arose confronting democratic liberalism. While the latter promoted the minimal State, the former tried by all means to establish a maximal State, i.e., a State that would promote and protect not only the negative rights but also the positive ones, establish a fair workday, prohibit exploitation of women and children, demand a minimum wage, and protect the unemployed, the sick, the retired, widows, and others from misfortune. Thus arose consciousness of everyone's right to education, adequate housing, well-paid work, unemployment compensation, a pension, and health care.

In the realm of health, economic, social, and cultural rights led people to conceive health as something that can be justly demanded. This in turn has prompted a radical change in the way governments deal with health problems; for in this light health can no longer be considered merely a private matter; rather, it becomes a matter of public concern and hence a political issue. This marks the beginning of 'health policy' as a chapter in social and welfare policy. The social Justice State, which in the Western countries has become identified with the Welfare State (or benevolent State), must have as one of its top priorities the protection of the right to health care. Otherwise, the development of the entire Western system of compulsory health insurance would be incomprehensible.

While liberalism discovered the *right to health*, socialism cast light on a new right, the *right to health care*. The first is a negative right preceding the social contract, and the State can do nothing but protect it; the second is a positive right that the State must actively promote. The first is a specification of the principle of freedom, while the second is deduced from the principle of equality.

The great democratic revolutions of the modern world - first the English Revolution, then the North American, and then the French - were carried out in the spirit of civil and political rights. Liberal revolutions of the 17th and 18th century applied and realised the human rights theory, affirming that all human beings are autonomous, subjects of the civil and political rights. With the liberal revolutions, paternalism disappeared from civil and political life. The English Bill of Rights of 1689, the American Bill of Rights of 1791, and the French Declaration of the Rights of Man and of the Citizen of 1789 were expressions of the new democratic style. The political assumption of the economic, social and cultural rights was made much more later, during the second half of the nineteenth century and the first half of the twentieth, after the so-called social revolutions, first of all the one of 1848.

It is impossible to understand the meaning of medical bioethics in isolation from this context. Bioethics is a necessary consequence of the principles that have been moulding the spiritual life of the Western countries for two centuries. If since the Enlightenment there has been affirmation of the autonomous and absolute nature of human individuals, in both the religious order (through the principle of religious freedom) and the political order (through the principle of democracy), it is logical that this should have led to what we might call the 'principle of moral freedom,' which can be formulated as follows: All human beings are autonomous moral agents, and as such should be respected by all those who hold distinct moral positions. Just as religious pluralism and political pluralism are human rights, so too should moral pluralism be accepted as a right. No morality can be imposed on human beings against the dictates of their own consciences. The sanctuary of individual morality is inviolable.

Pluralism, democracy, and civil and political human rights have been leading achievements of the modern era. The same is true of ethics in the strict sense, that is, of the moral in contradistinction to the physical. For this reason it should not seem strange that the development of ethics has been linked to the development of democracy and human rights. Indeed, all of the democratic revolutions, those that have taken place in the Western world since the eighteenth century, were organised to defend these principles.

Nevertheless, there is a curious circumstance - that this pluralistic and democratic movement, which had already been established in the civil life of Western societies centuries ago, only reached medicine very recently. Medicine has had no interest in autonomy. The theory of autonomous persons, autonomous actions and the evaluation of autonomy has been developed for legal reasons and needs. Only the autonomous person can be considered legally responsible, which is the reason why the evaluation of individual competency and capacity has been and still is legally essential. Medicine traditionally considered the patient to be incompetent. Roman Law recognised legal capacity and autonomy only where it concerned some privileged people. Autonomy, then, was a privilege. The relationship between the physician and the patient has obeyed the guiding principles set forth by Plato more than it has obeyed principles of a democratic cut. Specifically, within the framework of the physician-patient relationship the patient has been considered both physically and morally unfit, making it necessary for his physician to lead him in both areas.

In general, the physician-patient relationship has traditionally been paternalistic and absolutist. Pluralism, democracy, and human rights, in other words, ethics understood in the modern sense, has not touched this relationship until recently. It was only during the 1970s that patients began to be fully aware of their status as autonomous moral agents, both free and responsible, who had no wish to establish parent-child relationship with their physicians, but who instead sought adult relationships based on mutual needs and mutual respect. Since then, however, that awareness has caused the physician-patient relationship to be based upon the principle of autonomy and freedom for all the participating subjects, including both physicians and their patients.

The liberal revolution has entered the field of medicine in the last decades, when the rights of the patients were discovered as specifications of the civil and political rights. This has been the beginning of the end of medical paternalism. In the U.S.A. in 1969 the Joint Commission on Accreditation of Hospitals, a highly influential nongovernmental hospital accrediting association drawing its memberships from various medical and hospital groups, issued a new revised policy statement. Little was said in this policy about problems of patients, and the organisation was asked by various consumer groups to redraft the statement with an eye toward the concerns of patients. Leading these efforts was the National Welfare Rights Organisation, which in June 1970 drafted a statement with 26 proposals for the rights of patients. This seems to have been the genesis of the so-called patients' rights movement. After several months of deliberation and negotiation, the Joint Commission on Accreditation of Hospitals accepted some of the 26 proposals as parts of a new preamble statement in its 1970 *Accreditation Manual.* The American Hospital Association then began to debate the issue of patients' rights and adopted *A Patient's Bill of Rights* in late 1972. Shortly thereafter, in January 1973, a commission from the U.S. Department of HEW recommended that health care facilities adopt and distribute such statement in a manner that would 'effectively communicate' with patients.

The AHA Patient's Bill of Rights was published in 1973. Among its articles, the following ones can be found:

> The patient has the right to obtain from his physician complete current information concerning his diagnosis, treatment, and prognosis in terms the patient can be reasonably expected to understand ...
>
> The patient has the right to receive from his physician information necessary to give informed consent prior to the start of any procedure and/or treatment ...
>
> The patient has the right to refuse treatment to the extent permitted by law and to be informed of the medical consequences of his action.

> The patient has the right to obtain information as to any relationship of his hospital to other health care and educational institutions insofar as his care is concerned (American Hospital Association, 1973).

Notice what this signifies. When all the mature human beings who make up a social group live as autonomous adults, it is highly probable, not only in the world of politics but also in the world of morality and religion, that they will maintain different positions. This will have two results. The first is that a society based on the liberty and autonomy of all its members must by necessity be plural and pluralistic; in other words, its members will not only have distinct views in the areas of politics, religion, morality, etc., but will also commit themselves to respect the views of others, on condition that these others do likewise. And the second is that besides maintaining pluralism, the society will have to be secularised, since it will be practically impossible to achieve uniformity in religious matters.

Let us now return to medical ethics. During the many centuries in which the Greek philosophy of the natural order prevailed, a philosophy that was subsequently christianised by the theologians, medical ethics was drawn up by moralists and applied by confessors. The physician was presented with everything in completed form and asked or required to comply with it. There was no clear understanding that specific cases could provoke grave and substantial conflicts, since once the general, immutable principles had been established, the only things that might vary were the circumstances.

Expressed in other terms, over the course of all those centuries there was no true medical ethics, if by this is meant the moral autonomy of physicians and patients. What existed was something else, in principle heteronomous, which we might call 'ethics of medicine.' This explains why physicians have not generally been competent in questions of 'ethics,' their activity having been reduced to the sphere of 'asceticism' (how to educate the good or virtuous physician) and of 'etiquette' (what standards of propriety and civility should govern the practice of medicine). The history of so-called medical ethics offers effective proof of this.

Nevertheless, the current panorama is quite different. In a society where everyone, in lieu of evidence to the contrary, is an autonomous moral agent with distinctive criteria of good and bad, the medical relationship, being interpersonal, may involve inherent rather than accidental conflict, but also inherent rather than accidental moral creativity. Physicians and patients are not moral delegates or obedient subordinates, but free moral agents. They are autonomous persons, and in such way sources of morality. As Kant wrote, "heteronomy of the will [is] the source of all spurious principles of morality", and "autonomy of the will, the supreme principle of morality" (Kant, 1995, p. 1092-3).

5. THE POLITICAL TRADITION AND THE PRINCIPLE OF JUSTICE: THE SANITARY COMMON GOOD

Over the last century health has ceased to be a private matter concerning mainly individuals and has become instead a public problem, a political issue. The terms 'health' and 'politics', initially mutually exclusive, have become inextricably intertwined in the expression 'health policy', and today it is hard to find any aspect of health completely detached from the immense bureaucratic apparatus of health policy.

Many consider the interference of politics in health excessive, while for others it is still insufficient; but both sides justify their points of view by appealing to the concept of distributive justice. Thus it is not surprising that one of the liveliest and most polemic chapters of bioethics today is that of justice in health.

When should a health service be considered just or unjust? What resources must be allocated to comply with the obligation of justice? How should one proceed when available resources are less than those theoretically needed? How can insufficient resources be justly distributed? These are some of the questions policy-makers, health promoters, and members of the general public repeatedly ask themselves.

The primary and basic meaning of 'justice' is correction or adjustment of something in accordance with a model of what it should be. In this first sense, 'just' means 'adjusted', that which is adjusted to the model. Thus, we will say that an act is 'just' when it is in accordance with the law, and that the law is 'just' when it is an expression of moral principles. 'Unjust', to the contrary, is that which is not adjusted to the general principle, norm, or criteria being applied.

That general principle of justice with which all other criteria and acts of man must be brought into line was defined by the Roman jurists as *suum cuique tribuere*, 'to each his due'. An act is just when each is given his due, and unjust otherwise. The problem lies in spelling out precisely what this means. Throughout Western cultures there have been no fewer than five different interpretations, which to some extent contradict one another; these have variously interpreted justice as 'natural proportionality', 'contractual freedom', 'social equality', 'greatest utility', and 'equitable efficiency'. I will endeavour to characterise each of them as concisely as possible and to examine their impact upon the world of health.

1. Historically, the theory of justice that has been most widely applied is doubtless that which understands justice as 'natural proportionality'. Initiated by the Greek thinkers around the sixth century BC, it went unrivalled until the seventeenth century.

According to this notion, justice is a natural property of things, whose name need only be known and respected. This is the meaning that the Greek philosophers attributed to the term *dikaiosyne*. As natural entities, things are just, and any type of maladjustment constitutes a denaturalisation. Everything has its natural place, and it is just that it remain there (Plato, 1995, p. 147-8).

In addition to this 'general' justice, Greek philosophy distinguished other more concrete or partial meanings of the term. Aristotle differentiated at least two. They are called partial because they neither pertain to all of nature nor to the body politic as a whole, but are limited to relations among different members of society. One is 'distributive justice', which governs the relationship between the ruler and his subjects. The other, 'commutative justice', regulates relationships between private persons. In the health world both are important, but especially the first (Aristotle, 1960b, 1131b2-1132a2).

According to Aristotle, distributive justice regulates the distribution of public 'honours, or money, or anything else' among the private members of society (Aristotle, 1960b, 1130b31f). If by our nature we were all identical, and worked the same for the wealth of the State, there is no doubt but that such distribution would not be considered just if it were unequal. But given the 'natural' character of inequality and hierarchy in society, for the Greeks the distribution of public honours, wealth, etc. cannot and should not be done on an 'identical' basis, but rather 'proportionate' to one's natural abilities and actual works.

The repercussions of these schemes upon medicine have been tremendous. The patient cannot achieve the good of health other than through the general economy embodied in the physician. Therefore, the only virtue that should be demanded of the patient is obedience. In the relationship between patient and physician, as in the relationships between parishioner and priest, or subject and sovereign, there is no place for a perfect or complete commutative justice. The services of physicians, priests, and sovereigns are deemed so superior to those rendered by all other members of the community that it will never be possible to achieve equality in the exchange. Therefore, none of them is paid in accordance with the principle of commutative justice, but rather in 'honour'. The money they receive is an 'honorarium'.

The concept of justice as natural proportionality also has another health consequence of great importance. This derives from the fact that 'proportionality' should be reflected in medical care, in accordance with the individual's social task. This belief was already evident in Plato's *Republic*, which not in vain endeavoured to describe the order of a 'just' polis. There one can see how medical care should have a certain differential character, precisely by virtue of the principle of 'distributive' justice. Slave doctors attended slaves; artisans had no access to lengthy or costly therapeutic procedures; and only the rich had complete access to the world of health. Of these three groups, only those belonging to the latter fully benefited from the goods of the city, and only these members could have been and should have been just and virtuous. Perhaps that is why only they were beneficiaries of complete health care.

All of this, written in the fourth century BC, maintained its validity throughout the Middle Ages. Medieval society tried to follow platonic dictates insofar as possible and medical care basically accommodated to these norms. The only difference was that the Christian commandment of charity made room for the appearance of hospitals, places in which poor and ill people were attended, though more spiritually than medically.

2. At the end of the Middle Ages, and specially in Modern times, political science began to make basic changes in the concept of justice, and also began to insist increasingly on the importance of a social contract as the basis for all justice-related duties. In this way, justice was transformed from a mere 'natural adjustment' into a strict 'moral decision'. The relationship between the subject and the sovereign was no longer based on 'submission' but on free 'decision'. Man was seen as being above Nature, and as the sole and exclusive source of rights.

The rights described by John Locke in his *Second Treatise of Government* were thought of as individual and inalienable. But the defence of these rights was conceived much easier assuming the burdens of a social contract than in the pure state of nature. The state of nature, says Locke, "is full of fears and continual dangers: and it is not without reason that he [men] seeks out, and is willing to join in society with others, who are already united, or have a mind to unite, for the mutual preservation of their lives, liberties, and estates, which I call by the general name property" (Locke, 1823, p. 412). The 'common good' is now understood, therefore, as the achievement of the highest private good for everyone. In order to protect the private goods from all, or at least from the highest number of people, "they want an established, settled, known law, received and allowed by common consent to be the standard of right and wrong" (Locke, 1823, p. 412). This is the new idea of social justice proper of the liberal philosophy and politics. Just and unjust are now the same as right and wrong, understood as the system of laws established by common consent in order to preserve the lives, liberties, estates and properties of everyone. There is no substantive common good, as opposed to the thoughts of the ancient philosophy. The common good is only the lowest common factor of the goods of everyone.

For Locke, social and legal justice has no aim other than that of protecting the rights that man has already had from the beginning, in such a way that we can never transgress those limits or oppose them. The result is a minimalist notion of the State. Specifically, the State's only legitimate purpose is to facilitate people's exercise of their natural rights to life, health, liberty, and property. When the State does not do so, or does so poorly, i.e., when the laws do not respect the natural limits, or the State steps beyond its bounds and dictates laws that go beyond the powers granted to it in the social contract, such laws are unjust. The thesis of Locke is, for instance, that land property is a right of each individual and not of the State, and that the entitlement for property is work. Therefore, the concentration of land's property in the hands of kings, nobles and churches begins to be considered illegitimate, or immoral. From the first, says Locke, land property was not concentrated this way. At first, "right and conveniency went together; for as a man had a right to all he could employ his labour upon, so he had no temptation to labour for more than he could make use of. This left no room for controversy about the title, nor for encroachment on the right of others; what portion a man carved to himself was easily seen: and it was useless, as well as dishonest, to carve himself too much, or take more than he needs" (Locke, 1823, p. 367).

This was one of the most characteristic aims of the liberal revolutions, the land's reform, distributing the big properties of privileged people between the farmers. The new state would not be the landowner but only the governor or administrator of public affairs. In order to do so and due to its lack of resources, the liberal state must distribute proportionally the public charges between the private individuals. Consequently, the new function of the state will no longer be the distribution of its properties between private individuals, but the proportional distribution of public charges among citizens, and therefore the elaboration of a coherent theory of taxes. Later, the old thesis of proportionality came to be applied to the distribution of charges and taxes between individuals, instead of being used for the distribution of public goods, specially land property, between some positively privileged people.

All this has been and continues to be of enormous importance to the world of medicine. According to liberal philosophy, the health market should be governed, like all others, by the laws of free trade, without the intervention of third parties. This has been the guiding concept of the so-called 'liberal' medicine, which insists that the physician-patient relationship must accommodate itself to free-market principles, and therefore should not be mediated by the State. Everyone must secure his or her own health. And when private individuals are incapable of taking care of their own health, then civil society and the state can and should help them, but not due to the moral principle of justice but due to charity.

3. This was the point in which the third theory appeared. Its main thesis was that some basic social goods, like work, basic education, or health care, should be covered equally for all members of a civilised society. One of the most important goals of the state must be to secure universal and equal access for all individuals to those primary goods. And because the perfect society must be protected by the state, these are public duties that generate specific rights in citizens. As it is well known, there are two different types of duties, perfect duties or duties of justice and imperfect duties or duties of

beneficence. Liberalism considered that health care should be conceived of as an imperfect duty or duty of beneficence, because the moral duty of helping others when they were incapable of caring themselves could not be thought legal, and therefore compelled by force, but only moral. Perfect duties are those that generate rights in one or more people, and which, therefore, can be forced if necessary. As opposed to this, imperfect duties are those which do not generate rights in others, which is the reason why they cannot be obtained by force. Liberal thinkers considered that health care should be understood as an imperfect duty. As opposed to this, the socialist movement of the nineteenth century began to consider health care, like other primary social goods, as a perfect duty that generates rights in individuals and that must be secured by the state. This was the origin of the new group of human rights established in the second half of the nineteenth century, known as the economic, social and cultural rights. They were the expression of a new chapter of the theory of justice, the so-called social justice. According to this third theory, health care is or should be neither a problem of commutative justice, nor of distributive or redistributive justice, but of social justice.

In the realm of health, this attitude has led people to conceive of health care as something that can justly be demanded. This in turn has prompted a radical change in the way governments deal with health problems; for in this light health can no longer be considered merely a private matter; rather, it becomes a matter of public concern and hence a political issue. This marks the beginning of 'health policy' as a chapter in social and welfare policy. The social Justice State, which in the Western countries has become identified with the Welfare State (or benevolent State), must have as one of its top priorities the protection of the right to health care.

4. The problem is that it is not easy to define what kind of health is needed, and what should be covered by the state in virtue of the moral and legal principle of justice. Our society has identified health with 'perfect wellbeing', that is with happiness, and it is evident that this is an ideal goal which cannot be achieved by the state nor by the health care system, and that it cannot be performed as a right of justice. If we establish ideal goals, it is evident that not all of them can be achieved at the same time, and that we would be compelled by reality to define priorities. A way to do so is prioritising those interventions with a higher cost-effectiveness ratio, that is, those more useful and profitable. This is the reason why the utilitarian theory of justice has become so important throughout the last two centuries. In his *Fragments on Government*, Bentham established that the objective of any ruler can be none other than to bring about the greatest happiness of his subjects; and to do so he has no recourse other than to be guided by the principle that "the greatest happiness of the greatest number is the measurement of what is just and what is unjust" (Bentham, 1988, p. 393). There can be no other criterion of distributive justice. As a promoter of the *res publica*, the politician must seek

to achieve the greatest benefit at minimal cost, so as to maximise utility. This is very important today in medicine, since health has become a public matter, a political issue.

5. This brief review of the historical debate on justice in health illustrates just how complex the matter is. One reason for that complexity is that workable theories in this field, to explain events, must perforce articulate two moments, one deontological or principles-oriented and the other teleological or consequences-oriented. In the Western world, there seems to have been a certain convergence that the main deontological principle to be achieved is 'equity', and the teleological one 'efficiency'. Both are unavoidable in a comprehensive theory of justice. Economy looks for efficiency, and ethics for equity. But no one can deny or neglect the other. They must be articulated in a complex structure, which holds at the same time equity and efficiency. This is today, perhaps, the most frequent idea of justice, that which understands it as equitable efficiency.

The problem now is how to combine these two moments. How do they relate to one another? In some cases they 'complement' one another, and then there is no doubt that one's moral duty is to promote equity and optimise efficiency. Unfortunately, however, such cases do not abound and may well be exceptions. Indeed, what is most common in ethics is not complementarity but conflict, the conflict between principles and consequences, equity and efficiency. Efficiency always looks for the best or optimal allocation of resources. An optimum allocation is one that equalizes the returns of the marginal or last unit to be transferred between all the possible uses. The so-called 'law of eventually diminishing marginal productivity', ensures that such an optimum exists, and that after this point the marginal utility decreases progressively; the consequence is that the allocation of resources begins to be less efficient. In general, the efficient or optimum allocation of resources is not compatible with the extension of goods to *all* the individuals, but only to *some* of them. Efficiency is therefore more in accordance with the 'generalisation rule' (the Utilitarian maxim of the greatest efficient number) than with the 'universalisation rule' (that which affirms that a good must be distributed between all, despite possible inefficiencies). Ethics assumes preferably the universalisation rule, considering the full efficiency as inequitable; and economics handles the generalisation rule, thinking that universalisation is usually inefficient.

The theory of equitable efficiency affirms that both elements are unavoidable, but with different weight depending on the type of good. In the field of the so-called social primary goods, equity is prior to efficiency, and therefore the universalisation rule is the most important. This assures a decent minimum equal for all. The other goods, which are neither primary nor essential parts of the human dignity, can and perhaps should be distributed according to the efficiency rule. In any case, neither equity nor efficiency can be exclusive, abolishing the other principle. Both are necessary. But at the same time they are in a perpetual and unavoidable conflict with each other. Conflict is one of the most important categories of our moral life. Human reason can not avoid all conflicts. They are unavoidable. Life will always be conflicting. Our only moral duty is to resolve them, or to manage them in an equitable way, that is, respecting as much as possible all the duties and rights implied in the case. Justice, complete justice, is an ideal concept. The real duty of all human beings is not to be completely just but to work in order to diminish as much as possible the gap between the principle of justice and reality.

6. THE PRESENT SITUATION

Our situation is the first one in history in which the three traditions described above have converged in the health care arena. From the three, health care professionals are only accustomed to use the first. And this first tradition has also changed due to the interference of the autonomy of the patient. Really, the only principle of modern bioethics which health care professionals think they know is non-maleficence. The other three are seen as strange and the cause of an increasing number of conflicts. There are conflicts between autonomy and beneficence, autonomy and justice, beneficence and justice, non-maleficence and justice, non-maleficence and autonomy, and, finally, between non-maleficence and beneficence. There are also possible conflicts between three of them, or between the fourth. Probably there has been no other time in history in which medicine has been so conflicting. But the number of conflicts is not a good rate of moral quality. On the contrary, it can be thought that conflicts emerge when the participation in the process of decision-making is promoted, and in general when the respect for the rights of all participants is tampered with.

For instance, consider one of the most typical examples. A Jehovah's Witness is in a car accident and is taken to the emergency room suffering from severe hypovolemic shock. On seeing this, the emergency room physician makes a decision, based on the deeply rooted moral criterion of beneficence, to give the patient a blood transfusion. The patient's wife, who is at his side, informs the physician that her husband is a Jehovah's Witness and that he has repeatedly said that he does not wish to receive blood from other persons, even if this endangers his life.

In expressing her husband's views, the patient's wife asks for his moral criterion to be respected; she shares it, the doctor does not. Faced with the moral criterion of non-maleficence wielded by the physician, the wife in our example defends the criterion of autonomy, according to which all human beings, unless there is evidence to the contrary, are considered autonomous moral agents fully responsible for all their actions.

Here one can see how the simplest medical relationship, the one established between a physician and a patient, has been transformed into one that is autonomous, pluralistic, secularised, and characterised by conflict.

The potential intensity of this conflict is increased by the fact that others besides the physician and patient (nurses, the hospital administrators, the social security agency, the patient's family etc.) may intervene in the health relationship. However, all of these agents in the physician-patient relationship can be reduced to four: the physician, the patient, the relatives, and society. Each of these participants plays a particular moral role. By and large, the patient is guided by the moral principle of 'autonomy', the physician by that of 'non-maleficence,' the patient's family is guided by the principle of 'beneficence' relative to the patient, and society by that of 'justice.' The hospital administration, health insurance representatives, and judges have to look above all to safeguarding the principle of justice. Hence, these four dimensions are always present in the physician-patient relationship, and this is a good thing. If the physician or the family were to shift camps from non-maleficence or beneficence to justice, the health relationship would suffer irreversibly, as would also happen should the patient cease to act as an autonomous moral subject.

But the fact that these four elements are essential does not mean they must always be complementary, and thus never in conflict. The actual situation is more the reverse. It is never possible to completely respect autonomy without causing beneficence to suffer, or to honour beneficence completely except at the expense of justice, etc. From this arises the need to keep the four principles in play, weighing their importance in each specific situation. As David Ross would say, those four principles work like conditional or *prima facie* obligations, which must be weighed in each specific situation. Only then will it be seen how they might best articulate with each other, giving way to specific or effective duties (Ross, 1930, p. 19-22).

Thus, for example, despite the fact that all of us feel it necessary to scrupulously respect personal autonomy, we believe that in the case of a just war the State may compel individuals to give up their lives (that is, their autonomy) for others. Here it can be clearly seen how a primary obligation, respect for personal autonomy, may fail to coincide with the concrete and effective obligation, precisely as a consequence of the need to honour another primary obligation, justice, which in this specific case seems to be of a higher order.

Medical ethics has to do whatever is possible to scrupulously and simultaneously honour autonomy, beneficence, non-maleficence, and justice. There is an obligation to act in this way, even though the objective is very difficult and at times quite impossible to achieve.

The situation being thus, it is evident that the urgency of specific and daily problems cannot free us from the prescribed exigencies. Rather, very much to the contrary, these problems force us to take the utmost precautions and to find the strictest possible foundation for our decision-making criteria. When the issues are of such gravity that they determine the lives of individuals and societies, as frequently happens in medicine, then rationality must be honoured to its finest edge, and as much time as necessary must be dedicated to the problems involved in laying foundations.

In so doing, it is important to approach medical bioethics aided not only by logic but also by history, since human reason is simultaneously logical and historical. Hence, the history of bioethics should not be viewed as an erudite curiosity presented with no other purpose than to enlighten the reader. Rather, it should be seen as the best possible introduction to the study of bioethics, and as something that facilitates analysis of the problems involved in the lying of the discipline's logical and philosophical foundations. In this way it improves our ability to answer the question that serves as a kind of summary of all the other questions: What are the moral conditions that should attend upon what the Greeks called *téleios iatrós*, the Latins *optimus medicus*, and the Castilians *el perfecto médico*? Bioethics aspires to no greater task, nor to any lesser one.

7. BIOETHICS HERE AND NOW

The fact that bioethics began as a movement and had its first development in the US, leads some authors to think of it as a typical and specific product of the American culture, and that, therefore, outside the United States it can be spread out, applied, and particularised, but not enriched in its essence. Bioethics, hence, should be considered a genuine output of the American spirit. This was the general belief expressed in the talks by the American participants in the conference about *The birth of bioethics*, organised by Albert R. Jonsen in the University of Washington, in Seattle, in 1992 (Jonsen, 1998, p. viii and xv).

In this Conference, Daniel Callahan expressed his belief that "bioethics is a native grown American product, which did emerge elsewhere but finds uniquely fertile ground in the U.S." (Jonsen, 1998, p. 377). Three other participants, engaged in the study of the history of bioethics, David Rothman, Daniel Fox, and Stanley Reiser, debated why bioethics began in the United States. They all agreed that bioethics was "a response to the effects that technological medicine has on the relation between patients and physicians and between medicine and society" (Jonsen, 1998, p. 382); and because technologized medicine is typical of the developed world, and specially of the U.S., bioethics could have originated nowhere else.

Albert Jonsen had a wider interpretation. His thesis is that bioethics appeared in the United States because it is a typical product of what he calls the 'American ethos': "There is an American ethos that shapes the way in which Americans thinks about morality, and that ethos transformed the response of American medicine into a discipline and discourse called bioethics ... There are, I believe, three facts of the American ethos that exert a powerful influence on all American thought about morality. I call these three facets moralism, meliorism, and individualism" (Jonsen, 1998, p. 389f). Jonsen defines moralism as the "addiction to moralising", a behaviour closely related with the "pan-moralism" proper of the colonial Puritans (Jonsen, 1998, p. 391). Moral meliorism is the belief that "current situations can and should be made better" (Jonsen, 1998, p. 393). Finally, moral individualism is the idea that "from the beginning, each person counts as one and no more than one, a free agent, expected to become responsible and to make one's own way in the world" (Jonsen, 1998, p. 393). Jonsen concludes his analysis as follows: "Why an American bioethics and how does it differ from all other bioethics? The answer lies, I believe, in the American ethos: a destiny to make life better than it is and a conviction that it is possible to do so, a faith in the value of individuals and their capacity to reach consensual agreements, and a vague but genuine commitment to a conventional morality" (Jonsen, 1998, p. 395).

The conclusion of all these interpretations about the birth and first developments of bioethics is exactly the same: that this discipline has been a typical consequence of the American culture, and therefore an American product (Fox, 1989), which other countries and cultures can import and assimilate, but without the possibility of adding fundamental novelties. Bioethics either is American or is not at all.

Analyzing the history of bioethics from abroad, all these views seem to be extremely parochial and ethnocentric. A more accurate interpretation must give reason of at least two facts: first, why bioethics was born in America; and second, why it has been spread so quickly all over the developed countries, and to some extent also over the entire world. The theories about the beginning of bioethics described above provide an answer to the first question, but they do not have a convincing response to the second. Therefore, it is necessary to attempt a wider and more comprehensive explanation.

My hypothesis is that the success of bioethics is not directly related with the peculiarities of the American ethos, but with some more general characteristics, such as the secularisation of Western culture, and the emancipation of the decision-making process in the questions related with the human body, and therefore with life and death. These decisions were generally, until the second half of the 20th century, in the hands of priests, theologians and churches, and hence they were not open to public discussion.

In the field of moral problems related with the use of the body, normative ethics was until very recently in the hands of 'moralists', in general clerics of different churches, whilst 'ethicists' were only concerned with formal and abstract problems, like the logic, metaphysics and linguistics of the moral discourse. It is not casual that the word 'ethicist' did not appear in English until the end of the nineteenth Century, and with the restrictive meaning of 'specialist in ethics'. Until very recently, 'ethics' was conceived as the philosophical background of morality, and 'morals' as the discipline of behaviours. 'Ethics', rational ethics, had no normative role (Toulmin, 1997, p. 108), and 'morals', theological morals, had an important lack of interest about philosophical and foundational questions. Consequently, the word 'moralist' had the meaning of persons working in the field of normative ethics, but it was impossible to use the word 'ethicist' to designate the lay person working in practical ethics or in morals. Only during the sixties has this word enriched its content and covered the field of normative ethics. If a 'moralist' was traditionally a theologian engaged in normative questions, the new word 'ethicist' began to be used with the meaning of lay or secular person working in the field of moral norms; and 'bioethicist', that one who is working in normative questions related with the right use of life and body.

The turn began, as I have stated before, in the sixties. The debates about contraception and abortion opened the door to the public scrutiny in this set of questions. Instead of statements of authority, people began to ask for rational arguments. In this way, morality began to be secularised, and philosophers, professors and researchers in ethics, were progressively engaged in these kinds of discussions and started to be interested in the field of normative ethics. This is the process that Stephen Toulmin has described as the shift of philosophical ethics from meta-ethical questions to normative problems. This shift was crucial in the constitution of bioethics as a movement. It is neither an absolute coincidence that Toulmin was engaged in the works of the National Commission, nor that he was the main author of the influential Belmont Report (Jonsen, 1998, p. 103f). The works of the National Commission ended in 1978, and the article of Toulmin, 'How Medicine Saved the Life of Ethics' appeared in 1982 (Toulmin, 1997). The turning in the normative ethics from the hands of the 'moralists' to the hands of the now called 'ethicists' occurred, actually, between the sixties and the seventies. This change had two different origins. One was the new interest of philosophers and secular thinkers on normative questions, as Toulmin said. The other was the change made by moral theologians, which in a high number began to relegate authority arguments, in favour of reasons. As Sandro Spinsanti has shown, the birth of bioethics was directly related to this turn of moral theology (Spinsanti, 1995).

Looking at bioethics from this perspective, it is clear that it is not an internal American phenomenon. The above described revolution has taken place not only in America but also in Europe, and it has been the necessary consequence of the process of secularisation of life which has taken place in Western cultures. In the same way as in the liberal revolutions of the 17th and 18th centuries, the ethics of civil government was taken out from the hands of theologians and churches and became secularised; now a similar revolution is taking place in the field of the management of body and life. The right of liberty of conscience protects not only religious and political freedom, as it was clearly established by the liberal revolutions of the modern centuries, but also moral freedom, understood as the right to decide autonomously, without foreign constraints, about the government of the body and the life. This is the new revolution, and bioethics is its most significant expression.

No discussion is reasonable about the fact that bioethics has been, at least to some extent, the natural consequence of the process of emancipation of human beings from certain kind of tutelage or paternalism, traditional in Western culture. That is why bioethics has stressed so strongly the principle of autonomy, the respect of the different value systems and substantive ideas of good proper of each one and every culture, and the reach of moral agreements throughout common and deliberative procedures. In the same way as the political revolutions of the 18th century looked for a new moral legitimisation of public norms, different from the paternalism proper of the old regime, the new moral revolution also introduces methods and procedures for participation and deliberation in order to legitimate moral behaviours.

This has been, as I have stated before, a general phenomenon all through Western culture, and not only in North-America. The reason why bioethics was born in the US, is not due to some particularities of the American ethos, but because it showed the first manifestations of the general phenomenon: the Civil Rights movement after the Second World War, and the development of the new technology and medicine. They were the first, but they are not the only ones.

The question is now to determine what the other Western cultures, and in particular Europe, can contribute to the bioethical movement. And to my mind, the answer is that they can, and perhaps should add many substantive things, or at least the following.

1. Autonomy plays in European bioethics a less prominent role than in America, maybe due to the Latin sense of community, virtue, and shared values. Perhaps the European experience can offer something on this point to the general movement of bioethics (Reich, 1995a, vol. 3, p. 1556-1562).
2. On the other hand, Anglo-American democracy is far from being a morally clear concept (Manin, 1997). Many European thinkers are nowadays preferring participative democracy over the traditional representative democracy, and the universality rule to the classical majority rule, opening in

this way the door to a new kind of democracy, now called participative and/or deliberative democracy (Habermas, 1994, 1996). These recent ideas should have, with no doubt, many ethical and bioethical consequences.
3. Phenomenology and hermeneutics, two typical products of European philosophy, can play an important role in enriching the analysis of concrete facts (Levinas, 1971; Ricoeur, 1990).
4. Also the textual analysis and the theory of text deconstruction could perhaps improve the quality of narrative ethics (Derrida, 1967a, 1967b).

The last question is whether the bioethical movement, born in the US but typically Western, can and will be spread all over the world. It seems possible, due to the increased trend to globalisation. But a true globalisation can not be understood as the mere diffusion of the Western way of life to the other countries and cultures. Globalisation must not only be an economical and political concept, but also, and principally, a moral one. It is necessary to rethink ethics and bioethics from a global point of view (Küng, 1996, 1997; Potter, 1988). This new ethics will include in its judgements all actual human beings, not only those of the developed world, but also those of the underdeveloped countries; it will also include the human beings of future generations, plus the other living beings; and, finally, the environment. We are far from reaching a cogent global bioethics. This is not yet history, but only future.

REFERENCES

American Hospital Association: 1973, *A patient's bill of rights*. In: Annas, G.J.: 1992, *The rights of patients*. 2nd ed., Humana Press, Totowa, NJ, p. 40-42.

Aristotle: 1960a, *Politica*. In: Bekker, A.I. and Gigon, O. (eds.), *Aristotelis Opera*. Vol. II, W. De Gruyter, Berlin.

Aristotle: 1960b, *Ethica Nicomachea*. In: Bekker, A.I. and Gigon, O. (eds.), *Aristotelis Opera*. Vol. II, W. De Gruyter, Berlin.

Beauchamp, T.L. and Childress, J.F.: 1979, *Principles of biomedical ethics*. Oxford University Press, New York.

Bentham, J.: 1988, *A fragment on government*. In: Burns, J.H., Dinwiddy, J.R. and Rosen, F. (ed.): *The collected works of Jeremy Bentham*. Oxford University Press, London and Oxford.

Chambers, T.: 1998, Retrodiction and the histories of bioethics. *Medical Humanities Review* 12(1): 9-22.

Dell'Oro, R. and Viafora, C. (eds.): 1996, *Bioethics: A history*. International Scholars Publications, Bethesda, MD.

Derrida, J.: 1957a, *De la grammatologie*. Minuit, Paris.

Derrida, J.: 1967b, *L'écriture et la différence*. Seuil, Paris.

Dworkin, G.: 1983, Paternalism: Some second thoughts. In: Sartorius, R. (ed.), *Paternalism*. University of Minnesota Press, Minneapolis.

Fox, R.C.: 1989, The Sociology of Bioethics. In: Fox, R.C., *The sociology of medicine: A participant observer's view*. Prentice Hall, Englewood Cliffs, New Jersey, p. 224-276.

Goldman, A.H.: 1980, *The moral foundations of professional ethics*. Rowman & Littlefield, Savage, Maryland.

Gracia, D.: 1989a, *Fundamentos de bioética*. EUDEMA, Madrid.
Gracia, D.: 1989b, La relación médico-enfermo en España: Balance de los últimos veinticinco años. *Todo Hospital* 62(12): 23-26.
Habermas, J.: 1994, *Faktizität und Geltung: Beitrage zur Diskurstheorie des Rechts und des demokratischen Rechtsstaats*. Suhrkamp, Frankfurt am Main.
Habermas, J.: 1996, *Die Einbeziehung des Anderen: Studien zur politischen Theorie*. Suhrkamp, Frankfurt am Main.
Hippocrates: 1972, *Epidemics I and III*. In: Jones, W.H.S. (ed.), *Hippocrates*. Vol. 1, Harvard University Press, Cambridge, Mass.
Hippocrates: 1981, *Decorum*. In: Jones, W.H.S. (ed.), *Hippocrates*. Vol. 2, Harvard University Press, Cambridge, Mass.
Jonsen, A.R.: 1997a, Introduction to the history of bioethics. In: Jecker, N.S., Jonsen, A.R. and Pearlman, R.A. (eds.), *Bioethics: An introduction to the history, methods, and practice*. Jones and Bartlett Publishers, Sudbury, Mass, p. 3-11.
Jonsen, A.R.: 1997b, The birth of bioethics: The origin and evolution of a demi-discipline. *Medical Humanities Review* 11(1): 9-21.
Jonsen, A.R.: 1998, *The birth of bioethics. An historical memoir of the origins and evolution of bioethics in the United States, 1947-1987*. Oxford University Press, New York.
Jonsen, A.R., Veatch, R.M. and Walters, L. (eds.): 1998, *Source book in bioethics: A documentary history*. Georgetown University Press, Bethesda.
Kant, I.: 1995, *Grounding for the metaphysics of morals*. In: Cahn, S.M. (ed.), *Classics of Western philosophy*. Hackett Publishing Company, Indianapolis, p. 1092-3.
Küng, H. (ed.): 1996, *Yes to a global ethic*. SCM Press, London.
Küng, H.: 1997, *A global ethic for global politics and economics*. SCM Press, London.
Levinas, E.: 1971, *Totalité et infini*. Martinus Nijhoff, The Hague.
Locke, J.: 1823, *Two treatises of government*. In: *The works of John Locke*. Thomas Tegg, W. Sharpe and Son, G. Offor, G. and J. Robinson, J. Evans and Co., Dublin.
Manin, B.: 1997, *The principles of representative government*. The Syndicate of the Press of the University of Cambridge, Cambridge, England.
Mill, J.S.: 1995a, *Utilitarianism*. In: Cahn, S.M. (ed.), *Classics of Western philosophy*. Hackett Publishing Company, Indianapolis.
Mill, J.S.: 1995b, *On liberty*. In: Cahn, S.M. (ed.), *Classics of Western philosophy*. Hackett Publishing Company, Indianapolis.
Moore, G.E.: 1994, *Principia Ethica*. Cambridge University Press, Cambridge.
The National Commission for the Protection of Human Subjects of Biomedical and Behavioral Research: 1978, *The Belmont Report: Ethical principles and guidelines for the protection of human subjects of research*. DHEW Publication No. (OS) 78-0012, Appendix I, DHEW Publication No. (OS) 78-0013, Appendix II, DHEW Publication (OS) 78-0014, Washington.
Plato: 1995, *Republic*. In: Cahn, S.M. (ed.), *Classics of Western philosophy*. Hackett Publishing Company, Indianapolis.
Potter, V.R.: 1970, Bioethics, the science of survival. *Perspectives in Biology and Medicine* 14: 127-153.
Potter, V.R.: 1971, *Bioethics: Bridge to the future*. Prentice-Hall, Englewood Cliffs, New Jersey.
Potter, V.R.: 1988, *Global bioethics*. Michigan State University Press, East Lansing, Michigan.
Potter, V.R.: 1990, Getting to the year 3000: Can global bioethics overcome evolution's fatal flaw? *Perspectives in Biology and Medicine* 34(1): 89-98.
Potter, V.R.: 1995, Global bioethics: Linking genes to ethical behavior. *Perspectives in Biology and Medicine* 39(1): 118-131.
Potter, V.R. and Whitehous, P.J.: 1998, Deep and global bioethics for a livable third millennium. *The Scientist* 12(1): 9.
Reich, W.T.: 1993, How bioethics got its name. *Hastings Center Report* 23(6, Suppl.) S6-S7.

Reich, W.T.: 1994, The word 'bioethics': Its birth and the legacies of those why shaped it. *Kennedy Institute of Ethics Journal* 4(4): 319-335.

Reich, W.T.: 1995a, Medical Ethics, History of. In: Reich, W.T. (ed.), *Encyclopedia of bioethics*. Revised Edition, Macmillan, New York.

Reich, W.T.: 1995b, The word 'bioethics': The struggle over its earliest meaning. *Kennedy Institute of Ethics Journal* 5(1): 19-34.

Ricoeur, P.: 1990, *Soi-même comme un autre*. Seuil, Paris.

Ross, W.D.: 1930, *The right and the good*. Clarendon Press, Oxford.

Rothman, D.J.: 1991, *Strangers at the bedside. A history of how law and bioethics transformed medical decision making*. Basic Books, New York.

Spinsanti, S.: 1995, *La bioetica: Biographie per una disciplina*. Franco Angeli, Milano.

Toulmin, S.: 1997, How medicine saved the life of ethics. In: Jecker, N.S., Jonsen, A.R. and Pearlman, R.A. (eds.), *Bioethics: An introduction to the history, methods, and practice*. Jones and Burtlett Publishers, Sudbury, MA, p. 101-109.

Wear, A., Geyer-Kordesch, J. and French, R.: 1993, *Doctors and ethics: The earlier historical setting of professional ethics*. G.A. Rodopi, Amsterdam-Atlanta.

HENK A.M.J. TEN HAVE

THEORETICAL MODELS AND APPROACHES TO ETHICS

1. INTRODUCTION

Interest in the moral problems of medicine and health care has grown rapidly over the last three decades, not only in the U.S. but in the majority of European countries as well (Gracia, 1999). Although medical practice has always been the subject of ethical reflection and critique, two recent developments in particular have served to transform the traditional notion of medical ethics: (1) advances in medicine and health care due to the influence of biotechnology and technology-oriented medicine, and (2) the rapidly changing socio-cultural context marked by the prevalence of a plurality of values in Western countries, especially those values that bear on the provision of health care.

Since the middle of the nineteenth century, medicine and health care have advanced significantly; numerous life-threatening situations though common in daily clinical practice are now met with effective therapeutic interventions. However, advances in biotechnology and biomedicine have also stimulated the realisation that the medical enterprise as such is in need of reappraisal (Greaves, 1979). Not only have advances in medical technology become disvalued as impersonal and inhumane, but increasingly the goals, direction and effectiveness of technology-oriented medicine have come into question.

The socio-cultural context of medical practice has changed in many respects. During the last three decades, the influence of religious values in the resolution of moral problems in medicine has diminished whereas a non-religiously, secularly-grounded normative view of human life has become more influential. This view emphasizes personal autonomy and each patient's right to make his or her own health care decisions. At the same time, the writings of Zola, Illich, and Foucault focused on the power of health care professionals in present-day society, as well as on the so-called 'medicalisation' of post-modern culture. Such critiques have resulted in a change in attitude towards health care professionals and an increasing

H.A.M.J. ten Have and B. Gordijn (eds.), Bioethics in a European Perspective, 51-82.

demand by patients to participate in medical decision-making at virtually every level - not only in the physician-patient encounter but also within the health care system as such.

The social status of physicians has been affected significantly by these factors. Traditionally, 'medical ethics' referred to the deontology of the medical profession, to codes of conduct which consist partly of ordinary moral rules, partly of rules of etiquette, and partly of rules of professional conduct (Downie, 1974). In this sense bioethics has the following characteristics:

1. It is essentially a set of problems that focus on the internal morality of medicine, *viz.* those values, norms, and rules intrinsic to the actual practice of health care. Medicine is not considered a merely technical enterprise that can be morally evaluated from some exogenous standpoint. On the contrary, the professional practice of medicine always presumes and implies a moral perspective or point of view; therefore, what is judged to be sound medical practice is determined by the shared rules and standard procedures of the practice.
2. It is related to the professionalisation of medicine. Through this historical process emphasis is placed on the common good, and this was combined with an appeal to the self-interest of the members of the profession. Social recognition could only be gained on the basis of a strong internal organisation and self-imposed standards of behaviour. Self-regulation by physicians and a special style of life, structured in terms of high ideals, duties, and virtues, could promote the physician's image, and thus the power and prestige of each member of the medical profession.
3. Since it was primarily concerned with explicating norms and formulating standards of professional conduct, medical ethics and etiquette have been segregated for a long time from general intellectual history (Fox, 1979). Moreover, before the 1960s, medical ethics was not a subject frequently discussed in public fora and the extant literature. Apparently, there was a consensus of opinion concerning the moral commitments of those who provided medical care, and the explication or codification of these commitments was regarded as the principal concern of medical professionals.

Since the 1960s, professional medical ethics has gradually detached itself from its traditional deontology (ten Have and van der Arend, 1985), although there are important phase differences between the North-American and European countries; especially in some Eastern and South European countries the emphasis in medical ethics still is on 'medical deontology'. In North-western European countries professional medical ethics more and more is subsumed under 'health care ethics', or 'bioethics'. These new terms, as is shown by Diego Gracia in the previous chapter, tend to indicate that the discipline of ethics not only includes problems that arise in the physician-patient relationship, but also a number of moral problems posed by other health care professionals, as well as moral issues created by the health care system, and public policy issues engendered by biomedical advances and the results of research. This is also illustrated by the dramatic increase in the number of publications on moral problems in medicine and health care authored by non-physicians. Consequently, the range of problems that properly is subsumed under the rubric 'medical ethics' is considerably enlarged; there are ever new and more complex moral issues, and new participants emerge to participate in an intensified set of medico-ethical debates.

Thus, the result of the gradual transformation of medical ethics is two-fold.

First, it has produced a new professional - the health care ethicist or 'bioethicist' who possesses a specific body of knowledge and particular cognitive skills; who publishes in specialized journals, participates in newly-formed societies, and teaches in newly-established centres, institutes, and departments.

Second, it has produced a new socio-cultural interest in medico-moral matters of significant public concerns - particularly in those countries where advanced biomedical technology permeates public as well as private life. 'Bioethics' is a way of publicly addressing, explicating and debating problems generated by science and technology.

The outcome of the above-mentioned transformation process is more salient in the U.S.A. than in most European countries. Moral problems in health care are generally approached in U.S. bioethics in a specific way, - more analytic and applied as is usually the case in many Continental approaches.

The aim of the first part of this chapter is to identify and criticize the dominant conception of bioethics which seems to prevail in bioethical debates everywhere in the world. This conception originated in the American rather than in the European context. However, interest in alternatives conceptions and methods of bioethics is currently increasing. Criticizing the dominant conception, therefore, at the same time leads to the question whether there are specific characteristics of European approaches to moral problems in health care. Is it possible to identify typically European approaches in the area of bioethics? The second part of this chapter describes significant ideas and developments in European medical ethics.

2. EVALUATING THE CURRENT STATUS OF ETHICS

Today there is growing concern that the results of the transformation from traditional medical ethics to modern bioethics are unsatisfactory. Professionalisation and institutionalisation of ethics received an enormous stimulus because both the adequacy and the relevance of medicine's internal morality were put into question. Professional ethicists have placed more and more emphasis on the crucial role of external morality: the principles, norms, and rules operative in society that bear on medicine and are frequently codified in law. Thus, for some, medicine and health care are nothing more than interesting 'intellectual' phenomena with respect to which general ethical theories, principles, and rules may be applied.

This shift from internal professional to external morality and the predominant interpretation of medical ethics as 'applied ethics' encouraged physicians to criticize present-day medical ethics for its lack of attention to the practical vicissitudes of health care, for its theoretical biases, and its conceptual alienation from clinical reality (Editorial, 1990; Vandenbroucke, 1990; Fulford, 1994; Welie, 1998; Wulff, 1998).

It is also claimed that the conceptual ground of medical ethics is too limited and even reductive when seen from the perspective of the tradition of philosophical ethics itself. Must medical ethics be conceptualized as applied theory rather than as reflective practice? (Baier, 1985; Kass, 1990; Murray, 1994; Zwart, 1995; Evans, 1998).

In addition, it has been suggested that there is a serious discrepancy between the public's attention to moral questions and the actual impact of ethical analysis on the routine practices of medicine, as well as the current direction of medicine's development. Moral issues tend to appear every day, but how successful are 'bioethicists' in addressing these novel issues? To be sure, the media reflect a constant fascination with the myriad of moral problems in health care, but what concrete effect do these debates have on physicians' decisions in daily clinical medicine, on nursing practice, and on public health policies? Arguably, such 'discrepancies' result from the very conception of medical ethics in our time (ten Have and Kimsma, 1990).

3. THE DOMINANT CONCEPTION OF MEDICAL ETHICS

During the last thirty years, a popular and unique view of medical ethics as a new discipline separated from philosophy, theology as well as medicine has emerged. The growing appeal of this new discipline among public and scientific circles of opinion leaders can be attributed to the empowering combination of two traditional notions from the history of moral philosophy: 'application' and 'principle'.

The current conception of medical ethics reflected in the mainstream of scholarly literature is that of *applied ethics*. In Beauchamp and Childress' well-known textbook, biomedical ethics is defined as

> The application of general ethical theories, principles and rules to problems of therapeutic practice, health care delivery, and medical and biological research (Beauchamp and Childress, 1983, p IX-X).

Instead of the theoretical abstractions of traditional moral philosophy, applied ethics can contribute to analyse dilemmas, resolve complex cases and clarify practical problems arising in the health care setting. The practical usefulness of applied ethics not only manifests itself in biomedicine; it has a wider scope as well. In the *Encyclopedia of Applied Ethics*, the following definition is presented:

> Applied ethics is a general field of study that includes all systematic efforts to understand and to resolve moral problems that arise in some domains of practical life, as with medicine, journalism, or business, or in connection with some general issue of social concern, such as employment, equity or capital punishment (Winkler, 1998, p. 192)

A distinction is made between three major areas of applied ethics: biomedical ethics, business and professional ethics, and environmental ethics. However, the table of contents of the four encyclopedia volumes show a wide range of topics covered, such as archaeological ethics, censorship, divorce, electronic surveillance, gun control, nuclear power, vegetarianism, and wildlife conservation. Applied ethics can extend to almost any area of life where ethical issues arise. 'Application' here has a double connotation: it indicates that ethics is available for what we usually do, it applies to our daily problems; but it also is helpful, practical, in the sense that ethics is something to do, it works to resolve our problems.

The second characteristic of the dominant conception of medical ethics is the focus on *principles*. If ethics is conceived as applied ethics, then subsequent reflection is needed on what is being applied. The emerging consensus that principles should provide the answer to this query, is coherent with the moralities of obligation that have dominated modern ethical discourse, especially since Kant. Behaviour in accord with moral obligations is considered morally right. The morality of behaviour is a morality of duty. Morality is understood as a system of precepts or rules people are obliged to follow. Particularly in the early days of bioethics, when medical power was strongly criticized, and the rights of patients were vehemently emphasized as requiring respect, the moralities of obligation presented themselves as a common set of normative principles and rules that we are obliged to follow in practice. As Gracia (1999) pointed out, the *Belmont Report* in 1978 was influential because it was the first official body to identify three basic ethical principles: autonomy, beneficence and justice. A basic principle was defined as a general judgement serving as a basic justification for particular prescriptions and evaluations of human actions. From these principles, ethical guidelines can be derived that could be applied to the biomedical area. About the same time, Beauchamp and Childress, in the first edition of their book, introduced the four-principles approach, adding 'nonmaleficence' to the above three principles. In their view, principles are normative generalisations that guide actions. However, as general guides they leave considerable room for judgement in specific cases. Various types of rules are needed to specify the principles into precise action guides.

Although Beauchamp and Childress have considerably nuanced their theoretical framework in later editions, their work has contributed to the conception of medical ethics that is currently dominating the practical context, in ethics committees, clinical case-discussions, ethics courses, and compendia and syllabi. This conception is sometimes called 'principlism': the focus is on the use of moral principles to address ethical issues and to resolve conflicts at the bedside (DuBose, Hamel and O'Connell, 1994). Belief in the power of principlism is sometimes proselytizing. Raanan Gillon, for example, argues that the advantage of the four principles not only is that they are defensible from a variety of theoretical moral perspectives, but also that "they can help us bring more order, consistency, and understanding to our medico-moral judgements" (Gillon, 1986, p. viii). Later, Gillon used the principles-approach to develop a major scholarly project, the voluminous textbook *Principles of Health Care Ethics* (Gillon, 1994). Over 100 authors discuss in 90 chapters all possible ethical dilemmas in modern health care, employing the analytical framework of the four principles. In his Preface, Gillon confesses that he is inclined to believe that the four principles approach can encompass all moral issues, not only those arising in health care. Principlism apparently is a universal tool; it provides a method of resolving all moral issues in all areas of daily life, whatever the personal philosophies, politics, religions, moral theories of the persons involved.

The dominant conception of medical ethics, arising from the combination of the notions 'application' and 'principle', implies the following set of interdependent presuppositions (ten Have, 1990b):

1. medical ethics is application of ethical theory and ethical principles;
2. there is a body of available ethical theories and principles, and rules to be applied to a variety of practical, biomedical problems, in particular the framework of the four principles;
3. professional ethicists have a special expertise in applying ethical theories and principles, whereas non-ethicists (e.g., physicians) merely provide moral problems for applied ethics;
4. medical ethics is general ethics applied to medicine. That is, the context in which these problems arise is not unique in the sense of being characterized by specific values which generate special problems. Indeed, the medical context is viewed as a practice-ground for a new profession of biomedical ethicists; and

5. the aim of medical ethics is to proffer practical recommendations and prescriptions based on or deduced from ethical theories and principles.

This set of presuppositions to some degree clarifies why medical ethics is perceived by many as an independent discipline. For example, there is the view that ethics should perform four tasks: conceptual clarification, analyzing and structuring arguments, weighing alternatives, and advising a preferable course of action (de Beaufort and Dupuis, 1988, p. 19-20).

The central contribution of medical ethics is therefore restricted, but nonetheless powerful. It does not necessarily result in judgements regarding what we should do. The ethicist provides the topography of arguments, and objectifies the options. The ethicist regards himself as a disinterested and neutral observer of medical practice, who is in the best position to weigh moral alternatives.

4. THE DOMINANT CONCEPTION: CRITIQUE

Only recently have the presuppositions underlying the prevailing conception of applied ethics been critically questioned. Consider the following three arguments:

1. In daily medical practice, medical ethics focuses on *mid-level principles* - respect for autonomy, beneficence, nonmaleficence, and justice. These principles are applied to dilemmas, cases, and problems encountered in the practice of health care. From a specific principle, guidelines or recommendations can be derived in order to resolve various problematic situations. Yet there is no single rational criterion on the basis of which to decide which principle is overriding; there is no definitive scheme for ordering principles and for choosing between them. As long as the principles of applied medical ethics are not integrated into some broader theoretical framework they tend to lead to conflicting judgements about which actions and social policies one ought to carry out. Even if one proceeds from some articulated moral theory (e.g., consequentialism, contractarianism) one cannot evade the chaos of conflicting moral judgements (Brody, 1988). The lack of agreement on which moral theory to apply on concrete medical cases could make applied ethics counterproductive. Confronting physicians and medical students with a variety of conflicting but plausible theories, applied medical ethics may be seen to give no moral guidance but to reinforce the belief that whatever is done in problematic situations, some moral theory will condone it, another will condemn it (Baier, 1985). Thus the primacy of applied ethics and the deductivist model of applying general moral theories and intermediate principles can only lead to an inadequate way of conceiving the relation of ethics to medicine (Jonsen, 1990).

Because the dominant conception of medical ethics focuses on the application of principles, norms and rules, it is rather loosely embedded in philosophy, and lacks a more encompassing critical, theoretical perspective on its own practical activities.

2. Secondly, the dominant conception has developed within a *particular cultural context*. The fundamental ethos of applied medical ethics, its analytical framework, methodology, and language, its concerns and emphases, and its very institutionalisation have been shaped by beliefs, values, and modes of thinking grounded in specific social and cultural traditions. Nowadays, the medical ethics literature serves as one of the most powerful means by which to express and articulate these traditions. However, the medical ethics literature only rarely attends to or reflects upon the socio-cultural value system within and through which it operates. Scholars usually assume that its principles, theories, and moral views are transcultural. H.T. Engelhardt (1986), for example, distinguishes between *two* levels: that of secularized pluralistic society and that of the many particular moral communities with competing visions of the good life. Bioethics, in his opinion, should focus on the societal level, speaking across gulfs of moral discourse; it is a common neutral language, a secular moral grammar, guaranteeing a peaceable society. The most interesting task of ethics is on the first, societal level: promoting and defending, in the context of health care, the general secular moral language of mutual respect. Critics agree that this is an important task; but it flows from a rather thin or minimalist conception of ethics (Callahan, 1981). Ethics is conceptualized as procedural; it is the regulation of social relations through peaceable negotiation. In order to speak the language of mutual respect, all other moral languages must be pacified.

But why should we abstain from our particular moral language in favour of a neutral common language? This question points to an important problem: how neutral is the common neutral language? Is Engelhardt's language itself not the specific moral language of a specific moral community? Is this language itself not the expression of a commitment to a certain 'hypergood' (Taylor, 1989), in particular, the good of universal and equal respect and self-determining freedom, - primal values in the liberal tradition? Such questions assume that the value of mutual respect and rights to privacy are not decontextualized standards but themselves expressions of community-bound agreements.

Only recently has there been an increasing awareness that a critical examination of the socio-cultural context is necessary if we are to better understand the strengths and weaknesses of this currently dominant conception of medical ethics. The medical sociologist R. Fox (1989), for example, has shown how the political norms of liberalism and individualism are very much characteristic of North American bioethics. By stressing the autonomy and rights of individuals, other significant considerations (e.g., community and the common good, duties and responsibilities) have been neglected, as have critical philosophical questions concerning the value of medical progress and personal and public health in communal life. Although interest in the philosophy of medicine in Europe in general seems to emphasize the social aspects of medicine and the common good, rather than individual autonomy, the dominating conception of medical ethics in particular countries such as the Netherlands seems in many respects not significantly different from that in the U.S., where liberalism and personal autonomy are stressed.

3. Another criticism of the dominant conception of medical ethics is its *inattention to the particularities of the practical setting.* Moral theories and principles are necessarily abstract and therefore not immediately relevant to the particular circumstances of actual cases, the concrete reality of clinical work, and the specific responsibilities of health care professionals. By appealing to principles, norms or rules, applied ethics may fail to realize the importance of concretely lived experiences of health care professionals, as well as patients. The moral agent is taken to have an abstract existence. This point is critically elaborated by contemporary philosophers. Ethics, according to B. Williams (1988), does not respect the concrete moral subject with his personal integrity. It requires the subject to give up his personal point of view and exchange it for a universal and impartial point of view. This is, Williams argues, an absurd requirement, because the moral subject is requested to give up what is constitutive for his or her personal identity and integrity. The idea that knowledge of normative theories and principles can be applied to medical practice simply ignores the fact that moral concerns tend to emerge from experiences in medical settings themselves. A similar issue is raised by Ch. Taylor in his *Sources of the Self*, in which morality and identity are considered two sides of the same coin (Taylor, 1989). To know who we are is to know to which moral sources we should appeal. The community, the particular social group to which we belong, is usually at the centre of our moral experience. Even the use of ethical language depends on a shared form of life. The Wittgensteinian notion that our understanding of language is a matter of picking up practices and being inducted into a particular form of life is germane here.

In short, medical ethicists should become more appreciative of the actual experiences of practitioners and more attentive to the context in which physicians, nurses, patients, and others experience their moral lives, e.g., the roles they play, the relationships in which they participate, the expectations they have, and the values they cherish (Zaner, 1988). The physician-patient relationship is neither a-historical, a-cultural nor an abstract rational notion; persons are always persons-in-relation, are always members of communities, are immersed in a tradition, and participants in a particular culture.

From these points of view two conclusions may be drawn (ten Have and Kimsma, 1990):

1. Morality is something we all participate in; medical ethics in particular is not the result of esoteric knowledge; anyone involved in the medical setting is ipso facto a moral participant and 'expert' at least with regard to moral experience and intuitive knowledge.
2. The moral experience inherent in health care practices must be taken into account - more than the conformity of these practices with pre-existing ethical theories. From the perspective of applied medical ethics, abstracting from the reality of practices and appealing to moral principles and rules outside these practices, are necessary conditions to criticize health care practices. The problem, however, is not only how such a standpoint external to concrete practices is possible, but also whether appeals to external morality are not vain without intimate knowledge of the morality internal to the practices in question (Jensen, 1989).

5. ALTERNATIVES

Given the criticisms noted above, an alternative approach to and conception of medical ethics is clearly needed - a conception that provides a more comprehensive understanding of the nature, scope, method, and application of ethics in the contemporary health care context. It will be necessary to re-connect medical ethics with both a general philosophical standpoint and the concrete practice of medicine.

In order to achieve a more adequate understanding of the possibilities for such a re-connection, it may prove useful to outline promising new perspectives.

1. In response to the theoretical and methodological weaknesses of applied ethics, *new approaches* to medical ethics are available: phenomenological ethics (Zaner, 1988), hermeneutic ethics (Carson, 1990; Leder, 1994), narrative ethics (H. Brody, 1987; Hunter, 1988; Newton, 1995), and care ethics (Tronto, 1993). Furthermore, traditional conceptions have been revitalized, notably the new casuistry (drawing from the classical casuistic mode of moral reasoning) (Jonsen and Toulmin, 1988), and the virtue approach, emphasizing qualities of character in both individuals and communities (Drane, 1988; Pellegrino and Thomasma, 1993).
2. Other approaches have emerged due to the recent appreciation of the relevance of the *social and cultural matrix* in which medical ethics necessarily operates. For example, D. Callahan (1990) has argued that the ethical problems generated by the need for health care resource allocation and for the formation of a new health policy have forced us to explore the goals and ideals of medicine as well as the meaning of health in modern society.

However, thoughtful empirical research into the value systems relevant to the formation of moral issues in health care is relatively rare. For example, data acquired from sociological value research as well as the methodological approaches of social scientists (e.g., Halman, *et al.*, 1987; Inglehart, 1990) are virtually unknown and therefore ignored in medical ethics. The rigid distinction between descriptive and normative ethics could, in part, account for the absence of empirical value studies in medical ethics. Only recently, however, there are signs that a more positive interaction between medical ethics and the social sciences can be achieved (Weisz, 1990). What is particularly striking is the interest in so-called empirical medical ethics. The focus of medical ethical research is shifting from applying ready-made ethics toward studying ethics-in-action (Arnold and Forrow, 1993). A variety of research methods is used: participatory observation, questionnaires and interviews, decision analysis, quality assessment, preference polls. The common denominator is that qualitative and quantitative data are collected via the empirical study of ethical questions. Many of these studies are fascinating since they show the underlying value pattern of specific practices and the intrinsic norms which are operative in clinical work, for example in surgery (Bosk, 1979), genetic counselling (Bosk, 1992), intensive care (Zussman, 1992), neonatal care (Anspach, 1993), and nephrology (Lelie, 1999). Although empirical research in ethics can provide new and useful insights, and can be regarded as complementary to philosophical approaches (Hope, 1999), it is also troubled with fundamental problems (ten Have and Lelie, 1998). One of the basic questions concerns the moral relevancy of empirical data. Empirical research can help to explain and understand the attitudes, reasonings and motivations of the various actors in the health care setting, but empirical data in themselves can not justify how the actors ought to behave or what kind of decisions are morally justified (Pellegrino, 1995).

3. A relatively new conception of medical ethics is so-called *clinical ethics*. It has emerged in response to the criticism that applied ethics is too far removed from the realities of medical practice. Clinical ethics aims to reorient medical ethics within the daily health care setting (Jonsen, Siegler and Winslade, 1986).

The extent to which clinical ethics differs from the prevailing conception of applied ethics can be characterized as follows:

a. Interdependence of technical and normative dimensions of medical judgement. This interdependence which is at the basis of clinical ethics, is repeatedly underlined by recent work in philosophy of medicine. It is argued that clinical medicine is intrinsically a moral enterprise since it presumes a healing relationship between physician and patient. Since value judgements are pervasive in clinical decisions, moral concerns are inseparable from certain technical concerns, e.g., the correct diagnosis and the most effective treatment.

b. Insider perspective. The realities of clinical decision-making are crucial for the identification of ethical issues such that the ethical problems that arise in the practice of surgery are not identical to those that arise in pediatrics, obstetrics, or gynaecology. Moreover, they are not of the same nature 'medically', since they differ with respect to risks and benefits. Specifically, the insider's perspective allows for the determination of whether risks, in routine investigations, are low, or whether they are substantial with questionable benefits. Thus it is asserted that an insider perspective is not only required to direct attention to the ethical questions that arise in clinical encounters, but properly to acquire empirical data relating to the process and outcome of these clinical encounters: How do patients and physicians actually make decisions? What moral options are involved? What are the effects of personal and professional values in reaching clinical decisions?

c. Method of induction. Instead of utilizing a deductive method by which general theories and principles are applied to practical moral dilemmas, an inductive methodology should be utilized which begins with a careful analysis of specific empirical conditions. This view, in part, accounts for the renewed interest in classical casuistry (Jonsen and Toulmin, 1988). The casuistical method includes the search for paradigm cases in which a particular moral maxim for right action is clearly applicable. Analogies are then proposed regarding cases in which, due to different circumstances, other moral maxims appear less suitable. The casuist thus explores a range of cases and scenarios forming more or less plausible arguments. Thus the factual circumstances of a case are extremely relevant: by modifying them they reveal new insights. The casuist's task is to determine the degree to which relevant moral maxims 'fit' the particular circumstances. Even more: the casuist seeks to determine which factors, personal preferences, and social conditions and values are relevant enough to be judged as significant 'moral facts'.

d. Clinical ethics is an inherent function of medicine itself. This is a logical consequence of the points just mentioned. When physicians consider ethics as intrinsic to their craft, then the ethical analyses of medical decisions cannot proceed from an externally imposed system; essentially, they are an inherent, second-order function of clinical medicine itself.

From this survey of criticism and alternative approaches it is concluded that medical ethics is presently dominated by a limited conception of ethics - the *application* of moral theories and principles to cases. This conception depreciates the fundamental internal morality of the professional practice of medicine by stressing external morality. This conception also reveals a lack of interest in the empirical realities of clinical medicine and neglects the socio-cultural value-contexts in which medical care is provided. In short, a broader framework for a practicable medical ethics is needed.

6. EUROPEAN APPROACHES TO MEDICAL ETHICS

It is prima facie problematic to identify typically European approaches to medical ethics. Heterogeneous philosophical ideas and theories rule the Continent (e.g., postmodernism, hermeneutics, critical theory, to name a few) without any major and dominating school. The same is true for ethics. But this situation seems not too different from North America. Although in some European countries the principles approach is influential, the spectre of conceptions in medical ethics seems broader than in the U.S. On the one hand, in many countries ethics is very much under the influence of philosophical and theological traditions, and not dominated by analytical methodology. In Central and Eastern European countries, bioethics has emerged only recently, mainly from the former departments of marxist-leninist philosophy or social sciences. On the other hand, only in a very few countries, such as the United Kingdom, the Netherlands and the Nordic countries, medical ethics in particular is the specialized enterprise of a new profession; in most countries it is the recognized business of medical practitioners, who therefore dominate public debate. This is, presumably, also one of the reasons why the term 'bioethics' is not as frequently used as 'medical ethics' or 'health care ethics'.

However, another reason is that 'bioethics' is often not identified with a discipline of moral philosophy, but with a specific approach of moral problems. Sometimes, such identification leads to negative responses, accusing 'bioethicists' (not 'medical ethicists') of being facilitators of medical technologies attempting to soften moral resistance against innovations. For example, it has been argued that moral intuitions about the intrinsic dignity of human embryos do not in general favour the instrumental use of embryos for research; in order to undermine intuitive opposition to embryo research, bioethicists have introduced, according to this argumentation, the new terminology of 'pre-embryo' and new conceptions of personhood, thus connecting moral status and human development. These kind of arguments must be evaluated against the background of events in Germany. In 1988 and 1989, the philosopher Peter Singer was invited to lecture on the subject of euthanasia for severely disabled newborn infants in Germany. However, the invitation was cancelled. When trying to lecture at another university, protesters made it impossible for him to speak. A broad coalition of left- and right wing groups did not want issues such as euthanasia and the right to life of handicapped people to be discussed in Germany (Singer, 1990). Since then, other activities against bioethics have taken place, for example the cancelling of the annual conference of the European Society for Philosophy of Medicine and Health Care in Bochum in 1990, of the International Wittgenstein Symposium in Kirchberg in 1991, and the disruption of courses on practical ethics in several German universities. For the protesters, bioethics is an import-product supporting the U.S. medical-industrial complex.

In other European countries there is no such radical opposition against bioethics. But, what is noticeable, is a growing awareness that the dominant conception of medical ethics, though in a sense very effective in education and public debate, lacks attention to certain fundamental aspects. If there is a difference in the medical ethics literature between Europe and North America, it is that European authors put more emphasis on: (1) the historical perspective of ethical issues, (2) the sociocultural context, and (3) substantive normative viewpoints.

6.1. Historical Perspective

Present-day interest in medical ethics is regarded as the most recent phase of a tradition of theoretical reflection upon medicine. In the thematic development of philosophy of medicine over the last hundred years three phases can be distinguished: an epistemological, an anthropological and an ethical phase (ten Have, 1990c). It is remarkable that from 1870 onwards there has been a rapid growth of medico-philosophical literature, particularly in Germany, France and Poland. Initially, the identity of modern medicine was described in epistemological terms. Medicine was characterized as a natural science. In this scientific conception of medicine, the artistic element, the art of medicine, had been eliminated. But, at the same time, the unity and coherence of medicine were endangered through the successes of its scientific approach. In the philosophical literature, two problems were identified: first, medical knowledge is fragmented and medical practice one-sided because of specialisation; second, the patient as the object of medicine is no longer adequately addressed since the conceptual tools of medicine are insufficient and too simple. Solutions were sought by proposing more rigorous methodologies, synthesis of medical knowledge in grand theories, and re-interpreting medicine as an art.

The interpretation of medicine as an art evolved into a new conception of medicine as anthropological science, - influential from 1930 until 1960, particularly in Germany and the Netherlands. What is important in this conception is the tendency to idealize the doctor-patient relationship. The subject should be re-introduced into medicine; that implied acknowledgement of the subjectivity of the knowing and acting subject (the physician) but also that of the object (the patient). Medicine was considered a unique profession in systematically and methodically attending to the patient as an irreducible person.

Since the 1960s, this anthropological orientation has been rapidly superseded by a growing interest in medical ethics. However, there is a marked continuity between these two phases of philosophy of medicine. By concentrating on the subjectivity of the patient, anthropological medicine paved the way for the subsequent ethical phase. It opened up the moral dimension of medicine for public reflection, because it argued that medicine itself is a normative science of life.

The current preoccupation with ethical problems is, in this view, not discontinuous with earlier efforts to philosophize about medicine (ten Have, 1998a). In a certain way, it shares the same commitments and fundamental problems as earlier phases, although with different concepts and vocabularies. Medical ethics, therefore, is part of a long tradition of philosophical reflection on health care. However, what is new, is the tendency nowadays to phrase fundamental problems in the language of good or bad, right or wrong, acceptable or unacceptable. Furthermore, within such a traditional view, medical ethics is not so much focused on solving these problems, rather than on clarifying their value-context, analyzing, for example, the goals of medical practice and the subjectivity or personhood of the patient.

6.2. *Socio-cultural Context*

The genesis, impact and perplexity of medical ethical problems are analyzed in relation to the structure and organisation of the health care system in a particular country as well as the framework of social values in which the problems present themselves. For example, moral problems of neonatology are related to the rise of neonatology as an independent discipline, the use of increasingly sophisticated technology, and the development of a specific ethos in its practitioners. The sociocultural context is also considered important for both the perception and management of moral issues in medicine. In the euthanasia debate in the Netherlands, for instance, it is not only important to analyse the moral arguments pro and con active termination of human life, but also to examine the changing attitudes towards a good death, the rapid secularisation of a religion-based and organized society, as well as the fact that many patients have a long-standing relationship with a general practitioner who can 'manage mortality' at home. But contextualism also implies that the role of individual actors is related (and explained in reference) to sociocultural conditions. For example, although there is much discussion about the welfare state in the Netherlands, the basic notion of solidarity as well as state protection of the vulnerable are not really disputed (ten Have and Keasberry, 1992). Introducing libertarian and free market thinking into health care, has in many European countries only resulted in some degree of strongly regulated competition for, in most instances, marginal services. Two examples from the Dutch context can illustrate this point. When the government announced that it wanted to introduce competitive elements in health care financing, many insurance companies merged; from initially over 200 companies only a few dozen remained that have now divided the potential market among each other without much competitive risks. Another example are governmental proposals during the early 1990s to get hold on the pharmaceutical budget; finally, it was announced that only a restricted number of medicines would be fully reimbursed through health insurance. The response of the pharmaceutical industry almost annulled the effect of such proposals: although in the Netherlands the use of medicines is almost the lowest in Western Europe (8 units per capita in 1989, compared to 49 units in France and 27 in Italy), the consumer price per unit is the highest (162 per unit, compared to 90 in Italy and 62 in France) (*Kiezen en Delen*, 1991).

6.3. Substantive Ethical Issues

Although European nations are in principle as pluralistic as the U.S., procedural ethics seems less pervasive than the privileged solution to moral controversies. At least in politics, many countries have strong social-democratic and christian-democratic traditions, sharing many substantive normative ideas on communal relations, labour, social welfare, health care. Value research has shown that in post-modern societies there is de facto a lot of agreement and overlapping consensus concerning moral values (such as tolerance, equality, solidarity) (Halman, *et al.*, 1987).

Moreover, it is argued that for the most pressing problems in medical ethics a procedural approach is insufficient. How can scarce resources in health care be allocated without substantial ideas on essential or adequate care? And how can such ideas be developed without a philosophy of the kind of society we want, without a substantive conception of health and human life, without a politics of the good? An ethics of principles is too much focused on cure and technology. It is argued that a thin conception of ethics is unsatisfactory and that a broader conception is needed, for example an ethics of care concerned with meaningful life and filial morality.

7. TRANSCENDING PRAGMATISM

The focus on the historical background, the socio-cultural context as well as the substantive ethical issues in health care lead to a somewhat different agenda of bioethics. In addition to concentrating on attempts to analyse and resolve practical problems, European literature shows a desire to overcome pragmatism by raising philosophical questions concerning the human condition, the perfectibility of the human being, the impact of biotechnology, basic concepts such as health, disease, and disability, and the epistemology of medical science. The feeling is that the dominant conception of medical ethics has been loosely embedded in philosophy, thereby lacking a more encompassing critical, theoretical perspective on its own practical activities. The success of this conception flows from its applicability to practical problems, its educational value and its pragmatic concentration on elucidating and resolving dilemmas and problems. In doing so, medical ethics itself has been transformed into a more or less technical approach, technethics. This is a paradoxical result. Moral issues arise from an almost exclusively technological orientation to the world and a predominant scientific conceptualisation of human life; we try to address these issues with a conception of ethics, itself impregnated with scientific-technical rationality. The dominant conception of medical ethics still seems very much under the spell of the Marxist formula that philosophy should change the world, not interpret it. Unfortunately, through its emphasis on pragmatism and applicability, it cannot change the world of medical science and technology, since it is too much a part of it.

The paradox is exemplified by the role of the human body in bioethical discourse. Viewed from the tradition of philosophy of medicine, bioethics emerges from the anthropological criticisms of medicine, as argued above. Bioethics therefore originates from the recognition that medicine separates the individual person into subject and object, and that the human subject needs to be re-introduced into medical discourse. The best way to focus attention to the patient as a whole person and as an agent being in control of his or her own life, is to stress the autonomy of the individual subject and to demand moral respect for this autonomy. However, the emphasis on individual autonomy tends to neglect the significance of the human body. In most ethical discourse, there is no recognition of the special experiences of embodiment; it seems as if the autonomous subject is not embodied. Its body is merely the instrument through which the subject interacts with the world. The subject is in full control of its body. It is imperative that the integrity of the body should be respected, as it is the prime vehicle of the autonomous person.

The moral principle of respect for autonomy in health care ethics is apparently associated with a popular image of the body as property (ten Have and Welie, 1998). When the individual person is regarded as an autonomous subject, then the body is his private property. And the person is the sovereign authority with property rights over his or her body. Since autonomous individuals own their bodies, they have exclusive possession of it and they alone have it at their disposal. This concept of body ownership is increasingly important now in debates concerning transplantation, research, genetics and reproductive technologies. Property language in health care ethics is used to designate the locus of decision-making authority: the individual as owner is in control over his body. In view of the increasing medical possibilities to invade the human body as well as the potential of body parts for research and commercial purposes, it is necessary to protect the individual person against harmful and paternalistic interventions with and to his body. At the same time, the concept of body ownership is morally problematic. The distinction between person and body is contrary to the existential identity with our bodies and the self-experience of ourselves as embodied selves. In making such a distinction between autonomous subject, c.q. owner and a body, c.q. private property, bioethics seems to proceed from the same dualism which was criticized in the anthropological tradition. More so: it apparently uses a dualistic distinction between person and body, subject and object, - a distinction which has led to the emergence of bioethics itself. Whereas medicine tends to neglect the subject, bioethics tends to neglect the body (ten Have, 1998b; Zwart, 1998).

What has been one of the prime motivational sources for the growth of bioethical debate (a reductionist view of human beings as bodies without relevant subject) apparently is copied in bioethical discourse itself (a counterpart reductionist view of human beings as subjects without relevant body). Although the precise vocabulary has changed, in biomedicine as well as bioethics, a similar dualism of object and subject in regard to human beings can be recognized.

8. CONNECTING INTERNAL AND EXTERNAL MORALITY

The different emphases in European literature seem to have a common denominator: they focus on the dialectic connection between the internal and external morality of medicine, without reducing one set of norms and values to another. It is heuristically assumed that on the one hand there are specific values, norms and rules intrinsic to the actual practice of medical care (the 'internal morality'), on the other hand, there are values, norms and rules prevailing in social, cultural and religious traditions that function as external determinants of medicine (the 'external morality'). The dominant conception of medical ethics proceeds from a too strong distinction between these two sets of values, norms and rules, as well as an over-estimation of the relevancy and importance of the external morality. In order to obtain a better understanding of the interaction of both moralities, it is necessary to establish a theoretical framework relevant to medical practice in order adequately to take account of the norms and values inherent in the practice of medicine, but it requires at the same time sufficient detachment in order to provide a critical normative perspective on medical practice.

The problem is how this task can be accomplished. How to develop a theoretical perspective on medical ethical issues that connects philosophical reflection with the everyday realities of medical practice? Such a perspective not only aims at elucidating specific bioethical problems, but it intends critically to examine various conceptions of bioethics that purport to deal with such problems. It should also make clear why and how such bioethical problems appear, reappear, and even disappear in medical discourse; why certain problems emerge in various health care practices and others do not; and how such problems can be discussed and even resolved during daily interactions between physicians, nurses, patients, hospital administrators, and others.

In programmatic attempts and debates pursuing these objectives, at least four steps can be distinguished.

1. The first step is to examine the methods of clinical ethics in different health care contexts in order to obtain a better understanding of the *internal morality* of these practices. This will require to formally undertake both empirical research and philosophical investigations. A new theoretical perspective on bioethics can be developed only if we take seriously certain fundamental notions of clinical ethics (ten Have, 1990a). One of these notions is that there are internal standards and norms that govern professional medical practice. These internal norms are made salient by analyzing various health care practices. Recently, a revival of the concept 'internal morality' is noticeable. Especially changes in the health care system as well as financing structures have stimulated anew reflections on the nature of medical practice (Brody and Miller, 1998).

John Ladd introduced the concept to refer to norms governing medical practice; they determine what is good clinical medicine:

> Internal morality ... comprises moral norms relating to the clinical situations that depend on 'medical' considerations, such as diagnosis, prognosis, treatment plans, concepts of disease, and so on (Ladd, 1983, p. 212)

He contrasted it with the notion of 'external morality'; this notion includes the moral considerations that come from outside medicine; they are based on non-medical facts like social conditions, personal habits, and demands of individual patients and their families. The focus on the internal morality reiterates the view that medicine is a profession (as discussed earlier in this chapter; see also: Ladd, 1989). In this view medicine is not a morally neutral body of knowledge and technique; its moral content cannot be derived from the general morality of society. A full account of the content of the internal morality of medicine, as Brody and Miller (1998) acknowledge, requires further development of two constituents: the moral goals of medicine and the morally acceptable means for achieving those goals. The clinical practice of medicine is directed at a set of particular goals, a coherent range of good healing actions. As Brody and Miller point out these goals should not be too narrowly identified (interpreting 'healing' as 'curing a disease'); at the same time, even a comprehensive list of goals limits medical activities and requires particular moral values rather than others. Medical practice also requires internal standards of appropriate performance. Promotion of a particular goal alone is not sufficient; it should go with morally acceptable means. Brody and Miller suggest four standards, originating in the nature of medical practice:

> 1. The physician must employ technical competence in practice ...
> 2. The physician must honestly portray medical knowledge and skill to the patient and to the general public, and avoid any sort of fraud or misrepresentation ...
> 3. The physician must avoid harming the patient in any way that is out of proportion to expected benefit, and must seek to minimize the indignity and the invasion of privacy involved in medical examination and procedures ...
> 4. The physician must maintain fidelity to the interests of the individual patient ... (Brody and Miller, 1998, p. 388).

These examples from the literature indicate which direction to go for future developments. The first step towards a reconnection of practice-internal and practice-external moralities is to carefully examine daily health care practices. Surely clinical ethics requires such a reorientation; yet for many practitioners 'clinical ethics' does not suggest an alternative view of medical ethics, but only serves to further the application of ethical rules and principles to cases. That is, 'clinical ethics' simply means 'doing ethics in the clinical setting'. Clinical ethics under this construction is simply a special case of applied ethics. The disadvantages of this approach can be overcome, however, if by introducing ethical discourse directly into the clinic, thereby retaining the prevailing values.

Clinical ethics, under this view, involves a new approach to ethics that is relevant to all health care practices. Clinical ethics becomes a radically different interpretation of ethics because it takes place within the clinical setting. It is possible, therefore, to profit from clinical ethics without reducing clinical ethics to applied ethics.

2. The second step is to analyze and interpret the *external morality* governing health care practices. Making use of the results of recent social research and specific empirical investigations, this step requires the study of values, norms, and attitudes concerning medical-ethical issues. To date, value studies have only occasionally examined (patients') values regarding health, disease, dysfunction, disability, dying, illness prevention, and health care. These values in society need to be explored in order to understand more fully the value context in which current bioethical debates occur.

An example of this approach would be a research project that focuses on values regarding health, disease, dying, illness prevention and health care that are explicit or implicit in public policy documents concerning care for the chronically ill. What norms and values are reflected in public policy documents as well as actual public policy decisions for the chronically ill? Important values in this context are, for example, solidarity and justice. In the Dutch health care system, solidarity seems to imply not only that the

community will take care of the ill and helpless, but also that the weak will limit their claims to care when there is no longer any prospect of a meaningful life for them. The value of justice is significant as well, since we seem to lack a guiding vision of how a just and good society should accommodate the special needs of its chronically ill members. In view of the growing prevalence of chronic illnesses, traditional concepts of solidarity and justice will become more problematic. The question will be how much society wants to afford to care for the chronically ill. But the issue is also how chronic suffering is valued in a particular society. In order to have a better understanding of the current bioethical problems in chronic health care practices, the normative context of such practices (in social debate, in public policy decisions, and policy documents) should therefore be further explored.
3. The third step is the creation of *new theoretical perspectives* on health care practices. History of medicine as well as philosophy of medicine share a growing interest in the empirical realities of medicine. The so-called empirical shift in philosophy of science, some decades ago, has led to new approaches, e.g., several kinds of social constructivism (ten Have and Spicker, 1990). From this social constructivist perspective important contributions to medical theory have been made (Latour, 1987). From this point of view, diagnoses, diseases, medical knowledge, health care institutions are considered social constructions, which can be understood only in their *empirical* social and cultural context.

Ethics, philosophy and history of medicine may thus find common ground in creating new theoretical perspectives on health care practices. In any practice a complex set of activities guided by shared rules, cognition, action and normativity are inextricably linked. Focusing on the notion of *practice* as the common theoretical starting-point, the interdependence of the disciplines as well as the specificity of their expertise will become apparent. Philosophy of medicine analyzes the cognitive components of health care practices: concepts, methods, and ideas. Medical ethics examines the activities and action-guides embodied in health care practices as well as the values embedded in such practices. History of medicine studies the diachronical and synchronical construction and transformation of practices.

A critical evaluation of theories of medical practice is therefore necessary. The work of the Danish philosopher, Uffe Juul Jensen (1987), is a useful and interesting example of a philosophical theory of medical practice. Jensen's theory is a conceptual framework as well as a heuristic instrument to study the problems of modern health care - such as those arising in the care of chronic patients - from moral, philosophical and historical perspectives. The modern health care system is a complex network of practices based on different historical traditions, embodying different values and using different methods. Jensen specifically distinguishes three kinds of practice-orientations that are woven together in the modern health care system: the disease-orientated practice, the situation-orientated practice, and the community-orientated practice. Obviously, a critical analysis of Jensen's specific viewpoints is necessary; nonetheless, the focus of his model for the interrelationship of knowing, acting and valuing in health care practices seems to be a promising starting-point for analyzing and elucidating present-day moral problems in present-day health care.

4. The fourth step is to develop a *new conception of bioethics* that illuminates and clarifies the complex interactions between the internal and external morality of health care practices. As a particular domain of philosophy, ethics proceeds from empirical knowledge, *viz.* moral experience. The moral dimension of the world is first and foremost experienced. Moral experience is humanity's way of understanding itself in moral terms (van Tongeren, 1988). Ethics is therefore the interpretation and explanation of this primordial understanding. Before acting morally we must already know, at least to some extent, what is morally desirable or right. Otherwise, we would not recognize what is appealing in a moral sense. On the other hand, what we recognize in our experience is typically unclear and in need of further elucidation and interpretation.

In short, we approach the moral dimension of the world from a set of prior understandings; they form the basis of our interest in what at first seems odd and strange to us, requiring us to continuously reconstruct the moral meaning of our lives. Such an interpretive perspective will be helpful when integrating the experiences disclosed in the clinical-ethical studies, as well as utilizing the insights gained from describing the value-contexts of health care practices.

9. INTERPRETIVE BIOETHICS

Overlooking the theatre of competitive approaches, one of the challenges for contemporary medical ethicists is to formulate a new conception and practice of medical ethics that can bridge the gap between the internal and external morality of medicine, as well as between medical empiricism and ethical normativism. It requires the development of a theoretical framework relevant to medical practice so that it may adequately take account of the norms and values inherent in the practice of medicine, but with sufficient critical distance so that it may provide a normative perspective on these practices.

Graber and Thomasma (1989) developed the unitary theory of clinical ethics out of a concern with the problematic relationship between theory and practice in medical ethics. Having examined various models of theory-practice relation (for example, the model of applied ethics), they believe the new theory will avoid the weaknesses of these models and combine their strengths. The Unified Clinical Ethics Theory (UCET) therefore can incorporate elements of the virtue, deontological, and consequentialist theories of ethics. The theory is summarized as follows:

> Certain conditions (C) are present in this case such that the probability (x) exists that Value (V) A will be judged more important than B by (I) interpreters because the Principle (P) p' will be more likely to apply to the case than p" (Graber and Thomasma, 1989, p. 194)

However, in this summary statement it is not obvious that it does indeed combine such theories. It emphasizes the context of a case, the weighing of relevant values, and the role of interpreters but the normative justification for judging value A more important than B is that principle p' takes precedence over p".

At the same time, Graber and Thomasma consider UCET as a practical model of bioethical hermeneutics combining both theory and practice. The hermeneutic aspect is repeatedly mentioned by the authors: all cases require interpretation; interpreters are involved in profound ways in analyzing the case and balancing its important features (*O.c.*, p. 196); an essential part of making moral judgements is interpreting the fit between situation and principles (*O.c.*, p. 201).

However, the pragmatic orientation of UCET has possibly prevented a further elaboration of this interpretive point of view, so that it is unclear how radical the hermeneutic perspective really is: is it methodological hermeneutics, paying adequate attention to the interpretive components of medical practice, or is it hermeneutic philosophy, trying to develop a theory of interpretation and to explain medicine as a hermeneutic science? If the last focus prevails, the crucial question for ethics as a practical enterprise is not so much to clarify action guides and make moral quandaries controllable but rather to make them communicable.

Graber and Thomasma have not further developed their hermeneutic philosophy; it may seem that hermeneutics has simply been incorporated as a tool into a hybridisation of virtue, deontological and consequentialist theories. Even the name 'unitary theory' suggests an harmonious combination of different approaches, whereas in fact the authors are aiming at an encompassing ethical super-theory, absorbing specific ethical theories within a radical hermeneutic framework.

However, what really is innovative in their approach is the emphasis on the role of interpretation. The term 'hermeneutic' can be misleading. It may suggest that now a particular school of philosophy is applied in the context of health care whereas the philosophical point of view is the emphasis on interpretation.

9.1. Hermeneutics

Originally, hermeneutics refers to the art of interpreting and the science of interpretation. As such, it was used in theology, law and philosophy, all concerned with interpreting the meaning of texts. It has come into prominence in the last century as a methodology characterizing the humanities and social sciences. Philosophers such as Schleiermacher and Dilthey have shown that not only texts but all human products need interpretation, and that hermeneutics involves not only the interpretandum but also the interpreter. Finally in our century, through the works of Heidegger, Gadamer and Ricoeur, hermeneutics evolved into a philosophy of understanding and explaining human existence.

9.2. Medicine as Hermeneutics

Entering a new stage in the long-standing debate on the status of medicine, it has been argued that medicine has to be considered as a hermeneutical enterprise apparently presuming that medicine is not or not merely a natural science (Daniel, 1986; Leder, 1988; Svenaeus, 1999). The modern emphasis on information and empirical data has contributed to a new understanding of diagnosis and treatment as the physician's interpretation of what concerns the patient and what can be done to help the patient. And metaphorically, the patient is conceived as a text that may be considered on different interpretive levels. It is important to reflect upon the typical preconditions of interpretation in medicine. The patient is usually understood through an anatomico-physiological model. The patient's body is made 'readable' by the use of technology. The biomedical language of diagnosis and treatment reduces the overwhelming amount of information presented by the patient, so that the standard medical case report does not reflect the story of the patient's life but of the physician's relationship with the patient's illness (Poirier and Brauner, 1988).

It is also important to look at the effects of medical interpretation upon the interpreter. Interpretation seems to bring understanding and empathy. Interpreting symptoms involves understanding what is actually wrong with a patient and appreciating what he or she is going through. Interpreting the patient's illness arouses therefore an 'affiliative feeling' in the physician-interpreter (Zaner, 1988).

9.3. Bioethics as Interpretation

Some contemporary philosophers have argued that ethics is best considered to be a hermeneutical discipline. Ethics therefore can be defined as the hermeneutics of moral experience. Complex bioethical problems must be understood within the broader framework of an interpretive philosophical theory. Such a theory should concentrate upon four characteristic parameters.

9.3.1. Experience

The starting-point of medical activity is the moral experience of the patient. Through his illness he is confronted, in Zaner's words (1988, p. 65), with tears in the fabric of daily life. He presents himself to the physician as both puzzling and meaningful. The patient's symptoms are deeply textured by his biographical situation, with his beliefs, values, habits and life-style. To ascertain what is wrong requires an interpretation, the more so since there is an initial distance between patient and physician. The meaning of the individual human being who is the patient requires interpretation for two reasons: (1) intrinsic strangeness; the experience of illness in this particular patient is unique and unusual, (2) theoretical pre-understandings: the context in which the physician interprets the symptoms (e.g., the pathological models) is different from the context in which the interpretandum came into existence. It can reasonably be expected that moral experiences differ according to the interpretive models used in various health care practices and according to the specific complaints, illnesses and disabilities of the patients encountered in different health care settings. Different practices should therefore be examined and compared. Experiences are part of the text of life. But we never know whether our understanding of this text is adequate (van Tongeren, 1994).

9.3.2. Attitudes and Emotions

For ethics, the fundamental question is not so much "What to do?" but rather "How to live?". It is *praxis* not *poiesis* that is important (van Tongeren, 1988). The moral relevancy of our actions should not be reduced to their effects; it is also determined by an evaluation of what we do in executing our actions. For example: the problem of experimenting with human embryos should not be settled by reference to future results, but should also raise the question: Why are we interested in scientific research? This change of focus implies a re-orientation from activity to passivity, from acts to attitudes and emotions. Moral experience involves primarily feelings, for instance, of indignation, confusion or contentment; secondarily, these emotional responses can be made the object of moral thinking (Callahan, 1988).

A sharing of moral experiences of patients and physicians, and of the emotions and attitudes involved, is therefore required for elucidating the relevant ethical issues of the case or situation. Understanding and defining the morally relevant facts of a case do not involve the identification of relevant general principles and the deduction of a set of rules from which the correct response to the problem can be derived. The role of medical ethics is not so much to explicate and apply ethical theories and principles but to interpret and evoke what is implied in moral experience. The notion 'applied ethics' wrongly suggests that we already know which moral principles and rules to apply. However, rules and principles are in fact answers to what is evoked or appealed to in a particular case. First of all, we need to understand what the moral experience of vulnerability and appeal to assistance really mean in this case. We need to discover why particular principles will motivate us in this case; why is there a particular ideal, rule or obligation? It requires close scrutiny of the medical situation in all its complexity.

9.3.3 Community

The interpretive reading of a patient's situation is not an individual doctor's affair. The medical prior understandings that orientate the interpretation are the sediments of traditional cultural assumptions concerning the nature of the world and the body, and the results of a specific historical evolution of medical knowledge. Interpretation presupposes a universe of understanding. This is a consequence of the so-called hermeneutic circle; in order to interpret a text's meaning, the interpreter must be familiar with the vocabulary and grammar of the text and have some idea of what the text might mean (Daniel, 1986). For man as a social being, understanding is always a community phenomenon: understanding in communication with others. The continuous effort to reach consensus through a dialogue with patients, colleagues and other health professionals, induces us to discover the particularities of our own prior understanding, and through that, to attain a more general level of understanding. This seems to reflect the experience of hospital ethics committees: analyzing a case in terms of moral principles leads to a stalemate but interpreting the moral experience of the concrete participants involved in this particular case usually leads to a consensus. Since the interpretation of moral experience takes place within the context of particular social practices, intimate knowledge of the historical, medical and scientific components of those practices is essential to the task of moral criticism. Ethics can not be practised without a high degree of engagement in medical work.

9.3.4. Ambiguity

Ethics primarily aims at interpreting and understanding moral experience. But moral experience is complex and versatile. It implies that every interpretation is tentative; it opens up a possible perspective. Definitive and comprehensive interpretation is non-existent. An interpretive approach always has an ambiguous status: more than one meaning is admitted. As Zaner (1988, p. 272) points out: "Every life is linguistically inexhaustible, there is always a richer tale to be told that can never be wholly captured in words, no matter how evocative they might be". That means that moral judgements and decisions which must be framed on the basis of understanding the thematic moral ordering of a person's life are fundamentally uncertain.

9.4. Implications

Interpretive bioethics will have important consequences for the competence and role of medical ethicists. Since the interpretation of moral experience takes place within the context of particular social practices, intimate knowledge of the historical, medical and scientific components of those practices is essential to the task of moral criticism. Ethics cannot be practised without a high degree of engagement in medical work. Bedside medical ethics does not imply that only doctors can be ethicists. To comprehend the human terms in which actual moral dilemmas are experienced, the facts of medical practice must not be left to doctors alone. But being able to do that requires a sharing of moral experiences of both patients and doctors.

To implement the interpretive conception of medical ethics, more empirical study of actual decision-making processes is necessary. Like the empirical turn in philosophy of science, the current interest in empirical medical ethics combines medical sociology and anthropology, history of medicine and philosophy to construct a more sophisticated view of moral experience in medicine.

Interpretive bioethics also has consequences for ethics teaching. If it is assumed that ethics is part and parcel of the routine practice of medicine, and that ethics is not an abstract exercise of moral reasoning but characterized by the emotion, complexity and ambiguity commonly involved in real cases, then lessons should be drawn for the method and goals of medical ethics teaching (ten Have and Essed 1989). Preference should be given to team-teaching in the clinical period using the format of patient conferences and case review, generally accepted in routine clinical work. The objective of this problem-orientated ethics teaching is first and foremost to increase the students' sensitivity to moral problems in everyday medicine.

Finally, interpretive bioethics will require a new rapprochement between ethics and philosophical anthropology (ten Have, 1998c). As described above, during this century there has existed an undercurrent of philosophical criticism of modern medicine with very different manifestations: originally epistemological, then anthropological, now ethical. Particularly in health care, normative positions and moral theories are intimately connected with images of the human being. In the medical setting we cannot escape the question: what kind of human being do we want to realize in medical activities, what kind of person do we wish to respect, heal, inform, comfort in health care? However, the relationship between ethics and anthropology is a dialectical one. Instead of claiming the primacy of anthropology, with ethics based on anthropological theory and images of the human being, philosophy of medicine should also focus upon explicating the morality underlying anthropology itself. Ethical discourse can exemplify a particular image of man, but at the same time anthropological discourse itself is presupposing particular moral views. From the point of view of interpretive bioethics, morality is not something we choose, but a fundamental predicament we are already involved in before we even start to reflect upon it; such a predicament, however, at the same time is an anthropological characterisation of what is essential in human beings. Morality is choosing us, because we are primarily social beings. Ethical views are articulated and explicated because we are in a moral relationship with other human beings appealing to us. The face of the other makes us moral beings whether we like it or not, whether we choose to act accordingly or not. Morality is a social affair. Its interpersonal character makes it possible to scrutinize and criticize individual moral choices.

The dialectical interaction of anthropology and ethics is helpful in regaining a view of man as social being, and therefore restoring the idea of moral community. Currently, there is an interest in communitarian approaches to bioethics (Kuczewski, 1997). This emphasizes that cultural context and community are constitutive of the values and goals of individuals. Communal relatedness falsifies the idea of the unencumbered self, the idea of self-ownership assuming that the individual as an entity exists prior to the ends which are affirmed by it. Without societal culture our potential for self-determination will remain empty. Present-day individuals are not free of all moral ties. But neither are they bound in a universal community with clear encompassing loyalties. It is typical for the postmodern predicament that they are citizens who can think and act as "multiply-situated selves" (Sandel, 1996, p. 350). A similar point of view is stressed by Benjamin Barber (1995). Postmodern individuals are members of a world-wide community, so-called McWorld, the global theme park of MTV, Macintosh and McDonald's, a world tied together by communication, information, entertainment and commerce; in this world everyone is a consumer, defined by needs and wants. McWorld therefore is not really a community: the significant relations are exchange relationships among individual consumers and individual producers; society is privatized and commercialized. The question is how to reconnect individuals with civil society and civic culture, as the middle ground between big government and the private sector. Interposed between the state and the market is where community exists, where we are more than clients or consumers, where we are public beings having regard for the general good, where we as citizens relate in the cooperative, noncoercive pursuit of public goods. Barber defines a citizen as "an individual who has acquired a public voice and understands himself to belong to a wider community, who sees him/herself as sharing goods with others" (Barber, 1995, p. 286). But also Barber agrees that humankind depends for its liberty on variety and difference; we live in several spheres, in a many-sectored civil society. Whereas market choices are private and speak about individualistic goals and individual preferences, citizens speak about the social consequences of their private choices; they speak the public language of the common good; but at the same time, this public language is multiple and heterogeneous; civil society has many narratives about the common good.

The *universal* human condition of existence as a communal-cultural being can only be realized in *particular* ways; the communitarian self is constituted by particular cultural characteristics. A richer medical ethics can result from taking seriously the basic idea of moral community, and concomitantly, the various narratives about the particularities of people as communal beings.

10. CONCLUSION

The data, insights, and theoretical notions obtained and analyzed in the previous steps require integration through the development of a theory of medical practice with emphasis on its ethical dimensions, that can illuminate and clarify the complex interaction between the internal and external morality of various health care practices. Criticism of the dominant conception of applied ethics and principlism can help to articulate what kind of theory is needed and which conception of bioethics is most adequate in understanding medical practice.

REFERENCES

Anspach, R.R.: 1993, *Deciding who lives. Fateful choices in the intensive-care nursery*. University of California Press, Berkeley.

Arnold, R.M. and Forrow, L: 1993, Empirical research in medical ethics: an introduction. *Theoretical Medicine* 14: 195-196.

Baier, A.: 1985, *Postures of the mind. Essays on mind and morals*. Methuen, London.

Barber, B.R.: 1995, *Jihad vs McWorld*. Ballantine Books, New York.

Beauchamp, T.L. and Childress, J.F.: 1983, *Principles of biomedical ethics* (2nd ed.). Oxford University Press, New York/Oxford.

Beaufort, I.D. de and Dupuis, H.M.: 1988, *Handboek gezondheidsethiek*. Van Gorcum, Assen/Maastricht.

Bosk, C.L.: 1979, *Forgive and remember. Managing medical failure*. The University of Chicago Press, Chicago/London.

Bosk, C.L.: 1992, *All God's mistakes. Genetic counseling in a pediatric hospital*. The University of Chicago Press, Chicago/London.

Brody, B. (ed.): 1988, *Moral theory and moral judgments in medical ethics*. Kluwer Academic Publ., Dordrecht.

Brody, H.: 1987, *Stories of sickness*. Yale University Press, New Haven and London.

Brody, H. and Miller, F.G.: 1998, The internal morality of medicine: Explication and application to managed care. *Journal of Medicine and Philosophy* 23(4): 384-410.

Callahan, D.: 1981, Minimalist ethics. On the pacification of morality. In: Caplan, A.L. and Callahan, D. (eds.), *Ethics in hard times*. Plenum Press, New York, p. 261-281.

Callahan, D.: 1990, *What kind of life: The limits of medical progress*. Simon and Schuster, New York.

Callahan, S.: 1988, The role of emotion in ethical decision-making. *Hastings Center Report* 18: 9-14.

Carson, R.A.: 1990, Interpretive bioethics: The way of discernment. *Theoretical Medicine* 11: 51-9.

Daniel, S.L.: 1986, The patient as text: A model of clinical hermeneutics. *Theoretical Medicine* 7: 195-210. .

Downie, R.S.: 1974, *Roles and values. An introduction to social ethics*. Methuen, London.

Drane, J.F.: 1988, *Becoming a good doctor: The place of virtue and character in medical ethics*. Sheed and Ward, Kansas City.

DuBose, E.R., Hamel, R. and O'Connell, L.J. (eds.): 1994, *A matter of principles? Ferment in U.S. bioethics*. Trinity Press International, Valley Forge (Pennsylvania).

Editorial: 1990, Medical ethics: should medicine turn the other cheek? *The Lancet*, oct.6: 846-7.

Engelhardt, H.T.: 1986, *The foundations of bioethics*. Oxford University Press, New York/Oxford.

Evans, M. (ed.): 1998, *Critical reflection on medical ethics*. JAI Press, Stamford (Conn.).

Fox, D.M.: 1979, The segregation of medical ethics: a problem in modern intellectual history. *Journal of Medicine and Philosophy* 4: 81-97.

Fox, R.C.: 1989, *The sociology of medicine. A participant observer's view*. Prentice-Hall, Englewood Cliffs, N.J.

Fulford, K.W.M.: 1994, Not more medical ethics. In: Fulford, K.W.M., Gillett, G.R. and Soskice, J.M. (eds.): *Medicine and moral reasoning*. Cambridge University Press, Cambridge, U.K., p. 193-205.

Gillon, R.: 1986, *Philosophical medical ethics*. John Wiley & Sons, Chichester.

Gillon, R. (ed.): 1994, *Principles of health care ethics*. John Wiley & Sons, Chichester.

Graber, G.C. and Thomasma, D.C.: 1989, *Theory and practice in medical ethics*. Continuum, New York.

Gracia, D.: 2001, History of medical ethics. This volume.

Greaves, D.: 1979, What is medicine? Towards a philosophical approach. *Journal of Medical Ethics* 5: 29-32.

Halman, L., Heunks, F., Moor, R. de, Zanders, H.: 1987, *Traditie, secularisatie en individualisering. Een studie naar de waarden van de Nederlanders in een Europese context*. University Press, Tilburg.

Have, H. ten and Arend, A. van der: 1985, Philosophy of medicine in the Netherlands. *Theoretical Medicine* 6: 1-42.

Have, H.A.M.J. ten and Essed, G.: 1989, An experimental case-conference programme for obstetric and gynaecology clinical students. *Journal of Medical Ethics* 15: 94-98.

Have, H. ten and Kimsma, G.K.: 1990, Changing conceptions of medical ethics. In: Jensen, U.J. and Mooney, G. (eds.), *Changing values in medical and health care decision making*. John Wiley & Sons, Chichester, p. 33-51.

Have, H. ten: 1990a, *Ethiek tussen alliantie en dissidentie*. Inaugural lecture, University of Limburg, Maastricht.

Have, H. ten: 1990b, *Een Hippocratische erfenis; ethiek in de medische praktijk*. De Tijdstroom, Lochem.

Have, H. ten: 1990c, Verleden en toekomst van medische filosofie. *Scripta Medico-philosophica*, Schrift 7, p. 5-18.

Have, H. ten and Spicker, S.F.: 1990, Introduction. In: Have, H. ten, Kimsma, G., and Spicker, S.F. (eds.): *The growth of medical knowledge*. Kluwer Acad. Publ., Dordrecht, p. 1-11.

Have, H. ten and Keasberry, H.: 1992, Equity and solidarity: The context of health care allocation in the Netherlands. *Journal of Medicine and Philosophy* 17: 467-481.

Have, H.A.M.J. ten and Lelie, A.: 1998, Medical ethics research between theory and practice. *Theoretical Medicine and Bioethics* 19: 263-276.

Have, H.A.M.J. ten: 1998a, Philosophy of medicine and health care - European perspectives. *Medicine, Health Care and Philosophy* 1(1): 1-3.

Have, H.A.M.J. ten: 1998b, Health care and the human body. *Medicine, Health Care and Philosophy* 1(2): 103-105.

Have, H.A.M.J. ten: 1998c, Images of man in philosophy of medicine. In: Evans, M. (ed.): 1998, *Critical reflection on medical ethics*. JAI Press, Stamford (Conn.), p. 173-193.

Hope, T.: 1999, Editorial: Empirical medical ethics. *Journal of Medical Ethics* 25: 219-220.

Hunter, K.M.: 1988, Making a case. *Literature and Medicine* 7: 66-79.

Inglehart, R.: 1990, *Culture shift in advanced industrial society*. Princeton University Press, Princeton, N.J.

Jensen, U.J.: 1987, *Practice and progress. A theory for the modern health-care system*. Blackwell Scientific Publications, Oxford.

Jensen, U.J.: 1989, From good medical practice to best medical practice. *International Journal of Health Planning and Management* 4: 167-80.

Jonsen, A.R., Siegler, M., and Winslade, W.J.: 1986, *Clinical ethics: A practical approach to ethical decisions in clinical medicine* (2nd ed.). Macmillan Publ. Comp., New York.

Jonsen, A.R. and Toulmin, S.: 1988, *The abuse of casuistry*. University of California Press, Berkeley, CA.

Jonsen, A.R.: 1990, Practice versus theory. *Hastings Center Report* 20: 32-4.

Kass, L.R.: 1990, Practicing ethics: Where's the action? *Hastings Center Report* 20: 5-12.

Kiezen en delen: 1991, Rapport van de Commissie Keuzen in de Zorg. Den Haag.

Kuczewski, M. G.: 1997, *Fragmentation and consensus. Communitarian and casuist bioethics*. Georgetown University Press, Washington, D.C.

Ladd, J.: 1983, The internal morality of medicine: An essential dimension of the patient-physician relationship. In: Shelp, E.E. (ed.): *The clinical encounter*. D. Reidel Publishing Company, Dordrecht, p. 209-231.

Ladd, J.: 1989, The good doctor and the medical care of children. In: Kopelman, L.M. and Moskop, J.C. (eds.): *Children and health care: Moral and social issues*. Kluwer Academic Publishers, Dordrecht, p. 281-302.

Latour, B.: 1987, *Science in action*. Open University Press, Milton Keynes.

Leder, D.: 1988: The hermeneutic role of the consultation-liaison psychiatrist. *Journal of Medicine and Philosophy* 13: 367-378.

Leder, D.: 1994, Toward a hermeneutical bioethics. In: DuBose, E.R., Hamel, R. and O'Connell, L.J. (eds.): 1994, *A matter of principles? Ferment in U.S. bioethics*. Trinity Press International, Valley Forge (Pennsylvania), p. 240-259.

Lelie, A.: 1999, *Ethiek en nefrologie*. Dissertation, University of Nijmegen.

Murray, T.H.: 1994, Medical ethics, moral philosophy and moral tradition. In: Fulford, K.W.M., Gillett, G.R. and Soskice, J.M. (eds.): *Medicine and moral reasoning*. Cambridge University Press, Cambridge, U.K., p. 91-105.

Newton, A.Z.: 1995, *Narrative ethics*. Harvard University Press, Cambridge (Mass.).

Pellegrino, E.D. and Thomasma, D.C.: 1993, *The virtues in medical practice*. Oxford University Press, New York/Oxford.

Pellegrino, E.D.: 1995, The limitations of empirical research in ethics. *Journal of Clinical Ethics* 6: 161-162.

Poirier, S. and Brauner, D.J.: 1988, Ethics and the daily language of medical discourse. *Hastings Center Report* 18: 5-9.

Sandel, M.J.: 1996, *Democracy's discontent. America in search of a public philosophy*. The Belknap Press of Harvard University Press, Cambridge (Mass.) and London (UK).

Singer, P.: 1990, Bioethics and academic freedom. *Bioethics* 4(1): 33-44.

Svenaeus, F.: 1999, *The hermeneutics of medicine and the phenomenology of health. Steps towards a philosophy of medical practice*. Linköping University, Linköping (Sweden).
Taylor, Ch.: 1989, *Sources of the self. The making of modern identity*. Cambridge University Press, Cambridge.
Tongeren, P. van: 1988, Ethiek en praktijk. *Filosofie & Praktijk* 9: 113-27.
Tongeren, P. van: 1994, The relation of narrativity and hermeneutics to an adequate practical ethic. *Ethical Perspectives* 1(2): 57-70.
Tronto, J.C.: 1993, *Moral boundaries. A political argument for an ethic of care*. Routledge, New York/London.
Vandenbroucke, J.P.: 1990, Medische ethiek en gezondheidsrecht; hinderpalen voor de verdere toename van kennis in de geneeskunde? *Nederlands Tijdschrift voor Geneeskunde* 134: 5-6.
Weisz, G. (ed.): 1990, *Social science perspectives on medical ethics*. Kluwer Academic Publishers, Dordrecht/Boston/London.
Welie, J.V.M.: 1998, *In the face of suffering. The philosophical-anthropological foundations of clinical ethics.* Creighton University Press, Omaha (Nebraska).
Williams, B.: 1988, Consequentialism and integrity. In: Scheffler, S. (ed.): *Consequentialism and its critics*. Oxford, p. 20-50.
Winkler, E.R.: 1998, Applied ethics, overview. In: Chadwick, R. (ed.): *Encyclopedia of Applied Ethics*. Volume 1. Academic Press, San Diego, p. 191-196.
Wulff, H.R.: 1998, Contemporary trends in healthcare ethics. In: Have, H.A.M.J. ten and Sass, H.-M. (eds.): *Consensus formation in healthcare ethics*. Kluwer Academic Publishers, Dordrecht, p. 63-72.
Zaner, R.M.: 1988, *Ethics and the clinical encounter*. Prentice-Hall, Englewood Cliffs, N.J.
Zussman, R.: 1992, *Intensive care. Medical ethics and the medical profession.* The University of Chicago Press, Chicago/London.
Zwart, H.: 1995, *Weg met de ethiek? Filosofische beschouwingen over geneeskunde en ethiek*. Thesis Publishers, Amsterdam.
Zwart, H.: 1998, Medicine, symbolization and the 'real' body - Lacan's understanding of medical science. *Medicine, Health Care and Philosophy* 1(2): 107-117.

MARTYN EVANS

THE AUTONOMY OF THE PATIENT: INFORMED CONSENT

1.CASE 1.

"When God grew tired of salmon, He created the sewin,"[1] *announced Dr. Reeves to his patient, forty-five-year-old Paul D. The occasion was a general practice consultation, and Reeves had two things on his mind. First, how might Paul reduce the cholesterol content of his diet whilst continuing to enjoy his food? Hence the importance of a dish such as grilled sewin, delicious as well as rich in those oils currently approved by medical science for healthy eating. The second problem was the management of Paul's essential hypertension - Dr. Reeves was keeping an open mind on whether these two features of Paul's physiological condition were related.*

For this second problem Reeves had, four months previously, placed Paul on long term anti-hypertensive medication. He recalled the initial consultation now: it remained on his mind because Paul had turned up dressed rather conspicuously in a motorcyclist's protective suit and helmet. Paul had presented himself to the surgery for an unrelated problem, had agreed to a general health check and had exhibited raised blood pressure. The health check had been offered to Paul, and to many other patients who happened to consult their general practitioner around that time, as part of a programme of health promotion encouraged by the regional health planners.

Paul had agreed to come back to the surgery one week following the general health check, and they had discussed the problem of his raised blood pressure. Reeves had explained the risks of stroke and heart disease in later life, and the reductions in those risks which could be expected as a result of taking long-term anti-hypertensive medication. Paul had asked about the side-effects of this particular medication and Reeves listed the commonly reported ones: occasional tiredness or heaviness in the limbs, slight numbness or tingling in the extremities; nothing serious or unpleasant. Less commonly reported were headache, dizziness, sometimes a little nausea.

Dr. Reeves and Paul had chatted together for a short while and had pondered the seriousness of suffering that horrible affliction, a stroke, compared

H.A.M.J. ten Have and B. Gordijn (eds.), Bioethics in a European Perspective, 83-91.

to which some occasional heaviness in the limbs was almost laughably trivial - probably even reassuring, a sign that an active, modern, scientific remedy was working in the patient's body, managing his future protection. (In fact, though he said nothing of this to the doctor, Paul had been slightly amused by the thought that he now 'owned' a medical condition of just the right kind, something people had heard of and which required medical interest and attention, but which was not intrinsically unpleasant - indeed, in itself without symptoms entirely, so far as he could judge from what he had been told by Dr. Reeves - and which would be a useful resort in conversations that had lost their momentum. Moreover Paul had felt that he would have no difficulty adhering to the pattern of taking the prescribed tablets at regular intervals. It would be a sign both to himself and to others that he needed to be careful and, perhaps, to deserve a small amount of the care and attention and sympathy of others. But, of course, he said nothing of this.)

After some discussion they had agreed on the medication, the prescription had been written and presented to the pharmacist, and Paul had become a long-term patient.

"... He created the sewin," Dr. Reeves was saying. He did wonder whether Paul was entirely interested - in either the problem of his cholesterol levels (which had not fallen significantly in the four months between the two blood tests taken) or the wisdom of Reeves's gastronomic advice. Was his patient rather more lethargic than he had appeared in earlier consultations? There was no doubt that in general terms the medication was doing him good. His blood pressure had fallen nicely back within the normal range. But he must really do something about his dietary cholesterol intake. And a little more exercise - in fact, quite a bit more exercise - would be a good idea.

They discussed various sporting and leisure activities for a few minutes, and Paul acknowledged that he needed to "get around to being a bit more active" in his spare time. But right now he couldn't be sure what activities he would choose. At any rate, the main problems - heart disease and stroke - were surely being taken care of by the medication? Yes, rather well in fact, agreed Reeves. The blood pressure readings spoke for themselves. All in all Paul was responding well.

"Look, doctor, between ourselves I don't think Paul's at all his same old self since he's been taking those tablets."

Paul's wife Lucy had not intended to discuss her husband's case with Dr. Reeves, and had in fact come to the surgery for a routine examination in connection with a new job for which she had successfully applied. But Paul's general demeanour had certainly begun to worry her. He had always been rather unnecessarily interested in matters of health and illness, in her opinion, but now he was a downright hypochondriac - if anything she would have to say he was 'unhealthily' preoccupied with his state of health. And he seemed to have no energy, no vigour. He did next to nothing in the garden these days. He went to bed early - and certainly not for the reason that he used to suggest they went to bed early; his interest in sex had all but vanished. Lucy did not go into details in discussing this with the doctor, but she had found she could not help mentioning to her sister-in-law that Paul had actually appeared impotent on two or three occasions over the last eight months or so. Their sex life was, more or less, a thing of the past. Perhaps it was simply a matter of age. To the doctor she did mention Paul's habit of being somewhat forgetful or absentminded, something she thought she might have had to expect thirty years from now, but not, surely, at forty-six? Worst of all was his now-frequent periods of irritability, something that she was at a loss to explain and which she felt were directed, unreasonably, at herself personally.

Reeves listened sympathetically, and with interest, but with some discomfort. His proper concern in this consultation was not with Paul's general condition as reported by Lucy. These were things he should certainly discuss with Paul when they next met. But he did not have Paul's permission to get into detailed descriptions of confidential matters with a third-party, not even so intimate and special a third party as the patient's wife. Yet as a conscientious family physician he could not exactly ignore what he was being told either. He thanked Lucy appreciatively "for putting him in the picture" and made a mental note to himself that he must discuss 'quality of life issues' perhaps a little more fully with Paul - and also perhaps with other hypertensive patients. Although, he reflected, no-one could say that he ignored such things. After all, how many of his patients had he introduced to that noble fish, the sewin?

Reeves was generally a contented man, but in the 'middle' years of his professional career he had begun to ask questions of his clinical practice, questions about his goals as a physician, about the amount of good that he thought his medicine could achieve, and about the duties that professional practice brought with it. In effect, and surprisingly for him, greater experience brought with it the tendency towards greater self-doubt.

Sitting with his glass of whisky late in the evening of the day of Lucy's consultation, Paul's case somehow shouldered its way into his attention from amidst the background of his two-thousand-odd patients. Almost involuntarily he began to review Paul's case, to anticipate how he would approach the next consultation with Paul, to argue with himself whether or not he should himself initiate that consultation by asking Paul to come into the surgery, and to set out what he thought the ethical dimensions of Paul's case were.

It seemed to him that they ran like this. First, was Lucy describing the symptoms of a undiagnosed disease in Paul, or were these symptoms really just manifestations of the long-term medication to which he had encouraged Paul to commit himself? Probably, he thought, this was a straightforwardly 'clinical' matter which he could resolve by the appropriate questions and investigations within his own professional competence. But why had he not anticipated such manifestations? Some anti-hypertensive drugs had given rise to occasional reports of temporary loss of sexual potency in males, although so far as he was aware the particular drug prescribed to Paul had not been involved. Of course one might equally ask why Paul had not complained of these problems to him, his doctor. Indeed, why had Paul not complained of them to Lucy? - for it seemed clear that Lucy was reporting her observations, not Paul's descriptions.

Now, therefore, should he review Paul's medication? It seemed that he should, but he would be doing so on the basis of information disclosed to him outside the confidential relationship between himself and Paul, the patient in question.

Second, and more puzzlingly, how secure was Paul's agreement to take the medication? Sure enough, Paul had made a choice - on the basis of what he, Reeves, had described. But now he wondered whether that basis, namely the information reviewed and presented by him to Paul - was itself really secure. Reeves grimaced as he recalled to mind recent journal papers whose results, modifying the assumptions of existing best practice, confirmed the provisional nature of scientific knowledge: it was always liable to be amplified, qualified or even overturned by later experience. That was the nature of scientific enquiry, and one worked with the best information one had at the time. But - the question returned to his mind - should he have known more about the longer term effects of this kind of anti-hypertensive medication?

But, third, suppose that he had? Would it, should it, have made a difference to Paul's decision? That seemed to depend on what Paul's priorities were. Reeves had no doubt that the decision to take the medication was still essentially the correct one - the prevention of stroke was a major goal for the community as a whole and, surely, for individual 'at risk' patients in particular. Of course, Paul was at risk. But his uncontrolled blood pressure was only moderately, rather than excessively, high. The risks were there, but the extent to which they should be weighed was perhaps not itself a wholly scientific matter. He himself would choose the medication. At least, he thought he probably would, though admittedly some of the symptoms described by Paul's wife were, to say the least, unfortunate. That returned him to the question of attributing the symptoms, so he decided to leave this unresolved, as it had to be, until he could talk to Paul. The question of proportionately 'weighting' the gains and the losses of course remained.

The fourth question was this: had he really presented the information to Paul in a way that left Paul free to choose? For that matter, was it really Paul's choice to have undergone the general health check in the first place? In one sense it was, in that no-one had obliged Paul to do so. But people like to please doctors, especially those who wear white coats (Reeves never did so, and usually wore a sweater over a freshly-laundered rugby shirt to assure his patients that he shared what he supposed were their community's leisure interests.) Paul had seemed happy enough to have the health check. In deciding to take it, of course, he could have had no idea that it would lead to a commitment to long-term medication, for a condition which might never afflict him, and - perhaps - turning him from a man in apparently good health into a man whose life was, according to the person closest to him, in important ways diminished.

The fifth question was perhaps the most troubling so far: exactly how was this diminution to be assessed? There was no doubt that what was at stake was the present quality of Paul's life, and it was a long 'present', given that his blood pressure was stabilised by ongoing medication to which he was committed for the foreseeable future. But the problem was to know how to assess Paul's quality of life.[2] *And this was by no means a simple matter, he now realised. His own view of Paul's present quality of life had been that he presumed it was very good. He had no reason to think otherwise, apart from the slight lethargy he had noticed, and this he had attributed to the medication. Paul had not complained about it to him, nor indeed about anything else. Therefore why should he not conclude that Paul's quality of life was good? Attractive though the conclusion admittedly was, he recognised at once that it was superficial, and ignored any constraints on Paul's quality of life which Paul either chose not to report to him, or attributed to something other than his physical well-being. To that extent he was 'working in the dark'; but equally to that extent he might suppose Paul's quality of life was not the business of the doctor after all, except insofar as it was affected by the doctor's decisions and actions. Paul's quality of life was clearly Paul's business, of course, so what was Paul's own view of the matter? Again, Reeves didn't know, beyond what he was told by Paul or could infer from him. But of course that very day indirect evidence on the matter had been given to him by Paul's wife Lucy - a source who, if not altogether comfortable from the ethical point of view, might certainly be thought well-informed. From Lucy's point of view, Paul's present quality of life was clearly less than perfect. Indeed, Reeves had a feeling that Lucy had by no means told him the whole story. The question facing Reeves was how much weight to give to Lucy's views (leaving aside his discomfort over learning them from her outside the clinical relationship with Paul). Should the views of even someone as close as Paul's wife be taken more seriously than those of Paul himself - as far as they could be gauged?*

In short, and depending on how you looked at it, Paul's quality of life was apparently excellent (Reeves), satisfactory (the non-committal Paul himself), or frankly rather impaired (Lucy). Reeves disliked conundrums so late in the evening, finished his whisky and went to bed.

The following day was Saturday, the weather was vile, and Reeves himself had a mild sore throat. Disinclined to go out, and attracted by nothing on the television, he decided to try and square up to the previous night's conundrum in a more methodical fashion. He settled himself at his writing table and prepared to set out the issues on paper. The medical question - that of whether Paul's apparent symptoms (as reported by Lucy) were attributable to the medication was something he had to settle by clinical consultation. But the moral questions he could reflect on right now. And he regarded them as being centred around the matters of the patient's information, choice, and subsequent consent to treatment. He was mildly surprised to discover, as he elaborated his thoughts on paper, how even a routine, 'humdrum' matter - prescribing anti-hypertensive medication to an ordinary adult in the general practice surgery - could give rise to precisely those issues which receive such prominence when they are studied in the dramatic contexts of high-technology, life-saving interventions (often with patients whose capacity for reasoned judgement was impaired). This was the burden of what he wrote:

In the abstract, Reeves reasoned, Paul would make an autonomous choice to accept the medication if his choice were informed, if it were free, and if it were intended. This offered him a kind of template for considering the case, and he organised his thoughts under these headings.

2. WAS PAUL'S CHOICE INFORMED?

Did Paul choose on the basis of all the relevant information - was he 'fully informed', as one might say? That seemed to depend on the extent of 'full' information. Paul had no scientific or medical training, so had presumably only a lay understanding of what Reeves would regard as the relevant cardiology and neurology. But it was Reeves' job to bridge the resulting gap - to explain in lay terms the anticipated harms and benefits of the medication in relation to the risks Paul faced if his hypertension remained untreated.

Yet undeniably the relevance of any information given must be judged in relation to one's chosen goals. Had Paul chosen the goal of prevention of future stroke? This was rather less clear. Paul had not turned up at the surgery in search of such preventive measures. He had been directed towards them as a result of a health promotion exercise initiated by the health care professionals. He could be presumed to share the goal of preventing a future stroke - if it was presented to him in the right way. But could he be presumed to have made this choice in isolation from an agenda set by the health care professionals themselves? There was no doubt that Reeves had presented Paul with a *selection* of information which set up the choice Paul had to make.

Moreover, Paul had been invited to start long-term medication on the basis of information about how he would experience the medication. Paul's choice was to a greater or lesser degree an informed one according to the value of the information he had been given on that score. If Lucy were right, then the information had been deficient; Reeves had not allowed for the possibility that Paul would experience - or at any rate exhibit - all the features reported by Lucy. They were not part of the common profile of adverse reactions reported in the drug trial.[3]

3. WAS PAUL'S CHOICE FREE?

By 'free' Reeves meant, at first, being made without pressure or influence. But he immediately realised that this would not do. No genuine choice can be made without *some* influence - the influence of knowledge of those alternatives between which one must choose, for instance, and the influence of a genuine sense of the values which are at stake. If nothing else, one's choice must be influenced by one's own goals. A completely uninfluenced choice would not be a choice at all. Plainly then, Reeves needed to distinguish between what he might regard as proper or due influence and improper or undue influence. Once again the picture became more complicated than he had expected.

The main trouble in Paul's case was that it was not clear whether the goals which appeared to influence his choice were, in the final analysis, Paul's goals as such. If they were not, then could it truthfully be said that Paul's choice was free of undue influence?

Reeves was tempted to answer this question with a reluctant, but firm, 'No'. But he reflected that even this question was further complicated by the recognition that the perspective of the patient's doctor and, still more, that of his wife as the person closest to him did have some legitimacy of their own. His job as Paul's doctor *was*, to an extent, to influence Paul's choice of goals so far as they affected Paul's physical health. This did not, of course, mean obstructing choices Paul clearly wanted to make. It did, however, surely allow Reeves to ask Paul to review for example his general reluctance to undertake physical exercise in his leisure time. And when it came to things like excessive smoking or drinking - Paul indulged in neither, so far as Reeves knew - Reeves felt he would be justified in encouraging a patient who relied on such things for easy gratification or relaxation to reconsider his goals. (He reasoned that the justification could appeal to a wider view of the interests of such a patient himself or, failing that, to the interests of society at large whose health care resources would be jeopardised in the future by the consequences of a patient's present recklessness.)

If any of this were true for Reeves, it seemed still more true for Lucy who shared Paul's life and his limitations - whether these arose from medication today or from his suffering a stroke tomorrow. Lucy had entwined her life with Paul's; they had even taken traditional marriage vows addressing each other's future health and sickness. Such ties might reasonably bring with them a legitimate influence over each other's health care decisions (or so Reeves reasoned, from the somewhat detached position of a middle aged bachelor with no domestic entanglements of his own).

In short, it seemed to Reeves that there was no conclusively *unrestricted* supremacy to the autonomy of the patient; the patient could reasonably expect freedom from *undue* influence over his choices, but picking out the dividing line between due and undue influence was no simple matter. These considerations remained in Reeves' mind as he considered the final stage of his three-part analysis of Paul's choice.

4. WAS PAUL'S CHOICE INTENDED?

In one obvious sense Paul's choice was clearly not intended at all, in the sense that he had not intended having to make it. He had not initiated the general health check, and he had not foreseen the kinds of clinical problems to which the check might give rise. It followed that he had not intended to be confronted with the decision to undergo long-term medication.

But perhaps this was not the intention with which Reeves needed to grapple. The real problem was whether Paul was making the choice in what one might think of as authentically his own 'voice'. And Reeves felt that this was a genuine problem, given that he had already acknowledged the problem of whether the goals at stake in the choice were authentically Paul's own. Paul was somewhat hypochondriac, presented himself rather often at the surgery with generally trivial (indeed sometimes virtually undetectable) complaints, was disconcertingly deferential towards the physicians and appeared to take a minute interest in their views and advice. It had been very easy to persuade Paul to accept the prescribed medication. Of course from Reeves' point of view that was in one sense just as well - the working day did not permit elaborate discussions with every patient. But he nevertheless felt uneasy and made up his mind to review the decision more carefully with Paul when they next met - particularly in view of the possibility that Paul was experiencing significant symptoms which he had not yet reported.

This of course finally drew Reeves back to the observations made by Lucy. These observations after all were what prompted his decision to review the whole case. And now he faced the somewhat paradoxical possibility that if on reflection Paul were to change his decision then it might seem that the more authentic 'voice' in the decision was in fact Lucy's.

Reeves pushed away his pen and paper, marked the improvement in the weather and decided to go out walking after all. His analysis had been inconclusive and it had if anything intensified the conundrum of the previous evening. Yet he felt that it had also legitimised it. The questions of patient autonomy and consent were, perhaps, too-easily trotted out as mantras in the various practical discussions of ethical matters which he had read or encountered. He felt that autonomy and consent embodied not merely 'ethical' but also conceptual - even, philosophical - challenges as well. They marked the moral untidiness of his practice as a general physician, and they marked the inevitable interpenetration of the biological and the biographical dimensions of his patients.

NOTES

1. The sewin (or sewen) is a kind of trout found notably in the rivers of Wales, during certain months of the year.
2. Precisely this problem, in precisely this clinical context, is the focus of an important early paper on the problem of reconciling conflicting views on the impact of a medical treatment. See Jachuk, *et al.* (1982).
3. Jaschuk, *et al.* (1982) suggest that these features left no clinical *signs* to be picked up by the professional practitioners. Hence if patients do not volunteer information about them, we may conclude that it is not unreasonable for at least some of them to be missing from an essentially physiological profile of adverse reactions. I discuss this point in relation to medical evidence and measurement, see Evans (1998).

REFERENCES

Evans, M., Sweeney, K.: 1998, *The human side of medicine*. Royal College of General Practitioners, London, Occasional Paper 76: 1-16.

Jachuk, S.J., Brierley, H., Jachuk, S., Willcox, P.M.: 1982, The effect of hypotensive drugs on the quality of life, *Journal of the Royal College of General Practitioners* 32: 103-5.

LINDA S. SCHEIRTON

THE AUTONOMY OF THE HEALTH CARE PROVIDER: ADVERTISING BY HEALTH PROFESSIONALS

1.CASE 2.

Marcel Alles is an obstetrician and gynaecologist in a small upscale town on the outskirts of a large metropolitan area in the southern part of the country. The metropolitan area has a population of almost one million people with numerous educational institutions, colleges and universities, research facilities, hospitals, long-term care facilities, and rehabilitation centres. Coming from a long tradition of medical practitioners - his grandfather, father, and siblings having all been physicians - he decided to become a physician as well. Upon graduation from medical school, he embarked on an obstetrics and gynaecology specialisation program. Immediately thereafter, he entered into a partnership with his father.

Marcel's partnership with his father has been successful, although not always harmonious. Marcel and his father have divergent viewpoints on professional ethics and preferred patterns of practice. They have learned to tolerate and sometimes even accept each other's philosophical differences. Marcel knows that he has benefitted greatly from his father's years of experience and knowledge. Under his father's tutelage Marcel was able to develop and expand an interest in his own speciality area. Whereas his father has specialized in high risk deliveries, Marcel focuses on infertility issues. He studies trends toward longer life spans and delaying motherhood for career development as significant indicators of future fertility need. Although he does not have formal specialisation training in this area, he has routinely attended national and international continuing education programs and workshops to develop his skills in this area. Several certificates hang on his wall indicating continuing education training in microsurgery and in vitro fertilisation (IVF). He was recently approved for physician staff privileges at a new fertility clinic located in the metropolitan area. He also has physician staff privileges at the University Medical Centre.

After thirty-five years of distinguished practice - thirteen of those years in partnership with his son - the elder Alles retires. Six months have passed since his retirement. As a solo practitioner, Marcel Alles has started to implement his

H.A.M.J. ten Have and B. Gordijn (eds.), Bioethics in a European Perspective, 93-107.

own long range practice plan. Given the cosmopolitan atmosphere of the region and the number of young professionals in the area, Marcel feels he needs to attract an increasing number of patients to secure and, preferably, increase his income. Marcel hires a marketing consultant to assist him in reorganizing the business aspect of his practice and to help develop a sound marketing and advertising initiative. Together they develop a multifaceted marketing plan.

As the first step in this plan, Marcel hires a new office manager. The current manager was employed by his father 27 years ago. She has decided to retire as well. The marketing consultant has some knowledge of the practices of Dr. Marcel Alles' chief competitors in the local area. One such physician, Dr. Jan Vischl, has practised in the area for nearly 10 years. His office manager is well known for her astute business acumen. Upon the advice of the marketing consultant, Dr. Alles offers Dr. Vischl's office manager, Ms. Marie Contes, a large raise in salary and benefits if she will come to work for him. Ms. Contes gives her present employer, Dr. Vischl, a chance to meet Dr. Alles' salary offer. When Dr. Alles proves to be the highest bidder, Marie joins his office staff.

The second stage of his plan is to hire two young physicians to assist him with the increased patient load. He has decided not to look for a partner to take his father's place, but to hire young physicians to handle the bulk of routine medical procedures. He will offer them a fixed percentage of gross receipts from routine medical procedures performed. The marketing consultant suggests that Marcel mail out employment opportunity notices to recent medical school graduates. By having the young physician employees see routine patients, Marcel Alles hopes to concentrate and expand patient care in infertility medicine. He thinks this will generate a higher level of personal practice satisfaction as well as monetary returns.

The third part of the plan is aimed at patients directly. Marcel plans a 'discount' coupon mass mailing to a carefully selected number of residents in his town and the eight other communities surrounding the city. He also plans to put a display advertisement in the telephone directory and the leading area newspapers. To this avail, the marketing consultant secures the services of a leading advertising agency. After a meeting with Dr. Alles, the advertisement was created (see figure 1).

But when the display advertisement actually appears in the new metropolitan telephone directory and various area newspapers, Dr. Vischl and several other local physicians report Dr. Alles to the Provincial Medical Peer Review Committee. They argue that the advertisement is false, misleading, and unethical. Marcel seeks the elder Alles' support, but his father refuses.

2. CASE ANALYSIS

2.1. Introduction

Before we embark on a detailed analysis of the various ethical dilemmas embedded in this case, we may want to describe the core issue which is in some sense personalized by the two protagonists in this case, the elder Alles on the one hand and his son Marcel on the other. Although the case description does not specify the reasons of Marcel's father's refusal to support his son against the charged levelled, one may speculate that his refusal is directly related to the different views of Alles Sr. and Jr. respectively about professional ethics. Alles Sr. has always believed in the positive value of professionalism for the improvement of society. In his view, the patient's need for care has profound moral significance because it is the patient's need that produces the reason for the profession's existence in the first place. The elder Alles' views coincide with those of Nelson, Clark, Goldman and Schore (1989) who say "caring for the sick is not, and should not be, considered largely the same as commercial selling. There is a special relationship that links patient and health care provider which distinguishes the latter from the commercial sellers" (Nelson, *et al.*, 1989, p. 37).

We can easily imagine that the elder Alles is not at all surprised that Dr. Vischl and several other local physicians have reported Marcel to the peer review committee. When his son comes to him with this news, he is not sympathetic. He voices his arguments against advertising. Maybe, he even gets his old textbook *Law and Ethics for Doctors* (Hadfield, 1958) from the bookshelf. Quickly, he finds the section he once underlined:

> [I]t is accepted in the profession that any advancement in the profession shall be obtained only in the normal process of building up a good professional reputation. Anything else that the practitioner does, or causes to be done, or even knows is being done, which can be said to enhance his professional standing or reputation or to increase his practice, may be regarded as advertising. Anything which can be construed as a means of attracting patients to him whether indirectly by advertising or directly by canvassing, whether or not they are patients of another practitioner, is likely to incur the displeasure of the Disciplinary Committee and to result in erasure (Hadfield, 1958, p. 34).

The father feels strongly that advertising harms the professional nature of medicine. He points out to Marcel that in 1997, the World Health Organisation (WHO) in response to complaints from Belgium and other countries, convened an ad hoc committee to develop recommendations to curtail the escalating use of false and misleading advertisements by health care promoters on the Internet. His son objects that advertising stimulates competition among physicians, which in turn has a great number of beneficial consequences both for physicians and the consumer. In his opinion, advertising can encourage lower fees; it can stimulate the use of innovative systems; and it can force physicians to continually improve their practice. Marcel argues that advertising is becoming an important source of disseminating information to the health care consumer. He cites empirical evidence from his colleagues in dentistry indicating that those most likely to respond to advertising, for example, are utilizers of care who have moved to a new geographic area (Friedman, Jong, DeSouza, Burek and Kranz, 1988). The elder Alles is concerned that exorbitant advertising costs will be paid for by the unsuspecting patient and fees will not become lower. He mentions an article by Havard in 1989 pointing out that in countries having a National Health Service it will be the taxpayer who ultimately pays the bill for advertising as costs will form part of the practice expenses. More importantly, thus the elder Alles continues, even if the fee for most services is slightly reduced by advertising, if the advertisements are paid for by the physicians pushing more services, the savings from lower fees are illusory. When his son dismisses that allegation, the elder Alles gives up and drops the conversation while warning his son, "Marcel, you are more interested in becoming a wealthy businessman. You do not value genuine professionalism."

2.2. Profession or Business

This hypothetical conversation between the elder and the younger Alles illustrates what I take to be the core ethical issue in this case. There would seem to be a genuine incompatibility between treating the provider-patient relationship in economic terms and viewing it primarily as a moral relationship. Traditionally, the medical profession has eschewed advertising. The ethical conflict is between a health care provider's right to advertise for commercial gain and the right of the largely naive public to protection from false or misleading claims.

Agich (1990) has argued that it is really the utilitarian foundations of economics versus the deontological foundations of professional medical ethics that underlie this conflict. Dr. Marcel Alles takes the utilitarian viewpoint. From a utilitarian aspect, advertising serves two very distinct objectives: (1) the dissemination of information to the patient to assist in making informed choices, and (2) product differentiation, which economists define as public perception of differences between two products (or as in medicine, two physicians), even though such differences may not exist. Until recently, the profession has largely held that dissemination of information is acceptable, but that product differentiation or solicitation of patients is not. These tenets have now been somewhat relaxed or even abandoned in some countries. Table 1 chronicles ancient and modern codified changes in medical advertising prohibition.

Many critics (including the elder Alles) of relaxing the medical advertising guidelines feel that societal changes in modern medicine and the way health care services are financed are now transforming patients into consumers. To the elder Alles, health care is not strictly a commodity to be sold effectively for profit to the public. According to him, medical care is not - or should not be treated as - a product which can be evaluated and then accepted or rejected before buying. He points out that the patient cannot just return it to the store within 30 days if something is found unsatisfactory. He quotes Cunningham (1978), as saying:

> [T]hose who get sick or hurt have no sensible way to judge the value of medical services, let alone strike a balance of value and price, as they commonly do in the case of dresses and deodorants. So they have to trust the intentions as well as the competence of their physicians and institutions, and this trust imposes on physicians and hospitals the obligation that is the bedrock of the professional ethic. The obligation of the ethic emerges from it measure the difference between hospital and industry, and nowhere more obtrusively than in marketing (Cunningham, 1978, p. 81).

2.3. Truth in Advertising and Misleading Advertising

The critics of Dr. Marcel Alles do not simply charge him with advertising, but with spreading false and misleading information. The issue of falsehood and, hence, of truthfulness towards patients, is of crucial significance to the modern understanding of what the medical profession is all about. Granted, veracity has not always been as high on the medical ethical agenda as it is nowadays.

For example, neither the Oath of Hippocrates nor the Declaration of Geneva of the World Medical Association specify guidelines for veracity (Reich, 1995, p. 2632, p. 2646). The Principles of Medical Ethics of the American Medical Association made no mention of veracity until a 1980 revision. That revision as well as a revision to the 1983 International Code of Medical Ethics of the World Medical Association indicates that physicians should "deal honestly with patients and colleagues" (Reich, 1995, p. 2647, p. 2649).

As Beauchamp and Childress (1994) have observed "by contrast to this *traditional* disregard of veracity, virtues of candour and truthfulness are among the most widely praised character traits of health professionals in *contemporary* biomedical ethics" (p. 395, emphasis added). This new emphasis on veracity has impacted the communication between individual provider and his or her patient. In order to treat, a physician must obtain an informed consent, which in turn requires the physician to inform the patient truthfully. The same principle of veracity has also influenced the communications from physicians to the public at large. In 1996, the American Dental Association reorganized its Code of Ethics and placed all articles concerning such public communications under the heading of veracity. The 1996 behaviour rules for physician members of the Royal Dutch Medical Association underscore the principle of veracity by saying that "publicity ... must be factual, verifiable, and understandable". Both the British Medical Association and the College of Physicians and Surgeons of Alberta require that advertised information is verifiable and not misleading the public respectively.

According to the elder Alles, these codes of ethics provide normative guidance. Each form of ethical statement in a professional code implies a moral imperative, which binds the individual practitioner precisely because (s)he is a member and representative of the *profession*. Dr. Marcel Alles, on the other hand, feels that codes are just an ordered collection of injunctions and prohibitions. These injunctions and prohibitions are of historical interest and may be helpful to settle problematic cases. But they cannot bind the individual practitioner precisely because (s)he is first and foremost an *individual* practitioner. In Dr. Marcel Alles' view, the relationship between doctor and patient is not determined by generic, prefixed rules, but by the particular agreement negotiated between the individual doctor and his or her individual patient. Physician and patient "as equal bargainers who have no obligations to one another save a prohibition of coercion and an obligation to truthfulness and to keeping contracts once they are made" (Ozar, 1994, p. 28). In Dr. Alles' opinion, the obligation to truthfulness is an obligation to the truth, but not the whole truth. He knows that as a bargainer he may not lie, but he will not be interested in providing information unless it is advantageous to the sale. The elder Alles aims at a relationship with patients which is fiduciary in nature. He feels a patient should not have to root out ambiguities, material omissions, or confusing statements in a physician's information. He thinks his son is not meeting up to his professional obligation to maintain trust, respect the patient's autonomy, and adhere to high standards of truthfulness.

The disagreement between father and son Alles, so it shall be clear, is not about false advertisements. Marcel Alles grants his father that false advertisements are not morally acceptable. Their disagreement concerns the secondary charge against Marcel Alles, that of misleading advertisements. While falsehood can be determined by comparing the claim to the actual facts, whether or not a claim is misleading is much more difficult to ascertain because it involves the addressee of the advertisement as well, that is, the (potential) patients. Hence, "the moral adequacy of advertising or promotional literature cannot reside entirely on factual truth. ... The crucial point is that assessing the moral propriety of advertising requires an analysis of what the advertisement implies as well as what is says" (Banja, 1994, p. 1017).

Many codes of ethics indicate that advertisements should be factual and verifiable, such that the truth (or falsity) can be established in principle. But most codes also prohibit advertisements that are verifiable in principle, but not in practice. For example, even if an advertisement about a particular drug or medical intervention is factually correct, patients will not be able to assess the proper indications and uses of advertised products or services and may compromise their health when they fail to seek proper advice from a physician (Skolnick, 1997). Since it is virtually impossible for a patient to compare the clinical abilities of different practitioners, advertisements should not compare either directly or indirectly or by innuendo, the practitioners' services or ability with that of any other practitioner or clinic. The advertisement should not promise or offer more effective service with better results than those available elsewhere. Moreover, in order not to be misleading, advertisements should also be understandable and not create an unjustified expectation about the results the practitioner can achieve.

In the display advertisement by Dr. Marcel Alles, a number of potentially misleading statements are present regarding the inappropriate use of credentials, outcome measures as well as the potential to exploit fertility patients' vulnerability. These potentially misleading claims will now be discussed in more detail.

2.4. Inappropriate Use of Credentials

Dr. Alles' colleagues, who have reported him to the Peer Review Committee, are concerned that Dr. Alles is misleading the public when his advertisement says he is "Specializing in Infertility Medicine". Claiming you specialize does not guarantee that you have mastered all the latest techniques in the particular area. To be considered a specialist, the physician must be willing to commit to a course of study and training that leads to a mastery of these methods. Mastery is rarely attainable just by attending continuing education programs and workshops on infertility. Usually these courses are offered only as a means of exposing gynaecologists to new state-of-the-art reproductive endocrine/infertility practice. Most of these courses are short in duration, lack significant clinical training, and do not objectively evaluate the information or skill level attained by their participants. These short courses, however well taught, cannot be construed as subspeciality training. Argue Reade and Ratzan: "The determination of such expertise is problematic at best and certainly should not, in every instance be left to self-designation alone" (1987, p. 1318). Even if the advertisement agency hired by Dr. Alles did not intend to mislead or deceive the public, the advertisement most likely leads the reader to believe mistakenly that Dr. Alles possesses proven skills, and training, experience, and ability superior to others.

Another area of concern involves memberships in professional organisations. Dr. Alles indicates that he is a member of special societies such as the European Fertility Society. This does not indicate whether this is just a forum for the advancement of knowledge about infertility where anyone who expresses an interest can be a member (for a modest fee) or if membership is limited by some peer review process (Blackwell, *et al.*, 1987). If the physician displays that he is a member of the European Fertility Society, the public may once again read this claim to imply that the physician has special expertise or certification in this area. There is clearly a difference between interest and training and the general public are usually unaware of this difference. Making claims of superiority, abusing the trust of the patient or exploiting their lack of knowledge is always morally suspect.

2.5. Specifying Outcomes

The public can easily be misled and given false hopes by certain outcome measurements. In Dr. Alles' display advertisement he mentions that the clinic has produced over 100 pregnancies for intra cytoplasmic sperm injections. He also lists an in vitro fertilisation success rate of 30%-40%. As Dyer puts it, quantifiable results are "more than simply the numerator of successful pregnancies over the denominator of cost" (1997, p. 146). The statements given must not omit significant information that will allow the patient to make an informed treatment choice. After all, one can tell the truth and still mislead. For example, the same couple that utilizes the services of two different in vitro fertilisation clinics, could face live birth rates that are two, three, even fives times higher in one program than in the other. Unfortunately there is still no verifiable national or international outcome-based reporting system that achieves accountability in reporting clinic-specific success rates. An important question is the kind of statistics used; is it expected rather than actual births? Some IVF programs transfer more embryos than others. Higher success rates are achieved by increasing the number of implantations. There is also the increased risk of multiple pregnancies. Usually they report data on women younger than 40 years-of-age. This can inflate expectations for women over 40 if they are not informed of this fact.

The elder Alles probably thinks that his son is using puffery claims that are intended to deliberately influence the choice in favour of the advertiser. On the other hand, Marcel will counter that the claims are performance claims and not exaggerated hyperbole. Marcel feels that the one purpose of advertising is to bring accurate information to potential consumers. His father wants Marcel to substantiate every factual claim, especially performance claims. The father does not understand what a '30-40% success rate' means. If the success rate is not explained, he says, the claim should not be made. He also wonders how many patients were seen and during what time period to produce "over 100 pregnancies from intra cytoplasmic sperm injections". Did the "over 100 pregnancies" result in over 100 live births?, he asks. Informed consent is really the issue. To the elder Alles, having adequate and accurate information provided to the patient is a hallmark of ethical practice. In his opinion, a simple description of the types of services delivered by the physician and his associates without making success claims, comparisons or self-evaluations is appropriate. In response to his father's reproaches, Marcel quotes Irvine saying that "[t]o maintain a distinction between 'information-giving', on the one hand, and the full panoply of promotional techniques on the other inevitably involves making a subjective judgement as to where one ends and the other begins" (1991, p. 38). Although there has been considerable consensus within the profession in the past, "the diversity and complexity of modern medical practice has made it increasingly difficult to maintain a hard and fast line" (*O.c.*, p. 38). To illustrate his point, Marcel refers to recent changes and substantial relaxation of controls on advertising in major European codes of medical ethics and other international codes of ethics (see table 1).

2.6. Unfair Business Practices

Several physicians interviewing for the associate positions in Dr. Marcel Alles' practice are concerned about the financial reimbursement scheme which is outlined in an elaborate legal contract. According to the employer's contract, the associate will receive a fixed percentage of gross receipts in five categories of routine procedures. They feel this payment structure includes unethical financial incentives. Several others who have been interviewed are pleased that the arrangement is clearly outlined, nothing is vague or unattainable, and it is competitive and yet fair. They feel that if every detail is spelled out to avoid misunderstandings in the future and if both parties freely choose the exchange, then the resulting distribution of resources is ethical and just. These two divergent viewpoints just underline the distinctive nature of two different professional practice models.

Other areas of questionable but not necessarily unfair business practices involve marketing techniques that offer clip-and-save coupons, one-time specials, and package deals. An example in dentistry would be free tooth extractions with the purchase of an upper and lower denture. Having a coupon in the newspaper next to a coupon to have the oil changed in your automobile may not be considered unethical but may be, in actuality, viewed as demeaning to the medical profession. One genuine example of an unfair business practice would be to offer a discount fee of $45 for a routine gynaecology examination and then add on additional charges for items that would normally be included in a routine gynaecology examination. Other areas of concern can include statements guaranteeing the matching of competitor prices. In his display advertisement, Dr. Marcel Alles highlights in bold print that "we will match all competitor prices." Several of his physician colleagues and father find this statement extremely troublesome. Firstly, how can patients make the necessary comparisons to verify this claim? The patient is not offered a product with a price tag adhered to it. Even routine examinations can run a broad range from a cursory overview taking approximately 10 minutes or less to a more extensive process involving several diagnostic tests. Should a patient telephone other physicians in the area to ascertain examination time spent, the number of tests routinely run, and the fee charged. How will Dr. Alles have proof that he is matching competitors' prices? If other physicians' treatment fees could be substantiated, then this claim would not be suspect to as much criticism. However, even if verifiable, it would remain troublesome for the elder Alles because it puts the provider-patient relationship in economic terms rather than viewing it in an altruistic, self-effacing, and fiduciary orientation.

3. CONCLUSION

In summary, assessments of right and wrong advertising conduct in the provider-patient relationship depend on the way in which the relationship itself is understood from a moral perspective. Whether medicine is considered a profession or a business is decisive. If you think that medicine is a profession, there is no room for *creating* patients needs. Furthermore, it is unprofessional to compete with fellow professionals (colleagues) for patients because such competition does not have the patient's best interests at heart. However, if medicine is thought of as a business, the practices of physicians should be evaluated not from a professional perspective but from a commercial perspective. The same two behaviours that are ethically prohibited for professionals are key elements of a business relationship. Business is all about trading goods and making a profit in the process of doing so. Selling goods to clients who at first might not even have a desire for such goods is not immoral (provided clients are not lied to or otherwise forced into buying the goods). Advertising ones products in an attempt to draw new clientele is an intrinsic and necessary part of this process.

The previous examples show that these two models of the patient-provider relationship are at odds with one another in important regards. The question thus arises whether it is at all possible to unite these two models into one. On the other hand, maybe the quest for a model that perfectly merges the professional and business model is itself an unnecessary quest. This is still an ongoing debate in the bioethics literature.[1] In the ideal world the patient and physician would have equal moral standing within the relationship and advertising of physician services would not be an issue.

Table 1.

Various Medical Organisations and What Their Codes of Ethics Say on Advertising

The Hippocratic Oath

The oath did not mention advertising or self-promotion per se. It emphasized the principle of patient benefit, placing the patient at the centre of the physician's concern. "I will apply dietetic measures for the benefit of the sick according to my ability and judgement; I will keep them from harm and injustice" (Reich, 1995, p. 2632).

Percival's Medical Ethics

Percival apparently did not explicitly refer to advertising in his code. While is was reputed that Percival's medical ethics was drawn on utilitarianism (Reich, p. 1551), he developed a systematic view of medical ethics based on universal (and possibly deontological) truths about good professional behaviour and how it could be learned and applied by all medical practitioners. Percival stressed maintaining of harmony, dignity, humour, and reputation of the profession (Percival, 1803).

American Medical Association (AMA), Code of Medical Ethics

The 1847 AMA code was based upon Percivalian precepts. According to the 1847 code, Article II.1.3, "It is derogatory to the dignity of the profession, to resort to public advertisements or private cards or handbills, inviting the attention of individuals affected with particular diseases ..." (American Medical Association, 1995, p. 2642).

The 1980 revision of the AMA code deleted any reference to advertising when the Federal Trade Commission (FTC) successfully sued the AMA over the issue of restricting advertising through its code of ethics (Bates v State Bar of Arizona, 1977).

The 1996 AMA code lists seven principles of ethics. Principle number II. says "[a] physician shall deal honestly with patients and colleagues, and strive to expose those physicians deficient in character or competence, or who engage in fraud or deception" (*O.c.*, p. XVII). The current opinions of the AMA Council on Ethical and Judicial Affairs have twenty-two pages devoted to opinions on Confidentiality, Advertising, and Communication Media Relations. Section 5.02 states that "[a] physician may publicize himself or herself as a physician through any commercial publicity or other form of public communication ... provided that the communication shall not be misleading because of omission of necessary material information, shall not contain any false or misleading statement, and shall not otherwise operate to deceive" (p. 72). In addition, "[a]ggressive, high pressure advertising and publicity should be avoided if they create unjustified medical expectations or are accompanied by deceptive claims" (p. 72). Similarly, "[o]bjective claims regarding experience, competence, and the quality of physicians and the services they provide may be made only if they are factually supportable" (p. 73).

Australian Medical Association, Code of Ethics (1996)

"2.3.1 Do not advertise professional services or make professional announcements unless the chief purpose of the notice is to present information reasonably needed by any patient or colleague to make an informed decision about the appropriateness and availability of your medical services."

"2.3.2 Ensure that any announcement or advertisement directed towards patients or colleagues is demonstrably true in all respects, does not contain any testimonial or endorsement of your clinical skills and is not likely to bring the profession into disrepute."

British Medical Association, Handbook on Medical Ethics

As quoted by Dyer, the 1984 *Handbook of Medical Ethics* said that "[t]he medical profession in this country has long accepted the tradition that doctors should refrain from self-advertisement" (Dyer, 1985, p. 77). In 1990, the General Medical Council issued guidance allowing general practitioners to advertise to the public after the Monopolies and Mergers Commission recommended a relaxation of the advertising rules (Beecham, 1990, p. 1420). More recently, in 1996, British medical specialists have been allowed to advertise to the public in the same way as general practitioners. Doctors "may publish or broadcast verifiable information about the services they provide so long as it conforms with the law and with guidance from the Advertising Standards Authority. The information must not offer guarantees of cures, nor exploit patients' vulnerability or lack of medical knowledge" (Beecham, 1997, p. 1226). These changes were based on advice from over 50 medical and patient organisations.

Royal Dutch Medical Association: 1996, Behaviour Rules for Physicians, De arts in relatie tot publiciteit [The physician in relation to publicity]

V.1. Publicity by and for physicians must be factual, verifiable, and understandable; publicity may not be solicitous.

V.2. It is not permitted for a physician to support that third parties advertise on his behalf with the apparent intention to point attention in a solicitous manner to his professional practice.

V.3. It is not permitted for a physician to generate publicity in such a manner that his services are compared with the services of other identifiable physicians.

V.4. It is not permitted for the physician to release information about identifiable persons for the purpose of publicity unless they have granted consent explicitly for that purpose.

World Medical Association, International Code of Medical Ethics (amended 1983)

"A physician shall not permit motives of profit to influence the fee and independent exercise of professional judgement on behalf of patients." (Reich. 1995, p. 2647).

"Self-advertising by physicians, unless permitted by the laws of the country and the Code of Ethics of the National Medical Association" is deemed to be unethical conduct (*O.c.*, p. 2647)

Council of Europe, Convention for the Protection of Human Rights and Dignity of the Human Being with Regard to the Application of Biology and Medicine, Convention of Human Rights and Biomedicine (1997)

Article 4-Professional standards: "Any intervention in the health field, including research, must be carried out in accordance with relevant professional obligations and standards" (*O.c.*, p. 2).

College of Physicians and Surgeons of Alberta, Advertising Information Regarding Medical Services (1989)

"A registered practitioner or clinic may make information about himself/herself or itself and his/her or its medical services available to any patient, potential patient, or public generally, subject to the limitations contained herein", such as ... "marketing activities [not in] good taste, accurate and not capable of misleading the public", including ... "any conduct directly or indirectly or through any medium or agent that: (a) misrepresents fact, (b) compares either directly, indirectly or by innuendo, the registered practitioner's services or ability with that of any other practitioner or clinic, or promises or offers more effective service or better results than those available elsewhere; ... (d) creates an unjustified expectation about the results the practitioner can achieve" (*O.c.*, p. 1)

Norwegian Medical Association, Code of Ethics for Doctors (amended 1992)

"A doctor must not advertise medicaments, consumer goods, or methods. References to professional medical contexts in articles, lectures, and the like, and involving no pecuniary gain, are not regarded as advertisements (Reich, 1995, p. 2686)."

Conference Internationale des Ordres et des Organismes d'Attributions Similaires, European Code of Medical Ethics (1987)

Does not mention advertising per se, but Article 28 does say the rules of professional etiquette were designed to prevent patients becoming victims of dishonest manoeuvres between doctors (Reich, p. 2684-2685).

NOTES

1. For more information on models that attempt to incorporate the business and professional aspects of the provider-patient relationship see: Veatch, R.M.: 1972, Models for ethical medicine in a revolutionary age, *Hastings Center Report* 2(3): 5-7; Ozar, D.T.: 1985, Three models of professionalism and professional obligations in dentistry, *Journal of the American Dental Association* 110(2): 173-177.

REFERENCES

Agich778, G.J.: 1990, Medicine as business and profession. *Theoretical Medicine* 11: 311-324.

Australian Medical Association: 1996, *Code of ethics*. Advertising, [online]. Available:http://domino.ana.com.au/DIR0103P ... be8e64a2564e30025c79d? OpenDocument

American Medical Association: 1995 [1947], Code of ethics. In: Reich, W. (ed.), *Encyclopedia of bioethics*. Vol. 5, p. 2639-2644. Simon and Schuster Macmillan, London.

American Medical Association: 1996, *Code of medical ethics: Current opinions with annotations*. Chicago, IL, American Medical Association.

Banja, J.D.: 1994, Deception in advertising and marketing: Ethical applications in rehabilitation. *Arch Phys Med Rehabil* 75: 1015-1018.

Bates v State Bar of Arizona, 433 US 350, 53L.Ed. 2d 810, 97 S. Ct. 2691, 1977.

Beauchamp, T.L. and Childress, J.F.: 1994, *Principles of bioethics*. Oxford University Press, New York.

Beecham, L.: 1990, Advertising - no longer a dirty word. *BMJ* 300: 1420.

Beecham, L.: 1997, UK specialists to be allowed to advertise. *BMJ* 313: 1226-1227.

Blackwell, R.E., Carr, B.R., Chang, J.R., DeCherney, A.H., Haney, A.F., Keye, W.R., Rebar, R.W., Rock, J.A., Rosenwaks, Z., Seibel, M.M., Soules, M.R.: 1987, Are we exploiting the infertile couple? *Fertility and Sterility* 48(5): 735-739.

College of Physicians and Surgeons of Alberta: 1989, Advertising information regarding medical services. Alberta, Canada, [online].
available: http://www. cpsa.ab.ca/ policyguidelines/advertising.html

Council of Europe: 1997, Convention for the protection of human rights and dignity of the human being with regard to the application of biology and medicine: *Convention of human rights*. Oviedo, Spain, [online]. available: http://www.coe.fr/eng/legaltxt.164e.htm

Cunningham, R.B.: 1978, Of snake oil and science. *Hospitals* 52: 81.

Dyer, A.R.: 1997, Ethics, advertising, and assisted reproduction: The goals and methods of advertising. *Women's Health Issues* 7(3): 143-148.

Dyer, A.R.: 1985, Ethics, advertising and the definition of a profession. *Journal of Medical Ethics* 11: 72-78.

Friedman, P.K., Jong, A.W., DeSouza, M.B., Burek, M. & Kranz, S.: 1988, An investigation of the effect of advertising on dental consumers. *J Dent Pract Adm* 5: 166-173.

Hadfield, S.J.: 1958, *Law and ethics for doctors*. Eyre & Spottiswoode, London.

Irvine, D.H.: 1991, The advertising of doctor's services. *Journal of Medical Ethics* 17: 35-40.

Nelson, L.J., Clark, H.W., Goldman, R.L., and Schore, J.E.: 1989, Taking the train to a world of strangers: Health care marketing and ethics. *Hastings Centre Report* 19(5): 36-42.

Ozar, D.T.: 1994, Professional ethics and dental advertising. *CDA Journal* 22(10): 27-29.

Percival, T.: 1975 [1803], *Medical ethics: A code of institutes and precepts*, adapted to the professional conduct of physicians and surgeons. Krieger, Huntington, N.Y.

Reade, J.M., Ratzan, R.M.: 1987, Yellow professionalism. *N Engl J Med* 316: 1315-1319.

Reich, W.T. (Ed.): 1995, *Encyclopedia of bioethics*, Volume 5, Appendix: Codes, Oaths, and Directives Related to Bioethics. Simon & Schuster Macmillan and Prentice Hall International, New York.

Royal Dutch Medical Association: 1996, *Behaviour rules for physicians 'Blue Booklet'*. De arts in relatie tot publiciteit [The physician in relation to publicity], V1-4. Blue Book (translation: J.V.M. Welie).

Skolnick, A.A.: 1997, WHO (World Health Organisation) considers regulating ads, sale of medical products on internet. *JAMA* 278(21): 1723-1724.

Figure 1

Display Advertisement

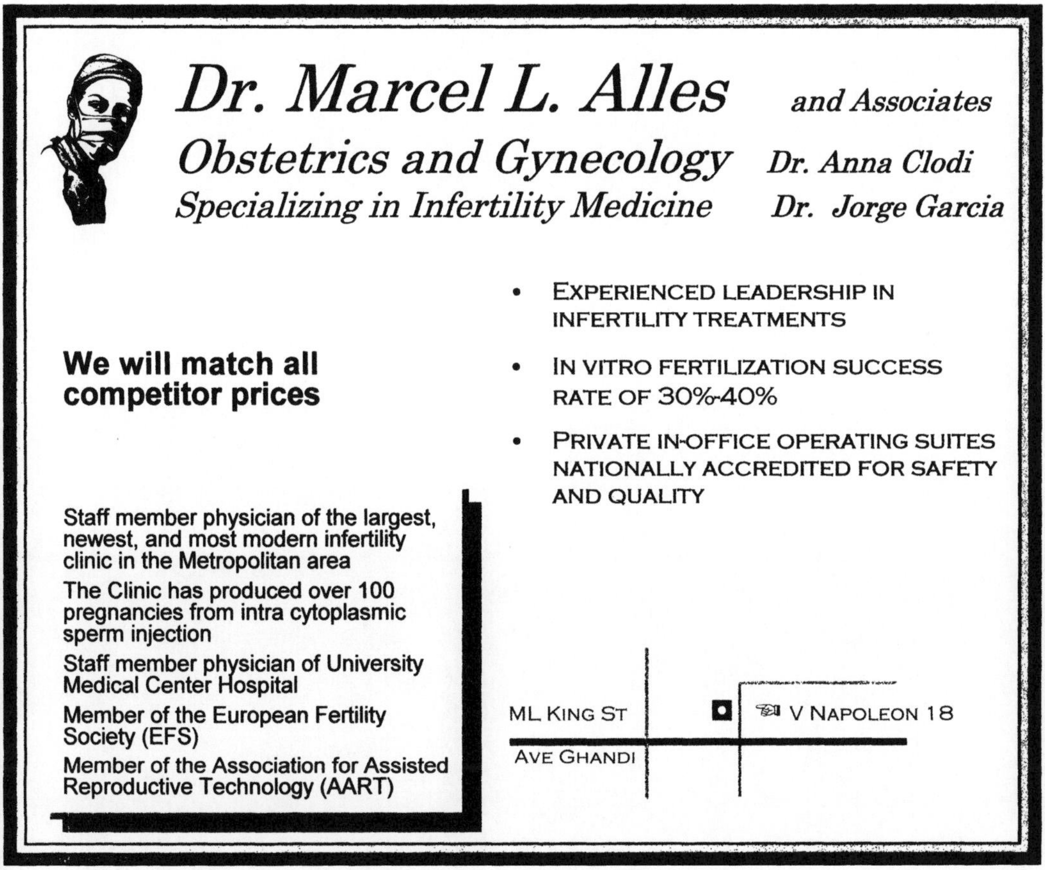

PART 2

PERSON AND BODY

HENK A.M.J. TEN HAVE

PERSON AND BODY

Introductory Comments

"Is my case very serious?" Ivan Ilych asks. But the physician ignores him. The irrelevance of such personal concerns only evokes the physician's contempt. The patient even accepts it, since he as judge has showed similar behaviour toward those who petitioned him in the courtroom. He recognizes in his physician the same method, reducing the most complicated case to an impersonal form.

In 1886, when Leo Tolstoy publishes his novel *The Death of Ivan Ilych*, medicine is in the midst of a tantalizing process of scientific evolution. Physiology and pathology are making enormous progress, due to the knowledge and methods of the natural sciences. Clinical medicine shows better diagnostic abilities (for example after the invention of the ophthalmoscope) and more successful therapeutic interventions (for example in surgery, following the introduction of anaesthesia and antisepsis). New disciplines such as bacteriology are developing. In short, medicine is transforming into the powerful science and technology of today.

Tolstoy, however, was very much aware of the possible drawbacks of this transformation. Every light has its shadow. The medical account of illness, though impartial and effective, is radically different from the experience of illness, the patient's story. Ivan Ilych's basic question whether his condition is very serious, - a matter of life or death -, was cruelly neglected by the physician. For the doctor the only relevant issue was whether this case was a matter of vermiform appendix or a floating kidney. The physician works as if diseases are real entities that have invaded the body, that can be recognized, localized and counteracted. The patient is identified with his diagnoses. More and more, he will have an abstract medical existence. Ivan Ilych is to his physician, then to himself, and finally for others the floating kidney (or perhaps the appendix) that is believed to be making him unpleasant and not feeling well. The diagnosis, the case is substituted for the person. The patient is impersonalized.

H.A.M.J. ten Have and B. Gordijn (eds.), Bioethics in a European Perspective, 111-113.

In Western medicine, physicians are believed to know everything about the body. Proceeding from the Cartesian presupposition that the person can be divided into two distinct realms of mind and body, medical science has concentrated on the mechanics of the human body. In order to have effective medical interventions it is taken for granted that a difference must be constructed between the subjective, experiential story and the outsider's objective, scientific interpretation.

The rise of bioethics, as analyzed in the previous part of this book, can be regarded as an incentive to medicine to enrich its one-dimensional interest in the de-personalized body. It should re-orientate itself towards the patient as a person. Not minds and bodies, but persons, are real entities existing in the world. What is needed, is attentive care that attempts to reconcile the subjective account of the patient's suffering with the medical version of illness.

More attention therefore should be given to the relevancy of subjective experiences of illness: uncertainty, anxiety, suffering, helplessness, fear of death, loss of control, loneliness. To complement medicine's focus on the patient's body, bioethics stresses the central importance of the person of the patient.

This part of the book addresses first of all the basic views of the human body as well as the notion of person. Subsequently, examples are discussed of moral problems regarding medical interventions directed to the bodies of persons.

Wim Dekkers from the Netherlands argues that medical ethical discourse as long as it conceives the human person as autonomous, rational agent, cannot take the body seriously. The body is simply viewed as an instrument for its owner. The significance of man's bodily nature for his or her moral experiences is largely ignored. Continental philosophical traditions, notably phenomenology and philosophical anthropology, assert the bodily nature of all experience. The human body is not merely object for moral reflection, but rather the source of moral experience. Human beings both have and are their bodies. Within the interpretive approach to bioethics this view implies that the human body itself is a text for interpretation; ethical reflection should begin with an interpretation of patients' experience of bodily suffering. Dekkers shows how concentrating on human experience can bridge the gap between human corporeality and morality.

In the following chapter, Martyn Evans examines various notions of personhood which are in use in bioethical discourse. It is obvious that in bioethics the term 'person' is not used in the ordinary sense, *viz.* 'individual human being'. Bioethicists use the term in a more sophisticated, technical sense, referring to particular qualities, such as rationality and self-consciousness. Such qualifying characteristics allow for distinctions to be made between persons and non-persons. These distinctions are useful for bioethical debate because they determine what is morally justifiable to what category of entities. Evans argues that the term 'person' is both descriptive and evaluative. In bioethics, different conceptions of personhood are used which exclude different categories of human beings. Each conception introduces a particular categorisation to treat human beings differently. Evans points out that instead of re-defining patients in or out the moral realm by focusing on conceptions of personhood, bioethics should concentrate on moral justifications for making choices among different people.

Transplantation, experimentation, and compulsory treatment are three major areas of contemporary debate in bioethics. In each area, medicine intervenes in the body of persons. The first case concerns living organ donation. Rita Kielstein from Germany discusses the moral issues involved in this case. Medical experimentation with human beings is one of the earliest controversial topics in bioethics, as reviewed by Wim Dekkers in the second case analysis. With the growing potential of effective medical intervention, the question has arisen what to do in cases where patients do not co-operate when medical treatment is indicated. Finally, Bert Gordijn from the Netherlands analyzes several cases of compulsory treatment. In all case analyses, notions of 'body' and 'person' are at stake in the bioethical evaluation.

WIM J.M. DEKKERS

THE HUMAN BODY

1. INTRODUCTION

It is a commonplace, but at the same time a statement that raises many philosophical questions, to say that man consists of a material part, that is the body, and an immaterial part, that is the soul, the mind, or whatever it may be called. In the long history of Western European philosophy the soul, the mind, and the consciousness have always received ample philosophical attention, while philosophical interest in the human body has been marginal. Concepts of the body have mainly come into play where the definition of the soul, the mind, or the mind-body relationship is at issue. The body as a problem could only become of real importance after the relationship of body and mind had become a philosophical problem of central concern (Verwey, 1990). From a historical perspective, it is in the twentieth century only that the body 'as such' has received philosophical attention. For two or three decades the body has called attention from other disciplines as well. Nowadays, a rather extensive literature about philosophical, social and cultural aspects of the human body exists.[1]

In this chapter I will primarily concentrate on the human body and try - as a heuristic method - to 'forget' the mind. It must be clear from the outset, however, that speaking about the body means speaking about the mind or the person. It is significant that in the English language the words 'somebody' and 'anybody' mean 'someone' or 'some person' and 'anyone' or 'any person' respectively. Thus, if one focuses on the body, one will inevitably come across the person, the mind and the mind-body problem. The reverse holds too. As may be clear from the chapter on the person in this volume, speaking about the person means speaking about the body (Evans, 2001). In philosophical and ethical considerations of the person one often refers to the (moral) status of unconscious bodily 'beings' such as embryos, brain dead people, and people in a permanent coma.

In my attempt "to take the body seriously", as Toombs (1997) put it, I will draw insights especially from three Continental traditions in philosophy and medicine: (1) philosophical anthropology, (2) the so-called anthropologi-

H.A.M.J. ten Have and B. Gordijn (eds.), Bioethics in a European Perspective, 115-139.

cal movement in medicine, and (3) what may be called the interpretive tradition, including phenomenology and hermeneutics. Insofar as these traditions are typical of Continental thinking, the approach in this chapter may be called 'European'.[2]

The argument is based on three assumptions (Dekkers, 1998b). The first assumption is the philosophical presupposition that 'corporeality' and 'morality' may be considered two fundamental anthropological categories, that is, that man is a corporeal and a moral being as well. By the (phenomenological) term *corporeality* I mean the fact of man's corporeal existence, the fact of his being a corporeal being. By *morality* I mean the fact of man's moral existence, the fact of his being a moral being.[3] In this context it is important to note that not infrequently 'corporeality' and 'morality' seem to be considered exclusive categories. If one surveys the history of philosophy, one can say that roughly speaking more attention has been paid to man's morality than to his corporeality (Zaner, 1995). If we consult philosophical literature in which attention is paid to the human body, we scarcely find anything about man's morality.

Secondly, if the human body plays a role at all in medical-ethical literature, it is foremost the 'body as object' to which attention is paid, for example, in the context of discussions on the moral acceptability of using dead body parts for transplantation purposes, medical experimentation, and the production of diagnostic tests and other industrial products. In many modern ethical debates the arguments are focused on the question of whether and to what extent it is morally acceptable to control the human body, which is primarily considered an instrument for the human person. Much emphasis is paid, in particular in the (neo)liberal tradition, to man as a free, self-sufficient, rational and autonomous subject (Beauchamp and Childress, 1994). The human person, conceived of as a conscious, rational subject, is considered the owner of the human body. In this view, the human body has an exclusively instrumental value only and seems to have no moral value by itself.

Thirdly, (moral) experiences of the human body establish an excellent starting-point for moral considerations and discussions about how 'to cope with the body'. The aim of this chapter could be summarized - though a bit schematically - as an attempt to bridge the gap between man's corporeality and his morality by means of focusing on human experience. Analyzing and interpreting our (moral) experience of the body is a sound basis for further (moral) deliberation. This is not to say that the appeal to (moral) experience is the end of moral deliberation. On the contrary, it is just the beginning of it. An appeal to moral experience must not be considered a search for a definite answer to moral problems, but rather "an invitation to conversation" (Lauritzen, 1996, p. 13). As Zaner has argued, "experience is the point of departure and return for theory: its ground and ultimate test" (Zaner, 1994, p. 230).

The central aim of this chapter is to emphasize the moral significance of the human body, the dead body and the living body as well, and to explore from a moral perspective what the human body 'has to tell us'. This implies an understanding of the similarities and dissimilarities between our moral attitudes toward living and dead bodies. As will be explained below, the phenomenal body is the subjective, lived body (Merleau-Ponty, 1994). The objective body as we know it from medicine and other objective approaches is only an impoverished image of this subjective, lived body. Though the dead body necessarily is an objective body, the argumentation will show that we can only understand the moral status of the dead body by referring to knowledge of the subjective, lived body.

I will concentrate on the notion of integrity of the body, taking the practice of anatomical dissections of the body, invasive surgery and cadaveric organ transplantation as examples (section 3). Regarding the living body I will focus on the notion of the subjective, lived body, in particular as it is experienced by patients with chronically disabled bodies (section 4). But first I will explain what I mean by a 'European approach' and elaborate on the metaphor of 'the body as a text', discussing three fundamental ways in which the human body can be thematized at all (section 2).

2. EUROPEAN APPROACH

The twentieth century philosophy of medicine and health care can be divided into three phases (ten Have, 1997, 1998): (1) the epistemological phase, (2) the anthropological phase, and (3) the ethical phase. This chapter has been mainly based on a retrieval of ideas and concepts which are typical of the second, anthropological, phase. The argument in this chapter can be considered as a contribution to the exploration of the continuity and discontinuity between the anthropological phase on the one hand and the ethical phase of modern health care ethics on the other hand. In particular, three important traditions which belong to the second, anthropological, phase are elaborated here: philosophical anthropology, the anthropological movement in medicine, and the interpretive tradition.

2.1. Philosophical Anthropology

Philosophical anthropology is the philosophical discipline that studies existing and desirable images of man, that is, an understanding of what man is and what he or she should be. Although the various aspects of being human have been the subject of philosophical thinking as long as philosophy has existed and although there is hardly any philosophy which is not concerned with man, it is in the twentieth century only that philosophers such as Scheler and Plessner laid the foundation of philosophical anthropology. Epistemology deals with human knowledge, linguistic philosophy with human language, philosophy of religion with human religion, and ethics with human actions. What is new in the development of philosophical anthropology is that all the questions which derive from our 'being in the world' converge to the one overwhelming question: what is man? (Marquard, 1991).

One of the most important themes of philosophical anthropology is corporeality or 'bodiliness' (Strasser, 1983).[4] In this context especially Sartre, Merleau-Ponty, and Ricoeur are worth mentioning. The common denominator of their view of the human body is the rejection of a strict, Cartesian dichotomy between the body, conceived of as a material, mechanistic structure, and the soul or spirit, conceived of as an immaterial substance. For Descartes the human body is an inanimate machine. He was fascinated by the automatons of his day which, while able to perform a variety of functions, even to imitate the behaviour of living creatures, were in fact driven by mechanical forces. According to Descartes, the living body can be treated as essentially no different from a machine (Leder, 1992b).

Ricoeur (1949, p. 12-20) refers to the human body as a mystery which cannot be deciphered. The very fact of man's bodily existence restricts his self-knowledge and self-insight. Man, as it were, is doomed to live with the mystery of the body. This is due to the fact that we cannot approach our body as if it were a thing which can be totally objectified. We are part of our body and our body is part of ourselves. In other words: we can neither identify completely with our body, nor can we detach ourselves completely from our body. By his well-known expression "to be the body and to have the body", Marcel (1935) means the body, which I am and which I have at the same time. I cannot choose one of the two possibilities. For example, when someone touches my arm, I can say to that person: "you touched my arm" or "you touched me". Both expressions are true. I can identify with my arm, but at the same moment I can detach myself from my arm. It is this ambiguity which constitutes one of the most important aspects of human existence.

The living body is commonly called by phenomenologists the 'lived body'. Merleau-Ponty argues that our lived body, our subjectively experienced body ('le corps-sujet'), is the only access to the outside world. It is through one's lived body that one manifests oneself to the world. The lived body is the body as it is given in direct experience. It is immediately felt, sensed, tasted, smelt, heard and seen. The lived body is the expression of one's existence and as such it is concretely lived by oneself. All structures and functions of the lived body (perceiving, moving, acting, sexual behaviour etc.) are modes of being of the person. Whether we are consciously aware of it or not, the lived body is present as a true companion in our personal existence. For Merleau-Ponty the lived body has its own knowledge of the world. This implies the existence of a 'tacit knowledge', a silent knowledge which functions without conscious control. On a subconscious level my body provides me with a lot of information about the world. Merleau-Ponty expresses one of his main theses as follows:

> The objective body is not the true version of the phenomenal body, that is, the true version of the body that we live by: it is indeed no more than the latter's impoverished image, so that the problem of the relation of soul to body has nothing to do with the objective body, which exists only conceptually, but with the phenomenal body (Merleau-Ponty, 1994, p. 431-432).[5]

As is illustrated in this quotation, the anthropological and phenomenological perspective on the human body constitutes a right counter-balance against the one-sided Cartesian concept of the body and also against current perspectives on the human body, in which the body is mainly considered an *object* with instrumental value. Many modern ethical debates concern the question of to what extent it is permissible for a person to control his body by means of surgery, the use of drugs, or intensive physical training. From the anthropological perspective the body is not primarily seen as an object of instrumental value, but as a lived, animated body. It is first and foremost considered a *subject.*

2.2. The Anthropological Movement in Medicine

The anthropological movement in medicine - with similar developments in biology, psychiatry and psychology - had its heyday between about 1925 and 1950. This movement consisted of a group of mostly German physicians who supported the idea of an anthropologically oriented medical science and medical practice. Some of their most important representatives are Von Gebsattel, Von Weizsäcker, Binswanger, Straus, Goldstein and Buytendijk (ten Have, 1995). If one attempts to summarize the motives and intentions of representatives of this movement in one sentence, this would be the following phrase: "to introduce the subject into medicine". It was their conviction that the human subject must have a place in the practice and theory of medicine. What this means can be explained by elaborating on some influential ideas of Buytendijk (Dekkers, 1995).

Buytendijk's significance lies primarily in an attempt to make the anthropological view on man and his body useful to medical science and medical practice. The purpose of his concept of an anthropologically oriented physiology was - following Von Weizsäcker - "to introduce the subject into physiology". Merleau-Ponty's notion of the body as a subject constitutes the foundation of Buytendijk's anthropological physiology. Buytendijk adopts the distinction made by Scheler and other phenomenologists between the objective thing-body (*Körper*) and the subjective lived body (*Leib*). The objective body is a complicated structure, an instrument, and as such the object of scientific anatomy and physiology. The lived body is the body from which one cannot be separated, the body, which one is and has at the same time.

Like most other anthropological thinkers Buytendijk rejects the Cartesian dichotomy of body and mind. This does not mean, however, that he has 'overcome' Cartesian dualism. Like Merleau-Ponty, he advocates another dualistic opposition, i.e., between the body-subject and the person-subject, although this distinction is far less radical than Descartes' dichotomy of two separate substances. The most essential characteristic of man is "existing in the world as polar unity of a mental and bodily subjectivity" (Buytendijk, 1974, p. 27). Buytendijk's concept of the subjectivity of the body means that our own body, even on a preconscious, prepersonal level, establishes a meaningful relation with the world. This bodily relationship accompanies our personal existence "as a provisional nameless draft of our being situated" (Buytendijk, 1974, p. 243) Buytendijk calls the lived body "a tentative sketch of our existence" (Buytendijk, 1974, p. 47), "the prereflexive entrance to the world" (Buytendijk, 1968, p. 302).

The slogan "the introduction of the subject into medicine" has epistemological-methodological and medical-practical implications as well. This means that the subject has to be introduced not only into the theory, but also into the practice of medicine. What this means will be dealt with in section 5.

2.3. The Interpretive Tradition: Experience and Interpretation

Ricoeur (1973) has described hermeneutics as "the science of interpreting texts". In hermeneutics it is explicitly recognized that every meaning is the result of an interpretation process, of a textual reading. What makes interpretation hermeneutical is the realisation of the historicity and conditionedness of any human action or human being. In the field of health care, the hermeneutical enterprise consists of three different levels (Dekkers, 1998a): (1) the level of being human, i.e., the very fact of our being in the world, (2) the level of (medical) ethics and (3) the level of clinical medicine. In this chapter I will explain what is meant by these three levels.

As hermeneutical philosophers like Heidegger, Gadamer and Ricoeur have argued, the very fact of our human existence necessarily implies interpreting and trying to understand the world and ourselves. In their view, interpretation is inherent to being human: man is a 'self-interpreting animal'. Being human in the sense of 'being-in-the-world' essentially is giving meaning to something. This meaning-giving process takes place at both a conscious and an unconscious level, in health and in disease. We cannot withdraw from this meaning giving activity because it is inherent to being human.

A promising conception of ethics is as "hermeneutics of moral experience" (van Tongeren, 1994). In this definition ethics is conceived of as a discipline which strives for knowledge of man as a moral being, for an interpretive self-understanding and for an answer to the question of what is happening to us as moral beings. Along the same line of thinking as Husserl, Heidegger and Gadamer, van Tongeren defines experience as something which calls for an answer and activates the process of interpretation. A moral experience can be described as an experience of the moral dimension of the world, an experience by which someone is challenged as a moral person. Moral experience is our way of understanding ourselves and the world in moral terms. A hermeneutical ethics tries to interpret daily experiences and real life events. Through interpretation we try to understand more thoroughly that is already - though provisionally and insufficiently - understood in experience.

Those who consider clinical medicine as a hermeneutical enterprise, argue that the 'object' of clinical medicine is a person's experience of bodily or psychic pain, dysfunction or other form of suffering. The patient's complaints are not to be seen as objective facts, but as phenomena to be interpreted. The meaning of a patient's experiences is not an absolute objective phenomenon waiting to be discovered by the patient or the doctor. The idea that clinical medicine is a hermeneutical activity has led to the well-known metaphor of 'the patient as a text' (Daniel, 1986). This phrase is a useful metaphor under one condition: it should include another metaphor, that is, the metaphor of 'the body as a text'. Not only verbal or written, but also bodily signs must be interpreted. Moreover, the lived body is a *text* as well as an *interpreter*. The lived body is a text to be interpreted, but also an interpreter on its own. These two aspects of the human body are inextricably linked.

2.4. The Body as a Text

If we connect the idea of the lived body as man's bodily 'being-to-the-world' to the metaphor of 'the body as a text', (at least) three forms of experience can be identified in which the human body plays a different role: (1) the human body as a pure means of perception, (2) the human body as a subject, that is as an interpreter (writer or reader) of texts and (3) the human body as an object, that is as a text to be interpreted.

All human experience has a bodily dimension in the sense that all experience is realized via the body. Via his own lived body, the person has access to the world and the world to him. In this sense the body is a *pure means,* the only means by which experience is possible at all. The notion of the body as our lived access to the outside world signifies that we are not aware of our body and its functions. In this view, the body may be seen as a 'mediator' between the person and the world. This is for example the case when I am totally absorbed in reading an exciting book or making love, thereby 'forgetting' my own body. In such situations the body itself does not speak. Because I am totally unaware of my body, for me (but not necessarily for someone else) no bodily text which can be interpreted exists.

The body is a *subject* of experience when it functions as an interpreter in its own right, when it - so to speak - speaks for itself. The body does not only interpret itself, but also everything in the outside world with which it is confronted via the various senses. This is what Merleau-Ponty means by the notion of prereflective, experiential, 'tacit' knowledge. The content of these bodily interpretations of the world does not necessarily need to be known (immediately) by the person. The body may be considered the author of a text, meaning, the author of bodily signs. But it also can be seen as a reader, meaning, as a reader of the text which is constituted by what is happening in the outside world. All kinds of experiences of the outside world can be - but do not necessarily have to be - experienced bodily. For example, a feeling of abhorrence can be accompanied by a bodily reaction of nausea. In section 3 and 4 I will elaborate on this notion of the body as a subject by exploring some (corporeal) experiences of dead and living bodies.

The body is an *object* of experience, when I experience my own body. In these situations, I am more or less aware of my own body. I can experience my body in a pleasant or unpleasant way. Unpleasant experiences of the body, of course, are often the start of medical investigation and treatment. Also when I see or feel the body of someone else, that body is an object of experience. In these cases the body can be described as a text to be interpreted. The person - the I-person or another person - is the reader. The notion of the body as object is basic to the question what might be done with living or dead bodies or body parts.

To conclude, in situations when the body is a pure means, it cannot be interpreted by the person to which that body 'belongs', because there is no available text for interpretation. However, the body as subject and object of experience is 'accessible' for the person. In the next section the body is considered an object. In section 4 the emphasis is on the body as subject.

3. THE INTEGRITY OF THE HUMAN BODY

From time immemorial the relationship between man and his or her body has been the subject of intense ethical debate, especially in medicine and health care. The structure and function of the human body pre-eminently exemplify the field of research of medical and biological sciences. Furthermore, it is especially in medicine as a therapeutic practice that the human body is subjected to all kinds of preventive, diagnostic and therapeutic interventions. Although since the beginning of medicine there have been discussions about the 'ownership' of the human body, it is especially since the development of modern medical technology that this question has gained importance (Andrews, 1986; Campbell, 1992; ten Have and Welie, 1998). The notion of 'ownership' means that there is somebody (a person) who can be seen as the owner of the human body and who can take control of the body and its parts. In this section another approach to the human body will be developed, concentrating on its integrity, the integrity of the dead and living body as well.

Obviously, the moral status of a dead body is generally considered to be lower than the moral status of a living body. This is not to say that we owe more respect to living bodies than to corpses *per se*. But it is generally felt that a violation of the integrity of dead bodies raises less aversion than a violation of the integrity of living bodies. On the one hand, basic to this intuition is that the living human body belongs to 'somebody' who can be harmed by those invasive actions. On the other hand, many good reasons exist to violate the integrity of the living body, the most important one being the welfare of the person concerned. In contrast, there is less reason to perform invasive actions on a dead body. Moreover, in the case of a corpse, the person to whom the body once 'belonged' no longer exists, at least not in this earthly life. Thus, one can argue, that there is no longer 'anybody' who can experience pain or who can be harmed otherwise, except in cases where the person concerned has consented to or explicitly asked for a particular invasive action to his or her dead body. But even in this latter case we feel hesitant to do so. There appears to be a taboo against violations of the dead human body. As does the living body, the dead body possesses an integrity which must be respected. This assumption may possibly explain why every day we can witness on television many technical manipulations of living human bodies, but manipulations of the dead body are seldom shown.

With regard to the dead body, its status raises questions such as: Why do we bury or cremate our dead fellow men instead of treating corpses like household rubbish? Why was there for example in Homer's *Iliad* such a violent struggle between the Greek and the Troians to possess the corpse of Patroclus, Achilles' friend, who was killed by Hector? Why is it difficult to accept that human corpses are used for trauma research, for the purpose of improving the safety of cars? Why may we experience a kind of hesitation to perform invasive actions on a dead body, for autopsies, transplantation purposes or the practice of embalming? It is far beyond the scope of this chapter to deal with all these questions. But one thing must be clear from the outset, i.e., that posing and answering questions like these have implications for our view of our own identity and of who (we think) we are. Kass writes:

> One of the most unsettling ... things about confronting cadavers, dead bodies, or the question of organ transplantation is that we are by practice *forced* to decide who or what we think we are ... How to treat dead bodies may seem to be a trivial moral question ... But ... few are as illuminating of our self-conception and self-understanding (Kass, 1985, p. 24).

Let me start with a recent example of treating dead bodies in a rather unusual way which has raised many ethical questions.

3.1. Body Worlds

From October 1997 till March 1998 an exhibition was held in Mannheim (Germany) at the Museum for Science and Technology: Body Worlds (*Körperwelten*). What was exceptional about this exhibition was that it marked the first time that prepared corpses were publicly displayed in a museum of art, that is, not in an anatomical museum. The corpses had been preserved through a method called 'plastination', developed by the German pathologist Von Hagens, which replaces the body's moisture with plastic providing for odourless body parts which retain their colour, shape and texture. Included as exhibits were whole bodies as well as organs and individual body parts. But more exceptional than the exhibition itself was the extraordinary number of people who felt compelled to see this exhibition, and the ethical debate it raised (Fielding, 1998). The most pressing point that was articulated was whether or not such an exhibition dishonours the dead. As the organizers pointed out, the corpses had all been knowingly donated before death for the purpose of plastination. Moreover, the organizers felt it was ethically important that the identity of each individual corpse, including cause and age at death, not be revealed.

At the entrance the visitor could read: "the sight of the human bodies and body parts exhibited may hurt religious feelings and may be an alienating experience". More than a few visitors were indeed rather shocked and experienced feelings of abhorrence and alienation. Their lived bodies provided them - often subconsciously - with knowledge about the dead bodies which were exposed. Obviously, a visit to an exhibition such as Body Worlds can be seen as a moral experience. The ethical debate raised by this exhibition is illustrative for the various ways in which the moral value of the human body is experienced. The exhibition of the corpse in 'unnatural' positions - that is unnatural for a dead body - would downgrade the human body to a mere piece of art. For example, from a religious point of view it has been argued that the upright position of a dead body goes against the idea of a final resting-place. A dead body must lie down.

It was Von Hagen's primary aim to bridge the gap between the living and the dead. The purpose of the exhibition was that one should learn more about one's own bodily existence than about a dead corpse. From this intention we can recognize the idea of the coexistence of the living and the dead which was an important item in late medieval art, finding its expression in the theme of the 'dance of death' or 'dance macabre'. A visit to the exhibition Body Worlds may be considered a meeting between living persons and dead bodies (persons). According to Von Hagens, modern man is totally alienated from nature and is surrounded by all kinds of artificial objects. In this artificial world man has forgotten that he himself, in particular his body, is the last piece of 'pure nature'. The human body is neither an artifact, nor a machine. Man should know that he - as a bodily being - is part of nature. He must be aware of the vulnerability and mortality of the human body.

3.2. The Dead and the Living Body

By showing dead bodies Von Hagens tried to emphasize the 'naturalness' and finiteness of the living body. However, this is possible only if we (to a certain extent) recognize ourselves and our living bodies in the dead bodies which are exposed. What, then, are the differences and similarities between a dead body and a living body seen from a moral perspective? What does death, conceived of as the transition from a living body into a dead body, mean from a moral perspective? This difference may be elucidated by referring to the metaphor of the body as a piece of art.

A considerable difference exists between living and dead bodies. Obviously, it was one of Von Hagens' intentions to show the aesthetic dimension of the human body. His practice of preparing the human body can be considered a combination of anatomy, technology and art. Von Hagen's corpses may be seen as pieces of art of the objective body. Merleau-Ponty also considers the human body a piece of art, but, in contrast, he focuses on the lived body. According to him the human lived body is not to be compared to a physical object, but rather to a piece of art (Merleau-Ponty, 1994, p. 150). What is expressed in a picture or a piece of music, he argues, is only recognizable by means of the display of colours and sounds. The same is true of a poem or a novel, although they are made up of words. Merleau-Ponty calls a novel, poem, picture or musical work 'individuals', that is, beings in which the expression is indistinguishable from the thing expressed, their meaning, accessible only through direct contact. It is in this sense that our own lived body is comparable to a work of art. The human body may also express something, a meaning, which is only understandable by direct contact, by interpreting the signs of the body. Merleau-Ponty calls the human body a 'nexus of living meanings' (Merleau-Ponty, 1994, p. 151).

Thus, on the one hand, a significant difference exists between dead and living bodies which can be elucidated by the different ways in which Von Hagens and Merleau-Ponty use the metaphor of the body as a piece of art. On the other hand, we must recognize similarities between dead and living bodies. A dead body necessarily refers to a living body. There are no dead bodies that have never been living bodies. A dead body is a body that once was a living body, that once was the body of 'somebody', a human person, or a potential human person. Even small body parts which are recognizable as originating from human origin refer to a particular person (whether or not known by name) who once was the 'owner' of these body parts. This holds for monstrous, non-viable malformations of the human body as well. It may be difficult to recognize in these malformations a strict human or personal element, but these defective human bodies refer to a 'promise of humanness'. They once had the potentiality of becoming/being the body of a human person, but this potential humanness has not been totally realized.

Thus, we owe respect to a body, living or dead, not only because it is or was the body of a particular person, but because it refers to mankind in general as well. The human body, dead as well as living, is in a way a symbol for all mankind.

3.3. The Closed and the Opened Body

Von Hagens' approach dates back to Renaissance time, when anatomists such as Da Vinci (1452-1519) and Vesalius (1514-1564) tried to grant the public insight into the ingenious structure of the human body. In those days anatomical theatres emerged at various centres in Europe. The famous anatomical theatre of Padua, where among others Vesalius, Fabricius ab Aquapendente and Harvey worked, dates from 1550. Particularly, Von Hagens' approach has to be placed in the context of the history of medical techniques to preserve the dead human body. In order to be able to use a corpse longer than about a week for dissections and public anatomical demonstrations, new ways of embalming corpses were developed. The Dutch anatomist and pathologist Frederik Ruysch (1638-1731) was one of the pioneers in this field. He was very skilful in preparing corpses of children and adults as well. He possessed a secret fluid which he injected into the bloodvessels. Corpses prepared by him were real pieces of art. Ruysch was not only an anatomist and artist, but a successful trader as well. In 1717 he sold his collection to tsar Peter the Great from Russia, for 30,000 Dutch guilders (Schwartz, 1998).

However, for an adequate understanding of what the exhibition Body Worlds brought about, we must return to the late Middle Ages. At that time, in which modern anatomy emerged, a paradigm shift concerning the view of the human body occurred. This shift can be summarized as a transition from abstraction to concreteness, from exteriority to interiority, from qualities to causalities, and from states to processes (Levin and Solomon, 1990). From a Lacanian perspective, one can say that it involves a transition from the imaginary body to the symbolical body (Zwart, 1998). Though this paradigm shift can be analyzed from several perspectives, its essential property is that it encompasses a shift in the moral attitude towards the body as well.

The changing moral attitude of Western man toward the dead body has been described by Van den Berg (1965a, 1965b). He distinguishes three periods in the history of the human body: the period of the closed body, of the opened body and of the abandoned body. The period of the *closed* body starts with Galenic medicine in the second century and lasts until the fourteenth century. In that period sections on the human body were seldom carried out. The theory and practice of medicine were based on knowledge of the closed body. The period of the *opened* body starts in 1316. In that year for the first time after centuries the anatomist Mundinus performed a dissection of the human body with the intention of seeing how the body looked from the inside. A well-known painting, dating from the year 1345, shows Guido de Vigevano, one of Mundinus' pupils, starting to open a human corpse. It is one of the earliest illustrations of an autopsy performed on a human body. De Vigevano gently embraces the body he has started to open with his lancet. Looking at the dead person's face, his facial expression reflects a sense of hesitation, even apology, for invading his fellow human being's bodily integrity. Obviously, it is a strong moral experience for Vigevano. Van den Berg has beautifully described Vigevano's ambivalence. On the one hand, Vigevano definitely wants to open the human body. On the other hand, he hesitates greatly. It is as if he is asking the dead person for permission.

It is precisely this ambivalence which anyone carrying out an invasive procedure on a corpse or a living body may experience. The same moral hesitation is experienced by the medical student who for the first time visits an anatomical theatre to dissect a corpse, or gives someone an injection, or makes a surgical incision (Kass, 1985). By making an incision and penetrating the human body a threshold has been crossed. The integrity of the human body is then at stake.

The portrait of Guido de Vigevano mentioned above has been put opposite the front page of volume I of Van den Berg's book on the human body. As a contrast - to demonstrate the radical changes in our attitude toward the human body and the medical-technical progress as well - a picture of an open heart operation has been put next to the front page of volume II. This picture shows the surgeon entering the human heart with his forefinger to examine the state of the valves between the left forechamber and left chamber. This picture represents man's capacity to perform technical invasions on the human body. In another book, called *Medische Macht en Medische ethiek*, Van den Berg (1969) shows us a picture which probably raises even more ethical questions. It concerns a case of a so-called hemicorporectomy, i.e., the surgical removal of half of the body. At the end of the fifties several cases of hemicorporectomy were reported in medical journals (Kennedy, *et al.*, 1960).

The case which Van den Berg refers to was described by Aust and Absolon (1962). It concerned a 29-year-old man who shortly after birth had a meningomyelocele repaired and had been paraplegic since that time, unable to move either lower extremity. There was no sensory or motor activity distally from the spinal level L-1. The lower half of his body was considered "useless, a hindrance to any activity due to its weight and deformity" (Kennedy, *et al.*, 1960, p. 756). Moreover, an extensive cancerous process had developed in a large decubital ulcer of long standing, overlying the entire sacrum. The authors considered a hemicorporectomy a possible solution to the otherwise insoluble medical problems of the patient. According to them the technical feasibility of a successful hemicorporectomy was no longer open to question. The patient accepted the surgeons' suggestion to perform a hemicorporectomy with equanimity. His body was literally cut through. The stomach, spleen, liver, kidneys and the upper part of the bowels were left. The bladder was elevated, turned upward into the anterior abdominal wall, and sutured to it. The other abdominal organs were removed. The patient was grateful for the surgical treatment which had been administered. The article does not mention how long the patient lived having only the upper half of his body in his possession.

As far as I know, hemicorporectomy is nowadays no longer carried out, since the progress of medicine has provided us with more effective and less mutilating ways to treat medical problems like the one discussed. But what we can learn from this (exceptional) case is that there exists something like the integrity of the body exists which from a moral perspective may function as a counterbalance to the wishes of the patient and the need to be operated on. Not every invasive action on the human body which is medically possible and which corresponds to the patient's wish can be allowed from a moral perspective.

3.4. Images of the Body

Admittedly, hemicorporectomy is an extreme case of invasive surgery and mutilation of the human body. Though the patient concerned gave his consent to be operated on, one may still ask the question of whether it is permissible to violate the integrity of the human body to such an extent. The same question holds for (often less invasive) violations of the integrity of the dead body, for instance in the context of transplantation medicine.

The practice of cadaver organ donation also calls for consideration of the integrity of the human body. Here also, moral experience may be the starting-point of ethical deliberations. In a multi-organ donation procedure, the human body, that is, the thorax and abdomen, are opened maximally:

from the clavicle up to the pubic bone. As never before in medical history we can now witness the vital functioning of an apparently living human body. Although the sight of the opened, still functioning human body would not cause much trouble for professionals such as transplantation surgeons, it would definitely be an intense moral experience for lay people. I wonder whether lay people would still consent to donate their own organs if they knew how a multi-organ donation procedure precisely proceeds.

The moral experience of physicians, nurses, patients and relatives (in various clinical cases) may be coloured by a completely different view of man and the human body. These different experiences and views of the human body may lead to different attitudes and decisions regarding organ donation (Dekkers, 1998b). Such a view of the human body is not based on a clearly defined, rational knowledge, but is rather to be seen as a more or less implicit 'image of the body'. On the one hand, people may have little difficulty with organ donation, because in their eyes a dead body is so to speak a Cartesian, objective body, which is merely to be seen as a collection of organs, tissues and cells, which retains few if any personal characteristics. For them the human being as a person is considered to be something actually separate from the body. On the other hand, people may experience strong hesitations to consenting to organ donation, because they view the human body in a totally different way. This was, for example, the case when the girlfriend of a potential multi-organ donor could not accept the decision of his parents to consent to donation. Because of her feelings of love for her boyfriend she could not accept that his heart, which had partly 'become her heart', would eventually beat in the body of someone else. The body image which is expressed by the girlfriend can be elucidated by means of the Merleau-Pontian notion of the lived body, that is, an individual, 'animated' body (Dekkers, 1998b).

Transplantation medicine offers a critical opportunity to reflect on the meaning of 'our bodies-ourselves' as well as upon the way modern medicine might enrich and/or threaten the sense of ourselves as bodily persons. Given the centrality of bodiliness to clinical practice in general and to transplantation medicine in particular, it also involves employing an understanding of corporeality which takes seriously the integrity of bodily life (McCormick, 1996). The examples above demonstrate that the integrity of the body is not to be seen as an abstract principle, the meaning of which is clear and undisputed, and which can simply be applied in practice. The experience of the integrity of the human body does not automatically lead to specific actions. What is meant by 'integrity of the body' must be explored by (hermeneutically) analyzing particular practices in which the integrity of the body is at stake.

Sometimes it is argued that there is no 'real difference' between several uses of dead bodies, for example, for trauma research, pathological examinations, autopsies, and organ donation. Such a distinction would be based on emotional or symbolic factors only. Symbolism, however, is the whole point of discussion, "the sole focus of concern and misgiving" (Feinberg, 1985). In my view, symbolism can be seen as the starting-point of ethics, at least if we conceive of ethics as 'hermeneutics of moral experience'. Thus, although one can generally say that the living and the dead human body has a moral dimension or possesses an integrity which must be respected, every violation of the body must be considered in its own context. Particular moral experiences are the starting-point for a hermeneutical interpretation.

So much for the integrity of the dead and the living body, which initially appears to be an objective body in the problems discussed. Now the subjective, lived body and the experiences of chronically disabled persons will be examined.

4. THE LIVED BODY

A fruitful approach to understanding the ambiguous relationship between body and person involves interpreting the lived experience of the chronically disabled body of patients with a chronic somatic disorder. A starting-point is the insight that illness is fundamentally experienced as a disruption of the lived body rather than as a dysfunction of the objective medical body.

This section will concentrate on multiple sclerosis (MS), because especially in view of 'chronicity' MS is an interesting disease. MS is a serious neurological disorder caused by pathological processes in the central nervous system. Although various types of MS can be distinguished, generally the disease is characterised by periods of exacerbation and complete or nearly complete remission. MS causes considerable external malformations and functional disabilities of the body. Due to the sheer variety of physical symptoms and their functional effects which may arise in the course of MS, the subjective experience of bodily changes in MS patients is a complex process. There appears to be a complex interrelationship between the disease course, the personal illness narrative and the sickness career, which influence each other strongly. In particular, it is important to note that there is often a discrepancy between the patient's perception of the disintegration of his/her bodily processes and medical explanations. For these reasons, it is to be expected that patients who suffer from MS can tell us a lot about the way they experience their own body.

I will describe some personal accounts of body experiences borrowing from auto-biographical self-descriptions of the American philosopher S. Kay Toombs (1990, 1992a, 1992b) and of the Dutch publicists Anneke Emmes (1987), Renate Rubinstein (1985) and Karin Spaink (1993). From their experiences two 'body images', that is, the alienated body and the unpredictable body, force themselves upon us. These body images may illustrate the troublesome relationship between the MS patient and his or her own 'chronic body', a relationship which can be expressed in the formula "to have a body or to be a body".

4.1. The 'Chronic' Body

The word 'chronic' may have a neutral meaning, i.e., referring to something which lasts a long period of time. However, since 'chronic' is commonly used in connection with unwanted things such as lack of money, unemployment or disease, it has in fact a pejorative meaning only. One may speak of a chronically ill person and not of a chronically healthy person. Both meanings of the word 'chronic' (neutral and pejorative) appear to be useful in exploring the meaning of the notion 'the chronic body'.

First, 'chronic' in the expression 'the chronic body' may have the same pejorative meaning as it has in the expression 'a chronic disorder'. 'Chronic' refers to something which is problematic, troublesome or painful, which bothers us, which cannot be cured, which we cannot get rid of. Second, at a more fundamental level, adopting the neutral meaning of the word 'chronic', the expression 'the chronic body' refers to human corporeality, i.e., the fact of man being a corporeal being. Although the paradoxical relationship between body and self explained in section II.1 is recognized in all forms of being-in-the-world and, thus, in all forms of illness, it is felt most profoundly in chronic illness. Patients with chronic physical disorders are aware each day of their dysfunctional body. Their disabled body may stand in opposition to the self. More than healthy people and more than patients with a temporary illness, they have a body with which they have to come to terms. For the chronically ill, the sense of alienation from and unwilling identification with their body is particularly profound. Self and identity, which are strongly connected to bodily aspects, are core aspects of everyday experience and of the everyday experience of illness. The experience of a loss, which is an important point in the self-experience of chronically ill people, in particular regards the loss of control over one's own body.

4.2. The Alienated and Unpredictable Body

Because of motor and sensory disturbances the body of MS patients does not function 'automatically' anymore. MS patients lose control over their bodies and feel disengaged from them. They can no longer trust their body. Many MS patients feel betrayed by their body. An MS patient describes in the following way how she painfully seeks to explain and understand what is happening to her body during the development of the disease:

> I hated my body at the time, I felt it had let me down by being inadequate, too weak to withstand living. I felt and still do a year later that this 'thing' which was taking over my body had nothing to do with 'me' inside it (Robinson, 1988, p. 42).

Toombs tells of a disassociation from her body as a result of an increasing loss of motor control. It is as if her inability to control the movements of her legs causes her to feel detached from them and they from her. She notes that when she sits in a chair and tries to raise her legs, she thinks to herself that 'these legs', rather than 'her legs' will not move. Since they are no longer under her control, she feels alienated from them (Toombs 1992a, p. 139).

Characteristic of MS is the unpredictability of the disease course. Disturbances of motor function and sensation may vary considerably in intensity and over time. Rubinstein (1985, p. 26) speaks of a 'capricious disease'. The bodies of MS patients do not fail constantly but only now and then. Their bodies are capricious because they may stop functioning suddenly, without warning. Toombs notes that while motor, sensory, bowel and bladder disorders all cause a loss of corporeal identity, alienation of the body may be experienced quite differently and uniquely. Motor disorders effect the establishment of the body as an oppositional force which places the body beyond the control of the self. The loss of tactile or kinaesthetic sensation leads to a feeling of gradual disengagement of the body from the self. Bowel and bladder dysfunctions such as incontinence, caused by the pathological process in the vegetative nervous system, represent an even more serious alienation of the body. They signify the most profound loss of control over one's body. The body is experienced not merely as oppositional, but as malevolent, posing a constant threat to one's dignity and self-esteem. The body appears inherently untrustworthy. It is capable of causing deep humiliation and shame. A patient writes:

> The worst thing about my MS is the problem I have with incontinence. It is utterly degrading to have to wear pads and to have to change them and not to know where the nearest loos are (Robinson, 1988, p. 44).

4.3. To Have a Body or to Be a Body

The notion of bodiliness means that one has a strictly personal body. My body is really mine: I could not have another body without being another person. And conversely, without just that body, I would not be me. This relationship between body and person may be clarified by analyzing various attitudes of MS patients toward their somatic problems. Some of them identify themselves with their illnesses. Others consider their disease as a strange entity which has nothing to do with their own personhood. Toombs appears to be a proponent of the first attitude. She argues that the different bodily dysfunctions have different existential meanings. According to her, the nature of MS is such that the disease process itself carries a particular existential significance. In her view, there is an intimate link between the various disease processes and a particular way of being-in-the world. She does not hesitate to relate this insight to her own situation, to her own being a patient. She writes:

> My once relatively orderly physical existence has been transformed into an uncertain and chaotic manner of being in the world ... this disordered manner of being which is my illness (Toombs, 1992b, p. 129).

Thus, from this point of view, patients with MS are persons living a disordered existence in very specific ways. They are not just persons who 'have' MS.

The notion of a personal existence in the illness, as argued by Toombs, does not seem to be the experience of many other MS patients. Often patients say: "I have MS, I am not MS". Expressions like this one are to be understood as a reaction to an all too drastic identification of the patient with his or her disease by others. Like many other chronically ill people, MS patients have the experience of being judged solely on the basis of their disease and handicaps. They heavily protest against such reactions by enlarging the distance between themselves and their illness. Spaink (1993) writes: "I am not ill, it is my body". She argues that illness causes a break between the self and the body, between the will and physical capacities. Chronically ill people may easily fall into this gap between body and self, unless they manage to construct a new bridge between the self and the body. They have to reach a new compromise with their own opposing body. Many chronically ill patients succeed in this. They maintain a certain distance to their body, which works positively: a light degree of disengagement from the body makes a renewed reconciliation with the body possible.

Descriptions like these suggest that besides the 'alienated body' and the 'unpredictable body' another body image exists. Apparently it is possible to accept a certain degree of alienation and capriciousness without losing the feeling of familiarity. Spaink is not fully on speaking terms with her body, but she cannot get along without her handicapped body. She and her body hang around together. She speaks of her body as a "drivelling chatterbox" (Spaink, 1993, p. 97).

These lived experiences of patients with a chronically disabled human body demonstrate (1) that the model of the lived body and the notion of bodiliness form an excellent guide to understanding the (experience of) the chronically ill body and (2) that - the other way around - a phenomenological-hermeneutical approach of body experiences of chronically disabled people can clarify and specify the rather abstract notion of bodiliness.

4.4. Significance for Medical Practice and Theory

The significance of the notion of the lived body for medical practice and theory lies - first - in the possibility of developing an anthropological, holistic, biopsychosocial or comprehensive medicine in which the Cartesian model is just one possible approach. To attend to the lived body is not to forsake the tools and learning that Cartesian medicine has provided. It is merely to refuse to grant this mechanical wisdom the status of ruling paradigm (Leder, 1992b). As the above examples of experiences of MS patients may demonstrate, the full significance of human disease and health necessarily eludes the model of body-as-machine.[6] The attempts of the representatives of the anthropological movement in medicine "to introduce the subject into medicine" must be seen in this context.

A second way in which the notion of the lived body in combination with the metaphor of 'the body as a text' may be fruitful to medical (and moral) theory is the following. If there is one thing which has become certain from bodily experiences of MS patients (some examples of which have been described above) it is the existence of many different ways of interpreting bodily signs. To recall the metaphor of 'the body as a text': a variety of texts available for interpretation exists.

The above experiences belong to (what Leder (1990b) calls) the *experiential* text, which consists of the patient's primary 'abnormal' bodily experiences which stand out as significant and disruptive.[7] Bringing about a disruption in the normal routine of life, these experiences of the subjective, lived body initiate a search for meaning: Why is this happening? What does it signify? Should I contact a doctor?

The experiential text is intimately connected to a second text described by Leder, i.e., the narrative text. The *narrative* text corresponds to the first part of the medical encounter, the taking of the medical history. It is, so to speak, the collaborative product of three different authors: the diseased body, the patient and the doctor. In fact, the experiential and narrative texts are two sides of the same coin. They are part of a complex set of meanings and are highly susceptible to different interpretations. The lived experience, which is the experiential text, is closely connected to the narrated experience, which is the narrative text. As Ricoeur (1984, p. 31) argues, all knowledge and all experience are mediated by language and therefore interpretive, narrated. Thus, there is an intimate relationship between a particular lived experience and a verbal, narrated, experience.

Things become even more complicated if we take into consideration two more possible texts, i.e. the physical text and the instrumental text. The *physical* text is written almost solely by the doctor while doing a physical examination. Here the objective body stands in the foreground. The *instrumental* text is the result of diagnostic testing. It is the text written by medical machinery.

Thus, if we take the notion of the body as a text seriously, the human body can be 'read' in (at least) four different ways: experiential, narrative, physical and instrumental. Then the following questions arise: What is the most important/basic text to read and what is the best interpretation? It is generally recognized that the view of patient and doctor not infrequently fails to correspond. In the literature, doctor's and patient's narratives are often seen in opposition to each other. However, one of the tasks of hermeneutics should be to explore and analyze that which is the common source of these two narratives, in other words, the precondition for a mutual understanding between doctor and patient. As Toombs (1992a) points out, a more fundamental world must exist, a world known, experienced or lived in by both patient and doctor. In the absence of such a shared world it would be impossible for doctor and patient to communicate with each other. From a hermeneutical point of view, this shared humanity may be founded in the fact that both doctor and patient are 'interpreting animals', and that they - as historical beings - share many common experiences and traditional interpretations. This fundamental shared humanity, this 'primary hermeneutical condition', may of course coexist with a complex variety of different and often opposite narratives. From a hermeneutical perspective one can say that it is this very difference in interpretation that primarily constitutes the moral dimension of clinical medicine.

5. HERMENEUTICS OF THE HUMAN BODY

The above considerations regarding the dead and the living body reveal a kind of paradox. Though the subjective body is more fundamental than the objective body and the latter is only an impoverished image of the former, as Merleau-Ponty argues, it is difficult to verbalize the lived body. The subjective, lived body can hardly be depicted or objectified otherwise. One can experience it, feel it, taste it etc., and afterwards try to explain in words how it is experienced or try to visualize how it feels. In this respect one can learn a lot from what literature, poetry and visual art can tell us about the human body. Thus, all in all, it seems possible to use Merleau-Ponty's phenomenology of the body, in particular his concept of the phenomenal body, as a starting-point to interpret experiences of dead and living bodies.

The metaphor of 'the body as text' may clarify that the living human body is an interpreter which has its own knowledge of the outside world. Further, this metaphor illustrates that the human body, living or dead, is not to be seen as an objective phenomenon, but must be considered a text requiring interpretation. Obviously, the way we approach and interpret the human body

- practically and theoretically as well - is not a neutral process. Every body practice and body theory is based on a normative stance and strongly interrelated to moral evaluations. If we analyze the predominant image of the human body in medicine and health care, it is - generally speaking - most of all the objective body that comes into prominence. Progress in scientific medicine is mainly based on the conceptualisation of the human body as an objective body. Anatomy and physiology, and most other medical disciplines as well, find their common ground in the notion of the human body as an external object. This process of objectification is inherent to medical theory and practice. Problems arise when this objectification of the human body exceeds certain boundaries.

Technology is a constitutive factor in modern medicine. Inherent to technology is the tendency to interfere with 'the order of things' and to manipulate the world. In a technologically-oriented medicine, a tendency exists to manipulate and transform the human body, a tendency which has altered the image of the body considerably. Transplantation medicine, for example, has contributed much to our view of the human body. The world of organ transplants is pre-eminently a world of technological enterprise. As a result there is an increasing tendency to consider the human body as a thing which can be transformed and manipulated, as a constellation of organs which can be replaced at any time we think necessary. The living or dead human body (or body parts) have a practical and economic value. It is increasingly difficult to see the human body as a subjective, lived body, as a body with its own integrity.

On the one hand, the images of the human body explored above appear to emerge from a morally significant layer of (implicit) meaning-attributions, habits and attitudes. On the other hand, they can be considered a moral source from which a particular action or decision can be understood. A hermeneutical analysis of these images may contribute to the exploration of their moral dimension. The 'hermeneutics of moral experience' should interpret and test concrete experiences of the body in the context of other experiences and traditional images, ideas and stories about the body. One of the tasks of health care ethics should be to explore what body image and what kind of approach is most appropriate in a specific situation. Instead of focusing solely on man as a rational, autonomous being, the ethicist should also investigate the moral dimension of the human body 'as such', young or old, beautiful or ugly, conscious or unconscious, healthy or disabled, living or dead. The living and dead human body bears marks of human dignity.

NOTES

1. See, for example: Zaner (1964); Spicker (1970); Leder (1990a); Harré (1991); Leder (1992a); Frank (1995); Cahill and Farley (1995); Komesaroff (1995).
2. I realize that naming a tradition 'European' may raise misunderstanding. At least one reason exists to relativize the distinction between a 'European' and a 'non-European' (Anglo-Saxon) approach, i.e., the fact that it is particularly some philosophers from the United States who recently have promoted a 'European' approach. See, for instance, the publications of Leder, Spicker and Toombs in the list of references.
3. This is a somewhat unusual notion of 'morality'. See for example Beauchamp & Childress (1994), who define 'common morality' - in its broadest and most familiar sense - as "socially approved norms of human conduct" (p. 6).
4. I prefer the term 'bodiliness' over the term 'embodiment' which is also commonly used, since 'bodiliness' indicates more clearly what is meant in phenomenology, i.e., an existential mode of being and not a process of becoming a body, of being incarnated.
5. A well-known criticism of the phenomenological approach of the human body is Lacan's understanding of the body. According to Lacan, we cannot say that the body as it is experienced in everyday life is more 'real' than the representations of the body which are produced by (objective) medical science. According to him, a phenomenological view entails a profound transfiguration of the body as well. The 'real' body is never experienced immediately (Zwart, 1998). With regard to this point, it must be noted that Merleau-Ponty, as far as I know, never speaks of the 'real' body and that phenomenologists will probably admit that the experience of the lived body is mediated by visual or linguistic entities. This insight does not contradict their claim that any knowledge of the body starts with a knowledge of the subjective body.
6. For a fine recent example of an attempt to criticize the Cartesian idea of the body as object focussing on disability, see Edwards (1998). Edwards argues that the body is properly viewed as a subject, that there are neither purely mental nor purely physical disabilities, and that selves are constituted, at least in part of their bodies.
7. Leder (1990b) distinguishes between a primary text, a series of secondary texts, and a tertiary text. The *primary* text that defines the clinical encounter is the individual patient, the 'person-as-ill.' The *secondary* texts through which the primary text expounds itself, are respectively the 'experiential,' 'narrative,' 'physical' and 'instrumental' texts. The *tertiary* text is the patient's medical chart.

REFERENCES

Andrews, L.B.: 1986, My body, my property. *Hastings Center Report* 16(5): 28-38.

Aust, J.B. and Absolon K.B.: 1962, A successful lumbosacral amputation, hemicorporectomy. *Surgery* 52: 756-759.

Beauchamp, T.L. and Childress J.F.: 1994, *Principles of biomedical ethics.* Oxford University Press, New York and Oxford.

Buytendijk, F.J.J.: 1968, *Woman. A contemporary view.* Newmann Press and Association Press, New York.

Buytendijk, F.J.J.: 1974, *Prolegomena to an anthropological physiology.* Duquesne University Press, Pittsburgh.

Berg, J.H. van den: 1965a, *Het menselijk lichaam. Een metabletisch onderzoek.* Vol I: *Het geopende lichaam.* Callenbach, Nijkerk.

Berg, J.H. van den: 1965b, *Het menselijk lichaam. Een metabletisch onderzoek*. Vol II: *Het verlaten lichaam*. Callenbach, Nijkerk.

Berg, J.H. van den: 1969, *Medische macht en medische ethiek*. Callenbach, Nijkerk.

Cahill, L.S. and Farley, M.A. (eds.): 1995, *Embodiment, morality, and medicine*. Kluwer Academic Publishers, Dordrecht/Boston/London.

Campbell, C.S.: 1992, Body, self, and the property paradigm. *Hastings Center Report* 22(5): 34-42.

Daniel, S.L.: 1986, The patient as a text: a model of clinical hermeneutics. *Theoretical Medicine* 7: 195-210.

Dekkers, W.J.M.: 1995, F.J.J. Buytendijk's concept of an anthropologically oriented physiology. *Theoretical Medicine* 16: 15-39.

Dekkers, W.J.M.: 1998a, Hermeneutics and experiences of the body. The case of low back pain. *Theoretical Medicine and Bioethics* 19: 277-293.

Dekkers, W.J.M.: 1998b, The interpretive tradition. The body as a text. In: Evans, M. (ed.), *Critical reflection on medical ethics*. Advances in Bioethics, Volume 4. JAI Press Inc., Stamford/London, p. 209-228.

Edwards, S.D.: 1998, The body as object versus the body as subject: The case of disability. *Medicine, Health Care and Philosophy* 1(1): 47-56.

Evans, M.: 2001, What is a person? This volume.

Emmes, A.: 1987, *MS in mijn ogen. Leven met multiple sclerose*. De Toorts, Haarlem.

Feinberg, J.: 1985, The mistreatment of dead bodies. *Hastings Center Report* 15(1): 31-37.

Fielding, H: 1998, Körperwelten: fenomenologie en ethiek. *Tijdschrift voor Geneeskunde en Ethiek* 8(3): 88-91.

Frank, A.W.: 1995, *The wounded storyteller: Body illness, and ethics*. The University of Chicago Press, Chicago.

Harré, R.: 1991, *Physical being. A theory for a corporeal psychology*. Blackwell, Oxford UK/Cambridge USA.

Have, H.A.M.J. ten: 1995, The anthropological tradition in the philosophy of medicine. *Theoretical Medicine* 16: 3-14.

Have, H.A.M.J. ten: 1997, From synthesis and system to morals and procedure: the development of philosophy of medicine. In: Carson, R.A. and Burns, C.R. (eds.), *Philosophy of medicine and bioethics*. Kluwer Academic Publishers, Dordrecht, p. 105-113.

Have, H.A.M.J. ten: 1998, Images of man in philosophy of medicine. In: Evans, M. (ed.), *Critical reflections on medical ethics*. Advances in Bioethics. Volume 4. JAI Press Inc., Stamford/London, p. 173-193.

Have, H.A.M.J. ten and Welie, J.V.M. (eds.): 1998, *Ownership of the human body. Philosophical considerations on the use of the human body and its parts in healthcare*. Kluwer Academic Publishers, Dordrecht/Boston/London.

Kass, L.R.: 1985, Thinking about the body. *Hastings Center Report* 15(1): 20-30.

Kennedy, C.S., *et al.*: 1960, Lumbar amputation or hemicorporectomy for advanced malignancy of the lower half of the body. *Surgery* 48: 357-365.

Komesaroff, P.A. (ed.): 1995, *Troubled bodies. Critical perspectives on postmodernism, medical ethics, and the body*. Duke University Press, Durham (NC).

Lauritzen, P.: 1996, Ethics and experience. The case of the curious response. *Hastings Center Report* 26(1): 6-15.

Leder, D.: 1990a, *The absent body*. The University of Chicago Press, Chicago and London.

Leder, D.: 1990b, Clinical interpretation: The hermeneutics of medicine. *Theoretical Medicine* 11: 9-24.

Leder, D. (ed.): 1992a, *The body in medical thought and practice*. Kluwer Academic Publishers. Dordrecht/Boston/London.

Leder, D.: 1992b, A tale of two bodies: The cartesian corpse and the lived body. In: Leder, D., *The body in medical thought and practice*. Kluwer Academic Publishers, Dordrecht/Boston/London, p. 17-35.

Levin, D.M. and Solomon, G.F.: 1990, The discursive formation of the body in the history of medicine, *Journal of Medicine and Philosophy* 15: 515-537.

Marcel, G.: 1935, *Etre et avoir*. Editions Montaigné, Paris.

Marquard, O.: 1991, Der Mensch 'diesseits der Utopie'. Bemerkungen über Geschichte und Aktualität der philosophischen Anthropologie. In: Nispen, J. van and Tiemersma, D. (eds.): *The quest for man. The topicality of philosophical anthropology.* Van Gorcum, Assen/Maastricht, p. 3-10.

McCormick, P.: 1995, Embodiment and organ transplants: On giving and taking our bodies more seriously. In: Cahill, L.S. and Farley, M.A. (eds.), *Embodiment, morality, and medicine.* Kluwer Academic Publishers, Dordrecht/Boston/London, p. 127-140.

Merleau-Ponty, M.: 1994, *Phenomenology of perception.* Translated by C. Smith. Routledge, London and New York.

Ricoeur, P: 1949, *Philosophie de la volonté.* Tome 1: *Le volontaire et l'involontaire.* Aubier, Paris.

Ricoeur, P.: 1973, The task of hermeneutics. *Philosophy Today* 17: 112-128.

Ricoeur, P.: 1984, *Time and narrative.* Volume I. The University of Chicago Press, Chicago and London.

Robinson, I.: 1988, *Multiple Sclerosis.* Routledge, London.

Rubinstein, R.: 1985, *Nee heb je. Notities over ziek zijn.* Meulenhoff, Amsterdam.

Schwartz, G.: 1998, Leren van lijken. In: Oderwald, A. and Groenewegen, H., *De verbeelding van het lijk. Essays over de dood in de geneeskunde.* Sun, Nijmegen, p. 11-24.

Spaink, K.: 1993, *Vallende vrouw. Autobiografie van een lichaam.* Van Gennip, Amsterdam.

Spicker, S.F. (ed.): 1970, *Philosophy of the body. Rejections of Cartesian dualism.* Quadrangle Books, Chicago.

Strasser, S.: 1983, Das Problem der Leiblichkeit in der phänomenologischen Bewegung. In: Tymieniecka, A.T. (ed.), *Analecta Husserliana.* Vol. XVI. Reidel Publishing Co, Dordrecht, p. 19-36.

Tongeren, P. van: 1994, Moral philosophy as a hermeneutics of moral experience. *International Philosophy Quarterly* 34: 199-214.

Toombs, S.K.: 1990, The temporality of illness: Four levels of experience. *Theoretical Medicine* 11: 227-241.

Toombs, S.K.: 1992a, *The meaning of illness. The phenomenological account of the different perspectives of physician and patient.* Kluwer Academic Publishers, Dordrecht.

Toombs, S.K.: 1992b, The body in Multiple Sclerosis. In: Leder, D. (ed.), *The body in medical thought and practice.* Kluwer Academic Publishers, Dordrecht, p. 127-137.

Toombs, S.K.: 1997, Taking the body seriously. *Hastings Center Report* 27(5): 39-43.

Verwey, G.: 1990, Medicine, anthropology, and the human body. In: Have, H.A.M.J. ten, Kimsma, G.K. and Spicker, S.F. (eds.), *The growth of medical knowledge.* Kluwer Academic Publishers, Dordrecht/Boston/London, p. 133-162.

Zaner, R.M.: 1964, *The problems of embodiment. Some contributions to a phenomenology of the body.* Martinus Nijhoff, The Hague.

Zaner, R.M.: 1994, Experience and moral life: A phenomenological approach to bioethics. In: Dubose, E.R., Hamel, R.P. and O'Connell, L.J. (eds.), *A matter of principles? Ferment in U.S. bioethics.* Trinity Press International, Valley Forge, Pennsylvania, p. 211-230.

Zaner, R.M.: 1995, Embodiment: The phenomenological tradition. In: Reich, W.Th. (ed.), *Encyclopedia of Bioethics.* Volume 1. Simon and Schuster MacMillan, New York, p. 293-299.

Zwart, H.: 1998, Medicine, symbolization and the 'real' body. *Medicine, Health Care and Philosophy* 1(2): 107-117.

MARTYN EVANS

WHAT IS A PERSON?

1. INTRODUCTION

Let us admit right at the start that this is by no means an easy question to answer. However, the reason for this is not that no straightforward answer can be given. I think a perfectly straightforward answer certainly *can* be given, and I will say in a moment what I think the answer is. No, the reason the question is not an easy one is that the straightforward answer is not likely to be what is wanted by anyone who actually *asked* the question. And the reason for this is that it is unlikely that the question itself is ever asked in a straightforward sense. If someone asks me (as happened the other week), "What is a camellia?", there seems no reason why the question should not be a straightforward request for information. That is why the simple and straightforward answer, "It is a flowering shrub related to the tea bush", is just the sort of answer I can expect to satisfy whoever asks. It is perfectly reasonable for someone who happens not to be particularly interested in gardening not to know what a camellia is. Then it becomes perfectly reasonable for them to ask what a camellia is in the simple spirit of asking for information. But it is hard to think that the question, "What is a person?", could be asked in this spirit. You could ask for *information* in this sense only if you did not know what a person was. There could be few competent language-users who really did not know what a person was. The question "What is a person?" is, I would confidently say, almost certainly never a request for information. But then if it is not such a request, what sort of question *is* it? It is certainly an unusual, even odd, question.

In thinking about what *sort* of question it is, it is useful to think about who is asking the question, and about what it is they have in mind in asking it. We shall pursue this in a little while. But first let us try out the following answer to this odd question "What is a person?". Let us suggest that the answer depends on who is speaking about persons, and on what they have in mind when doing so. For instance:

> "Mrs Fairley informs me that she saw her only this morning talking with a person." Mrs Poultney used 'person' as two patriotic Frenchmen might have

H.A.M.J. ten Have and B. Gordijn (eds.), Bioethics in a European Perspective, 141-155.

> said 'Nazi' during the occupation. "A young person. Mrs Fairley did not know him" (Fowles, 1977, p. 92-3).

At first sight, admittedly, this approach seems discouraging. Ordinarily we want to know what *as a matter of fact* a word or an idea really does mean, and not what it might happen to mean in the mind of an individual (particularly an individual as loathsome as the perverse and sadistic Mrs Poultney). However, the world is not so simple. Words and ideas are, after all, used by individuals to say what they want to say. Certainly, they have to do this within limits; there is no point in your using words in ways which your hearers will simply fail to recognise. But a suitably-prepared audience may allow you considerable flexibility. Mrs Fairley would have known precisely what Mrs Poultney meant.

2. ORDINARY AND EXTRAORDINARY USAGES

Something rather similar is true in the world of bioethics, where - like Mrs Poultney and Mrs Fairley - we too have our characteristic concerns, our ambitions, our expectations and our unspoken assumptions. Consider, for instance, what the following authors might mean by the word 'person'.

> A human being is a person to the extent that they are a rational self conscious agent with the capacity for the distinctive human emotions and affective ties. So there are some humans who are not persons (Carruthers, 1989, p. 234).

> [T]he justification for an upper-brain-death criterion would be better enunciated thus: the individual's essence consists in the possession of a conscious, yet not necessarily continuous, mental life; if all mental life ceases, the person ceases to exist; when the person ceases to exist, the person has died. Upper-brain death destroys all the capacity for a conscious mental life, and it is therefore the death of the person (Gervais, 1986, p. 157-8).

> I shall use the term person to stand for any being who has what it takes to be valuable ... whatever else they are like. Although in normal use 'person' is just another (and usefully gender-neutral) term for 'human being', as I shall use it from now on it will also be species-neutral as well (Harris, 1985, p. 9-10).

It does not appear that any of these authors is much closer to an ordinary, *matter of fact* usage of the word 'person' than was Mrs Poultney. I will say in a moment what I think their usages amount to (though let me emphasise that anyone who wants to be clear in their own mind what these authors are doing needs to go and read the whole of their discussions of this subject, and not just the short extracts which I have selected). Meanwhile we need to have some idea of the ordinary usage in order to decide whether or not these authors come close to it. Let me now therefore suggest, as I promised, just such an ordinary, matter of fact usage of 'person':

> Would the last person to leave the room please turn off the lights?

Outside the world of bioethics and its contributory disciplines - philosophy, theology, law, perhaps psychology - virtually everyone would use the word 'person' like this. It means, more or less, 'individual human being', though it would ordinarily seem very odd to be *asked* what we meant by the word. Indeed, to be asked at all would probably make us suspicious - suspicious that, for instance, we were being tricked, or that we had accidentally stumbled into an obscure and technical discussion or word-game.

Unfortunately, this might be precisely the right suspicion to carry with us into the world of bioethics. At this point I ought to make it clear that in this chapter our aim is to bring out what is 'going on' in certain discussions in bioethics, rather than to attempt a definition of the word 'person' or to produce a comprehensive classification and analysis of the various ways in which the word is used, or *could* be used, in bioethics. This is because an awareness, that words are liable to be used in ways that suit their speaker's or the writer's purposes, is a more useful achievement than a catalogue of what you can find in the bioethics literature. Anyone can go and look at that literature at any time.[1] But it will be a more useful and profitable activity if one can do so critically.

I should also make it clear that I have my own understanding of what the word 'person' means, and how it should be used. It is that the ordinary sense above, i.e. where "the last person to leave the room" is nothing more than the simple and natural way of saying "the last individual human being to leave the room", is exactly the right sense. Whenever I use the word 'person' that is exactly what I mean, whether in ordinary conversation or in writing philosophy. But I obviously have to recognise that - in bioethics and its contributory disciplines - some people use the word 'person' in quite different ways. Therefore, because we should naturally expect the word to be used in an ordinary, familiar way, it becomes extremely important to find out if anyone is using it to mean something else. If we do not find this out, we might end up agreeing to statements, claims and positions to which we would not wish to agree at all, once we had discovered what they actually amounted to.

3. ARE THERE 'QUALIFICATIONS' FOR BEING A PERSON?

Looking again at the three authors quoted above, let us consider how closely their usages conform to this ordinary sense of 'person'. Carruthers' usage clearly does not conform to it - he explicitly makes a *contrast* between some individual human beings and the class of beings who could be regarded as a person. Any human being who fails as a "rational self conscious agent with the capacity for the distinctive human emotions and affective ties" fails as a person. But it does seem clear - at least from this passage - that he thinks only humans can be persons. Gervais' usage has something in common with Carruthers'. She too is committed to the idea that a person must have an essentially mental dimension. We cannot tell from the passage I have quoted whether she thinks that a person must also be human, as Carruthers does. So her view *might* be more radical than Carruthers'. But at any rate she is clear that there can be individual human beings who fail to make the grade as persons, because they lack the mental characteristics she thinks persons must have. In this respect at least, she does not align with ordinary usage. And Harris's usage is still more striking. He probably agrees to some extent with Carruthers' and Gervais' insistence on rationality and self-consciousness. But he certainly does not think you have to be human to be a person. He explicitly claims that the term 'person' does not tie us down to *any* given species, human or otherwise. Moreover he even takes the trouble to remind us that he knows full well that 'person' *ordinarily* means 'individual human being'. It is just that for his purposes - the conclusions he wants to draw and the arguments by which he means to reach those conclusions - that ordinary sense will not be of any use to him. It will not take him where he wants to go.

It is both considerate and shrewd on Harris' part to take this trouble - to remind us of the gap between ordinary usage and his own. Considerate, because the careless reader, or the reader lacking in confidence, might perhaps miss what is going on and, as a result, come to a misleading or incomplete understanding of Harris's position. Shrewd, because it avoids the risk that unfriendly criticism might accuse him of deliberately taking advantage of the confusion between different senses of what is normally thought to be a perfectly ordinary term. This last point is particularly important, to my mind. For me it is this potential confusion which carries with it the danger of genuine moral harm in clinical policy and practice. This is a serious and uncomfortable suggestion for me to make, and I will develop and try to defend it as we go on. For one thing, not all writings in bioethics make their controversial assumptions as explicit and visible as Harris makes his. At least in the passages quoted, neither Carruthers nor Gervais does, for instance.

Carruthers here writes as though anyone who looks around with her eyes open must just *notice* that persons are as he defines them: "*Note that* ... a psychopath - an 'inhuman monster' - is quite literally less of a person than the rest of us", he declares (Carruthers, 1989, 234, my emphasis]. Of course we cannot show by this short passage whether or not he appreciates the essentially metaphorical way in which we speak of a sufficiently evil man as 'inhuman'. It is hard to imagine that he does not appreciate this. And he confirms here that, for him, being human is an essential part of being a person. But any claim that this re-classification - i.e. of psychopathic individuals as being less human than the rest of us - is a matter merely to be 'noted' rests on the prior belief that it is also a matter of *observation* that all persons must have the usual capacity for distinctively human feelings. Well, it seems to me that, on the contrary, it is a matter of simple observation that there are far too many persons who *do not* have this usual capacity. Now there may well be problems about what to do with such persons (or, as we would more naturally say, 'such people'). But these are moral problems - not problems of classification. We will return to this question.

For her part, Gervais also tends to write as though it were just obvious that her definition of 'person' should be accepted by all. She states (Gervais, 1986, p. 157-8), as though it were self-evident, that "if all mental life ceases, the person ceases to exist" (and, as though these were equivalent things, "when the person ceases to exist, the person has died"). In much the same spirit, though with more difficult ideas in mind, she implies that the 'metaphysical' dimensions of the category of personhood are simply there for all to see and acknowledge: "we commonly and most easily defend conclusions about moral personhood *by appeal to features of* metaphysical personhood" (Gervais, 19866, p. 181, my emphasis). This is not the place to try to explore what might be meant by metaphysical personhood. But at least we can say that 'features' can be 'appealed to' only by someone who is sure both that those features are really there for inspection, and that they will be seen as such by everybody else (specifically, those whom the individual is trying to convince).

Elsewhere in the field of bioethics, we can find quite a variety of different usages. They range from something very like the ordinary usage we have already noted to a variety of versions requiring the ability to think and reason - 'cognitive capacity' - and in one instance to a version involving a kind of social *awarding* of moral status.

Capron, for instance, reports (and seems to endorse) a usage which effectively amounts to the idea that a person is an individual human being:

> The accepted criterion for being considered a person ... [is] live birth of the product of a human conception (Capron, 1987, p. 10).

This is a rather technical way of putting it, probably to satisfy the lawyers. But it amounts to the same thing in the end. However Capron does not tell us *who* accepts such a criterion. He implies that it would be a majority view. I think it would indeed be a majority view - at least in substance if not precisely in those rather dry terms - outside the field of bioethics. But within bioethics it is not clear that it could be taken for granted as an accepted criterion. Indeed Gillon suggests that the accepted view is something quite different - namely, that it is usual to build a requirement for at least the potential for consciousness into a concept of the person.

> Whatever one's concept of a person is, one feature widely acknowledged as *necessary* for being a person is a capacity - or at least the potential for a capacity - for consciousness (Gillon, 1990, p. 4).

In this Gillon is certainly reflecting an influential trend *within* bioethics. The views of Harris, Gervais and Carruthers, noted above, all align with this trend. It is grounded in an equally influential tradition in the wider discipline of philosophy, well summarised by Lizza:

> Aristotle claimed that man is essentially a rational and social animal; Descartes, that thinking is essential to the nature of a person; Locke, that a person is an object essentially aware of its progress and persistence through time; Hume, that persons are bundles of psychological characteristics; Kant, that persons are rational agents who, among other things, can synthesize experience and act on moral principles; and Sartre, that persons are self-conscious, intentional beings.
>
> What all these philosophers have in common is the belief that some type of cognitive function is necessary for something to be a person. Any being devoid of the capacity for cognitive function would by implication lack each of the particular characteristics that these philosophers use to define persons (Lizza, 1993, p. 355).

This looks like an impressive consensus. But certainly not all writers agree. One who does not is Crosby, writing about the application of the idea of a *person* to the human embryo:

> Those who deny the personhood of the human embryo typically reduce the person to personal consciousness, and they argue that, since there is no personal consciousness in the embryo, there can be no person. Their inference is irresistible if they are right in reducing the person to personal consciousness. I will try to show that they are not right, and will try to do this by bringing out the distinction between *being* and *consciousness* in ourselves as persons. I will try to show that as persons we are not all consciousness, or all self-consciousness, or self-presence, or conscious acting and experiencing, but that we have a personal being which has to be distinguished from personal consciousness, which can even exist apart from consciousness. Then I will try to determine the assumptions which, given this

> distinction, we have to make about the status of the human embryo (Crosby, 1993, p. 400).

Crosby's enterprise is an ambitious one, though also somewhat puzzling. He is obviously trying to defeat the idea that being a person depends on having self-consciousness or even just plain consciousness. But to do this he wants to try and argue in terms which his philosophical opponents will find familiar and, presumably, difficult to deal with. He wants to put forward some other kind of *property* - he calls it 'personal being' - which makes the holder of that property into a person. I can see why he does this. His opponents use precisely this strategy, so they can hardly complain if he uses it himself. He plans to use it in connection with a property quite different from the one that most of them use. Where they appeal to rationality or some other feature of conscious mental life, which is a fairly narrow idea that is supposed to exclude certain kinds of human beings (the early foetus, people in a coma or a persistent 'vegetative' state, and so on), Crosby's idea of *personal being* is much wider, and is meant to include these various kinds of human beings who get left out by the narrower idea.

Insofar as Crosby is doing this as a means to trying to secure greater moral regard for those human beings, then I find myself in sympathy with him. But I have doubts about whether the best way to secure this moral regard is to try to show what kind of an *entity* the individual in question is. Crosby's opponents - certainly the ones he takes himself to be confronting - are all playing a rather similar game. They want to show how much, or how little, moral regard we should pay to certain groups of individual human beings. They plan to do this by showing what sort of *entity* these individuals constitute - specifically, by showing that they are not 'persons' in some strange, technical sense. Crosby tries to resist them by showing that on the contrary they really are persons - but in some other, apparently equally strange, technical sense. So, morally speaking I happen fully to share Crosby's wish to secure for these groups of people the moral respect that he believes all individual human beings should enjoy. But the way he tries to do this is in danger of making it seem that the general approach his opponents take is the right one - and that they just happen to have picked the wrong technical criteria (rational consciousness instead of 'personal being') to apply. So it will be clear that to my mind Crosby is ultimately playing what I think is the same misleading game as his opponents. Like them, he is asking "What are the qualifying characteristics for a person?" Of course I prefer his answer to theirs. But I think the question is the wrong one to put.

4. USING WORDS FOR SOCIAL PURPOSES?

One of the most sophisticated suggestions, about how we should define the concept of the person, is that we should accept from the outset that it is socially useful to have different senses of the term 'person'. According to this view, one of these senses is a bit of a trick - a kind of convention, but nothing more - which nonetheless supports the interests of *proper* persons:

> It is difficult to determine specifically when in human ontogeny persons strictly emerge. Socializing infants into the role *person* draws the line conservatively. Humans do not become persons strictly until some time after birth. ... This ascription of the role *person* constitutes a social practice that allows the rights of a person to be imputed to forms of human life that can engage in at least a minimum of social interaction. The interest is in guarding anything that could reasonably play the role *person* and thus to strengthen the social position of persons generally.
>
> It should be stressed that the social sense of person is primarily a utilitarian construct. A person in this sense is not a person strictly, and hence not an unqualified object of respect. Rather, one treats certain instances of human life as persons for the good of those individuals who are persons strictly. As a consequence, exactly where one draws the line between persons in the social sense and merely human biological life is not crucial as long as the integrity of persons strictly is preserved. Thus there is a somewhat arbitrary quality about the distinction between foetuses and infants. ... One might retort, Why not include foetuses as persons in a social sense? The answer is, Only if there are good reasons to do so in terms of utility (Engelhardt, 1982, p. 97-98).

Engelhardt here suggests that our usages of the word 'person' are rather untidy, but this reflects the fact that the world is a rather untidy place. So long as it is morally useful to do so, he thinks we can put up with a bit of arbitrariness in language. Engelhardt - like Harris - accepts from the outset that the word 'person' must be used to do some moral work. He also seems to accept that people could disagree over exactly where to draw the line between proper persons and merely conventional or, as we might say, 'honourary' persons. Again it seems clear that this is a far cry from what I have called the ordinary, familiar sense of 'person' - where 'person' just means an individual human being, no more, no less. Of course, most of the persons Engelhardt is talking about will turn out to be persons in the ordinary sense. But plainly Engelhardt will say that at least some persons in the ordinary sense are not, for him, strictly persons at all. He is happy for us to go on calling them persons. But that is because he thinks that it is socially useful to treat them *as if they were* persons - whereas I would want to say that we should treat them simply *as the persons they actually are*.

5. DESCRIBING AND EVALUATING

Now superficially we might think that this is a dispute over whether the term 'person' is a way of describing someone or a way of awarding moral value to them. It certainly looks as though the ordinary sense of 'person' is just descriptive. Either someone is, or is not, an individual human being (though people manage to disagree over even this, in some admittedly rather strange contexts: I am thinking of disputes about the very early human embryo, for instance). By contrast Engelhardt seems clearly to be saying that the term 'person' is a kind of moral insurance document, guaranteeing safety for those to whom it is awarded - and that sooner or later the award has to be earned. On this view the term 'person' picks out something especially valuable. So is the question primarily one of whether the term 'person' is a description or an evaluation? One writer who seems to see it like this is Michael Tooley:

> It seems advisable to treat the term 'person' as a purely descriptive term, rather than as one whose definition involves moral concepts. For this appears to be the way the term 'person' is ordinarily construed. Second, however, it seems desirable that the descriptive content assigned to the term 'person' be guided by moral considerations, in order to have a term that can play a certain, very important role in the discussion of moral issues (Tooley, 1983, p. 51).

To begin with, Tooley seems to agree that 'person' is a descriptive term. But he does not here tell us what he thinks the term describes; for instance he does not here tell us that the term refers to individual human beings, though he notices that it is a term with an *ordinary* sense or use, and we might expect him to agree that this ordinary sense does just refer to individual human beings. However, Tooley also seems to agree with Engelhardt that we can decide for ourselves how the term *ought* to be used. He thinks that we should decide on what the term is to mean, the kinds of things it is to refer to, and be guided in our choice by moral considerations. This, I think, is exactly Engelhardt's view (though perhaps Engelhardt is willing to put up with more untidiness than is Tooley). In a curious way Crosby also tries to argue like this, though of course he is more concerned to endorse the traditional or ordinary way of using the word 'person' than any narrower or more selective sense, as would be preferred by all the other writers we are considering.

We noted that there was a *superficial* appeal about the idea that this whole dispute rests on whether the word 'person' is a description or an evaluation. I do indeed think the appeal is no more than superficial. It seems to me that really we should not try to choose between these alternatives. For I think that the word 'person', as ordinarily used, is fairly obviously *both* a description *and* an evaluation. In ordinary use I do not think anyone is in any sincere doubt about how to use the word as a description - and, as I have suggested, it certainly looks as though at least Tooley would agree with this. If we return to the world outside bioethics debates for a moment, I am confident that the word 'person' in its daily use is just a description, referring to any individual human being, as in "I saw that person on the bus yesterday". But the interesting point is that this does not stop it also being an evaluation. Consider the following: "It does not matter if we put the goalposts here." "Yes, it does - there is a person reading on the bench over there." Here, as always, *precisely because* 'person' refers to an individual human being, it refers in the process to a set of interests that must be taken seriously in the moral sense. Putting the goalposts there is bound to lead to disturbing and annoying the person quietly reading on the bench. So pointing out that the person was sitting there has evaluative as well as descriptive content - moral constraints follow from it. And obviously 'person' is by no means the only word like this. As an example of an aesthetic evaluative constraint, think of using the term 'weed', which refers to a list of plants that you can look up in a gardening book but which also has the negative connotation of being a plant that is unwanted, in the wrong place. Returning to moral evaluative constraints, think also of transparent biological descriptions such as 'father', 'mother', 'sister', 'brother' (and so on), terms which nonetheless have equally clear moral content and importance; as the words of the song put it, "He ain't heavy, he's my brother".

So if the dispute is not after all about whether a word is descriptive or whether it is evaluative, what *is* the dispute about? I think it is about whether we *discover* what words mean by looking at how people actually use them, or whether we can *decide* for ourselves how words are to be used. Now if this is really what is at stake, how can we settle the matter?

6. DECIDING AND DISCOVERING MEANINGS

Let us go back to the question that has concerned us throughout, the question "What is a person?". It looks like a request for information, but we noted right at the start that it is probably never seriously asked in that spirit. People who ask it are in effect asking whether the ordinary way we use it is of any use for their particular purposes - they make the question a rhetorical one, an invitation to get into a special kind of conversation with them, in the course of which they will try to persuade us to accept a new, unfamiliar, unordinary way of using the word 'person'. Why they do this will depend on the particular kind of conversation they want us to join. There are a number of these, each with its own characteristic interests. For instance within the philosophy of mind, the question "What is a person?" might throw light on the problem of personal identity. In traditional theology and metaphysics, the question might be important in working out Mankind's relation to God and to the other living beings in the created universe. In moral philosophy it might be an essential part of deciding the characteristics of a moral agent. And of course in bioethics it is typically used to decide the characteristics of a full moral *subject* - the sort of subject enjoying the full range of constraints and protections concerning how he or she is to be treated. (This is exactly the role which the question plays in the authors I have quoted above.)

In most of these conversations, as I have called them, I think that the question "What is a person?" is essentially being treated as one of classification. Questions of classification are resolved by agreeing what things will be included within, and what will be excluded from, a particular category or group; or by agreeing what tests we will carry out - what characteristics we will test for - in order to find out whether something can be included in the category or group. So when treated as a question of classification, a question like "What is a person?" is not regarded as being something which we settle by going out and looking and discovering how words are used. Rather it is regarded as being settled by a kind of *decision* - a decision about how words *ought* to be used within a particular conversation.

7. UNFAMILIAR CONCEPTIONS THE CONSEQUENCES OF DISAGREEMENT

It is easy to see that, whilst most of the authors I have quoted seem happy to establish a new and unfamiliar conception of 'the person', they certainly do not agree on what that conception should look like. For instance, we have seen how even those authors who agree that it has something to do with rational consciousness do not agree on whether you have to be human to be a person. For someone interested in bioethics, the crucial thing about this disagreement is that these different conceptions of 'the person' have *different exclusions*. They do not agree on whether newborn babies, anencephalic newborns, people with advanced progressive dementia, people in a persistent vegetative state, unborn children, people in a coma, or people who are simply "unable to value [their] own existence" are persons.

This lack of agreement is remarkably inconvenient for these authors. For it seems pretty clear that their main purpose is to provide a way of deciding how different kinds of individuals, who are almost always patients in one of the groups I have just listed, ought to be treated - whether or not they should be given life-saving or life-sustaining treatment, whether or not they should be born alive rather than being aborted, whether or not they should be made the subjects of medical research, whether or not research involving them should end in their destruction, and so on. And virtually all of the authors we have looked at assume that the way to decide how these individuals should be treated is to establish first of all whether they are persons. Now since these various kinds of patients are (with the possible exception of the early human embryo) undisputedly individual human beings, it is obvious that the ordinary sense of 'person', which simply means 'individual human being', applies to them all and can not be used as a way of choosing which of them should be treated in this or that way. Any sense of 'person' which *is* meant to help us choose among them must be narrower, more exclusive, than the ordinary sense. As discussed, nearly all the definitions that have been considered are certainly narrower than the ordinary sense. But they do not lead to general agreement on how we choose among patients, or on how different patients ought to be treated precisely because they involve different exclusions from the class of persons. The very dispute which these authors intended to settle, by means of their various conceptions of 'the person', remains as far from resolution as ever.

8. DISGUISING MORAL QUESTIONS AS MATTERS OF SCIENCE

I think what this all shows is just how deep the oddness of the question "What is a person?" really goes. The question is not one of information, as we have seen. But nor is it a simple question of classification either. Because the classification is one of *moral status*. The reasons why people put forward their various answers to the question, at least in the conversation we call bioethics, are that they want to produce a test which will tell us whom we ought to treat, whom we ought to keep alive, whom we can experiment on without consent, and so forth. In other words, the question "What is a person?" concerns not a scientific classification but rather a moral classification. The question turns out to be a moral question in disguise. Of course, no one should complain if moral questions are raised in the context of bioethics! These are exactly the sorts of questions we should expect. If I am complaining about anything, it is that some writers in the subject are presenting moral questions as if they could be settled by some more scientific-looking procedure, of classification and of setting tests and criteria which individuals must satisfy if they are to count as members of the class of 'persons' in some unfamiliar, narrow and technical sense.

Back at the beginning of this chapter I admitted that simply looking at how people happened to use a word was an unsatisfactory way of deciding what the word meant - not least, because of the sheer variety of the possible uses of the word. The trouble is that this variety has to be taken seriously. Not only the unpleasant Mrs. Poultney but the no doubt highly likeable participants in bioethical debate seem to use the word 'person' in ways that suit their purposes. Some of them - perhaps unlike Mrs. Poultney - take the trouble to tell us how, and why, they use the word in the way that they do. At least then we are given the opportunity to decide whether we can agree with them. But of course we do finally need some criterion for deciding this - some grounds for thinking that the way *we* use the word is the right way, or at least the most appropriate way; only then we can decide if the way that someone else uses the word is appropriate. So what, finally, is our criterion? My answer to this has been that the ordinary and familiar use is the right way, and that we have a variety in use only because some people *choose* to depart from the ordinary and familiar use.

9. CONCLUSION: RE-AFFIRMING THE TRADITIONAL CONCEPT

Let me finish by facing up to the principal challenge which might be put to any conception of the person offered within bioethics - including the traditional conception. The intention of this challenge is to dispute the belief that all human beings are persons. It shares this feature with most of the various accounts of 'the person' which we have reviewed in this chapter, of course, and as such can be taken to stand for the moral challenge which all those accounts are meant to pose to us. It is this: how can we maintain that a human being who has permanently lost - or who has never attained - any possibility of cerebral function, and who is permanently condemned to a merely vegetative 'condition', is a person? My own answer to this question is to turn it back on the questioner. Why should we *not* regard someone in this condition as a person, albeit a person in dreadful circumstances? The assumption behind this challenge seems to be that only if we can classify different patients as different sorts of *entities* can we offer any moral justification for treating them differently. Indeed, this seems to be the programme underlying most if not all of the various accounts we have reviewed. To my mind this assumption looks like the result of a kind of moral panic - as though we never had morally defensible grounds for making choices and discriminations among different people, or as though such grounds are so difficult to find, so difficult to defend, that metaphysical reclassification is somehow easier! I suggest that this assumption is complete nonsense. We make moral choices all the time; we judge people as deserving or as undeserving, as guilty or innocent, as worthy or as unworthy; and daily life requires us to make such judgements, and to treat people differently according to how we judge, and finally to stand by our choices and to take responsibility for them. The fact is that such choices are especially uncomfortable in the clinical arena, and perhaps we think they are even improper - few would wish to admit to distributing health care resources according to whether or not patients deserve them, for instance. But the moral challenge of deciding whether or not to withdraw life-sustaining treatment from someone who is in a permanent coma is precisely that - a moral challenge - and it will not disappear just because we try to disguise it as an exercise in re-classification. Why not instead argue openly about whether someone who lacks the possibility of conscious experience has a lower claim on scarce resources, or whether it is kinder to the family to put an end to an intolerable situation, or whether the indignity of being maintained unconscious on a machine ought to stop? This seems far better than attempting to re-define the patient in question out of the moral realm of persons (and can a non-person really suffer indignity?!). Judgements on such questions will then be plainly moral judgements, and they can be disputed and defended in moral terms. This seems to me to be more honest and more realistic than manoeuvres in metaphysics, and far less likely to allow us to forget our responsibilities.

NOTES

1. In addition to the authors quoted in this chapter, the reader might look at, among others: Foot (1986); Glover (1997); House of Lords (1993); Kuhse and Singer (1985); Lamb (1988); Lockswood (1985); McCullagh (1993); Ruyter (1996); Tooley (1983).

REFERENCES

Capron, A.M.: 1987, Anencephalic donors: separate the dead from the dying. *Hastings Center Report* 17(1): 5-9.
Carruthers, P.: 1989, *Introducing Persons*. Routledge, London.
Crosby, J.F.: 1993, The personhood of the human embryo. *Journal of Medicine and Philosophy* 18(4): 399-417.
Engelhardt, H.T.: 1982, Concepts of personhood, in: Beauchamp, T. and Walters, L. (eds.), *Contemporary issues in bioethics*. Wadsworth, Belmont, p. 94-101.
Foot, P.: 1986, Life and death. *London Review of Books*, 7th August.
Fowles, J.: 1977, *The French lieutenant's woman*. Triad, Granada/London.
Gervais, K.G.: 1986, *Redefining death*. Yale University Press, New Haven and London.
Gillon, R.: 1990, Death (editorial). *Journal of Medical Ethics* 16(1): 3-4.
Glover, J.: 1977, *Causing death and saving lives*. Penguin Books, Harmondsworth.
Harris, J.: 1985, *The value of life*. Routledge, London.
House of Lords: 1993, *Airedale ZNHS Trust v. Bland*. House of Lords 1 all ER 821.
Kuhse, H. and Singer, P.: 1985, *Should the baby live? The problem of handicapped infants.* Oxford University Press, Oxford.
Lamb, D.: 1988. *Death, brain death and ethics*. Croom Helm Ltd., Beckenham.
Lizza, J.: 1993, Persons and death. *Journal of Medicine and Philosophy* 18(4): 351-374.
Lockwood, M.: 1985, When does life begin. In: Lockwood, M. (ed.), *Moral dilemmas in modern medicine.* Oxford University Press, Oxford.
McCullagh, P.: 1993, *Brain dead, brain absent, brain donors.* John Wiley & Sons, Chichester.

Ruyter, Z.W.: 1996, Embryos as moral subjects and limits of responsibility. In: Evans, D. (ed.), *Conceiving the embryo*. Martinus Nijhoff Publishers, The Hague.
Tooley, M.: 1983, *Abortion and infanticide*. Clarendon, Oxford.

RITA KIELSTEIN

TRANSPLANTATION MEDICINE

1. CASE 3.

Mr. Schmidt (46) has been a dialysis-patient for a year now and still extremely uncomfortable with his new situation. He cannot adjust to the new living conditions. One day he asks for a doctor's appointment and appears with his son Peter (19). Peter informs the physician that he can no longer accept his father's suffering and that he is ready to help his father and his mother in offering to donate one of his kidneys for transplantation. During the discussion, the physician becomes quite suspicious whether or not Peter acts out of free choice and free will, in particular as the father gets actively involved in the debate and Peter's body-language and other aspects of behaviour suggest that he might be under some kind of direct or indirect pressure from one or both of his parents. Peter requests, however, that he be given six weeks to have a good time with his friends prior to the explantation. Mr. Schmidt, who is sometimes depressive, sometimes seemingly angry during the course of the conversation, adds that he might not survive another six weeks. Finally, both father and son request that the necessary tests be done. What should I have done?

But then, who was my patient, Mr. Schmidt or the son? If Mr. Schmidt feels uncomfortable to be dialysis-dependent and a donor organ is available, the transplantation would be an excellent solution, especially since - in general at least - quality of life and patient's autonomy as well as clinical and allocational benefit-cost calculations are excellent for transplantation. On the other hand, should I have become a partner in the exploitation of a dependent fellow human being if Peter, indeed, was pressured into accepting an explantation against his will? Would it not be better to help the patient to accept the new conditions of his life or to wait till a post-mortem kidney becomes available? There was no urgent need for transplantation from the clinical point of view at that time.

H.A.M.J. ten Have and B. Gordijn (eds.), Bioethics in a European Perspective, 157-166.

2. FROM BIOMEDICAL TO BIOETHICAL CHALLENGES

In 1954 the first successful kidney transplantation with living donors was performed on identical twins in Boston. In the late fifties it became possible to keep bodies 'alive' under new ventilation and circulatory support technology, a clinical situation first called *coma dépassé*, an irreversible biological state of the body below the level of traditional permanent coma and without any activity of the brain, since 1968 defined as *brain-death*: irreversible coma, missing activity of the brain stem, loss of spontaneous breathing. Since the sixties drug development for immunosuppression had been increasingly successful and had allowed for transplantation of tissue and organs without instant rejection and with long-term organ survival. Today, good tissue-compatibility between the implanted organ and the recipient still is of great value and will result in low rates of rejection and long life of the recipient and the organ. In Germany, the first kidney was transplanted in 1963. The following organs were transplanted in Germany during 1997: 2249 kidneys, 762 livers, 562 hearts, 146 pancreases, 120 lungs (Smit, *et al.*, 1998).

Over 16,000 patients had been waitlisted: many will die before an organ becomes available for them based on allocational criteria which include tissue type, waiting time, and other data (Deutsche Stiftung für Organtransplantation, 1997). EUROTRANSPLANT, headquartered in Leiden, the Netherlands, provides instant and full information on potential recipient's tissue type, health condition, waiting list time, urgency, and location; it matches that information with data from post-mortem organs as soon as they become available.

But as these and other biomedical and organisational hurdles have been solved or eased, still not enough organs are available to prevent death of those who badly need them or for those whose lives would be more enjoyable (Kielstein, 1991) as Mr. Schmidt's case shows. There are not enough donors: a bioethical, cultural and moral challenge, not a medical problem. As more powerful immunosuppression drugs have become available in recent years, transplantation will become an additional choice among living persons, who are not closely related by blood and tissue type and therefore not an easy tissue match, but who are closely related by friendship, marriage, or otherwise. The pool of potential living donors therefore has tremendously increased over the last years. But, again, we have not seen an increase in the availability of organs yet. On the contrary, it has been mentioned that undue pressure, challenges and even exploitation might occur within families and among friends or within moral communities to 'voluntarily' donate. The legal parameters in some countries seem to be aware of those new probabilities for exploitation and have written closely worded regulations for living donors, among them the requirement for extended psychological evaluation.

If we look into the global situation, we have to recognize that today over 85% of end-stage renal disease (ESRD) patients live in Western Europe, North America and Japan (Davison, 1994). The point has been made that the rest of the world will never be able to pay the high costs of keeping all ESRD patients everywhere alive or pay for transplantation, even if organs would be easily available. Xenotransplantation (using organs from transgenic animals) or artificial organs (such as totally implantable rotary pumps replacing human hearts) have been mentioned to ease the morally uncomfortable situation that many in the 'rich world', but none except the super-rich in the 'poor world', will be able to benefit from efficacious modern replacement therapies (Kielstein and Sass, 1995).

3. POST-MORTEM EXPLANTATION

It has been argued that the donation of organs after one's death is a *moral obligation* of each and every civilized and morally concerned citizen, as well as being a right of those who otherwise would die or live a much more miserable life (Thiel, 1996). It has also been mentioned that in particular the Christian tradition of 'love your neighbour' seems to mandate that organs be made available after one's death when they are no longer of any use for the owner, except for acts of charity and beneficence (Sass, 1998). Other religious and philosophical positions, such as Taoism, Shintoism, and all animistic religions, on the other hand, cannot even formulate the concept of giving parts of the own body to someone else nor accepting body parts from other people. Explantation and implantation of organs, also of tissue and even the transfusion of blood, are assessed differently by different cultures, religions, families, and individual persons, who might or might not agree with the prevailing majority position in their culture or legal environment. This makes organ transplantation a very individual, very personal issue. Legal parameters or official default positions for those who have not voiced their dissenting position are of minimal help only.

Additional to religious and philosophical controversies over *concepts of body and person*, even within positions such as the Protestant Christian religion, there is a variety of positions pro and contra post-mortem donation (Jörns, 1992; Kielstein, 1994). Some of the controversies are related to the definition of death: when is a person dead, and when will it be acceptable to open their bodies and harvest tissue and organs? A clinical definition of death, the so-called *brain-death criterion* replacing previous cardiovascular criteria which are not significant any more under conditions of intensive care ventilation and cardiological support, has since 1962 found wide clinical, religious and philosophical support (Angstwurm, 1994). But, again, such a support is only possible within worldviews, capable of differentiating between a mortal body and a soul separable from the deceased or dead body, therefore not acceptable to the other positions holding different opinions on the human body and person. Schockenhoff quotes Matthew 7,12 - do to others what you expect from them. He holds that the donation of organs can be a final and last personal answer to the query what life is and what it means to be a human being:

> Life and death are close to each other; driven by solidarity with fellow-humans and by a spirit of Christian neighbourly love each of us can by donating an organ help an ill fellow human, beyond one's own death, who then will regain health and live a newly donated life within the family and the workplace (Schockenhoff, 1997, p. 17).

He also mentions the ethical and social principle of reciprocity. This principle would suggest that only those who are willing to donate organs would be among those who would be given a priority in receiving organs.

The legal situation regarding post-mortem transplantation is different from country to country in Europe. Some countries assume that the prospective donor after her or his death would be willing to donate (position of agreement), others assume that the prospective donor would not be willing to donate (position of refusal), if not expressively stated otherwise. All national laws and regulations will respect individual choices contrary to the default position. Some countries have a so-called 'extended solution', allowing the family to make decisions in the absence of choices made by the deceased. In Germany, after a lengthy public and parliamentary debate during which the numbers of prospective donors decreased by a high margin, organs may only be explanted if the deceased has agreed in writing or by donor card or if the surviving relatives agree (Deutsche Stiftung für Organtransplantation, 1997).

4. REVIEWING EXPECTATIONS OF PHYSICIANS, NURSES, AND PATIENTS

In May 1997 we asked physicians, nurses, patients in dialysis or with donated kidneys, and their relatives at the Otto-von-Guericke University in Magdeburg whether or not they would agree to donating or receiving organs. Of 574 physicians 17.2%, of 1197 nurses 49.1%, of 253 patients in dialysis 68.6%, of 40 patients living with donated kidneys 100%, of 109 partners of dialysis patients 68%, of 320 children over the age of 21 of dialysis patients 25% responded. One of the results is that 25.8% of physicians, 10.7% of nurses, 1.5% of partners of dialysis patients and 4.5% of children of dialysis patients had carried a donor card. Among those who responded, 92% of physicians would be willing to donate and 92% to receive organs. The figures for the nurses were 91.6% and 93%, for the dialysis patients 72% and 68%, for patients living with kidney-transplants 98% and 95%, for the partners of dialysis patients 72.7% and 70.5%, for the partners of patients with kidney-transplants 90.9% and 90.9%, for children of dialysis patients 54.1% and 59.6%, and for children of transplant patients 61.1% and 61.1%.

There was a reluctance to donate or to receive certain organs; among the physicians and nurses 5.3% would not donate and 2.5% would not want to receive cornea transplants, 2% would not donate and not receive lungs, 2% not donate and not receive a heart, 2% not donate and not receive a liver, 0.7% not donate and 0.4% not receive a kidney, 0.3% not donate and 0.1% not receive bone marrow, 3% not donate and 1.5% not receive cochlea transplants. The difference between prospectively donating and receiving organs among responding physicians, patients in dialysis and with transplants, partners and children was insignificant; among nurses 1.2% wanted to donate only, 1% to receive only.

Overall, 42 persons would accept organs, but not donate any; some would not want to donate a liver for alcoholics or previously addicted persons; others would only donate to those who themselves carry a donor card. Many preferred a clear-cut legal situation allowing for either a strict model of refusal or acceptance; they would not want at all to have their relatives make hard choices in situations of stress or allow them or others to question or interpret choices made previously by themselves. Other individual answers included: why should I have a donor card, if my relatives are given the legal or moral right to have a final word (13 different respondents); only those who will donate, should be among the recipients (physician); organ transplantation should be discussed in schools so that at maturity everyone would be capable of having her or his choice recorded when applying for a passport (nurse); parliament should provide for a solution requiring everyone to make a choice (nurse); organs should be retrieved, except in cases where written statements by the deceased request otherwise (nurse); persons carrying a donor card might feel that their care will be driven by the need for organs (nurse); I had a donor card, but destroyed it after experiencing that organs got lost and destroyed in the process of poor transportation and organisation management (physician). The following results need more and detailed interpretation:

1. The number of prospective donors is higher than the number of those carrying a donor card.
2. The number of those who would reject an organ is surprisingly high and has never played a role in public debate or religious or bioethical discourse.
3. Surprising was also the reluctance to either receiving or donating specific organs such as cornea; donor cards therefore should allow for making such choices (Kielstein, 1998).

5. LIVING DONORS

The number of potential renal transplant recipients far exceeds the number of cadaveric donors. For this reason living related donors and, more recently, living unrelated donors have been used to decrease the cadaveric donor shortage. Living related and unrelated renal donation continues to be an important source of kidneys for patients with end-stage renal disease. As mentioned, recent breakthroughs in immunosuppression widen the possibilities of organ donation among living donors. Outcome in emotionally related living kidney donor transplantation represents a valuable option. Recipients and graft outcomes were superior to cadaver kidney transplantation. The 2-year recipient and graft survival were 100% and 91% respectively, compared to 99% (recipients) and 93% (graft) in the non-HLA-identical living related kidney transplant group, and to 93% (recipients) and 83% (grafts) in the cadaver kidney transplantation group, according to the recent report of Binet, *et al.* (1997).

Easy availability of kidneys from living donors allows for a shorter waiting period, planning and timing of transplantation, omission of haemodialysis and its positive clinical, social and quality-of-life side-effect for ESRD patients. Also donation among living spouses is a final expression of love, allowing for a continuation of spousal love, sex, devotion, valued forms of lifestyle and social habits (Thiel, 1996). However, as in the case of Mr. Schmidt, familial pressure and unwarranted social and ethical complications following transplantation cannot be excluded. In Switzerland, a pool of partners, friends, and spouses of prospective recipients of donor kidneys has been formed in order to reduce some possible ethical conflicts of direct spousal donation and to match those recipients and donors within the pool based on good tissue compatibility. Sells (1997) from the International Forum for Transplant Ethics, Liverpool proposed to support the model of paired kidney exchange. Others argue that donating to and receiving from a pool is ethically preferable if compared with an exchange of pairs of donor-recipients (Ross, *et al.*, 1997a, 1997b). It has been proposed to increase the number of potential donors based on the principle of reciprocity. It was also suggested that an inner circle be formed among donor card carriers who would be entitled to be in a first group of prospective recipients, while non-card-holders would get organs only if no one is found in the reciprocity group, - a proposal ethically supported by the data discussed above, namely that some of those who do not want to donate might have their very personal reasons not to receive any either (Sass, 1998). These and other ethical challenges of living donor transplantation will become even more pressing as clinical possibilities in immunosuppression progress.

6. HUMAN ORGANS FOR SALE, XENOTRANSPLANTATION, AND ARTIFICIAL ORGANS

Given the scarcity of organs, gratuity models for transplantation are discussed (Land, 1991; Sass 1998). Unfortunately, there is a multitude of anecdotal evidence that in the Indian subcontinent those who have sold their kidneys for financial benefit have been exploited and given poor medical care (Sever, *et al.*, 1994). A federal law in India outlaws organ trade, but many Indian states have not yet and may not implement that law into their own. If all others parties - physicians, nurses, hospitals, insurance companies - profit from the transplantation business, why should the 'donor' be the only one who takes a high risk without any gratuity or compensation? Gratuity must not necessarily be expressed in financial terms, but in better health care or a good job offer, educational vouchers, etc. There are many highly risky enterprises in life, including in medicine such risky forms of clinical trials with questionable benefit for the individual research subject. These risks are highly regulated but not criminalized. The gratuity model, the moral challenges of which are big and risky, has not found an extensive ethical analysis and assessment yet and might for some time still be treated as a taboo. As long as we have not had a critical debate, we will not be able to determine whether or not and under which conditions it might be unacceptable or acceptable.

Given the bioethical and cultural problems associated with post-mortem and living donor transplantation, there is not much hope that even in European countries there will be a sufficient supply of organs in the foreseeable future to care for all those who very much need them for survival or for less suffering in life. Two other alternatives to post-mortem donation and living donor transplantation therefore might give these fellow humans a better chance than the still open solution to a debate on a gratuity model for organ giving: xenotransplantation and artificial organs. Especially the need for organs in the less developed countries calls for an increase in research and development in these two areas.

The *breeding of transgenic animals* for organ transplantation (Kamstra, *et al.*, 1996) and tissue farming has caused controversy and rejectionist statements in the media and from healthy adults - politicians, theologians, philosophers, commentators, and journalists - most of them not vegetarians. Prospective recipients, patient groups, dying or suffering patients on much too long waiting lists have not yet been asked for their opinion, hope, and *angst* regarding the implantation of organs from animals into their body. In my opinion, the moral rejection of farming animals for medical purposes, especially for saving lives, has a weaker justification than the rejection of breeding and farming animals for food or for supporting gourmet living. In a culture of eating sausage, cutlet, chicken, eggs, and cheese and of drinking milk, it seems to be either an unreflected prima facie reaction or an unacceptable moral double standard to have one's cutlet but to refuse dying and suffering fellow humans animal organs for survival.

Prostheses for teeth, joints, bones, cochlear implants, heart valves and vessels, pacemakers, haemodialysis machines and prototypes of totally implantable rotary heart pumps provide hope for further progress in developing even better and less expensive *artificial organs*, the availability of which will allow to avoid all or most of the above discussed bioethical problems of human organ transplantation. Also, better and less expensive artificial organs might be the best solution for reducing rates of death and suffering among fellow humans in countries less rich and less fortunate than our European countries (Kielstein and Sass, 1995).

7. CONCLUSION

We have discussed bioethical challenges of giving one's organs to other people in various biomedical and bioethical scenarios of post-mortem donation, donating or selling by living donors, also the bioethical challenges to the recipients of human organs post-mortem or from living donors or accepting transgenic organs from animals. Different as these challenges are, they will have to be answered by the moral agents, primarily involved as recipients, donors or prospective donors. The legal and cultural environment can only provide general support for ethical decision-making by those involved. Two principles seem to be important for providing an ethically acceptable framework for organ transplantation: (1) a consensus-oriented continuation of the public debate concerning collective and individual values, wishes, fears, and hopes involved and (2) the ethical identification of preferred decision-makers for giving or receiving organs. Based on our modern understanding of human dignity and the respect for conscientious choice-making, prospective recipients and prospective givers of organs should be the prime moral agents to decide whether or not and under which conditions they would take or give or refuse. The challenge to the public discourse and to those involved in it, would therefore be to support and respect individual values and wishes and choices made on their basis. This would require the ethics of the transplantation debate to change from confrontational and paternalistic attitudes towards information, communication, understanding, and supporting.

In the ethical setting it will be important to provide information and to be available for advice and assessment if requested, and to respect individual choice without pressure, coercion, or misleading statements. In particular, there are the differences between (a) explicit and clear consent by the individual person herself, (b) information and consent given in more general terms, (c) presumed consent by a person based on indirect evidence or authority of previous oral statements, conversations, or recollection of persons representing the donor or recipient with various degrees of authority for substituted judgement, (d) consent given by representatives, designated by authority of the individual represented or those chosen by courts. Each case will be different and clinicians should be critical in evaluating the authority and legitimacy of the consent given (Faden, 1986).

Back to Mr. Schmidt and his son Peter. Their story shows how closely connected request for consent, coercion or various forms of 'persuasion' can be. We did the tests; they were both told that the match was less than ideal and that the probability of immunological rejection was high. Mr. Schmidt was lucky enough to receive a post-mortem kidney less than six months later.

REFERENCES

Angstwurm, H.: 1994, Der vollständige und endgültige Hirnausfall als sicheres Todeszeichen des Menschen. In: Hoff, J. and In der Schmitten, J. (eds.), *Wann ist der Mensch tot? Organverpflanzung und Hirntod.riterien.* Rowohlt, Reinbek, p. 41-50.

Binet, I., Bock, A.H., Vogelbach, P., Gasser, T., Kiss, A., Brunner, F., Thiel, G.: 1997, Outcome in emotionally related kidney donor transplantation. *Nephrol Dial Transplant* 12(9): 1940-1948.

Davison, M.A.: 1994, Commercialization in organ donation. *Nephrol Dial Transpl* 9: 348-354.

Deutsche Stiftung für Organtransplantation (ed.): 1997, *Bundesrepublik Deutschland: Transplantationsgesetz. Gesetz über die Spende, Entnahme und Übertragung von Organen* [vom Bundestag beschlossen am 25. 6. 1997], Neu-Isenburg.

Faden, R.R. and Beauchamp, T.L.: 1986, *A history and theory of informed consent.* Oxford University Press, New York.

Hamstra, A., Sass, H.M., Westphal H., *et al.*: 1996, *Transgene Tiere*. Basel, Ciba Communications.

Jörns, K.P.: 1992, Organtransplantation: Eine Anfrage an unser Verständnis von Sterben, Tod und Auferstehung. *Berliner Theologische Zeitschrift* 9: 15-39.

Kielstein, R.: 1991, *Klinisch-ethische Probleme der Nierenersatztherapie*. Bochum, Zentrum für Medizinische Ethik.

Kielstein, R.: 1994, Inkompetenz und Demagogie. Replik auf die 'Berliner Initiative für die Zustimmungslösung'. *Der Dialysepatient* 1: 31-39.

Kielstein, R.: 1998, Ihre Meinung zur Organspende. *Deutsche Ärztezeitung* 15: 7.

Kielstein, R. and Sass, H.M.: 1995, From the wooden limbs to biomaterial organs. *Artificial Organs* 19(5): 475-80.

Land, W.: 1991, Das belohnte Geschenk? Überlegungen zur Organspende von gesunden Menschen. *Merkur, Dt. Zeitschrift für Europäisches Denken* 45(2): 120-123.

Ross, L.F., Rubin, D.T., Siegler, M., *et al.*: 1997a, Ethics of paired-kidney-exchange program. *New Engl J Medicine* 336(24): 1752-1755.

Ross, L.F., Woodle, E.S., Siegler, M.: 1997b, To the editor: Paired kidney exchange programs. *New Engl J Medicine* 337(19): 1393.

Sass, H.M.: 1998, Philosophische und ethische Probleme bei der Organtransplantation. *Nieren- und Hochdruckkrankheiten* 27: 184-192.

Schockenhoff, E.: 1997, Angstfreie und verantwortliche Entscheidung für die Organspende. In: Deutsche Stiftung für Organtransplantation (ed.), *Bundesrepublik Deutschland:*

Transplantationsgesetz. Gesetz über die Spende, Entnahme und Übertragung von Organen. Neu-Isenburg, p. 16-19.

Sells, R.A.: 1997, Paired kidney exchange programs. To the editor. *New Engl J Medicine* 337(19): 1393.

Sever, M.S., Eclert, T., Ayden, A.E., *et al*: 1994, Living unrelated (paid) kidney transplantation in third world countries: High risk of complications besides the ethical problem. *Nephrol Dial Transpl* 9: 350-354.

Smit, H., Sasse, R., Zickgraf, Th., Schoeppe, E., Molzahl, M.: 1998, *Organspende und Transplantation in Deutschland 1997*. Dt Stiftung Organtransplantation, Neu-Iserburg.

Thiel, G.: 1996, Ethische Probleme der Leichen- und Lebend-Nierentransplantation. In: Kielstein, R. (ed.), *Ethik in der Nephrologie*. Bochum, Zentrum für Medizinische Ethik, p. 81-91.

WIM J.M. DEKKERS

EXPERIMENTATION WITH HUMAN BEINGS

1. INTRODUCTION

Experiments on human beings are as old as medicine (Rothman, 1995). Medical research, especially research on human subjects, is integral to medicine as a science and as a practice. Over the past few centuries there has been a gradual, almost imperceptible, shift away from the Hippocratic absolute of medical beneficence to a new relativism that allows research to be performed on human subjects (Fagot-Largeault, 1988). Scientific medicine and clinical research in the modern sense are scarcely a hundred years old. Systematic controlled experimentation is even more recent. The randomized clinical trial (RCT) was used for the first time in medical practice in Britain in the 1940s to evaluate streptomycin in the treatment of tuberculosis. The RCT is now recognized as the most powerful research tool - the gold standard - in medical-biological research. Modern medical research involves "systematic design and analysis of interventions involving human subjects in order to develop generalisable knowledge regarding the prevention and treatment of disease" (Ackerman, 1994, p. 874).

The recent history of the ethics of experimentation on human beings reflects the deeply felt aversion to the Nazi-experiments which manifested itself in the *Nuremberg Code* (1947). Perhaps the most important effect of the Nuremberg Code was that it raised the consciousness of people everywhere about experiments on humans (Bankowski, 1993). The Nuremberg Code formed the basis of the "Recommendations guiding physicians in biomedical research involving human subjects" of the World Medical Association, the so-called *Declaration of Helsinki* (1964, last revision in 1996). This declaration is now the most important guideline for those engaged in medical research on human beings. In the late 1970s the Council for International Organisations of Medical Sciences (CIOMS) and the World Health Organisation (WHO) undertook an examination of the applicability of the Nuremberg Code and the Declaration of Helsinki to the special circumstances of developing countries. In 1982, these organisations issued their *Proposed International Guidelines for Biomedical Research Involving Human Subjects* (last revised in

H.A.M.J. ten Have and B. Gordijn (eds.), Bioethics in a European Perspective, 167-178.

1993). The three general principles for biomedical research on human subjects are: respect for persons, beneficence, and justice (Levine, 1986). In the United States these principles were formulated by the National Commission for the Protection of Subjects in Biomedical and Behavioural-scientific Research in the influential *Belmont Report* (1978). These three principles are generally accepted as the basis from which further, second-order principles, rules and norms may be derived (Bankowski, 1993). Respect for persons incorporates two further principles: autonomy (leading to the rule of informed consent) and the protection of those of impaired or diminished autonomy. The principle of beneficence means that predictable benefits must outweigh predictable risks. The principle of justice includes the rule of distributive justice, which requires a fair and equitable sharing of both burdens and benefits.

Beecher's 'Ethics and Clinical Research' is a classic article in modern research ethics. He presents 22 examples of medical experiments which can be classified as 'unethical'. In many cases investigators risked 'the health or the lives of their subjects,' without informing the subjects of the risks or obtaining the subjects' permission (Beecher, 1966). One of Beecher's conclusions is that the statement that 'consent has been obtained' has little meaning if the subjects or their guardians are incapable of understanding what is to be undertaken, or have not been clearly informed about all the hazards involved. Beecher regards the intelligent, conscientious, compassionate, responsible investigator as a more reliable protection. There is also a well-known book by Katz (1972) dealing at length with many cases of 'ethical' and 'unethical' experimental research on human beings.

The prevention of 'unethical' research on human beings was one of the reasons for establishing medical ethics committees. Ethics committees emerged in the 1960s. By the mid-1970s the United States had a number of functioning hospital or health care ethics committees (HEC's). Research Ethics Committees (REC's) - also called Institutional Review Boards (IRB's) or Local Research Ethics Committees (LREC's) - for research involving human subjects were also set up in this period (Levine, 1995). Though nowadays most HEC's deal only with the ethics of clinical treatment and institutional care policies, both types of committees share closely related histories, ethical presuppositions and procedures. One of the most important functions of both HEC's and REC's is to protect patients and subjects from abuse. Their charge is to sustain the autonomy and interests of subjects and patients (Tristram Engelhardt Jr., 1999).

For about four decades, ethical aspects of medical research on human subjects have been one of the main areas of activity and reflection in the philosophy of medicine and bioethics. It has been argued that the issues and problems of medical experimentation have led directly to the emergence of modern bioethics (Rothman, 1991). Medical ethicists are sometimes viewed with suspicion by physicians and medical researchers. Many would agree with an editorial in *The Lancet* of October 6, 1990, which decried the interference of "professional medical ethicists, latterly joined by lawyers specialising in health matters" - the so-called 'strangers at the bedside' (Rothman, 1991). The offending ethicists are perceived as propagating the dogma "that whatever is done for the sake of medical science is alien to the treatment of the individual, and should therefore be labelled an 'experiment', necessitating informed consent by the patient and adjudication by an ethics committee" (Editorial, 1990, p. 846). The increasing emphasis on ethics, patients' rights and autonomy is seen as having a negative impact on medical research. Further, it is argued that national research ethics committees will lead to bureaucratic restriction of research in the name of medical ethics, and eventually to the total alienation of research from bedside medicine.

Despite the criticisms, the debate about ethical and legal issues has substantially changed the practice of experiments on human beings. There are a growing number of international ethical standards and legal regulations for research involving human subjects. In addition to the *Declaration of Helsinki*, one must also mention the guideline *Good Clinical Practice* (GCP: European Recommendation Concerning Good Clinical Practice for Drug Research in the European Community) of the International conference on the harmonisation of technical requirements for the registration of pharmaceuticals for human use (1996). GCP is an international ethical and scientific quality standard for designing, conducting, recording, and reporting trials that involve human subjects. There is also *the Convention for the Protection of Human Rights and Dignity of the Human Being with regard to the Application of Biology and Medicine* of the Council of Europe (1996). Most countries have now enacted legislation controlling experiments on human beings which includes guidelines for Research Ethics Committees (REC's). Although the expression 'experiments on human beings' may suggest otherwise, most laws and regulations in this field also cover non-experimental observational and epidemiological research (Coughlin, 1995; Coughlin and Beauchamp, 1996). The recent Dutch legislation 'Regulations on Medical Research involving Human Subjects' (Medical Research involving Human Subjects Act) uses the following broad definition of an experiment: "medical-scientific research which involves subjecting persons to certain procedures or imposing particular behavioral requirements." This definition includes not only (invasive) medical experiments on human beings, but also experiments with their body parts. Epidemiological research in which human subjects have to provide personal data also falls under this law.

The existing body of literature in the field of research ethics focuses primarily on experiments on human subjects as a whole, that is, on experimentation on persons and their (entire) bodies (Katz, 1972; Spicker, 1988; Capron, 1989). In today's medicine, however, human organs, tissues and body fluids are increasingly being used for research purposes. The availability of human body parts is of undeniable importance for basic research, for research aimed at improving therapies and for research aimed at developing new treatments. Despite this, relatively little attention has been paid to biomedical research on human body parts (Dekkers & Ten Have, 1998; Van Agt, *et al.*, 1999).

Ethical problems that may arise in the field of experimental research on human beings include the selection of research subjects, the use of vulnerable, incompetent or dependent people (children, elderly people, mentally handicapped or demented people, students, employees, prisoners) as experimental subjects, ethical aspects of the scientific methodology such as the randomisation procedure and the use of placebos, the use of human body parts, the assessment of risks and benefits, informing the research subjects, and the reimbursement of costs. Some of these problems will be elucidated in an analysis of the following research protocol.

2. CASE 4.

2.1. Clinical Context

The clinical context of the research protocol 'A Activity in B Induced Skin Response in Normal Volunteers' consists of endeavours to develop a new drug against psoriasis. Psoriasis is one of the most common chronic skin diseases, with a prevalence generally estimated at between 1.4 and 2.9% of the population. It is a non-infectious disease, characterised by sharply marginated areas of affected skin which appear thickened, red and scaly, and may itch. The symptoms reflect a markedly elevated rate of epidermal proliferation. Dermal blood vessels are dilated and there is infiltration of the skin with immunologically active cells. The pathogenesis is not well understood. The disease cannot be cured in the strict sense of the word, and the current therapy for psoriasis is strictly symptomatic.

The experimental drug to be tested is substance A. A can possibly play a role in the development of a new drug against psoriasis. A will be tested with the help of the chemical substance B which, if applied on the skin, causes a lesion of the skin. The local application of B results in an experimentally reproducible inflammation of the skin which is more or less comparable to the lesions caused by psoriasis. Drug A is a potent and specific antagonist of B receptors. In vitro, it specifically inhibits chemotactic activity and some other chemical reactions in a specific category of white blood cells, the so-called neutrophils, which are stimulated by B. Drug A has to be taken orally. In previous studies drug A has been administered at several doses in rats and in monkeys. It has also been given to thirty healthy male subjects in a dose-finding study. Furthermore, in a multiple oral dose study three groups of eight healthy subjects received either drug A or placebo in a single subject blind manner. A few adverse events were reported in both the human volunteer studies, but no serious reactions.

2.2. Objectives

The objectives of this research project are:
- to assess the suppression of local neutrophil response with oral A to epidermal B challenge in healthy volunteers, using samples obtained from skin biopsies.
- to assess the effect of oral A on inflammation and epidermal proliferation and inflammation induced by B in healthy volunteers.

2.3. Scientific Background

There is already considerable expertise with this research model. The experimental inflammation induction by the application of B on the skin is an approved test model. The effect has many characteristics in common with lesions caused by psoriasis. Animal testing is inappropriate because there is no adequate animal research model for psoriasis. Psoriasis does not occur in animals and it is not possible to induce the experimental inflammation in animals.

Existing psoriasis lesions in patients with psoriasis cannot be used for this study. If one were to use these lesions for research purposes, it would be necessary to prolong the administration of drug A considerably. It would take 4 to 6 weeks (instead of 10 days) to assess the effect. Nor can unaffected areas of the skin of patients with psoriasis be used. The skin of psoriasis patients is not 'normal' in the sense that it contains a 'natural' low level of B, and there is a higher risk of complications if a biopsy is taken from the (unaffected) skin of psoriasis patients.

Thus, in healthy volunteers the experimental model is more 'pure' than it is in psoriasis patients or research animals.

2.4. Methodology

Eighteen healthy male volunteers between 18 and 45 years will participate in this randomised, double-blind, parallel group study. All will be randomised to one of the possible three treatment sequences:
- 10 days of treatment with drug A: 48 mg bid.
- 10 days of treatment with drug A: 200 mg bid.
- 10 days of treatment with placebo bid.

Volunteers will be subject to epicutaneous applications of B to their (normal) skin (on the shoulder) before treatment, and after 10 days of treatment (with drug A or placebo). They will have skin biopsies before B application (day 1), after 24 and after 72 hours. On day 8 they start medication. Following 10 days of medication or the use of placebo, B will be applied for a second time. Skin biopsies will again be taken after 24 and 72 hours. Thus, every volunteer will be subjected to 5 punch biopsies. The B induced inflammation will be analyzed with immunohistochemical techniques.

A punch biopsy involves removing a small area of skin under local anaesthesia. If the wound has a diameter of about three mm, a suture is seldom required. If the wound has a diameter of 4 mm or more, one or two sutures may be necessary. The biopsies seldom cause visible scars. Although volunteers with a predisposition to keloid formation are excluded from participation, keloid formation is still a possible complication.

Each research subject is paid 250 Dutch guilders.

3. CASE ANALYSIS

3.1. Clinical and Scientific Significance

In order to assess the clinical significance of research protocols it is helpful to distinguish between therapeutic and non-therapeutic research. While in therapeutic research the individual patient will possibly benefit from the experiment, non-therapeutic research is purely scientific and is conducted without direct diagnostic or therapeutic value to the person subjected to research. The above protocol is clearly non-therapeutic, though psoriasis patients will hopefully benefit from it in the future. If this study contributes to the development of a new drug against psoriasis in the long run, the clinical importance of the research protocol is considerable. Psoriasis is a serious skin disease which affects patients not only physically, but also psychologically. At the moment there are no drugs available that can cure the patient and treatment is strictly symptomatic.

It is now generally accepted that a research study that is scientifically inadequate is automatically also 'unethical'. The scientific value of the protocol depends both on how well the method can work, and to what extent the method can contribute to our understanding of the pathophysiological mechanisms which cause psoriasis (and thus to our capacity to therapeutically influence these abnormal mechanisms). The possible contribution of this study to our knowledge of psoriasis involves questions such as: is the aim of the study clearly defined and scientifically realistic? Will it provide useful information? Do the study objectives conform to the overall aim of finding a new drug against psoriasis? These questions are taken for granted and will not be discussed here. A few things can be said about the method used.

3.2. Methodology

The most important question is whether the method is adequate, and in particular, whether there are other (less invasive, less burdensome and less risky) ways of addressing the scientific problem. One could ask whether it would be possible to use research on animals or even *in vitro* studies instead of experimenting on healthy volunteers. This question invokes the principle - which is questionable! - that it is not right to perform scientific research on human beings that could alternatively be done on animals. The researchers convincingly argue that studies like these can only be done with living human subjects because no adequate research model in animals exists. Other ethical questions are: is the randomisation procedure correct, are the numbers of healthy volunteers justified, are the inclusion and exclusion criteria correct, what are the criteria for evaluation of the study protocol?

3.3. Beneficence

The most important (but not the only) motive for regulation and legislation is concern about the welfare of people as research subjects. The first two sentences of the *Declaration of Helsinki* read as follows: "It is the mission of the physician to safeguard the health of the people. His or her knowledge and conscience are dedicated to the fulfilment of this mission." Thus, the welfare of the research subjects comes first, but there is often a tension between the interests of the research subject, and of the interests of the researcher, of science in general, and of society. One of the most difficult questions for research ethics committees to answer is whether there is a reasonable balance between the clinical and scientific value of a particular study, and the burden (risks and inconveniences) it causes to the research subjects. The principle of beneficence means that predictable benefits must be weighed against predictable risks. Risks can be divided into physical, psychological, social, and economic risks. In the study being considered, the ethical discussion will probably mainly concentrate on physical risks to the volunteers, although physical risks cannot be completely separated from other forms of risk. An ethical review of the above protocol must focus on a weighing process at an *inter*-individual level, because the healthy volunteers themselves will not benefit from the results of the study. The burden for eighteen volunteers must be weighed against the possible benefits for others. However, while it is relatively easy to assess the burden for the volunteers here and now, it is far more difficult to weigh the possible benefits for psoriasis patients in the long run.

3.4. Non-maleficence: Integrity of the Human Body

Apart from the requirement of the volunteers' informed consent (see below) one can ask whether it is morally permissible to mutilate the healthy, intact human body by inducing an inflammation and by taking five biopsies. Do these violations of the volunteers' human body outweigh the possible benefit to others? If this is a justifiable assault on the integrity of the human body, is the degree of inflammation induced, the size of the scar after a biopsy and the exact place on the body morally relevant factors? Scars on the shoulder are more visible than, for example, on the inside of the upper leg. One could also ask whether it is morally appropriate to relate the extent of the injury inflicted to the scientific value of the research. Is it, for example, right to give less weight to the integrity of the human body, if the protocol being considered is excellent and promising? Conversely, are the injuries caused by the application of B and by the biopsies more reprehensible, if the research protocol has little scientific value?

Questions like these presuppose that the human body is not just an instrument of the person and that the 'owner' of the body has no absolute right to misuse it. The principle of integrity of the human body implies that although the person has given explicit consent to a particular manipulation of their body, one must still ask whether that manipulation is morally acceptable or not. The human body is not just an instrument, but has its own moral value (Dekkers, 2000). The integrity of the human body is an excellent - *prima facie* - ethical principle that may counter-balance other current principles. No ethical principle, including the principle of respect for autonomy, is considered to outweigh other moral principles, duties or goals in either medical care or research (Faden and Beauchamp, 1986, p. 19).

3.5. Respect for Persons

Despite the importance of the notion of integrity of the human body, respect for persons remains a crucial principle. This principle incorporates two further principles: autonomy and protection of those with impaired or diminished autonomy. The Nuremberg Code states that the first requirement for 'ethical' research is the autonomous, voluntary, informed consent of the research subjects. Informed consent is often considered the most important ethical principle in medicine and health care. In order to be recognized as valid, the consent of the subject must have four essential attributes. It must be competent (legally), voluntary, informed and comprehending. Through informed consent, the investigator and the subject enter into a contractual relationship which requires communication. According to Levine (1986, p. 99), the most appropriate single word for the communication between researcher and research subject is 'negotiation.' The negotiations must include four different components: (1) exchange of information, (2) assessment of the prospective consentor's comprehension, (3) assessment of the prospective consentor's autonomy and (4) consent.

In the study being considered the research subjects are healthy adult volunteers, and it is unlikely that there will be problems in the assessment of their comprehension of the study, in the assessment of their autonomy, or in registering their consent. A potential problem does arise when recruiting the volunteers - how can we be sure that they feel absolutely free to choose whether or not to join the project? The researchers planned to recruit the eighteen volunteers via announcements on notice boards for students. Respect for persons includes a reasonable compensation for research-induced injuries. In the study at hand the compensation is 250 Dutch guilders. Is this a reasonable amount of money? What would be too little? What would be so much that it would influence the free choice of potential volunteers who were short of money?

A fully informed consent may be ideal, but what does this mean in this situation? Do the volunteers have to know all the scientific details of the study? The biopsies seldom cause visible scars, but the possibility of keloid formation cannot be excluded. What do the volunteers have to know about this possible complication? Would it be appropriate to provide them with pictures of the scars which they might acquire?

3.6. Justice

The principle of justice includes the rule of distributive justice which requires a fair and equitable sharing of both burdens and benefits. In general, the research subjects should be representative of the class of persons the research is designed to benefit - by developing knowledge - so that the class of persons that carry the burden should receive an appropriate benefit, and the class primarily intended to benefit shares a fair proportion of the risks and burdens. This general rule is difficult to satisfy in studies with healthy volunteers. Although the eighteen volunteers carry the burden of the research, they will never profit from it, excluding the unlikely case that they later develop psoriasis and benefit from a new drug against psoriasis that had been developed as a result of this particular study. But we can apply the rule of distributive justice to bring in further considerations.

In the case description it was mentioned that neither psoriatic lesions of psoriasis patients nor unaffected parts of their skin can be used because their skin contains 'natural' substance B. Seen from the perspective of a fair distribution of burdens and benefits, however, one could argue that it would be better to use patients with another chronic skin disease (for example, eczema) than healthy volunteers, assuming that the unaffected parts of the skin of patients with chronic eczema do not contain substance B. If patients with chronic eczema were to be used as research subjects, the burden of the research would be carried by a *category* of patients that possibly benefits from the research, namely the group of patients with a chronic skin disease. However, although this may seem a fair solution from an ethical point of view, many new problems arise. It is extremely complicated to assess the burden-benefit-ratio of a research study, if we take into account not only the burdens and benefits of individual volunteers and patients, but also of categories of people (patients with psoriasis or with another chronic skin disease) and of society as a whole.

3.7. Solidarity

Against the argument that the burden of experimental research must be shared by those who benefit from it, one could argue that patients with psoriasis (or with another chronic skin disease) already carry enough burden in suffering from that particular skin disease. Thus it is healthy people who have an obligation to participate in biomedical research. Based on principles such as solidarity and altruism healthy volunteers have to take part in the burden of psoriasis (and other chronic skin diseases) and contribute to the development of new drugs against psoriasis. As Caplan (1988) argues on the basis of the principle of 'fair play,' if one expects benefits from modern research, one is obliged to offer to participate in that research. It is morally reprehensible for the rest of society to avoid this obligation, and thus to take a 'free ride.' Protocols like the example should provoke ethical debate on the notions of altruism and solidarity.

4. CONCLUSION

In the past four decades a number of ethical guidelines for REC's have been developed at national and international levels. These guidelines have been based on current ethical principles such as respect for persons, beneficence, and justice, and establish an internationally acceptable standard for the review and conduct of medical research. The relevant ethical principles constitute a coherent model that is useful for discussing ethical aspects of experimental research on human beings. However, as this case analysis shows, no single general ethical principle - no matter how important - can be applied in isolation to a real case. In an ethical review of any particular research protocol the first task is to assess which ethical principles and values are at stake. Second, ethical principles must be weighed against each other, for example, respect for autonomy versus integrity of the human body. Third, ethical principles are abstract notions which must be interpreted in concrete cases and situations. The above case analysis raises, for example, the question what 'solidarity' exactly means. As a subdiscipline of bioethics, research ethics should primarily be regarded as an interpretive discipline (Ten Have, 2000).

Thus, while we have attained an international consensus on fundamental guidelines for research on human beings, every REC is left to interpret the guidelines and devise their own means of implementing them. Moreover, the international guidelines leave many unanswered questions about the ethical review of research, such as to whom are REC's accountable, how many members should be enlisted to an REC, how should the members be selected and trained, what sort of consent forms should be used etc. (Macpherson, 1999). And the RECs have to reckon with national traditional values, local regulations and institutional care policies as well as the internationally accepted standards.

REFERENCES

Ackerman, T.F.: 1994, Medical research, society and health care ethics. In: Gillon, R. (ed.), *Principles of healthcare ethics.* John Wiley & Sons, Chichester etc., p. 873-884.

Agt, F.M. van, Dekkers, W.J.M. and Riel, P.L.C.M. van: 1999, Medisch-wetenschappelijk onderzoek met mensen: waarin de wet niet voorziet. *Nederlands Tijdschrift voor Geneeskunde* 143(1): 45-50.

Bankowski, Z.: 1993, International ethical considerations for research on human subjects. In: Cheney, D. (ed.), *Ethical issues in research.* University Publishing Group, Inc., Frederick, Maryland, p. 177-188.

Beecher, H.K.: 1966, Ethics and clinical research. *The New England Journal of Medicine* 274(24): 1354-1360.

Caplan, A.L.: 1988, Is there an obligation to participate in biomedical research? In: Spicker, S.F., *et al.* (eds.), *The use of human beings in research, with special reference to clinical trials,* Kluwer Academic Publishers, Dordrecht/Boston/London, p. 229-248

Capron, A.M.: 1989, Human experimentation. In: Veatch, R.M. (ed.), *Medical ethics.* Jones and Bartlett Publishers, Boston, p. 125-172.

Coughlin, S.S. (ed.): 1995, *Ethics in epidemiology and clinical research. Annotated readings.* Epidemiology Resources Inc., Newton, Massachusetts.

Coughlin, S.S. and Beauchamp, T.L. (eds.): 1996, *Ethics and epidemiology.* Oxford University Press, New York and Oxford.

Council for International Organizations of Medical Sciences (CIOMS): 1993, International ethical guidelines for biomedical research involving human subjects. In: Bankowski, Z. and Levine, R.J. (eds.), *Ethics and research of human subjects: international guidelines.* CIOMS, Geneva, p. 231-283.

Council of Europe: 1996, *Convention for the protection of human rights and dignity of the human being with regard to the application of biology and medicine: convention on human rights and biomedicine.* Directorate of Legal Affairs, Strasbourg (Explanatory Report: Strasbourg, 1997).

Cox Macpherson, C.: 1999, Research ethics committees: a regional approach. *Theoretical Medicine and Bioethics* 20: 161-179.

Dekkers, W.J.M.: 2001, The human body. This volume.

Dekkers, W.J.M. and Have, H.A.J.M. ten: 1998, Biomedical research with human body parts. In: Have, H.A.M.J. ten and Welie, J.V.M. (eds.), *Ownership of the human body. Philosophical*

considerations on the use of the human body and its parts in healthcare. Kluwer Academic Publishers, Dordrecht etc., p. 49-63.
Editorial: 1990, Medical ethics: should medicine turn the other cheek? *Lancet* 336: 846-847.
Engelhardt Jr, H.T.: 1999, Healthcare ethics committees: Re-examining their social and moral functions, *HEC Forum* 11(2): 87-100.
Faden, R.R. and Beauchamp, T.L.: 1986, *A history and theory of informed consent.* Oxford University Press, New York and Oxford.
Fagot-Largeault, A.M.: 1988, Epistemological presuppositions involved in the programs of human research. In: Spicker, S.F., *et al.* (eds.), *The use of human beings in research, with special reference to clinical trials.* Kluwer Academic Publishers, Dordrecht/Boston/London, p. 229-248.
Have, H.A.M.J. ten: 2001, Theoretical models and approaches to medical ethics. This volume.
International conference on harmonisation of technical requirements for registration of pharmaceuticals for human use: 1996, *Good clinical practice.* European Recommendation Concerning Good Clinical Practice for Drug Research in the European Community.
Katz, J.: 1972, *Experimentation with human beings.* Russell Sage Foundation, New York.
Levine, R.J.: 1986, *Ethics and regulation of clinical research* (second edition). Yale University Press, New Haven and London.
Levine, R.J.: 1995, Research ethics committees. In: Reich, W.Th. (ed.), *Encyclopedia of bioethics.* Simon & Schuster Macmillan, New York etc., Vol. 4., p. 2266-2269.
National Commission for the Protection of Human Subjects of Biomedical and Behavioral Research: 1978, *The Belmont report*: Ethical principles and guidelines for the protection of human research subjects. U.S. Government Printing Office, Washington, DC.
Rothman, D.J.: 1990, Human experimentation and the origins of bioethics in the United States. In: Weisz, G. (ed.), *Social science perspectives on medical ethics.* Kluwer Academic Publishers, Dordrecht/Boston/London, p. 185-200.
Rothman, D.J.: 1991, *Strangers at the bedside. A history of how law and bioethics transformed medical decision making,* BasicBooks.
Rothman, D.J.: 1995, Research, human: historical aspects. In: Reich, W.Th. (ed.), *Encyclopedia of bioethics.* Simon & Schuster Macmillan, New York etc., Vol. 4, p. 2248-2256.
Saunders, C.M., Baum, M., Houghton, J.: 1994, Consent, research and the doctor-patient relationship. In: Gillon, R. (ed.), *Principles of health care ethics.* John Wiley & Sons, Chichester etc., p. 457-469.
Spicker, S.F., *et al.* (eds.): 1988, *The use of human beings in research, with special reference to clinical trials,* Kluwer Academic Publishers, Dordrecht/Boston/London.
World Medical Association: 1996, *Recommendations guiding physicians in biomedical research involving human subjects (Declaration of Helsinki).*

BERT GORDIJN

COMPULSORY MEDICAL TREATMENT

1. INTRODUCTION

One of the major developments in medical ethics after World War II has been the increasing importance of the principle of respect for autonomy. At present, for example, it is widely accepted that informed consent is a necessary condition for starting a morally justified medical treatment of a competent patient, meaning a patient in power of a substantial decision-making ability. According to this view, morally justified medical treatments of competent patients can only be performed after the patient has given his consent (having been informed of all relevant aspects of the proposed treatment in an understandable way). Generally, the right of competent patients to refuse medical treatment has to be respected.

However, there are situations in which good moral reasons can be brought forward to perform a medical treatment, even if the patient does not want to co-operate. Of course, this can be the case if the patient is not competent and medical intervention can avoid harm, either to the patient himself or to others. Moreover, even in the case of competent patients, there are situations in which good moral arguments in terms of avoidance of harm can be developed for overruling the patient's unwillingness or reluctance to be medically treated. Accordingly, four different categories of such situations can be distinguished on a rather abstract level:

1. Not treating the incompetent patient is harmful to the welfare of others.
2. Not treating the competent patient is harmful to the welfare of others.
3. Not treating the incompetent patient is harmful to his or her own welfare.
4. Not treating the competent patient is harmful to his or her own welfare.

Of course, most real cases of compulsory medical treatment will rarely fit into just one of the above categories because, on the one hand, competence does not seem to be an all-or-nothing phenomenon and, on the other hand, harm to the welfare of others often implies harm to the patient's own welfare and vice versa. Nevertheless, the above systematisation can be helpful in ordering some reflections about the moral aspects of exerting coercion in medical treatment. Accordingly, four cases of compulsory treatment - each

H.A.M.J. ten Have and B. Gordijn (eds.), Bioethics in a European Perspective, 179-192.

more or less exemplifying one of the above-mentioned four categories - will be presented and discussed.

2.CASE 5.: SILENCING A PATIENT[1]

Mr. A., an 84-year old married veteran of World War I whose working years were spent as a carpenter, had always been a mild-mannered apolitical man of few words, who did not drink or smoke. But in the past decade his cognitive capacities had continuously declined. At the same time, Mr. A started yelling. In fact, the patient's yelling grew so incessant and loud that it disrupted his stay at home with his wife, but also his stay (for a variety of smaller medical problems) on the medical floor of the hospital, and in the nursing home. He was eventually admitted to an acute-care psychiatric unit where a host of pharmaceutical and behavioral interventions were totally unsuccessful. At one point Mr. A was observed yelling 45 times per minute, non-stop for seven hours.

One intervention into the yelling problem involved the application of bilateral hearing aids connected to a microphone which was placed in front of the patient. Such 'amplified auditory feedback', along with various other creative interventions, were unsuccessful and the patient's yelling continued unabated. Involuntary commitment to a long-term psychiatric hospital was briefly considered, but rejected as inappropriate for this severely demented individual who was now incontinent of bowel and bladder, losing weight, and subject to recurrent pneumonia's. Mr. A's wife stated she could not take him home, and all the nursing homes in the area refused to accept him because of his incessant yelling.

The staff in the psychiatric unit of the hospital were beginning to weary of Mr. A's relentless yelling. Feelings of discouragement, frustration and helplessness grew. After two months of exasperation and thwarted attempts to stem the yelling, an ad hoc *advisory committee consisting of psychiatrists, internal medicine physicians, an ear-nose-and-throat surgeon and other staff, recommended that a single, recurrent laryngeal nerve be crushed. The committee indicated that the surgical procedure could be done under local anaesthesia and would not compromise the patient's ability to clear secretions or to breathe, but would render his maximum voice level to be a very acceptable soft tone.*

Substituted judgement was impossible to reconstruct in the absence of written evidence attesting to Mr. A's wishes in such a situation. In retrospect however, his wife indicated the belief that her husband would have opted for the surgery, as "he was a kind and quiet man who never wanted to bother other people". The committee unanimously endorsed the procedure, as did Mrs. A, who by this time felt guilty, ambivalent and anxious.

2.1. Benefits and Beneficence

The benefits that first and foremost would be served by the surgical intervention on the incompetent yelling patient are clearly those of the other patients being treated and the persons working on the psychiatric unit of the hospital. They are the ones who have to cope with the incessant yelling. As a side effect, however, the yelling man himself would perhaps also benefit from the surgery. Although the operation might not prolong Mr. A's life, the quality of his life might be enhanced. In general, a person who, be it deliberately or not, causes a detrimental effect on others, will usually be harmed by a reaction of negative social feedback. As a quieter individual, the old man's presence would presumably produce a much less negative effect on others. As a consequence, the latter would probably be more inclined to give him positive attention, or at least not to avoid him. With the yelling eliminated, he could perhaps live in a less restrictive setting, possibly even in his own home. Consideration of these benefits probably led the advisory committee to give a positive advice on the surgery. The advisory committee decided according to the so-called principle of beneficence; they felt the moral obligation to act for the benefit of others.

Considerations of beneficence have played a central role in different ethical theories, for example in utilitarian theories. There are different views on what beneficence exactly involves. Frankena (1973, p. 47), for example, holds that beneficence implies four different obligations: (1) One ought not to inflict evil or harm, (2) One ought to prevent evil or harm, (3) One ought to remove evil or harm, (4) One ought to do or promote good. Beauchamp and Childress (1994, p. 190) distinguish between non-maleficence and beneficence, isolating the first above mentioned obligation that one ought not to inflict evil or harm, and categorising it as an obligation of non-maleficence. Be this as it may be, obligations of beneficence or non-maleficence are not absolute. Considerations of benefits being conferred and harms being prevented or removed result in determining *prima facie* obligations to act in some way or another. These *prima facie* obligations, however, can be overruled by other moral arguments.

In the case of the yelling man, therefore, it must be asked whether in view of the arguments against the surgery the above mentioned benefits can really justify the suggested intervention. First of all, since the yelling man seems to be severely demented, it can be doubted whether he himself would really experience any benefits after the operation. The benefits seem to lie mainly, if not exclusively, on the side of the others. Furthermore, it can be argued that, though it is all too understandable that the personnel on the psychiatric ward would rather work in a quiet environment without yelling patients, being able and willing to cope with these kinds of behaviour should be regarded as part of their professional duties and virtues. Finally, one could also refuse to approve of the procedure on the grounds that it involves non-medical surgery which is not without risk. It does not seem morally right to operate on healthy tissue of one person in order to solve a social problem of other persons.[2]

2.2. Competence and Incompetence

Judgements about competence and incompetence are of great importance in health care, because they *prima facie* decide on whose decisions should be respected and whose not. An incompetent person's decisions cannot in principle be autonomous. Consequently, these kind of decisions need not be respected from considerations of respect for autonomy. The decisions of a competent person, on the other hand, need not necessarily be autonomous decisions. For example, they can be based on false information. Therefore, a competent person's decision ought to be respected from considerations of respect for autonomy, only if they meet additional criteria making them also autonomous decisions.[3]

The severely demented old man seems to be a clear-cut case of an incompetent patient. In medical practice, however, it is often not easy to determine whether a patient is competent or not. Therefore, it must be asked on a more general level under what conditions a patient can be regarded as incompetent. There seems to be no single accepted definition of incompetence. In general, a competent person can be regarded as having a sound decision-making capacity, whereas in an incompetent person this capacity has been seriously distorted. However, opinions differ with regard to the degree or level of impairment of decision-making capacity being necessary when categorising somebody as being incompetent. Consequently, there are various competing standards of incompetence that all require different inabilities as a necessary condition for incompetence. In this list (Beauchamp and Childress, 1994, p. 137), the standards for incompetence are ranked from the one requiring the biggest inability to the last requiring the least inability as a condition for incompetence:

1. Inability to express or communicate a preference or choice
2. Inability to understand one's situation and its consequences
3. Inability to understand relevant information
4. Inability to give a reason
5. Inability to give a rational reason (although some supporting reasons may be given)
6. Inability to give risk/benefit related reasons (although some rational supporting reasons may be given)
7. Inability to reach a reasonable decision (as judged, for example, by a reasonable person standard)

The result of choosing for the standard last mentioned would lead to a very broad concept of incompetence covering a rather large group of persons. Following this standard, most persons would be incompetent with regard to a great amount of decisions they make in every day life. If, on the other hand, the first standard in the list would be held as a measure for incompetence, the resulting concept of incompetence would be a rather narrow one that could be applied only to a relatively small set of persons. Following this standard, the demented yelling man, for example, would probably have to be categorised as being competent because he is presumably able to express some of his preferences or choices, be it in a very primitive way. Already if the second standard in the list would be applied to determine the competence of the demented patient, he would turn out to be incompetent. Since choosing a standard either at the beginning or the end of the above list seems to lead to contra-intuitive results, some standard in between would be best to select.

Evidently, some forms of incompetence rest on reversible causes, e.g., certain forms of depression, pain or overmedication, whereas in other cases the disabilities in decision-making cannot be restored. In some cases competence can vary from hour to hour. Moreover, the quality of one's decision-making capacity seems to be relative to the particular decision to be made. Competence, therefore, seems to be a capacity that can be intermittent and specific.

2.3. Coercion

Another relevant question with regard to the issue of 'compulsory treatment' brought up by the unusual case of the severely demented, incessantly yelling old man is what kind of coercion would be exerted if the team would proceed to surgical intervention? It seems safe to suppose that the whole procedure of crushing a single, recurrent laryngeal nerve under local anaesthesia would involve different kinds of discomfort. It is also likely that the severely demented old man would not be able to understand the reasons behind the uneasiness he is confronted with. Therefore, he probably would be scared and accordingly unwilling to co-operate. Since trying to enhance his willingness to co-operate by presenting to him the reasons and goals of the surgical intervention would most certainly not be successful, the only effective form of coercion would be physical force.

This brings up the question of what are the different forms of coercion? What actually counts as compulsory treatment? First of all, it is necessary to distinguish between varieties of coercion on the one hand and more and less dramatic medical interventions on the other hand. The kinds of coercion exerted with regard to more dramatic interventions, e.g., sterilisation of mentally handicapped, forced feeding of people involved in an hunger strike, blood transfusions of Jehovah Witnesses, life-saving measures after attempted suicide, need not necessarily be stronger than those with regard to interventions of a less dramatic nature, e.g., restraining a patient to keep him from falling out of bed, forcing an older, depressed patient to get out of bed in the morning, reinserting a feeding tube that a patient keeps on pulling out, getting a patient to undergo one more session of chemotherapy.

Generally, three different forms of coercion can be distinguished: compulsion, pressure and persuasion (Gezondheidsraad, 1996). *Compulsion* is the strongest form of coercion. It leaves the patient no alternative but to comply with the proposed intervention. For example, it can be regarded as compulsion when police force is used to isolate a person with an open tuberculosis. In comparison with compulsion, putting pressure on somebody is a weaker form of coercion. *Pressure* means that the person the pressure is being put on is presented an alternative, in case she is not willing to co-operate. This alternative, however, is a highly unattractive one. Pressure can be exerted in two ways. On the one hand, it can be promised that, only if one complies with the proposed intervention, something of a highly advantageous nature will happen or be given (positive pressure). For example, when a drug addict is being told that he will only get his substitute drug under the condition that he will participate in a screening programme for certain infectious diseases, positive pressure is being put on him. On the other hand, it can be said that something very negative will happen as a consequence of not co-operating, e.g., when a political refugee will be deported unless he co-operates with a certain medical intervention (negative pressure). Finally, the weakest form of coercion is *persuasion*. It neither implies physical violence nor limiting the possibilities of choice. Nevertheless, when a health care professional exploiting all the powers of his rhetorical talent tries to convince a layman to undergo a certain medical treatment, persuasion can be rather coercive.

3. CASE: THE ENDANGERED FOETUS[4]

Mrs. Z. is expecting her first child. After a pregnancy of almost 25 weeks she starts to get contractions, which would force her to go into premature labour. When she is admitted in the hospital she already has a dilatation of the cervix of 2 cm. The chances for the baby are not very good if she were to give birth at this stage of her pregnancy. The majority of such babies die shortly after birth. Half of the babies surviving the premature birth will be multiple handicapped, despite intensive treatment.

In consultation with the patient it is decided to start with medication (via a drip) which will restrain the contractions. The therapy turns out to be successful and the contractions can be fully suppressed; however, Mrs Z. suffers a lot from the side effects. She complains notably of a high pulse rate and heart palpitations. She also has trouble with the prescribed bed rest.

After a few days it turns out that every now and then the patient loses little bits of amniotic fluid; apparently there is a small malfunction in the amniotic and chorionic membranes. Because of this there is a rising chance of an intrauterine infection. Also the possibility of long-term postponement of the birth will be restricted. This is obviously a serious blow for the patient and the team treating her.

Because in this stage the chances of survival for the baby are increasing considerably with every week, it is decided that it is in the child's best interests to continue the medication which restrains the contractions. Through 'sterile nursing' the team tries to limit the chance of an infection. However, it turns out that Mrs. Z. has become so discouraged through this latest setback and the discomfort caused by the therapy that she does not want any further treatment. She insists on the drip being removed and wants to leave the hospital.

This is against the medical advice. Should she leave the hospital, the chances of survival for the baby will decrease and if it should stay alive, the chances of a handicap will increase. It can also be damaging to her own health. Her husband asks the physician to do everything he can to convince his wife to stay in the hospital and if necessary to continue her treatment under duress.

3.1. Respect for Autonomy

The pregnant woman in the case insists on the drip being removed and wants to leave the hospital. She obviously does not want any further treatment and under normal circumstances a wish like this being put forward by an autonomous person should be respected. However, in this case there are two factors that make one hesitant in resolving the matter. First, it could be doubted whether the decision the woman has taken is indeed an autonomous one. The decision has a strong irrational aspect and the woman seems at least to be strongly influenced by emotions of discouragement through the latest setback and all the discomfort caused by the therapy. Second, even if the decision could be regarded as an autonomous one, there seem to be other persons who would probably be severely damaged by the consequences of the decision. If, on closer consideration, abstaining from any further treatment and leaving the hospital would indeed present itself as having a seriously detrimental effect on other persons, should the autonomous decision of the woman still be respected? As the concept of autonomy plays a central role in these two questions, resolving the case above has to start with a closer investigation of this pivotal concept.

The concept of autonomy dominated the bioethical debate, especially Anglo-Saxon discussions, since its beginnings in the sixties. This also had important repercussions for biomedical research and health care practice. The change of attitudes with regard to informed consent, informed refusal and truth-telling that we have witnessed in the last decades all have a conceptual background in which autonomy plays a major role.

There are different theories with regard to autonomy. Many theories, however, seem to agree that three conditions are essential to autonomy: first, some kind of independence from controlling influences, second, the capacity for intentional action, and finally, true information and understanding. Most theories, however, do not concur in their answers to the question of what kind of independence from what sort of controlling influences (e.g., influences from outside or also certain internal irrational influences), what kind of capacity for intentional action and how much information and understanding is needed to be autonomous. Opinions differ also with regard to the question whether, and if yes, what additional conditions need to be met. It seems to be for sure, though, that since independency from controlling influences as well as understanding and information obviously are properties that admit of different degrees, autonomy is not an all-or-nothing phenomenon.

Connected with the problem of the defining properties of autonomy there is the question of what respect for autonomy involves. Again, with regard to this question different theories exist. Already, Immanuel Kant (1724-1804) and John Stuart Mill (1806-1873), the two founding fathers of contemporary thinking about autonomy, held different views on the matter. According to Kant, respect for autonomy requires that autonomous persons (being persons intentionally following the moral law) ought never to be treated merely as means. Since autonomous persons are ends in themselves each having the capacity to determine their own destiny, respect for autonomy implies that it is morally forbidden to instrumentalize autonomous persons to serve exclusively the goals of other persons.

Mill regarded respect for autonomy as implying not to interfere with the choices and actions of autonomous persons, as long as they in their turn do not conflict with the freedom of other autonomous persons. Sometimes, however, when other persons act on the basis of erroneous beliefs we ought to interfere by way of persuasion. Mill's point of view exemplifies that there are two different duties that both seem to be somehow implied by respect for autonomy. First, there is the negative duty of not interfering with the decisions and actions based on the considerations, beliefs and values of an autonomous person. The liberty and freedom to act and think of autonomous persons ought not to be restricted arbitrarily. Second, respect for autonomy implies the positive obligation to interfere when possible in order to enhance or restore the autonomy of persons. For example, when somebody wants to commit suicide because he falsely believes that his family was killed in a car-accident, respect for autonomy demands giving him true information concerning his family thereby enabling him to reconsider his decision in the light of a better and more true understanding of his situation.

Returning to our case it can be doubted whether the woman's desire to stop the treatment has indeed resulted in a fully or even only substantially autonomous decision. Though she does seem independent from controlling influences from outside her behaviour appears to be strongly controlled by internal irrational influences, namely, strong emotions of discouragement. It looks like her mood is fairly regressive and self-destructive. From the case description, it is not clear to what extent the woman is able to understand her situation and its consequences. Has she really understood the relevant information concerning her decision? Is she able to give a rational reason for her behaviour? Questions like these would have to be considered more closely to determine her autonomy.

Supposing, however, that her decision were substantially autonomous, would in view of the detrimental effect on the welfare of others her desire have to be respected? Stopping treatment and leaving the hospital would decrease the chances of survival for the baby and if it should stay alive, the chances of a handicap would be increased. Moreover, it can also be damaging to the woman's own health. Finally, through this all the husband's welfare could be seriously affected. Should, for example, the foetus die or be gravely handicapped as a result of abstaining from treatment, the woman's husband would have to cope with feelings of mourning and perhaps self-blame. That is probably why he asks the physician to do everything he can to convince his wife to stay in the hospital and if necessary to continue her treatment under duress. In view of the possible harmful effects on the well-being of others it seems reasonable that the physician tries to convince the woman to rethink her decision. Therefore, persuasion seems to be an appropriate instrument of coercion in this case.

The question of whether more coercive measures like pressure and compulsion would also be morally justified to force the woman to continue the treatment is not easy to answer. Again supposing that her decision were substantially autonomous, it does not seem right to use physical force to compel the woman to refrain from abstaining medical treatment and to stay in the hospital. Moreover, it must be doubted whether such measures could be put into practice at all.

4. CASE: DEMENTED MAN WITH A SWOLLEN BLADDER[5]

Night-time has fallen on the internal medicine ward of a general hospital. There is snoring to be heard on the ward, a troubled sleeper turns around noisily. In a corner lies a demented old man, he moans and is agitated. He looks at the night nurse with big eyes full of fear. When the doctor tries to examine him he keeps him away, saying: "No, no ...", though it seems that he does not understand what is happening. The doctor discovers that his bladder has swollen enormously. It appears the old man cannot urinate anymore. This can be caused by lying in bed or by the disease he suffers from. The doctor catheterises him and the urine is drained off. Half an hour later the old man is quietly asleep.

4.1. Weak Paternalism

The case of the demented old man suffering from a swollen bladder describes a common situation in clinical practice. In this case, the medical intervention is clearly to the benefit of the patient. Inserting the catheter is a very effective method of alleviating the distressing symptoms caused by the swollen bladder. In doing this, however, the physician is obviously overruling the resistance of the old man. This behaviour is called medical paternalism.

In general, paternalism is the intentional interference with or refusal to conform to another person's preferences, desires or actions with regard to his or her own good to avoid harm or to benefit this person. It involves the overriding of the wishes and desires of one person by another person being justified by the goal of preventing harm or doing well to the person whose will is being overruled. The paradigmatic model of paternalism is the way parents take care after their children. Usually, parents take almost all the important and consequential decisions with regard to the welfare of their children and try to do so in accordance with what they regard as their interests.

The principles of beneficence and non-maleficence are being referred to as providing a basis for the justification of paternalistic interventions. Evidently, if the person whose will is overridden is substantially non-autonomous, justifying paternalism is easier than it would be if the person's preferences and decisions would be substantially autonomous. Therefore, the most widely accepted form of paternalism is the intervention in the decisions or actions of a substantially non-autonomous person to serve his or her own benefits. This latter form of paternalism is also called *weak paternalism*, whereas intervention with a person's substantially autonomous preferences and decisions is called *strong paternalism*.[6]

Clearly, the case of the demented old man who does not understand what is happening, exemplifies an instance of weak paternalism. The treatment of the old man involves coercion in the form of compulsion. The whole procedure evidently implies a harsh intervention in his desires: his unwillingness is being overruled by physical force. However, the resistance of the old man is the product of not being able to understand and rationally decide on his situation. The demented old man is obviously a substantially non-autonomous person. Furthermore, the benefits for him and the harms prevented by the intervention outweigh the sense of invasion caused by the intervention. Finally, it could also be argued that the old man himself, if rational, probably would consent to the medical intervention. Therefore, the compulsory medical treatment in this case can be regarded as morally justified.

5. CASE: A SUICIDAL PATIENT[7]

Mr. B. is 28 years old and has been admitted to a general psychiatric hospital for the third time. He has agreed to his admittance voluntarily because he could not manage at home anymore. He had become increasingly frightened over the last few weeks and he had been hearing 'voices', addressing him in a condescending way. Moreover, he had been preoccupied with ideas which related to his parents, in which cases of mistaken identity as well as sexual and sadistic motives played an important role. From time to time he was swamped by these ideas and panicked, but at other times he managed to distance himself somewhat from them.

The anamnesis shows that Mr. B. became psychotic for the first time when he was 19 years old. After six months in a psychiatric hospital he had managed to get rid of his psychotic experiences, but he could not function anymore at his previous level. He had become slow and inactive, he avoided approaching other people, he suffered from lack of concentration and complained of 'muddled' thinking. As a result of this he failed his studies, his relationships disintegrated and he became increasingly socially isolated. There were times when he was deeply depressed and fully realised the hopeless situation he was in. About five years later he was admitted suffering from a psychosis relapse, and was again mentally very upset. A treatment with medication helped in removing the most alarming symptoms and experiences.

The patient now seems to come to his senses. During a conversation he still expresses some psychotic experiences, from which it is clear that he still lives, in many ways, outside 'the' real world. At the same time however, he expresses in a very understandable and realistic way his despair about his hopeless situation. He seems fully to understand the different aspects of his situation and says he has thoroughly reflected on his predicament. There are no realistic options left for him to cope with his condition. Therefore, he tells the psychiatrist that he intends to commit suicide.

The psychiatrist now faces a very difficult decision: he takes the patient's suicidal ideas very seriously and subscribes to the view that the patient presents of his situation. He is then faced with the decision to start the process of compulsory admission or compulsory treatment, at the cost of a damaged trust and with the vague but not very well-founded hope that he can avert the calamity. After considering this the psychiatrist decides not to opt for compulsory admission. He makes an appointment with his client about frequent ambulant contacts and makes himself available for this, but he cannot prevent the patient from committing suicide two days later.

5.1. Strong Paternalism

The case above is a paramount example of balancing respect to the patient's autonomy and the protection of his bodily and mental integrity, versus the avoidance of harm or his long-term well-being. Of course, in the case above the psychiatrist, after having reflected on the different options, decided to abstain from compulsory admission. Was he morally right in doing so?

It has already been pointed out that, in general, the quality of one's decision-making capacity seems to be relative to the particular decision to be made and that, furthermore, there are cases in which competence can vary from hour to hour. Competence, therefore, has to be regarded as a capacity that can be intermittent and specific. This certainly seems to be the case here. The young patient seems to be competent while speaking to his psychiatrist with respect to the decision to commit suicide. Moreover, he also seems to be substantially autonomous with regard to his decision. He does neither seem to be controlled by outside influences, nor are irrational inner coercive factors, e.g., strong feelings of depression, forcing him to his resolution. Moreover, his decision is based on true information concerning his condition. Of course, more information would be needed to judge on the degree of autonomy his decision shows. Supposing, however, that the decision as presented to his psychiatrist is indeed a substantial autonomous one, could then compulsory admission or compulsory treatment be morally justified?

Overruling a substantially autonomous decision to the benefit of someone is a form of strong paternalism. In general, the moral problem with strong paternalistic acts is that they disrespect the autonomous agent being overruled. Strong paternalism displays a failure to treat an other autonomous agent as a moral equal. It can only be justified, if at all, under very special circumstances. Justifying a strong paternalistic act, for example, is easier if the beneficiary is at risk of serious and preventable harm. Furthermore, it must be relatively sure that the paternalistic intervention will indeed prevent the harm. Finally, compared with the conceived benefits the restriction of autonomy of the patient has to be insignificant.

In the case above it appears that none of these conditions is given. First, it can be doubted whether suicide would be a serious harm for the patient as he himself regards it as being the solution to his problems. Here, the more general question is: can the quality of life decrease to such a measure that not living no longer involves a harm but becomes a benefit? Also, the suicide does not appear to be preventable because the motives do not rest on reversible causes. Second, it is absolutely not sure and not even probable that the paternalistic intervention will indeed prevent the harm. Third, compared with the doubtful benefits the restriction of autonomy of the patient is not insignificant at all. Compulsory admission and compulsory treatment would probably entail physical violence and would as such display a lack of respect of autonomy with regard to a very important decision. Therefore, it would not be right to use compulsory admission and compulsory treatment. So it seems that the psychiatrist refraining from compulsion acted in a morally justified way.

6. CONCLUSION

A few general remarks with regard to the ethics of compulsory medical treatment can be put forward. In general, compulsory medical treatment of substantially autonomous patients is more in need of moral justification than coercion with regard to competent persons who are substantially non-autonomous (for example through lack of true information). Again, coercion with regard to the latter is easier to justify than constraining incompetent patients for their own benefit. Likewise, the need for justifying coercion grows inversely proportional with the detriment caused by the patient's refusal or unwillingness. Finally, stronger and more dramatic forms of coercion in treatment and care are from a moral point of view more problematic than coercive measures of a less dramatic nature.

NOTES

1. This case has been adapted from a case published by Gafner (1987).
2. In Gafner's case the Surgery Department later refused the procedure on similar grounds (Gafner, 1987, p. 196).
3. See section 3.1. for further discussion.
4. This case has been described by Essed (1992, p. 31-32).
5. This case has been adapted from a case published by Janssen, de Jonge and Pols (1995, p. 19-20).
6. This distinction was introduced by Joel Feinberg (1971).
7. This case has been adapted from a case published by Kuilman (1990, p. 18-19).

REFERENCES

Beauchamp, T.L. and Childress, J.F.: 1994, *Principles of biomedical ethics*, (4th ed.), Oxford University Press, New York and Oxford.

Essed, G.G.M.: 1992, Obstetrie. Vakinhoudelijke aspecten. In: Have, H.A.M.J. ten, *et al.* (eds.), *Ethiek en recht in de gezondheidszorg*. Kluwer, Deventer, XI 31-41.

Kuilman, M.: 1990, Over dwangmaatregelen in de psychiatrie. In: Have, H.A.M.J. ten, *et al.* (eds.), *Ethiek en recht in de gezondheidszorg*. Kluwer, Deventer, XXI 11-25.

Feinberg, J.: 1971, Legal paternalism. *Canadian Journal of Philosophy* 1: 105-124.

Frankena, W.K.: 1973, *Ethics* (2nd ed.). Prentice-Hall, Englewood Cliffs, New Jersey.

Gezondheidsraad, Commissie Ethische en juridische aspecten van TBC-bestrijding: 1996, *Dwang en drang in de tuberculosebestrijding*. Den Haag, nr. 1996/07.

Gafner, G.: 1987, Surgery to quieten the yelling of a demented old man. *Journal of Medical Ethics* 13: 195-197.

Janssen, M., de Jonge, M. and Pols, J.: 1995, *Dwang in de psychiatrie. Dilemma's, feiten, ervaringen en alternatieven*. Babylon de Geus, Amsterdam.

PART 3

PERSON AND COMMUNITY

HENK A.M.J. TEN HAVE

PERSON AND COMMUNITY

Introductory Comments

In 1516, Thomas More published his philosophical novel describing an ideal state on the island of Utopia. The book has two parts, one criticizing the unjust contemporary social and economic conditions, the other outlining an idealized society. Precisely the contrast between the two parts is interesting: the injustices of his own world in comparison to the glories of egalitarian Utopia; on the one hand a situation of commercial exploitation where greed for gain and accumulation of wealth in the hands of a few leads to indigence, oppression and malady; on the other hand the ideal state where private property is abolished, the means of livelihood are assured to all, and all citizens have perfect health.

Justice and injustice are recurrent themes in the history of ethics. More than two centuries later David Hume analyzed the origins of justice in *A Treatise of Human Nature* (1739). He argues that justice takes its rise from human conventions. These conventions are intended as a remedy to some inconveniences, resulting from two sources: certain qualities of the human mind, such as selfishness and limited generosity, and outward circumstances, such as the easy change of external objects and their scarcity in comparison to the wants and desires of men. These circumstances of justice are the pecularities of the human conditions which at the same time facilitate and endanger human cooperation. Human nature, as Hume points out, can not subsist without the association of individuals; that association could never take place were no regard paid to the laws of equity and justice.

One of the central topics in bioethics is the just organisation of human life in general, and the fair distribution of goods in particular. According to several bioethicists it is precisely this topic that brought modern bioethics into being. With an increasing supply of effective technologies, the major question became: Who shall be saved when all cannot be saved? The case of dialysis for end-stage renal patients and resuscitation technology for patients with respiratory failure made the headlines in the early 1960s. However,

H.A.M.J. ten Have and B. Gordijn (eds.), Bioethics in a European Perspective, 195-197.

1980s, when rising health care costs necessitated allocation of scarce resources.

In 1912 the French sociologist Emile Durkheim proposed the concept of *homo duplex*. In a human being there are two beings: an individual being which has its foundation in the organism, and a social being representing the highest reality in the intellectual and moral order, *viz.* society. Man is a member of nature by virtue of being a bodily organism as well as a member of society by virtue of culture. This double aspect of the human being is the theme of the third part of this book.

In his chapter, Eugenijus Gefenas from Lithuania explicitly deals with issues of social justice in health care. Particularly for the countries in Central and Eastern Europe the basic question nowadays is which fundamental principles should guide the structure and organisation of the health care system. The precarious economic condition of these countries dictates to reform the health care system which until recently has been primarily based on the principles of equality and solidarity. However, Gefenas shows that the present-day necessity to fairly distribute resources is not merely the consequence of a weak economic situation. Scarcity is a relative phenomenon. As Hume already mentioned, scarcity is related to the wants and desires of human beings. Economic needs and societal medicalisation however, have both the same effect: they imperil traditional equality-based solidarity. Having examined various theories of justice, Gefenas finally analyzes the two-tier model of health care, which is considered by many to be the best translation of the idea of distributive justice in the present circumstances.

In bioethics, the issue of justice is under debate in three different contexts. First, it is at stake in health policy, regarding the questions of health care reform and health care system, as discussed by Gefenas. Second, in every health care system, just distribution is an issue if resources are scarce. In European countries, its most visible sign is the existence of waiting lists for a range of health services. Third, justice is problematic on the micro-level of doctor-patient interaction when health professionals have to select among candidates for treatment.

The second context is analyzed in the chapter of ten Have. He argues that the waiting list as a rationing instrument is morally problematic, although it is organized on the basis of ethical selective principles. A different approach to the allocation problem is needed. This approach should take seriously Durkheim's double nature of men, and especially the social dimension of human beings. As long as the focus is on individual needs and wants, it will be impossible to adequately resolve the issue of fair allocation of resources. The focus should shift towards the question what, from the perspective of the community of individuals, is necessary to provide. A more communitarian approach is proposed in order to safeguard the basic notions of equal accessibility and solidarity in health care. A priority setting process focused on patient categories and community needs rather than individual patients and individual needs is necessary to determine what are essential services that must be provided without restrictions to all citizens.

A striking phenomenon in many countries is the establishment of health care ethics committees. These committees operate as a platform to mediate between individual and community concerns in health care. Christopher Melley from the U.S.A., working in Germany, addresses the relevant aspects of this phenomenon. He argues that ethics committees are the preferred mechanism of resolving moral conflicts in health care. He carefully analyzes the strengths and weaknesses of ethics committees.

Practical consequences from the above chapters are elaborated in the case analyses. Bela Blasszauer from Hungary illustrates how in his country basic human rights are not easily recognized. Justice and community considerations can only play a significant role in health care if the conditions for participatory democracy are not only created but also working in practice. The second case is located on the above-mentioned microlevel of doctor-patient interaction. Ten Have reviews various criteria for patient selection. He examines in more detail the now popular criterion of personal responsibility for health. In the third case analysis, Hub Zwart from the Netherlands focuses on mechanisms of consensus formation in ethics committees. He portrays these mechanisms as a practice of moral experimentation within today's health care.

EUGENIJUS GEFENAS

SOCIAL JUSTICE AND SOLIDARITY

1. INTRODUCTION

It would not be an exaggeration to say that social justice and just distribution of health care resources is probably the most fundamental issue in bioethics. The perspectives on social justice shape the model of the health care system, which a particular country aims to establish. These perspectives also determine, for example, the line between health care measures freely available for the whole population and those which have to be purchased privately. The kind of health care resources distribution also provides the context for the interpretation of different ethical dilemmas, such as the tension between autonomy and paternalism or providing and withdrawing extraordinary health care interventions. What is regarded as extraordinary treatment in Lithuania could be an ordinary procedure in the Netherlands, simply due to a huge difference in the health care resources available.

The issue of social justice in the field of health care is very important in all European regions. However, the socio-economic and cultural differences make the discussion on social justice in different parts of Europe to some extent incommensurable. For example, within the Central and Eastern European (CEE) countries, the discussions on the topic of 'justice' quite often concentrate on such issues as bribery, medical malpractice or negligence. Western societies, however, have already established legal and procedural rules to deal with these topics and therefore do not conceptualise them as 'ethical' ones.

At the same time, there is a solid basis for the common denominator in European social justice debates. Despite their cultural and socio-economic differences most European countries still base their health care policies on the principles of equality and solidarity. These principles dominate both international documents as well as academic discussions and papers dealing with the question of just distribution of health care resources. The so-called Bioethics Convention, which was recently signed by the majority of the Council of Europe countries and which in a few years time will probably be legally binding in Europe, explicitly refers to "equitable access to health care".

H.A.M.J. ten Have and B. Gordijn (eds.), Bioethics in a European Perspective, 199-218.

The Convention states, however, that "equitable means first and foremost the absence of unjustified discrimination" and is "not synonymous with absolute equality" but "implies effectively obtaining a satisfactory degree of care" (Explanatory report, 1997, p. 9). The meaning of *equitable access* as opposed to an *adequate* or *a satisfactory degree of care* is however rather ambiguous and, as we will try to prove, related to fundamental practical and theoretical controversies.

Firstly, these controversies arise because of the tendency to escape explicit procedures of limiting the access to some types of health care services. Secondly, even if it is accepted that it will be necessary to set limits to approaches based on equality and solidarity, there is still no easy way to reconcile and balance the different paradigms of distributive justice. Therefore, the main core of this chapter is to analyze the *limits* of the equality and solidarity based perspective on social justice in the European context, even if these ideals remain to be the leading ones in health policy issues (ten Have, 1998).

2. THE FACES OF HEALTH CARE CRISIS AND REFORM

The prevalent health policy in Western as well as Central and Eastern European countries has been based on the principle of solidarity and the right to equal access to health care in the sense that everybody is entitled to every health care intervention available for others. However, this ambitious or 'absolute' equality based paradigm appears to be unrealistic and even morally controversial in the context of the transformation of modern medicine.

Several interrelated factors, which have marked the transformation of modern medicine, should be mentioned in this respect. Firstly, the pattern of morbidity in modern societies has shifted from acute life-threatening conditions to chronic diseases prevalent in the ageing population. Secondly, modern medicine has been marked by the proliferation of new and often very expensive biomedical technologies to cope with chronic conditions. Thirdly, increasing supply of new technologies and interventions has fuelled the constantly rising patients' expectations to launch a desperate "fight" against ageing and mortality, to use these technologies and receive every treatment available, even if it is highly expensive and marginally beneficial (Callahan, 1987). These tendencies have also been enforced by increasing valuation of health in modern societies as well as the inability to accept death and suffering as integral components of human life (ten Have, 1993).

The interactions between these factors have created a huge health care inflation. For example, in the Netherlands the relative investment of the Gross National Product (GNP) in health care raised from 3.8% in 1958 to almost 10% in 1986 despite considerable increase in GNP (ten Have, 1993). What is also important is that the largest proportion of health care resources in modern societies is consumed by a small minority of the ageing population suffering from chronic diseases. For example, in the USA 1% of the population consumed 29% of all health care expenditures; and 10% accounted for 70% in 1980 (Blank, 1992).

If the proportion of social resources used for marginal health care services continues to increase, it would drain resources from other areas of social welfare like housing, education or environmental control, which are much more influential for the health of the population than high-tech medical interventions and narrowly defined health care. On the other hand, if the share of GNP allotted to health care is kept constant, for example, approximately 10% of the GNP as is the case in affluent European countries, the ideology of unrestricted use of marginally beneficial medical interventions would, sooner or later, cause a relative scarcity of resources within the health care system as a whole, including basic services.

That is why almost all modern societies nowadays face the dilemma of restructuring their health care systems. The health care reform debate is on the political and public agenda all over the world. It is also clear that the escalation of health care costs mirrors a very complicated phenomenon of medicalisation of human life, which is a tendency to reformulate existential and social problems of human life in terms of medical complaints and symptoms. Another aspect of medicalisation is the claim to 'fix' these problems by medical interventions, which is, of course, irrelevant when fighting human mortality and prolonging life by any measures available (ten Have, 1993). Therefore, the crisis is caused paradoxically by 'successful' development and expansion of biomedical technologies in human life. The implementing of these technologies is, however, directed by inadequate ideas about the role of health care in human existence and well-being. Such an "expansionary vision of medicine" (Callahan, 1993) inevitably leads to a critical situation in health care.

In contrast to Western countries, the health care crisis within a transition society is usually presented not as the paradox of 'medical success' but rather as a failure of the health care system to cope with the deteriorating health of the population. This is understandable taking into account the recent socio-political transformation of CEE countries where the economic crisis has had an enormous detrimental effect on living conditions, health care systems and the health status of the population (Knudsen, 1996). Therefore, within the transition societies the arguments used in the discussions on health care crisis and the necessity of health care reform are first of all based on the improvement of such quantitative health indicators as life expectancy, mortality and the like.

What is even more important for our discussion, is that the high-tech enclaves such as the centres for transplantation or intensive care units for low birth-weight babies continue to function, even when health care resources amount to less than 10% of the health care budget in Western countries. For example, annual expenditure per person amounts to approximately 100 USD in Lithuania as compared with 1700 USD in Sweden (Brody and Lie, 1993). The justification for using these enclaves is, of course, based on the ideal of equitable access, which is still declared as the governing principle of health care. That is why, despite a shortage of resources for primary care or emergency medicine, a centre for heart transplantation is being run in Lithuania. In a similar way, ignoring the fact that many cases of infant mortality are due to poor prenatal care and lack of training to deal with the complications of labour, the units for low birth-weight babies are expanding their activities in the hospitals of many transitions societies.

Therefore, even if the pictures of health care crisis are rather different within the affluent and the post-communist countries, the mechanism of 'expansionary vision of medicine' is as relevant in the transition societies as it is in economically developed countries. This very feature of modern medicine raises the necessity to set health care priorities and restrict the devastating influence of marginally efficient expensive biomedical technologies, which as a rule are distributed according to the same principles as primary or basic health care services. The undiscriminated use of these technologies makes the post-communist health care crisis comparable to the Western one. However, in this respect the transition societies are even more vulnerable because in the context of financial scarcity the use of marginally efficient technologies has stronger detrimental effects on what could be regarded as a basic level of health care.

3. TRADITIONAL AND MODERN HEALTH CARE ETHICS

In order to understand the obstacles and driving forces of the transformation of health care practice it is useful to analyze these ethical dilemmas in the context of the tension between traditional and modern health care ethics. For example, in the sphere of individual health care, the traditional paternalism in the health care provider - patient relationship has been replaced or at least substantially supplemented by the principles of respect for the patient's self-determination and transparency of health care decision-making (even dealing with life-threatening diseases). In the European context, these changes were most clearly expressed in Northern and Western Europe (Thomsen, *et al.*, 1993).

The shift from the traditional perspective on resource allocation to the modern interpretation of social justice is, however, more complicated. The transparency and explicitness with which health care services are prioritised - the feature of modern health care ethics - is still a rather avoided and tacit question even in Western democratic societies.

Health care professionals have been forced to reconsider the traditional individualistic ethos of seeking for the exclusive good of the individual patient because of the scarcity of extremely expensive biomedical technologies (Blank, 1992). The difficulties related to the selection of patients for kidney dialysis drew the attention to the moral dilemma how to select patients for expensive and scarce (at that time) treatment in the early sixties.[1]

In such a way, the developments of biomedical technology have brought into conflict the individualistic one-to-one doctor - patient relationship and the social aspects of health care decision-making, based on the physician's obligations to a group of patients or even to a broader community. The discrepancy between individualistic and social aspect of health care decision-making is not an easy one to solve. It stems from the tension between traditional and modern perspectives on health care. This tension could also be seen as a controversy between medical and health care ethics because a physician at the bedside of a terminally ill patient can not and probably should not make public and societal choices leading to an equitable and efficient health care system (ten Have, 1988).

That is why even if the issue of just allocation of scarce biomedical resources marked the emergence of modern health care ethics and showed the necessity to take into account the social dimension of ethical decision-making a few decades ago, this dimension is, as we will see, still a neglected and avoided issue in health care reform strategies.

4. ESCAPISM FROM SETTING PRIORITIES

The necessity to set priorities and limits to health care is denied because very often the consequence of such a decision is a painful awareness of having to sacrifice the care of one human being for the sake of another or for the common good. As we have seen, such a policy contradicts the traditional attitude of seeking for the benefits of the individual patient.

On the other hand, it is quite difficult to make the process of setting priorities explicit because health care probably has a symbolic function to compensate for the dramatic inequalities in many other spheres of human life. It is a paradox to see that while some people live in luxury houses or have access to expensive sports, others can hardly afford to live in a polluted environment, and that these situations are more easily accepted than inequalities in technologically oriented medical treatment (Quintana and Infante, 1995).

The tendency to deny or escape the necessity to set limits on some health care interventions appears in the context of strong adherence to the principles of equality and solidarity and simultaneous inability to implement egalitarian ideals, due to financial scarcity. The most grotesque and hypocritical form of escapism from distributive issues took place in the former Eastern block countries where a constitutional principle explicitly entitled everybody 'to free and comprehensive health care'.

The outcome of declaring free and equal access to comprehensive medical care has been, however, the paradox that the so-called socialist countries have never implemented the ideals of equality and solidarity to the extent of free market oriented societies, especially Nordic countries, which until very recently never questioned the justifiability of their one tier public health care system.

Even during the 'golden' time of the Soviet Empire (not to mention the decline period in the eighties) the principle "for everyone according to his or her needs" served simply as an ideological cliché. Statistics which became available after the collapse of the old political system revealed that the indicators of morbidity, mortality and health status in the CEE were much worse compared with Western countries. Those who had the greatest health care needs - the mentally ill, the disabled, and the handicapped - found themselves at the bottom of the health care pile. They were usually kept in 'special institutions', places isolated from the 'egalitarian society' and well away from the sight of the foreigners (Gefenas, 1995).

On the other hand, in contrast to the most vulnerable part of the population, the Soviet *nomenclatura* always had elite clinics and hospitals, which were not available to the general public. In fact, this was a second tier of health care, which never existed in such a form in Western Europe. Finally, the so-called 'black market' of medical services has always been a part of the Soviet health care system. It enabled those seeking for high-tech specialised care to bypass the waiting lists or to be simply more confident that health care staff would do their job properly in child delivery clinics.

Within the democratic societies, the escapism from distributive issues takes a different form. The most popular strategy to escape the issue of setting limits is to present the problem of scarcity as a result of poor management and inefficient use of health care resources. This is a common mechanism in Western as well as post-communist European countries, which are now also following a democratic development. For example, nowadays many CEE countries concentrate their efforts to replace the former public health care system by a mandatory insurance one, leaving the question of limiting high-tech medical interventions almost untouched.

Introducing a number of cost containing schemes, different types of reimbursement, fee-for service etc. is another universal step to escape a radical decision to cope with financial scarcity (Hanson, *et al.*, 1994). This step, however, contradicts in itself the principles of equality and solidarity because it could be regarded as a punishment for lower socio-economic groups, which have, as a rule, the greatest health care needs.

5. REDEFINING SOLIDARITY AND PLURALITY OF MORAL IDEALS

A characteristic feature of solidarity as a principle of health policy is its attachment to the principle of equality. Even when solidarity is mentioned as a separate principle of health or social policy, it is still presupposed that we are talking about reducing inequalities in the living conditions or the health status of the population. In such a way, solidarity could be characterised as a group oriented responsibility to care for the weaker and more vulnerable members of the community. The popular expression that the healthy pays for the sick and the young for the old conveys the basic idea behind our traditional understanding of solidarity.

It has been argued, however, that the principles of self-determination and subsidiarity are no less important to modern health care than the traditional ideal of solidarity. Self-determination and personal responsibility for one's health are getting more important as long as *acute care medicine* is supplemented by *long term care, non-acute prediction and prevention* of diseases. This shift of the health care profile requires a redirecting of the patient's moral virtues from traditional compliance and hope to health literacy and partnership with the physician (Sass, 1995).

Another theoretical strategy to cope with the failure of traditional understanding of solidarity to meet the challenges of modern health care, is to reinterpret or transform the traditional meaning of solidarity, which has been used to justify every medical intervention available without the reference to its benefit and cost. The attempt to 'rescue' the notion of solidarity is based on harmonising its traditional meaning with the ideal of self-determined and autonomous limitation of one's own expectations and corresponding claims for health care interventions. This is in a sense a *mutual* or *reciprocal solidarity* because a person not only receives support from others but he himself restrains his or her claims to marginally efficient health care services for the benefit of others. This type of solidarity does not stimulate the redistribution of income but is rather based on the acceptance of the responsibility for organising self-care and shifting the burden of care from the state to the individual (ten Have, 1993).

6. RHETORIC OF SOLIDARITY

It is important to take into account that the plausible interpretation of solidarity in the context of modern health care refers to different and very often controversial moral principles. This is because solidarity is the notion which receives a specific content and normative meaning due to its use in a particular historical and social context (Holm, 1993). There is a danger therefore that, because of its emotional component and flexible meaning, the appeal to solidarity could be used as a rhetoric measure to bring a message that is outdated or irrelevant to changing circumstances.

Let us think of solidarity in such different contexts as health care, national liberation movement or mafia's activities. It seems that in all these contexts the notion of solidarity implicates rather different contents. What is, however, common for all those diverse contexts and what could be defined as a more or less constant core of the concept, are three essential features. Firstly, solidarity as a group concept presupposes sufficient emotional bonds among the members of the group (Sass, 1992). Secondly, it is also essential that the group be united by common goals and/or ideals. Thirdly, in order to reach a common goal, the members of the group are committed to sacrifice some of their welfare (or even their life in extreme circumstances), which in itself is a sign of strong emotional involvement.

It is important to stress that if either the goal is achieved or, in contrast, the ideals are not any more unanimously agreed upon among the members of the group, the appeal to solidarity becomes a simply rhetorical one. It looses its significance and original meaning. The liberation of the former Eastern Block countries is an example of such a fragmentation of solidarity. In the early nineties, the feeling of solidarity united not only people seeking for independence within a particular country but also roused remarkable international initiatives. As an expression of such an international solidarity, people from the three Baltic states organised what is known as the 'Baltic Way' action, during which they formed a "live chain" stretching hands along hundreds of kilometres through the roads and seashore of Lithuania, Latvia and Estonia. This feeling of solidarity was replaced by the attitudes of competition between three countries as soon as the common goal - political independence - had been achieved. Nowadays, an appeal to the former international solidarity is often regarded as futile and irrelevant.

The fragmentation of solidarity in health care is to some extent similar to the example described above. The rhetoric of traditional equality-based solidarity is a misleading one in the context of financial inflation of modern health care. The necessity to limit both the unrealistic patients' expectations to fight mortality as well as marginally efficient health care interventions makes the traditional notion of solidarity self-defeating (ten Have, 1993).

That is why in order to 'rescue' the notion of solidarity in health care from being a simple rhetoric word, we have to make its new meaning explicit. The reinterpretation of solidarity means *limiting or supplementing its traditional equality-based core* with (1) the principle of responsibility for one's life and health choices, as well as with (2) the ideal of seeking the benefit of the whole community. Firstly, the principle of self-determination to accept responsibility for one's own life-choices including private initiative to organise and finance one's own health care corresponds to libertarian ideas. Secondly, such ideals as promoting the common good and self-exclusion from care for the benefit of others are akin to utilitarian ethics.

7. THEORIES OF JUSTICE

Usually, considerations of justice are divided into *distributive* and *criminal* domains (Buchanan, 1992). The latter one has only marginal relevance for health care. On the other hand, theories and principles of distributive justice dealing with the ways of distribution of benefits and burdens are of high importance in the health care context.

A *formal principle of justice* claims that equals are to be treated equally and unequals unequally. This definition is a rather uncontroversial one because it does not implicate any content. At the same time it is hardly applicable to resolve practical dilemmas of how to distribute scarce resources. In order to bring to our world situation a certain model of distribution, we need what is called a *material principle of justice*, namely, the criterion of classifying people into groups receiving an equal share of benefits or burdens (Beauchamp and Childress, 1994). This is exactly the point where the rival theories of justice come into the scene because such material principles as distributing benefits according to needs, effort, contribution or free-market exchange provide different solutions to the same problem.

For this reason, there is no theory of justice, which is universally acceptable and applicable without controversies within the health care context. It is not surprising therefore that different encyclopedias and textbooks on bioethics provide us with rather diverse lists and prioritising criteria of the most influential theories of justice. For example, the *Encyclopaedia of Applied Ethics* puts an exclusive emphasis on Rawlsian theory ('Theories of Justice', see Chadwick, 1998). On the other hand, the *Encyclopaedia of Bioethics* provides a short description and analysis of such alternative approaches as libertarian, socialist, contractarian, utilitarian, communitarian and feminist justice (Sterba, 1995). It is not the aim of this chapter to provide a comprehensive overview of these theories. We would rather concentrate on those aspects of three influential theoretical approaches, which seem to be the most relevant ones to deal with the changing values in distributive issues; egalitarianism, utilitarianism, and libertanianism.

We have chosen these approaches because they illuminate different and inseparable moral perspectives, which are essential to reconsider the meaning of solidarity. Firstly, the emphasis on equal access to health care for all leads us to an egalitarian theory of justice based on the traditional ideals of *equality* and *solidarity*. The necessity for providing the best possible health care for the largest part of the population, which is *utility maximisation*, is the basic idea of the utilitarian theory of justice. Finally, the personal responsibility to maintain and finance one's own health care, and the unrestricted freedom to purchase every health care service available are based on self-determination and *liberty* - essential features of libertarianism applied to our health care context.

8. EGALITARIANISM

The principle of equality in the field of health care, which is the basis of the egalitarian approach, is usually formulated as a right to 'equal access to health care'. One plausible interpretation of this right is that those who are in equal need to health care are to receive the same level of care. This is in a sense an expansive or strong egalitarian approach because it entitles a person to every intervention available for others including even the most expensive ones. On the other hand, a qualified egalitarian approach entitles a person to the so-called adequate level of care, which is based on satisfying only basic or fundamental health care needs.

The egalitarian approach is closely related to the idea that people should be compensated for the disadvantages they are not responsible for. A metaphor of life's lottery 'distributing' people their health status has often been used by egalitarians to convey the idea that we are not responsible for most of our diseases, disabilities and other health problems. The idea of distributing health care resources according to the needs implicates also that priority should be given to those who have the greatest need of health care, - the most seriously ill. This is a criterion of severity of disease (Hanson, *et al.*, 1994). The rule of rescue, which is a requirement to save first of all those suffering from life-threatening conditions, is another example of egalitarian prioritisation (Hadorn, 1991).

Egalitarian ideas have very deep roots in health care. First of all, they correspond to the traditional medical ethos of seeking the best for a particular patient. Second, the central place of equality when considering health care policy issues stems from the idea of equal moral worth of all human beings as well as the belief that health is a very important component of human well-being. Value research in modern societies reveals a significant increase in the valuation of health in human life in the last few decades (ten Have, 1993).

The most comprehensive egalitarian account in the field of health care is based on the Rawlsian principle of 'fair equality of opportunity', which has been explicitly applied in the health care context by N. Daniels (1985). The distribution of health care resources should allow each person to achieve a fair share of the normal range of opportunities present in a given society. The normal range of opportunities means the possibility to pursue life-plans corresponding to every person's talents and skills. Health in a sense of 'normal functioning' is regarded as an important condition for a person to pursue his or her life-plan.

The main objection to this egalitarian approach stems from what Buchanan called a 'black hole' phenomenon. This means that unlimited use of goods and services that are provided for the disadvantaged in the process of ameliorating the effects of life's lottery (and trying to provide equal opportunities for everybody) can easily drain unlimited quantities of social resources (Buchanan, 1989). This drainage of all available resources could take place in both macro- and micro levels of allocation.

Another objection called 'egalitarianism of jealousy' has been raised against the egalitarian position of prohibiting private purchase of health care services, because such a prohibition does not aim at improving the status of the disadvantaged but at lowering the status of the advantaged ones (Engelhardt, 1994). It could also be argued that such a policy is an incongruent interference with the person's liberty to use his or her income to buy health care services the same way this person is permitted to buy antique cars and other luxury goods (Buchanan, 1989).

These objections convey the idea that *qualified egalitarianism*, which requires only some basic equalities and permits inequalities that benefit the least advantageous, is a more plausible perspective.

9. UTILITARIANISM

Utility maximisation is an important consideration in common sense thinking as well as in most ethical theories. Utilitarians, however, make utility maximisation an exclusive principle of their theoretical approach. "The greatest amount of health for the greatest number of people" could be regarded as a paraphrase of the famous utilitarian principle applied in the health field.

Utilitarian reasoning in health care is based on cost-effectiveness evaluation of different health care interventions. This procedure allows us to compare different interventions or the same intervention applied to different patients in terms of their cost and the effects, which could be evaluated in quantitative terms of *saving lives* or *years of life* as well as the *quality of life*. Quantity of life measurements are more adequate in acute care medicine, which deals with such life-threatening conditions as serious traumas or infectious diseases. On the other hand, quality of life considerations rather than quantity of life years saved, are becoming relatively more relevant measurements of health care interventions in modern societies because of the ageing population and prevalence of chronic diseases. In these circumstances, the effect of treatment is evaluated either in terms of the ability to function in the basic social roles or as a self-evaluation of one's own well-being (Musschenga, 1997). The attempt to integrate *quantity* or the length of life and *quality* of life considerations into a single framework has been achieved through a methodology known as QALY, i.e. quality-adjusted life-years (Williams, 1985).

It is important to make a distinction between trivial efficiency improvements and utilitarian reasoning. Utility maximisation becomes an ethical problem when it comes into conflict with other moral ideals, for example, equality. Setting health care priorities brings exactly this type of controversy between competing moral ideals. The conflict between them becomes increasingly dramatic in the context of limited resources.

The most general objection to utility maximisation is a tendency to take into account only utilities (whatever they are defined) but not the persons who are the subjects of these utilities. That is why QALY-based rationing could lead to what Harris (1987) called a 'life-threatening device', which places the priority on life-years rather than on individual lives. For example, a draft Oregon plan which was based on utilitarian QALY's calculus, ranked such life-threatening conditions as appendicitis below splits of temporomandibular joint disorder (Hadorn, 1991). Another controversial aspect of utility maximisation is that it leads to a discrimination of a society's sickest and most vulnerable section of the population.

Even though there are other serious moral objections to ground health care on the utilitarian theory of justice, none of them supports the extreme conclusion that utilitarian considerations of maximising overall utility are irrelevant in allocating scarce health care resources. These considerations are an inseparable component of the decisions made at different levels of resource allocation.

10. LIBERTARIAN APPROACH

Libertarians deny all the welfare rights including the right to health and argue that strictly speaking there is neither a right nor a corresponding obligation to provide health care for others. A characteristic feature of libertarian thinking is the emphasis on personal responsibility for health care decisions. Only specific disadvantages in human life are called unfair, while the rest also including most of the health problems are regarded to be simply unfortunate. That is why even if it is virtuous to care for the unfortunate, at the same time it is not unfair not to provide health care if it cannot be regarded as a rectification of a former injustice, for example, treating war veterans. Individuals have the right to the income and wealth they earn in a free market and nobody has the right to take even a part of this income to provide health care for others. This theory is based on the material principle of ability to pay and the free market model of distributing health care resources (Engelhardt, 1996).

Libertarian ideas have been criticised from several points of view. Firstly, in contrast to the 'egalitarian jealousy' argument it could be argued that the private market of health care gradually reduces the quality of care in the public or compulsory insured tier. Other objections to the libertarian model of distribution of health care resources rise from the assumption that the patient and health care provider are not equal partners in the free market interactions. The patient and the doctor interacting in the clinical setting do not possess a comparable understanding of the services provided (especially in the field of high-tech medicine) and very often the patient's decision-making process is interfered by pain or suffering.

11. TWO-TIER SYSTEM: THE RESOLUTION OF DISTRIBUTIVE TENSIONS?

Theoretical considerations help to elucidate the basic reasoning behind the alternative moral ideals of distributive justice, as well as to point out the main objections to each alternative approach. The approaches which have been sketched are coherent in some contexts, however, at the same time in some cases they provide opposing answers to the same question. This situation of competing approaches raises a scepticism regarding the usefulness of philosophical theories to resolve public policy issues where intuitive controversies are simply reformulated in a more abstract theoretical language.

It seems, therefore, that the limits to equal access to health care as well as the definition of what 'an adequate' or 'a satisfactory level of care' is, should be examined in the concrete socio-economic context of a particular society. The most suitable framework to re-examine theoretical arguments seems to be a two tier model of health care, which is increasingly accepted as an unavoidable way to organise health care in a modern society (Sass, 1995).

The move towards this model from the traditional one tier system also reflects a shift in the patterns of distributive justice. Firstly, a two-tier health care system is an attempt to set health care priorities *explicitly*. Such a system is based upon the distinction between high priority and low priority services, which is in a sense the attempt to define what could be regarded as a satisfactory degree of care in a particular society. Secondly, such expressions as 'adequate' or 'satisfactory degree of care' refer to the tendency of limiting the strong egalitarian ideals which strive for the equal access to all health care services.

The two-tier system can be regarded as the attempt to balance and reach a compromise between egalitarian, utilitarian and libertarian approaches. Such an attempt reflects a democratic process of health policy reform in a pluralistic society (Beauchamp and Childress, 1994). It also avoids the consequences of adopting one of the moral ideals as a sole basis for a health care system. The services included in the first tier guarantee an adequate package of health care on the premise of equal access for all. The distinction between the tiers is based, first of all, on utility maximisation, using the technique of cost-effectiveness, which is the main feature of utilitarian reasoning. Those who emphasise the necessity to purchase private health care services also find within the two-tier system an opportunity to do so in the framework of the second tier. According to this 'optimistic' scenario, the development of the two-tier system is a way of solving the most pressing dilemmas of just distribution of health care resources.

12. CONTROVERSIES OF THE TWO-TIER SYSTEM

However, a more critical perspective on the process of implementing a two-tier system uncovers theoretical controversies between competing perspectives on justice as well as some practical difficulties. One of the main difficulties arising in the way of defining the health care package is to find the balance between giving priority to the worst-off, who by definition have the greatest health care needs (egalitarian position) and assigning priority to the services with the lowest cost-effectiveness based on utilitarian ethics. The exclusive application of the principle of cost-effectiveness is related to counterintuitive results because it sometimes leads to ranking life-saving procedures below the routine ones. As we have seen, a draft Oregon plan, which was based on utilitarian QALY's calculus, ranked such life-threatening conditions as appendicitis below splits of temporomandibular joint disorder. That is why QALY calculations have to be balanced with the egalitarian rule of the rescue argument.

The balancing of these different perspectives is a very complicated process which depends on several factors. Firstly, on the value a particular society assigns to the egalitarian ideal of reducing inequalities. Secondly, on the methodology by which the costs and benefits of medical interventions are measured. The quality-of-life criteria and the methods employing those criteria are very diverse and hotly disputed matters.

Another difficulty that arises when defining the distinction between the first and the second tiers of health care is related to the necessity of taking into account the peculiarities of individual cases. If the procedure is based on the average of cost-efficiency for an intervention in a group of patients, and not on the efficiency of treatment given to a particular patient, the results could also be counterintuitive.

The justifiability of the second tier forms yet another group of major ethical controversies related to the establishment of a two-tier system. The services which are neither included in the basic nor in the comprehensive health care packages are supposed to form a second tier of services distributed on a private basis. Nevertheless, the controversies of the libertarian distribution of resources do not disappear due to the fact that the private market is allowed to regulate only low priority interventions. Even in this case health care does not 'behave' as ordinary goods in the market. The assumptions of marketability (certainty in outcome, symmetric information) for low priority services, which are considered to be experimental and having an uncertain outcome, are violated to an even greater extent than for well-known and beneficial services included into the basic tier (Norheim, 1995).

The second tier of private services should also be balanced with the principles of equality and utility maximisation. In this respect, a private tier of health care is justified if people using private health care facilities are made better off, without making those who use only first tier services worse off. The danger of gradual deterioration of the first tier is based on the assumption that the quality of a universal tier of health care depends on the involvement and efforts of resourceful people and of those having political power (Hanson, *et al.*, 1994).

13. 'SATISFACTORY CARE' IN THE TRANSITION AND AFFLUENT SOCIETY

It is important to stress the relativity of the distinction between low and high priority health care interventions. This distinction as well as the expression 'satisfactory' or 'adequate' level of care are crucial for the drawing of the line between the first and the second tier of health care. At the same time they are very much context-dependent distinctions. What would be regarded as a satisfying level of care in Lithuania may be evaluated as a very basic minimum level of health care services in Norway.

Let us compare the evaluation of health care priorities in Norway with the one in the context of the transition society. Norwegian guidelines for priority-setting assign the *first priority* to the measures which are necessary to cope with immediate life-threatening consequences (e.g. emergency medicine, neonatal care). The *second priority* is given to those interventions which prevent catastrophic or very serious consequences in the long run. These are, for example, such severe and chronic diseases as cancer, heart failure or care for the elderly. For this group of patients there is a guaranteed limited waiting period. The *third priority* is assigned to measures with documented effect, obviously undesirable consequences of non-intervention, but less serious than those of the first or the second priorities (e.g. moderate hypertension). The *fourth priority* measures still have some health- and life-furthering effect, however, the consequences of not carrying them out are rather insignificant, like when abstaining from medical treatment of a common cold. The fifth or *zero priority* group is defined as health services, which are in demand, but have no documented effect (Norheim, 1993).

Even if these guidelines are too vague in many cases to indicate which particular interventions should be given priority, they are a suitable device to compare the interpretation of what could be regarded as a satisfactory level of care and the distinction between high and low priority services in different socio-economic contexts. For example in an affluent society the cut-off point for satisfactory or adequate package of care would probably be drawn somewhere around the fourth priority level of services.

It seems therefore, that in affluent countries the problem of defining the adequate package of health care and the distinctive line between the tiers is more appropriately defined as the separation of "upper limits or lateral fringes of the comprehensive health care package" (Holm, 1995). The second tier would include in this case such costly and marginally beneficial interventions as cosmetic surgery, experimental transplantation or novel cancer therapies.

On the other hand, in most of the CEE countries the upper limit of the health care package available would probably hardly encompass the second priority group of interventions, which amounts only to a bottom portion of what is regarded as a comprehensive and "adequate" package of health care in the affluent society. This is understandable taking into account that, within the transition society, health care resources amount to less than 10% of the resources allotted to health care in the Nordic countries.

It seems therefore that establishing a private tier in the context of the transition society is more controversial than in the affluent one, because the list of services, which a society affords to include into the first tier (ranked according to the cost-efficiency technique) would probably be limited to primary care interventions. Including the rest of available services in the private tier of services would result in a dramatic increase of inequalities in the access to many beneficial health care services, which would have a detrimental effect on the quality of the first or basic tier of health services.

However, the consequences of disallowing a private tier of services are no less controversial. Even if the services which are on demand are not available in the framework of the first tier of health care, neither patients who want to get these services nor health care providers who provide them will be satisfied with the restrictions imposed. They will try to interact in the second tier on the basis of free market exchange justifying mutual interaction by a libertarian approach. In many post-communist societies a second tier will continue its existence on a 'black market' basis, while in affluent countries the rich will try to use the services in foreign countries.

14. INTERNATIONAL JUSTICE

The transformation of Europe, which has been taking place since the collapse of the so-called Eastern Block, has brought a growing co-operation between different European countries, including relationships in the field of health care. For example, since the early nineties CEE countries have been participating in international multicentre clinical trials sponsored by Western pharmaceutical companies. It is not difficult to imagine the situation and mechanisms by means of which these countries could be used as a cheap polygon for the development of new drugs using methods sometimes not acceptable in Western countries. The international organ procurement and donation programmes could raise similar questions.

It seems that in such a context considerations of international justice in health care, which until recently have been a neglected area of justice debate, are getting increasingly important. These considerations help to highlight once again the question raised in the beginning of this chapter: what are the criteria of classifying people into groups receiving an equal share of benefits or burdens?

If we endorse the ideals of equality and solidarity in the domestic matters of justice, what are the criteria we use to explain striking inequalities between different countries? For example, if people in an affluent society would be horrified if adequate care and treatment were not available for all children, irrespective of income or social status (Holm, 1995) how should we react being aware that similar children in Romania or Russia have very often no access to decent primary care at all?

The theoretical approach, which justifies the existing differences in availability of health care services is the libertarian one. The libertarians would simply say that Romanian children are unfortunate children. Even if it would be morally virtuous to donate them vaccines and toys, it is not unjust to use the resources for one's own leisure activities.

However, if the egalitarian arguments are followed we have to admit that if inequality is bad, it is far greater, and so far worse between the richest people in the richest countries and the poorest people in the poorest countries, than the inequality between the most advantaged and least advantaged in a particular affluent society (Arneson, 1998).

This line of argumentation throws a serious challenge to the international policies, which are so different from the domestic ones. It seems that the same double standard is implicated even in the international documents, which employs the language of 'tolerating' inequalities. Equitable access to care, which is compromised by the 'satisfactory degree of care', means hardly commensurable things in different European countries.

In order to bring some elements of egalitarian perspective (which is so strong and influential within the domestic health care policy debate) to the international context we have to assume that at least some form of moral cosmopolitanism and a minimal view of justice as a commitment not to victimise others, and improve the protection of those whose vulnerability is constituted by ill health should be accepted. A realistic account of international justice aims not to abolish the borders between the states, but only to reduce the boundaries or to create 'just gaps' of the policies and laws, which limit and reduce the forms of vulnerability, which otherwise make people ready victims of destruction and damage (O'Neill, 1993).

It seems that the tendency to fragmentize the justice debate into smaller communities is also the sign of universal escapism from distributive issues. In this context it is a denial of the reality that we live in a world where resources used for someone's cosmetic surgery of neck wrinkles would be sufficient to save tens or hundreds of children's lives by giving them the necessary vaccines. Even if this is a sentimentalisation of the debate it helps us to think critically about the concepts and moral ideals we analyze and sometimes defend in our discussions, papers or even international conventions. What is 'unjustified discrimination', how could we understand 'satisfactory degree of health care' not only in the domestic but also in the European context? These are the questions we will have to deal with in the near future if we want to live in Europe with 'just gaps' in the borders between the states.

15. CONCLUDING REMARKS

A realistic and morally acceptable system of health care has to be based on the balanced plurality of different approaches to social justice. Each of these alternative approaches illuminate a particular aspect of just distribution of health care resources. At the same time taken in isolation as the sole basis of health care system, each of them creates a grotesque picture of health care.

The socio-economic and cultural context is also important for the evaluation of different distributive decisions and policies. Whatever context we choose - traditional or modern, domestic or international, transition society or affluent country - all of them imply diverse combinations of alternative ideals and perspectives of justice. Plurality of perspectives, however, also implicates unavoidable controversies and tensions because it is impossible to reconcile different and competing paradigms. What we are trying to do when searching for a decision in every concrete situation, is to change the balance between competing approaches and to reduce these tensions.

NOTES

1. A multidisciplinary commission was established to deal with this moral dilemma. The importance of the commission activities helped to reorganize it into the first institute on health care ethics in the world in 1969, which was later called The Hastings Center.

REFERENCES

Arneson, R.J.: 1998, Equality and egalitarianism. In: Chadwick, R. (ed.), *Encyclopedia of applied ethics*, Vol. 2. Academic Press, San Diego, p. 115-125.

Beauchamp, T. and Childress, J.: 1994, *Principles of biomedical ethics* (4th ed). Oxford University Press, New York.

Blank, R.: 1992, Introduction. In: Blank, R.H., Bonnicksen, A.L. (eds.), *Emerging issues in biomedical policy*. Columbia University Press, New York, p. 3-12.

Brody, B.A. and Lie, R.K.: 1993, Methodological and conceptual issues in health care system comparisons: Canada, Norway, and the United States. *Journal of Medicine and Philosophy* 18: 437-463.

Buchanan, A.: 1989, Health care delivery and resource allocation. In: Veach, R.M. (ed.), *Medical ethics*. Johns and Barlett Publishers, Boston, p. 292-327.

Buchanan, A.: 1992, Distributive justice. In: Baker, L.C., and Baker, Ch.B. (eds.), *Encyclopedia of ethics*. Garland Publishing, New York and London, p. 655-661.

Callahan, D.: 1987, *Setting limits*. Simon & Schuster, New York.

Callahan, D.: 1993, Health care reform and the goals of medicine. *Phi KAPPA PHI Journal*: 14-16.

Chadwick, R. (ed.): 1998, *Encyclopedia of applied ethics*. Academic Press, San Diego.

Daniels, N.: 1985, *Just health care*. Cambridge University Press, London.

Engelhardt, H.: 1994, Health care reform: A study of moral malfeasance. *Journal of Medicine and Philosophy* 19: 501-516.

Engelhardt, H.: 1996, *The foundations of bioethics* (2nd ed.). Oxford University Press, New York.

Explanatory report to the convention for the protection of human rights and dignity of the human being with regard to the application of biology and medicine: Convention on human rights and biomedicine: 1997. Council of Europe, Strasbourg.

Gefenas, E.: 1995, Health care in Lithuania: From idealism to reality? In: Seedhouse, D. (ed.), *Reforming health care: The philosophy and practice of international health reform*. John Willey & Sons Ltd., Chichester, p. 121-127.

Hadorn, D.C.: 1991, Setting health care priorities in Oregon: Cost-effectiveness meets a rule of rescue. *Journal of the American Medical Association* 265: 2218-2225.

Hanson, L.F, *et al.*: 1994, Equality, explicitness, severity, and rigidity: The Oregon plan evaluated from a Scandinavian perspective. *Journal of Medicine and Philosophy* 19: 343-366.

Harris, J.: 1987, QALYfying the value of life. *Journal of Medical Ethics* 13: 117-123.

Have, H.A.M.J. ten: 1988, Ethics and economics in health care: A medical philosopher's view. In: Mooney, G., McGuire, A. (eds.), *Medical ethics and economics in health care*. Oxford University Press, Oxford, p. 23-39.

Have, H.A.M.J. ten: 1993, Physicians' priorities - patients' expectations. In: Szawarski, Z. and Evans, D. (eds.), *Solidarity, justice and health care priorities*, Health Service Studies 8. Linköping University Press, Linköping, p. 42-52.

Have, H.A.M.J. ten: 1998, Philosophy of medicine and health care - European perspectives. *Medicine, Health Care and Philosophy* 1: 1-3.

Holm, S.: 1993, Solidarity, justice and health care priorities. In: Szawarski, Z. and Evans, D. (eds.), *Solidarity, justice and health care priorities*, Health Service Studies 8. Linkoping University Press, Linkoping, p. 53-64.

Holm, S.: 1995, 'Socialized medicine', Resource allocation and two-tier health care - The Danish experience. *The Journal of Medicine and Philosophy* 20: 631-637.

Knudsen, K.: 1996, *Lithuania in a period of transition*, Fafo Report 186. Falsh Hurtigtrykk, Oslo.

Musschenga, A.W.: 1997, The relation between concepts of quality-of-life, health and happiness. *Journal of Medicine and Philosophy* 22: 11-28.

Norheim, O.F.: 1993, Guideline for priority-setting in the Norwegian health care system. In: Szawarski, Z. and Evans, D. (eds.), *Solidarity, justice and health care priorities*, Health Service Studies 8. Linköping University Press, Linköping, p. 65-77.

Norheim, O.F.: 1995, The Norwegian welfare state in transition. *Journal of Medicine and Philosophy* 20: 639-655.

O'Neill, O.: 1993, International justice and health care. In: Szawarski, Z. and Evans, D. (eds.), *Solidarity, justice and health care priorities*, Health Service Studies 8. Linköping University Press, Linköping, p. 120-128.

Quintana, O. and Infante, A.: 1995, Setting priorities in the Spanish health care system. *Journal of Medicine and Philosophy* 20: 595-606.

Sass, H-M.: 1992, Introduction: The principle of solidarity in health care policy. *Journal of Medicine and Philosophy* 17: 367-370.

Sass, H-M.: 1995, The new triad: Responsibility, solidarity and subsidiary. *Journal of Medicine and Philosophy* 17: 587-594.

Sterba, J.P.: 1995, Justice. In: Reich (ed.), W.T., *Encyclopedia of bioethics* (rev. ed.). Macmillan, New York, p. 1308-1315.

Thomsen, O., Wulff, H.R., Martin, A., Singer, P.A.: 1993, What do gastroenterologists in Europe tell cancer patients? *The Lancet* 20: 473-476.

Williams, A.: 1985, Economics of coronary artery bypass grafting. *British Medical Journal* 291: II 326-329.

HENK A.M.J. TEN HAVE

CHOICES IN HEALTH CARE: WAITING LIST, RATIONING AND PRIORITIES

1. INTRODUCTION

The phenomenon of the waiting list is one of the most visible symptoms of scarcity of resources in health care. It is a very common experience for most citizens that they cannot immediately use the health care facilities when they are needed. Long waiting lists have become normal in many health care systems, particularly in Europe. However, waiting for appropriate treatment, diagnosis or care is also a source of deep concern. Individual patients may experience considerable stress or anxiety when they have to wait for treatment since they have a medical indication and are referred to specialist care. Patients may also suffer from uncertainty when they cannot have the diagnostic examination when there is a suspicion that they have a serious disease. Family systems may break down because of psychological burdens when the condition of the mentally handicapped child or the dementing parent they are caring for is deteriorating and the waiting lists for admission to professional facilities are very long. Furthermore, waiting lists are a source of frustration for health care professionals; they cannot provide the treatment and care that is indicated on medical criteria. Waiting lists give also rise to public concern and political debate, particularly when they have negative consequences, such as patients on the waiting list who died before they could receive adequate treatment, or serious accidents in private homes or in neighbourhoods which could have been avoided with earlier placement of a demented person in a nursing home. Recently, in the Netherlands, there has been a intensive public debate on the issue of waiting lists since the news was published that 90 patients die annually while waiting for cardiac surgery (*NRC Handelsblad*, 1996). Cardiologists argue that these patients could have been saved if they had had surgery in time. The waiting period for cardiac surgery is now between 3 and 6 months, while on the other hand the physicians' estimation is that a maximum of 6 weeks is acceptable. These waiting patients spend their waiting time at home; many of them are seriously handicapped, limited in mobility, anxious and under severe stress. Of course,

H.A.M.J. ten Have and B. Gordijn (eds.), Bioethics in a European Perspective, 219-239.

this situation is a waste of health insurance funds, human productivity, and personal happiness. Therefore, the interpretation of this phenomenon usually is straightforward: waiting lists exist because treatment capacity is limited. Because the available care facilities cannot adequately cope with the demand for health care not all patients in need of care can be helped in time. The solution advocated on the basis of this analysis is equally simple: expansion of treatment capacity is necessary. This conclusion will often be connected with political implications: when doctors who can really help patients are denied more resources, then some patients, and sometimes an exactly specified number of patients, will be sacrificed for budgetary reasons.

2. THE HETEROGENEITY OF THE WAITING LIST

This established interpretation of the waiting list in public debate is problematic. We do not really know what practices and activities are reflected in the phenomenon of the waiting list. The explanation for a long waiting time can be that quality of care is good. General practitioners may perceive this to be the case and refer more people. A long waiting list in this example reflects excellent care. Alternatively, it could be result of outmoded care being practised, for example because patients have to stay longer in hospital than necessary when up-to-date treatment schedules have been used. Likewise with short waiting times. A short waiting list may reflect either the use of effective day care or low quality care with resulting low GPs' referral rate. Not much research has been carried out on the interpretation and meaning of waiting lists; it is therefore difficult to know what situation is reflected; long or short waiting times can occur for good or bad reasons (Donaldson, 1993).

It is obvious that waiting lists do exist throughout the health care system. Notorious, for example in the Netherlands, is the waiting problem in the care for the mentally handicapped. There are waiting lists for admission to a professional institution, for transfer among departments within an institution, and for release from a facility, for example to special homes within the community. Although these lists use several urgency categories, in practice only those within the top priority urgency class stand any chance. In this class, several thousand people are indicated for admission to an institution. The average waiting time is three years. The list is growing faster than the number of places becoming available.

A similar situation exists for nursing homes. In psycho-geriatric nursing home care the shortage of beds is a continual problem. According to recent research 5130 patients were on the waiting list for admission in 1991. In that year the average waiting time was 23 weeks. Approximately 600 waiting patients deceased before admission (Meiland, *et al.*, 1994). The indication for admission is primarily a socio-medical one. Not only the severity of the dementia in itself, but also the insufficiency of care facilities at home or the excessive burden on the lay caregivers, usually the family members, determine the estimation of the necessity of admission.

The waiting lists for institutions of chronic care usually do not receive the same public attention as, for example, the waiting problem for cardiac surgery patients. Feelings of indignation because of risks to a patient's life naturally are stronger than because of risks to his social network or deterioration of the quality of his existence. Perhaps there is also a feeling that some conditions do not involve medical determinants only, but are of a more social nature, associated with the willingness of others to assist in care and with the capacity of the family system to cope with seriously disabled members.

Cardiac failure is more easily regarded as a biological disaster that an individual cannot overcome unless medically treated. Mental handicap or dementia is at the same time a more or less social phenomenon; its impact on the person is at least partly dependent upon the compensative support of others.

It is remarkable that when the waiting list issue is on the agenda in public and political debate, waiting problems in cardiosurgery rather than chronic care catch the eye. At the background, various kinds of valuations play a role, but they are seldom explicated and analyzed.

Even in the hospital setting the situation is diverse. It is estimated that on average one of three patients to be admitted to hospital on medical indications spends too long on a waiting list (Government Committee, 1992). Long waiting lists exists for orthopaedics, cosmetic surgery and general surgery, ophthalmology and otorhinolaryngology. Between regions and among hospitals the length of the waiting list differ considerably. Long waiting lists also exist for outpatient consultations, especially in ophthalmology, stomatology, plastic surgery and orthopaedics. A well-known problem is the waiting list for kidney transplantation. In the Eurotransplant area the average waiting time in the period 1989 - 1995 was 2.1 years. The percentage of dialysis patients waiting for transplantation longer than 3 years is 28; 11% waits longer than 5 years (Vanrenterghem and Persijn, 1996).

3. THE AMBIGUITY OF THE WAITING LIST

One of the reasons for the ambiguity of the waiting list is its definition. Immediate access to complicated systems such as health care is impossible. There is always a need to manage the input, throughput and output of the system. Especially in health care, the use of facilities is not constant, but can vary considerably. Workload, workforce and available beds call for a stable and efficient use. From this managerial perspective, a waiting list is, up to a certain extent, not only an unavoidable, but rather a rational instrument of planning. For the patient such limited waiting time has advantages because he can plan to have treatment at the time that is most convenient for himself. When there is no acute need many patients probably do not want instantaneous hospitalisation. As long as the waiting time is within certain limits, the acceptability of waiting for health care is not problematic. Using waiting lists for the rational planning of services is not morally objectionable since it is focused on promoting the benefit of the patients as well as the optimum use of available facilities.

Waiting, however, will only be an indicator of good management as long as the average waiting time is limited. It has been argued that hospitals using waiting lists to plan the flow of patients should not exceed 2% of the total admissions; this would imply an average waiting time of 14 days (Nationale Ziekenhuisraad, 1989). Sometimes this period is included in the definition of a waiting list: it is defined as a list of patients with a medical indication for hospital treatment, who by constraints outside of their own sphere of influence have to wait for admission longer than two weeks (Van Willigenburg, 1994). A waiting period of 14 days is regarded so normal that it is not labelled as a waiting list. Others, however, use extension of admission for longer than 4 weeks as a decisive criterion for defining a waiting list (Kimsma, 1993).

Besides being an instrument for rational planning, waiting lists can have other functions. Both functions are inappropriate.

First of all, they can work as a filter. Having to wait some time before hospital admission can make patients realize that the intended medical intervention may not be as necessary as initially thought. Perhaps they may learn to adapt to the condition, when the prospect of immediate relief has involuntarily vanished, and they may prefer to continue their existence in a medically not optimum state, without wishing to take the risks involved in medical intervention. However, in general, when a patient is on the waiting list, he is medically indicated for treatment; in this case, it is questionable when medical intervention is delayed with the deliberate aim of discouraging him or her to undergo treatment which is necessary; when, on the other hand, treatment is not medically indicated, the patient should not have been placed on the list in the first place. There may be additional benefits of waiting, particularly when the natural course of a condition is uncertain. For example, studies showed that women registered on the waiting list of a hospital providing in vitro fertilisation wrote to cancel their appointment because they were pregnant (Laborie, 1993). These benefits are unintended effects of waiting. They may, however, be a reason for cautiousness when arguing that waiting lists as necessary evils should be eliminated.

A second function of the waiting list is its possible use as an instrument of pressure or lobbying. The existence of a waiting list and the increasing length of a list can be effective, particularly in the hands of health care managers, to put pressure on social and political bodies. In the absence of explicit criteria for access to the waiting list as well as consistent rules for selection from the list, it is hard to objectively assess the existing list and make comparisons with other lists.

4. THE WAITING LIST AS A MORAL ISSUE

The proper use of a waiting list as an instrument of rational planning depends on a critical time frame. When the number of patients on the list grows and the waiting period becomes longer, the waiting list loses its function as a planning instrument. Critical her does not so much refer to the length of the list. If there is a rapid patient flow and therefore an adequate capacity for treatment and care, the number of patients on the list is not really a problem. What is critical is a long average waiting period. When the period between referral and placement on the list and the actual treatment is long, on average, for every patient, the waiting list is problematic. In these circumstances the list is used as an instrument of distributive policy and selection. When treatment capacity is scarce and resources are limited, waiting lists are used for rationing with at least two selective moments: the selection of patients to be placed on the list, and the selection of patients from the list for treatment or care facilities.

When waiting lists are longer than is strictly necessary for planning purposes, questions arise about the moral acceptability of the waiting list as a way of rationing. Does a waiting list contribute to a just distribution of scarce resources? Are the selection criteria for access and ranking of individual patients morally acceptable?

Upon analysis, waiting lists are apparently organised around three selective principles: (1) first come, first served, (2) medical need, (3) medical success (Nationale Ziekenhuisraad, 1989).

4.1. First Come, First Served

The first principle leads to a procedural criterion: those patients with the longest waiting period, are the first to receive treatment or care. The moment of referral and placement on the list is crucial. A fair distribution of waiting time among patients is the result. From the point of view that everybody is equally entitled to treatment, every patient counts for one. The advantage of the application of this criterion is that each patient has a clear overview of the waiting period ahead of him or her; he or she slowly moves through this period towards the moment of treatment.

The assumption is that the time of referral and placement is decisive because it cannot be manipulated. Nobody knows when he will become ill and in need of treatment. Placement on the waiting list, and the resulting waiting period will be a kind of 'natural lottery', beyond human influences.

4.2. Medical Need

The second principle implies a material criterion: decisive is the severity of the present condition of the patient. It is argued that it is fair to allocate health care resources on the grounds of medical need. Those who need treatment most, should be given priority. This criterion operates with a different understanding of equality. When patients are treated in the order of admission, we disregard that not everyone needs equal treatment since not every need is equal. What is important is that patients with equal needs should be given equal consideration (Chadwick, 1993). More than a system of randomness in allocation by natural lottery, we should take care that patients' equal interests are given equal consideration.

4.3. Medical Success

The third principle leads to the criterion that those patients with the greatest capacity to benefit will have priority. The potential outcome of treatment, the chance of medical success should be a decisive factor when choosing among patients. The implication is that the actual need of patients is more or less ignored; priorities are set in the light of what is expected to lead to the best possible outcome. It can be the case that those patients who are in the worst condition are those providing the least opportunity for achieving medical success.

The structure of the waiting list depends at the moment on a mixture of these three principles. Various combinations of selection criteria are applied. In many cases, a distinction is made between different classes according to urgency. Patients are therefore assigned to categories on the basis of their need for treatment or care. Priority within these urgency categories then depends on either the order of presentation or the chance of success.

However, in practice many difficulties arise. It can be, for example, a complicated matter to predict the potential benefit of a treatment. Should priority be given to patients in the most serious condition, those who are furthest below the minimum acceptable level? Or should priority be given to those patients with the greatest capacity to benefit? The concept of 'urgency' is also not always easy to apply. The interpretation of 'urgency' may differ between and within various settings and institutions. In the context of kidney transplantation, urgency refers to the serious medical condition of the patient now undergoing dialysis treatment. However, urgency as a criterion is not often applied; the major criterion is immunological: the matching on broad HLA antigens and the mismatch probability, - which is in fact the criterion of medical success. On the other hand, there is variety in the point of time where a patient is referred and placed on the waiting list. In the context of psychogeriatrics, urgency refers to the immediate risk of family breakdown due to excessive burdens on the caregivers. It is very difficult to determine who needs help most, since the medical judgement is influenced by the capacity of caring within the informal network supporting the patient as well as by the willingness of family members, friend or proxies to continue with informal care.

What is important, however, is that the above set of criteria, although difficult in practice and complicated in use, safeguardes the moral acceptability of the waiting list as an instrument of selection of patients and distribution of scarce resources. The waiting list can be regarded as a contribution to a just distribution of health care resources, as long as the moral criteria regulating its existence and use are actually applied in practice. Precisely at this point there is a major problem.

5. THE MORAL BREAKDOWN OF THE WAITING LIST AS RATIONING INSTRUMENT

Waiting lists have a critical length. If patients have to wait too long, they feel that they can no longer expect to be treated within a reasonable period of time. In these circumstances waiting is equivalent to endless postponement of treatment or admission to a health care institution. Waiting is without perspective of help. The very harm that is planned to be prevented or eliminated in the foreseeable future because of appropriate treatment or care within an institution, will probably occur during a long waiting period.
Waiting itself can bring risks to the individual and his or her family members because of stress and anxiety. There is also potential harm to the community because of risk of chronic disability and loss of working days.

When the waiting time for a health care facility exceeds a critical limit, it cannot be argued anymore that the waiting list helps to distribute fairly the scarce resources available. In fact, we have a situation where care and treatment are irresponsibly delayed or not provided at all to particular categories of patients. In this situation, the moral criteria for selection do not work anymore. With such a long waiting time, the principle 'first come, first served' has become meaningless. Patients do not have any prospect of treatment in the near future; they cannot have any expectation to slowly approach the moment of treatment. Especially when combined with a system of urgency categories, they run the risk that every time more urgent patients will have priority. When the facilities for care and treatment are so limited, the principle of medical need will also become unrealistic. Every patient will be in urgent need, which makes a classification of categories of urgency futile.

The conclusion then is that the waiting list can be regarded as an instrument of just distribution of resources *under conditions of moderate scarcity*. In these circumstances, the moral principles which underlay the existence and use of the waiting list can be applied. Under conditions of excessive scarcity, the discrepancy between demand for care and facilities to provide care has so much increased that too many patients have too long a waiting period. The moral principles underlying the waiting list have lost their differentiating power.

It is of course difficult to pinpoint exactly where and when the moral principles are transformed into futile abstractions. One signal is a general perception that the waiting list should no longer be considered a fair instrument. Waiting for many people has become equivalent to not being treated or cared for; waiting implies that some patients have to live with significant disability and suffering without reasonable prospect of relief. Waiting in fact implies that the health care system in some areas has become inaccessible. This moral experience of unfairness will on the other hand stimulate patients to seek other solutions. Other signals of transformation therefore are the use of informal criteria of selection or suspicions that the waiting list is manipulated. Inappropriate strategies may be used to place patients on a waiting list or to give priority to particular patients on the list. The general practitioner, for example, may attempt to prioritise some of his patients through directly and personally contacting the supervising specialist, through presenting social reasons as medical arguments, or through shopping among the health care facilities in the region and referring the patient to the one with the shortest waiting period (Kimsma, 1993). The problem with these informal strategies is that it is unclear what selection criteria are used: it is therefore publicly uncontrollable. It is also problematic since the interests of the individual patient are preferred for unknown reasons over the collective interest of all patients who are waiting on the list. The basic idea that patients with equal need and equal capacity to benefit are given an equal consideration is completely sidetracked. The best thing patients can do in such circumstances is to register with an assertive, influential and well-connected physician.

6. POSSIBLE SOLUTIONS: MORE RESOURCES?

When the crucial point has been reached where the moral principles regulating the waiting list have become futile, we have two options for a solution. The first is to expand the resources, so that the mismatch between demand and supply in health care is removed. The second option is to restore the discriminating power of the selection principles, either by re-interpreting the existing principles or by introducing new principles.

The first option is expanding the budget for health care in order to increase the capacities for treatment and care so that the number of patients waiting can be significantly reduced.

It is questionable whether this option would really give a solution for the problems of the waiting list.

One argument is that not all waiting lists are a matter of inadequate financial resources. For example, the waiting list for kidney transplantation is primarily caused by shortage of donor organs. Some waiting list problems are apparently due to inadequate management. In a Dutch survey of reasons for the development of long waiting lists, it turned out that in 20% of the cases hospitals could not even mention any reason for the waiting list (Government Committee, 1992, p. 99).

A second argument is that policies to reduce the waiting list through increasing the health care budget, in the past did not give adequate results. The Dutch Government Committee on Choices in Health Care refers in its report to Norwegian experiences. Explicit policies in the 1980s to allocate enormous extra funds to reduce waiting lists did not result in a decreasing number of patients on the lists. Paradoxically, many more patients were treated than before. Now that health care facilities have been expanded, demand has also increased. A similar Dutch example is the introduction of percutaneous transluminal coronary angioplasty (PTCA) which was initially intended to substitute coronary bypass surgery. The result was not that the waiting list for bypass surgery had been reduced, but that a new waiting list came into existence for PTCA which is now longer than the list for surgery ever was.

Besides these practical arguments that increasing resources will not automatically lead to a solution of the waiting list problem, there are also significant theoretical considerations that are related to an alternative interpretation of the resource allocation problem in general.

7. A DIFFERENT INTERPRETATION OF THE ALLOCATION PROBLEM

The Dutch Government Committee on Choices in Health Care, reporting in 1992, argued that even if more resources would be available for health care, making explicit choices and identifying core health services will be inevitable.

Essential to the argumentation of the Committee is that it developed an interpretation of the allocation problem that differs from current opinion.

Significant for the present state of the allocation debate is the notion that scarce resources are not just a financial, managerial, or organisational problem. It is also, and perhaps first of all, a socio-cultural problem. Both doctor and patient are participants in a cultural process, and actors with a fundamental post-modern mentality, overvaluing the contributions of medical science and technology to the pursuit of human happiness and wellbeing, and believing medicine's promise to eliminate human suffering and mortality. Rising health care costs should therefore give an impulse to critically examine the unprecedented power of medicine, not only because of its promises and its actual contributions to diminishing morbidity and suffering but also because of its interference with human dignity and its transgression of the boundary of meaningful life. The recently acquired medical power should be counterbalanced by a new medical ethics and a new awareness of a more critical use of medicine's technologies. Purely economic arguments do not suffice; there should be philosophical arguments for setting limits. Moreover, individual actors cannot be blamed since their actions are guided by common value orientations.

This basic interpretation is reflected in three components of the Committee's analysis: (1) a terminological change from 'limits' to 'choices', (2) a perception that the basic problem is a lack of public awareness that choices are unavoidable, (3) an emphasis on the priority setting process rather that on the product (i.e. a priority list).

7.1. Change of Terminology

Usually a distinction is made between the limits *of* care and the limits *in* care, referring, respectively, to limiting the health care system which is regarded as a responsibility of the government, and to limiting care for individual patients or patient categories which is considered to be the responsibility of health care professionals. This distinction implies that *medical* ethics as a discipline primarily concerned with individual welfare can only function within a more encompassing framework of *health care* ethics, which is primarily concerned with the general welfare (ten Have, 1988). It is often argued that intrinsic factors underlie rising health care costs: specialisation, professionalisation, medicalisation and above all, technological innovation. Within the present system such factors generate almost unlimited claims. Notions such as 'customary in the profession' and 'whenever a medical indication is prevalent' seem to be the only criteria available to differentiate between claims which are supposed to be justified and those which are not. On this basis a health care policy of equal access and financial solidarity produces an almost uncontrollable system.

One way suggested to regain control is by setting *external* limits: budgeting systems, review organisation, technology assessment. However, it is unclear whether this external approach will be successful as long as there is no evolution from external to *internal* limits. The argument is that we must learn autonomously whether or not to restrict our claims to health care, and whether or not to withdraw from the system of scarcity which is to a great extent maintained by an obsession with longevity in the sense of 'surviving others', living longer than, and thus outliving others as a consequence of an inability to integrate and accept death and suffering as an integral component of life. Since the concept of autonomy may be, and in many European countries is linked with the concept of solidarity, self-determination may be considered to imply an ability to restrain our claims for the benefit of others. In its literal (Kantian) interpretation, 'autonomy' denotes the capacity of setting limits to one's own behaviour and one's own resolutions: my health may not necessarily be a prime value compared to the needs of others. Self-determination should involve responsibility for the self-realisation of other members of the community. It is necessary to develop a new ethos of critical use, moderation and temperance in health care.

The rhetoric of setting limits is therefore inadequate. Introducing the terminology of 'making choices' represents a more positive attitude towards the problem of scarce resources, emphasizing that growth of medical knowledge is the result of deliberate options chosen by individual scholars, policy makers and subsidizers, recognizing that governments have in fact encouraged some developments and discouraged others by regulations and financing policies, and inducing every actor to select positive opportunities from everything modern medicine makes possible.

7.2. Necessity of Making Choices

Research data show that the majority of the Dutch population is opposed to making choices in health care: 55% agrees that every treatment should be available regardless of its costs and regardless of the probability of it having a curative effect. Moreover, 51% is prepared to pay double the current health insurance premium if that could guarantee the availability of every treatment possible. An even larger majority (78%) disagrees with choosing amongst expensive medical technologies making those technologies only affordable for higher income groups. The major problem identified in health care is the shortage of personnel (and not the increasing costs of health care) (NSS Marktonderzoek, 1991).

A similar questionnaire among Dutch health care professionals has produced more or less the same results: for 50% of the respondents a further increase of health care costs is acceptable; 66% agrees that people should pay more for health care. At the same time a majority of the physicians acknowledges that there are too many treatments with low or marginal benefit (63%), that the use of diagnostic procedures is overrated (86%), that public expectations concerning medical technology are too high (82%). One of the major problems in health care identified by health professionals, is consumerism (Tijmstra, Busch and Scaf-Klomp, 1991).

These data point out that the crucial issue in the resource allocation debate today is the perception that making choices, at least in health care policy, is not really necessary. Such a perception is influenced through a specific way of interpreting the problem at stake as *financial scarcity*. In this case, three different types of analysis can be distinguished. Policy-makers are used to refer to economic constraints as a hard fact; financial resources are always limited; there are competing social goods and the opportunities to satisfy all desirable goods are restricted; making choices is therefore unavoidable. A second analysis aims at showing the relativity of the phenomenon of scarcity: scarce resources are not an objective reality leading to the necessity of choices, but they are themselves the result of implicit a priori choices; scarcity itself is a human construct resulting from deliberate human limitations and decisions. A third analysis introduces a distinction between real and fictitious scarcity. Scarce resources that are problematic now, are in fact the result of inefficient use of available resources; for the time being, scarcity is not a real problem; what is needed before we start making painful choices, is a large-scale operation of making the delivery of health care more efficient.

In contradistinction to this interpretation, the Committee proceeded from another interpretation: the issue of scarce resources is first and foremost a *cultural problem*. Allocative questions will not be less problematic if more resources, money and manpower are available. Scarcity will not disappear if every health service is delivered in the most efficient way possible. The reason is that the basic problem is not scarcity of resources but the infiniteness of human needs. The issue of allocation of resources is in fact a result of the prevailing value-system of post-modern societies. This system is a conglomerate of values that sustain the ever-increasing health care structures: health, the right to health care, equal access, and solidarity.

For modern man, health is apparently one of the most important values in human life. Value research in the Netherlands, for example, indicate that approximately 60% of the population identifies health as the most important value in life. Many have come to feel that they have a *right* to adequate health care and medical treatment, and that by virtue of this right the provision of services in this area is a task morally incumbent on the community and therefore on the various agencies of government.

Concern for the maintenance of equal access to health care leads to an ongoing critical assessment of every innovation and new service; innovative treatment which has experimentally been received by some, will soon be demanded by the public to be made available to all. The value of solidarity, understood as a collective obligation to care, is still generally endorsed in Dutch society (ten Have and Keasberry, 1992).

Given this value-system of modern societies, particularly those in Western Europe, choices in health care will be inevitable. The problem is not so much a matter of finances; it concerns more the self-perpetuating system of infinite patients' needs and rising expectations on the one hand, and newer technologies and interventions, as well as medicine's ever greater promises on the other (Callahan, 1990).

7.3. Priority Setting Process

The Government Committee studied the Oregon Health Care Plan as an example of a priority setting procedure. In a European context, the Oregon health policy is several steps too far. First, it assumes that there is broad public awareness that making choices is unavoidable, - and apparently the lack of such awareness is a central problem in many Western-European countries. Second, it assumes that there is consensus on the moral desirability of approaching the issue of scarce resources by setting priorities, rather than by rationing, waiting lists, and patient selection. In many European countries it is too early to notice such consensus, although there is growing dissatisfaction with the current practice of rationing. In fact, this practice leaves the solution of distributive problems to the individual health care professional.

From a moral point of view, it is desirable that allocative issues are approached from a health policy, *viz.* macro-level, perspective. This is the level of 'first-order determinations' that settle the scope of individual possibilities (Calabresi and Bobbitt, 1978). Prefering the macro-level is connected with its specific characteristics:

1. Decisions have a bearing not on individual persons but on patient categories;
2. Decisions require explicit criteria, equally applicable to and for everyone;
3. Decisions are made within a public process of deliberation;
4. As many actors and groups of actors as possible are involved in the decision-making process.

Because of these characteristics, decision-making on the macro-level of health policy will give *prima facie* better guarantees for equal treatment of individuals than on the micro-level where specific and idiosyncratic factors may determine the outcome of the individual doctor-patient relationship. The same is true for fairness of distribution since the allocative criteria and procedures are more open and controllable through public inspection than on the micro-level. Next, the macro rather than the micro-level requires the development of a procedure on the basis of democratic involvement of all actors. Finally, developing a priority setting procedure on the macro-level underlines that the ultimate responsibility for allocative decisions has been accepted by society. That will be a significant revision of the current practice in which individual health care professionals are involuntarily attempting to solve problematic situations that they have not individually created.

8. OTHER SOLUTIONS: A COMMUNITARIAN PERSPECTIVE ON THE WAITING LIST PROBLEM

Based on the above analysis of the allocation problem three approaches have been suggested for a solution of the waiting list problem: (1) increased public accountability, (2) prioritising among waiting lists, (3) re-interpretation of selection criteria.

8.1. Increased Public Accountability

Because many rationing mechanisms are implicit, choices made are not monitored and evaluated; those who ration are therefore not accountable for the ways in which funds are deployed and choices are made. The practice of patient selection in the context of waiting lists is no exception. In order to restore the moral acceptability of the waiting list as an instrument of fair rationing, it is necessary to have explicit selection criteria and clear selection procedures. The idea is that a set of formal decisions scrutinised by the public is preferable to the informal and covert decisions that currently govern health care decisions about who is treated for what and when (Hancock, 1993). Explicit rationing will allow open and public debate, and it will also remove rationing decisions solely from the hands of the medical and health care professionals and make them into a community responsibility.

The Government Committee on Choices in Health Care presents several recommendations to increase public accountability. Health care institutions should have a clear and standardized registration of all patients waiting for treatment or admission. Criteria and procedures for access to the waiting list as well as selection from the list for treatment should be publicly known. Health care facilities should report annually about the relevant waiting lists as well as about the policies regarding these lists. Facilities should also develop quality control systems, which can be assessed by supervising authorities. Only when greater efficiency, responsible substitution and strict admission criteria have not significantly resolved the waiting problem and existing treatment or care capacity is still insufficient, expansion of financial support can be taken into consideration.

8.2. Prioritising Among Waiting Lists

A second approach follows from the argument that priority setting on the level of health policy is preferable over patient selection on the level of individual patient care.

Even if one agrees that it is morally preferable to develop priority setting procedures, it is in a European context difficult to implement such preference. Priority setting implies making distinctions between more or less important health care services, between essential and non-essential care, between necessary and unnecessary treatment options. To many people, the idea of making such distinctions (though flowing from moral considerations) is in itself not consonant with traditional moral notions as equal access and solidarity.

Therefore, after having argued that choices in health care are unavoidable and that the best way to make choices is through priority setting on the health policy level, the next step in a European context is to show that making choices is at least compatible with equal access and solidarity. In its report, the Dutch Commission went further: the best way to safeguard the realisation of these moral notions in future health care practices is through making choices between what is more important in health care and what is less.

In times of scarcity the notion of *equal access* is inadequate: it furnishes little or no guidance on which rationing policies should be applied and which health care settings they should be applied to. By making choices in a priority setting process, equal access for everyone can be guaranteed to every service or treatment that is regarded as important or essential. Not all waiting lists are equally important; an order of priorities among waiting lists must be established which corresponds with the priority of the health care services involved.

The same holds for *solidarity*. By asking solidarity for every health care service possible and every medical treatment available without any reference to their necessity and benefit, the notion of solidarity will be stretched beyond reasonable and affordable borders, and thus it will be self-defeating. Making choices in health care can revitalize the concept of solidarity and endow it with new meaning. Health policy today is in many ways involved in attempts to shift the burden of care from the state to the individual. In doing so, a new type of solidarity might be promoted: solidarity not in the sense of an endorsement of redistribution of income, but in the sense of a disposition to accept responsibility for one's own life and one's own choices in life. In its latter sense solidarity may become a reason for self-exclusion from care as well as a reason for private initiative in organising and financing self-care in new social support systems. The latter concept of solidarity would imply that the autonomy of the individual consists in a recognition that one's own interests may be best served by promoting the common good.

In its proposals to differentiate among various waiting lists, the Dutch Government Committee argues from a communitarian perspective. Starting-point for the Committee's argument is the proposition that everyone who needs health care must be able to obtain it. Health care is a communal good. However, equal access to health care should not be determined by demands but by needs. In order to have a just distribution of services, it is not important *that* all services are equally accessible, but crucial is *what* services are accessible. Not every health care service is equally relevant for maintaining or restoring health. Thus it is important to identify 'basic care', 'essential services' or 'core health services' that are focused on basic health care needs in contradistinction to individual preferences, demands or wants. Relevant needs should be distinguished from all the things we can come to demand or want. In his theory of health-care needs, Daniels argues that needs are distinct in relation to their object, *viz.* health (Daniels, 1985). The concept of health is therefore the most appropriate standard for characterizing health care needs. This focus also illuminates the fact that health enables persons to maintain a normal range of opportunities to realize their life plans in a given society. Since it is not health care services as such that are 'basic' or 'essential', the Committee prefers the expression 'necessary' because it implies a relationship between the particular kind of care or service with a particular goal ('necessary for what?').

The Committee defines health in general terms as the ability to function normally. However, 'normal function' can be approached from three different perspectives.

8.2.1. The Individual Approach

Here, health is related to autonomy and self-determination. It is the "balance between what a person wants and what a person can achieve" (Government Committee, 1992, p. 51). Defined as such, health can vary according to various individuals; its content depends upon individual preferences. But then, no distinction is possible between basic needs and preferences; what is a basic need for one will not be for another. This approach therefore is not helpful in determining on a societal level what is necessary care that should be accessible to all. Even if through a democratic decision-making process (such as in Oregon) the largest common denominator or the smallest common multiplier of individual demands could be determined, we would lack criteria to identify necessary care.

8.2.2. The Medical-professional Approach

Usually it is the medical profession that defines health, *viz.* as the absence of disease. This approach is defended by Daniels. He interprets health as "normal species-typical functioning"; disease is defined as "deviation from the natural functional organisation of a typical member of a species" (Daniels, 1985, p. 28). Basic functions of the human species are survival and reproduction. Health care is more necessary as it prevents or removes dangers to life and enhances normal biological function. In this approach, necessary care could be distinguished according to the severity of illness; this was in fact proposed as a criterion by a Norwegian Committee (Royal Norwegian Ministry of Health and Social Affairs, 1990). Nevertheless, this approach has a tendency to neglect the psychosocial functioning of individuals. It is also questionable whether normal species-typical functioning can be defined regardless of the social circumstances.

8.2.3. The Community-oriented Approach

In this approach, preferred by the Dutch Committee, health is regarded as the ability of every member of the society to participate in social life. Health care is necessary "when it enables an individual to share, maintain and if possible to improve his/her life together with other members of the community" (Government Committee, 1992, p. 54). 'Necessary' care is what the community thinks is necessary from the point of view of the patient. This approach is not utilitarian because what is considered to be in the interests of the community is dependent on its social values and norms. Every community exists because it presupposes a normative, deontological framework defining the meaning of its interests. In most European societies at least three normative presuppositions define the communal perspective: (1) the fundamental equality of persons (established in the Constitution), (2) the fundamental need for protection of human life (endorsed in international conventions) and (3) the principle of solidarity (expressed in the organisation and structure of social systems, particularly the health care system).

Given this normative framework, it could be specified from the perspective of the community what should be regarded as necessary care. The Committee distinguishes three categories of necessary care: (a) facilities which guarantee care for those members of the society who cannot care for themselves (e.g., nursing home care, psychogeriatrics, care for the mentally handicapped), (b) facilities aimed at maintaining or restoring the ability to participate in social activities when such ability is acutely endangered (e.g., emergency medical care, care for premature babies, prevention of infectious diseases, centres for acute psychiatric patients), (c) care depending on the extent and seriousness of the disease; priority among facilities in this group depends not only on need, but additional criteria decide whether a facility would be included in the basic package of health care provided to every member of society. From a community-oriented perspective, the first category is more important than the second or the third, and the second more than the third.

On the basis of its community-oriented approach it is argued that absolute priority should be given to care for people who cannot care for themselves, such as the demented elderly and the mentally handicapped. Waiting lists for these health care facilities should have priority over, for example, a waiting list for PTCA.

8.3. Re-interpretation of Selection Criteria

The third approach that may lead to a solution for the waiting list problem is to restore the discriminating power of the selection principles, either by re-interpreting the existing principles or by introducing new principles.

Some authors argue that additional discriminating criteria are needed in order to have a more stringent distribution of resources (e.g., van Willigenburg, 1994). Age, life style and individual responsibility, personal characteristics, life situation, social value, or the importance of the individual for family members are mentioned as candidates for new criteria. The Committee argues that these non-medical criteria are not acceptable. There are indeed many practical difficulties regarding the use of such criteria; they need specification and articulation when implemented as selective instruments. There are also significant moral objections. But the point is that at least for some criteria it will be hard to argue from a communitarian perspective that they should not be used at all. For example, there is an intensifying debate now about the question whether or not patients who as a result of disease cannot work, should have priority over those who are not employed. Employers and insurance companies are making arrangements with hospitals for quicker treatment schedules in order to significantly reduce waiting time. They even want to do so outside the official working hours of health care professionals, making more efficient use of available facilities. Because private businesses and insurance companies will pay for these extra efforts, it is argued that they in fact allocate extra money to the hospitals which can help to shorten the waiting list for other patients. On the other hand, these initiatives have led to strong political responses; they are rejected because they could lead to disadvantages for other patients on the waiting list, and because they could hinder attempts to create more structural solutions of the problem for everyone (*NRC Handelsblad*, 1996). The interests of the individual person may not outweigh the collective interest of all patients waiting on the list.

However, from a communitarian perspective, productivity seems to be a significant way of contributing to and participating in social life. The only reason why having a job is not acceptable as a selective criterion within this perspective is that it is not compatible with the fundamental equality of persons as a basic principle of Dutch society.

The significance of equality is probably also the reason why the Government Committee wants to consider the re-interpretation of the existing moral principles underlying waiting lists, rather than introduce new principles. It argues that the criteria of medical need and medical success should be more strictly applied in reference to its redefinition of necessary care. Only when explicit and public criteria are applied, when a priority order among various waiting list is determined, and when a strict re-interpretation of the existing selection criteria according to a community-oriented perspective has been accomplished, only then is expansion of resources required.

REFERENCES

Calabresi, G. and Bobbitt, P.: 1978, *Tragic choices*. W.W. Norton, New York.

Callahan, D.: 1990, *What kind of life. The limits of medical progress*. Simon and Schuster, New York.

Chadwick, R.: 1993, Justice in priority setting. In: Smith, R. (ed.): *Rationing in action*. BMJ Publishing Group, London, p. 85-95.

Daniels, N.: 1985, *Just health care*. Cambridge University Press, Cambridge (Mass.).

Donaldson, C.: 1993, Economics of priority setting: let's ration rationally! In: Smith, R. (ed.): *Rationing in action*. BMJ Publishing Group, London, p. 75-84.

Government Committee: 1992, *Choices in health care*. Ministry of Welfare, Health and Cultural Affairs, Rijswijk.

Hancock, C.: 1993, Getting a quart out of a pint pot. In: Smith, R. (ed.): *Rationing in action*. BMJ Publishing Group, London, p. 15-24.

Have, H.A.M.J. ten: 1988, Ethics and economics in health care: a medical philosopher's view. In: Mooney, G. and McGuire, A. (eds.), *Medical ethics and economics in health care*. Oxford University Press, Oxford, p. 23-39.

Have, H.A.M.J. ten, and Keasberry, H.J.: 1992, Equity and solidarity; the context of health care allocation in the Netherlands. *Journal of Medicine and Philosophy* 17: 463-477.

Kimsma, G.K.: 1993, Hoe gaat de huisarts met wachtlijsten om: de toename van informele selectie en oncontroleerbare systemen van voorrang. In: Have, H.A.M.J. ten, *et al.* (eds.): *Ethiek en recht in de gezondheidszorg*. Kluwer, Deventer, VI 21-26.

Laborie, F.: 1993, Social alternatives to infertility. In: Stephenson, P. and Wagner, M.G. (eds.), *Tough choices. In Vitro Fertilisation and the reproductive technologies*. Temple University Press, Philadelphia, p. 37-50.

Meiland, F.J.M., Varekamp, I., Hoos, A.M., Danse, J.A.C., Krol, L.J. and Wendte, J.F.: 1994, *Selectie en wachtlijsten in de gezondheidszorg*. Instituut voor Sociale Geneeskunde, Amsterdam.

Nationale Ziekenhuisraad: 1989, *Selectie van patiënten, wachtlijsten en modellen voor een rechtvaardige verdeling van gezondheidszorg*. Nationaal Ziekenhuisinstituut, Utrecht.

NRC Handelsblad: 1996, Vooral hartpatiënten gedupeerd door wachtlijsten. Rotterdam, 21 juni 1996, p. 3.

NSS Marktonderzoek: 1991, Keuzen in de gezondheidszorg; Een opinie-onderzoek. Conclusies NSS-Marktonderzoek 1990. In: Commissie Keuzen in de Zorg, *Kiezen en delen 1*. DOP, Den Haag, p. 305-308.

Royal Norwegian Ministry of Health and Social Affairs: 1990, *Health Plan 2000*. Ministry of Health and Social Affairs, Oslo.

Tijmstra, Tj., Busch, M.C.M. and Scaf-Klomp, W.: 1991, Keuzen in de zorg: Meningen van beroepsbeoefenaars. In: Commissie Keuzen in de Zorg, *Kiezen en delen 1*. DOP, Den Haag, p. 309-316.

Vanrenterghem, Y. and Persijn, G.G.: 1996, The implementation of the new Eurotransplant kidney allocation system, *Eurotransplant Newsletter*: 4-7.

Willigenburg, T. van: 1994, Wachten op zorg: wie heeft er voorrang? In: Have, H.A.M.J. ten, *et al.* (eds.): *Ethiek en recht in de gezondheidszorg*. Kluwer, Deventer, VI 131-142.

CHRISTOPHER D. MELLEY

HEALTH CARE ETHICS COMMITTEES

1. INTRODUCTION

Sincere attempts at becoming more ethically responsive to individual and community needs, both in terms of increases in personal choice of individual patients and more general concerns in providing better health care to the public, sometimes take the form of an institutional addition to the existing health care structure. Health care ethics committees (HECs) are one such addition to the structure. Health care ethics committees have enjoyed rising popularity in North American, European, Commonwealth countries, and elsewhere, as an institutional means of helping to resolve moral problems in a variety of settings, primarily in direct medical care of patients and evaluation of research proposals. What are these committees? What are their various functions, methodologies, composition, and problems? How are we to understand and evaluate the emergence of these committees within the larger historical and intellectual context of applied philosophical ethics?

2. EXISTING SOLUTIONS

There are already institutional ways in which bio-ethical issues are currently debated and resolved:
(1) the courts, (2) direct government control of specific medical practices, and 3) the traditional doctor/patient/family relationships. There are, however, problems with each.

2.1. Courts

Courts, although acting according to a body of law based upon principles of justice, although acting or attempting to act in a disinterested manner, are simply incapable of handling the potential case load that would find its way to the already overburdened court dockets. Also, given the speed at which procedures - emergency operations, accidents - occur, that is, oftentimes

H.A.M.J. ten Have and B. Gordijn (eds.), Bioethics in a European Perspective, 239-259.

quickly and without notice, a court system would prove too lethargic, too slow to respond to what are often dynamic and fluid situations.

2.2. Central Government

Direct government involvement also poses problems. Policies developed and broadcast from a national government would create uniformity and consistency. Yet centralisation of the decision-making and policy-writing processes can take much time in reacting to immediate conditions at the regional and local level. Once guidelines and laws have been written, there is still the problem of application to particular cases.

Further, Alan Fleischman of the Albert Einstein School of Medicine points out that central government involvement "would be intrusive and extremely inefficient" (Fleischman, 1987, p. 387). He insists that the increased central government involvement does not ensure better health care. On the contrary, such intervention would reduce the morale of health care workers as well as result in over-treatment to guard against criticism.

Government regulation does not always mean compliance. In the Netherlands, for instance, the Dutch government stipulated in 1990 that physicians involved in euthanizing a patient must supply a "certificate reporting a non-natural death" (ten Have and Janssens, 1997, p. 394); however, there was significant non-compliance, leading the same government, recently, to propose retrospective regional committees as a supplement to existing laws.

2.3. Doctor-Patient-Family Triad

Traditional participants in the decision-making process have been physicians, patients, parents, other family members, and guardians. Why not let them continue in that tradition? The relationship of trust that is often found between the patient, family, and attending physician can positively influence the patient's treatment. Though this trust may not always be achieved in every case, the interpersonal relationships built up between the members of this triad are often strong and make for established lines of communication. Still, there are difficulties.

Although family members may be and often are close to the patient in need, the very proximity can have a blurring effect upon their judgement. Emotions may cloud their ability to think rationally. Ulterior motives may divide the family and the patient about what is in the best interests of the patient.[1] Although the attending physician is often that person aware of the technical options for that patient, given the patient's situation, the physician too has only one perspective, the clinical/technical one. The clinical perspective is important. Still, the clinical perspective may overshadow other moral and psychological views of what should be considered. HECs are said to ameliorate some of these problems by bringing morally-laden issues in patient treatment to an interdisciplinary committee that would open the issue to several non-familial, nonclinical, and non-judicial views.

3. HISTORY OF HEALTH CARE ETHICS COMMITTEES

HECs have existed in nascent or limited forms in some US, Canadian, and European hospitals in the form of Hospital Review Boards. HECs emerged as identifiable entities in significant numbers in the latter part of the 1970s. As John A. Robertson points out in his article 'Committees as Decision Makers: Alternative Structures and Responsibilities,' HECs have some precursors. One precursor was the institutional committees convened to determine the feeble-mindedness of patients as decisional aids for questions of sterilisation. Another one was the committees convened during the 1950s and 1960s to determine acceptable cases of therapeutic abortions. Yet another paradigm for HECs came about when committees were first formed in 1973 to decide the acceptance of government funding of all dialysis treatment in the US (Robertson, 1984, p. 86).

There is an important distinction between these historical committees and today's HECs. These precursors were constructed for limited, finite aims, whose completion brought the end of those committees. Today's HECs, though, have gained permanent or semi-permanent institutional status in what has been estimated at "sixty percent of the medium and large-sized hospitals in the country [US]" (Bouton, 1990, p. 62), and growing numbers of committees in European countries.

4. HEALTH CARE ETHICS COMMITTEES IN EUROPE AND ELSEWHERE

The use of HECs is of course not a uniquely American experience. In contrast to the American HEC models, in Europe at least, the function of clinical ethics committees is primarily to provide approval of research projects involving patients.

The first medical ethics committees in Europe emerged in 1966, in the UK and Sweden, with other countries following the Helsinki Declaration revision of 1975. The UK utilizes some 250 "local research ethics committees (LRECs)" that function similarly to the advisory institutional review boards in the US, though there are some isolated HECs that have expanded their agenda into clinical practice (Thornton and Lilford, 1995, p. 667; Nicholson, 1998). In 1997, centrally controlled French ethics committees number over

130, Belgium possesses 112, while Germany lags behind with 65 committees, most of which operate within a university setting (Nicholson, 1998).

The Swiss Academy of Medical Sciences officially established its own private central ethical committee in 1979. According to Gerhard Stalder, "all medical faculties, major hospitals and all pediatric clinics affiliated with a university have consulting bodies of this type" (Stalder, 1981, p. 120). Denmark has established HECs since 1982, with presently seven committees in operation (Nicholson, 1998).

Although it has been commented that "Japanese bioethics is ten years behind the world-wide movement" (Murase, 1989, p. 225), Japan established its first ethics committee in 1981, at the Medical Institute of Tokyo University. Today, virtually all Japanese medical schools, among other professional schools, have some form of ethics committees (Murase, 1989, p. 222). Gambia too possesses an ethics committee, concentrating on research proposals and risks to human research subjects (Gambia Gov., 1998).

5. HECs IN NORTH AMERICA

Committees in the US tackle multiple issues in health care, from case-specific do-not-resuscitate orders to research proposals and procedures for advance directives and allocation questions. Those specializing in research proposals are sometimes referred to as institutional ethics committees (IECs), institutional review boards (IRBs); others dealing with specific cases are variously called ethics committees (ECs), human values committees, medical-moral committees, and ethics consultation services (President's Commission, 1983, p. 161; Toulmin, 1988, p. 12). Some are specialized according to a clinical speciality and take on names that reflect that speciality, like pediatrics ethics committee (Michaels and Oliver, 1986, p. 566) and infant bioethics committees. For purposes of clarity - unless otherwise specified - we will refer to such collectives as health care ethics committees or HECs.[2] The diversity of names reflects a diversity of opinions of how committee members view their position in the greater scheme.

In the last 25 years, the growth of HECs that have institutional or semi-institutional status in the United States has been dramatic. A survey taken in 1982 shows only 3 percent of 602 hospitals surveyed had HECs (Youngner, *et al.*, 1983). Another survey in 1984 shows that roughly 50 percent of American hospitals had formed HECs. The American Hospital Association's (AHA's) National Society for Patient Representatives Survey plotted the growth of HECs between 1983 and 1985 and found that 59 percent of American hospitals have committees. Another report by the AHA puts the figures even higher: "60 percent of acute-care institutions and 80 to 90 percent of major medical centers" have HECs (Kurp, 1988). The state of

Maryland has since 1987 required by law that every hospital in the state develop HECs at their facilities. To date, Maryland is the only state to do so (Allen, 1990). According to Samuel R. Sherman, Chairperson of the Judicial Council of the American Medical Association, "Before long, all hospitals will have ethics committees" (Sherman, 1984, p. 131).

The growing frequency of HECs in Canada is similar. According to E.H.W. Kluge, Director of the Division of Ethics and Legal Affairs in Canada, "most of the provincial medical associations and ministries of health care [in Canada] have either guidelines for their use or are developing them; and most of the major institutions have them in place.[3]

6. HEALTH CARE ETHICS COMMITTEES: A PROMISING ALTERNATIVE?

As Michael Yeo of the Westminster Institute, in London, Ontario, wrote, "Medical ethics has become something of a growth industry in the last two decades" (Yeo, 1989, p. 23) and health care ethics committees are expressions of this growth. An article sharply critical of HECs' continues the 'economic growth' analogy by suggesting that "Bioethics is the latest buzz-word and institutional ethics committees have become a growth industry that rivals fast-food outlets" (Gerber, 1988, p. 229). Another critic contends that "In the US, bio-ethicism is a growth industry; posts are being created all over the country, and much heat but no great light is being generated" (Waterston and Sanders, 1984, p. 387). The growth of such committees is not uniform, nor do they share similar functions, and their emphases differ significantly between American and European models. In the US, such committees have enjoyed substantial growth in the last 16 years, while British and other European counterparts have moved more cautiously.

6.1. Varying Functions of HECs

There is no uniformly accepted mandate for an HEC. In the US, the President's Commission for the Study of Ethical Problems in Medicine and Biomedical and Behavioral Research (1983) provides four open-ended functional paths individual HECs could take. According to the *Report*, HECs provide an institutional setting for (1) consultation, (2) discussion of actual cases in relation to ethical theory, (3) multi-disciplinary discussions as well as (4) a forum for continuing education of moral themes in medicine.[4] Sister Corrine Bayley, one of the authors of the *Handbook for Ethics Committees*, reiterates and reinforces the functions of the President's Commission: (a) education, (b) policy and guideline development, and (c) consultation (Bayley and Cranford, 1986, p. 193).

The American Hospital Association (AHA) strongly suggests the advisory function of HECs:

> Ethics committees should not serve as professional ethics review boards, as substitutes for legal or judicial review, or as 'decision makers' in biomedical ethical dilemmas. An ethics committee should not replace the traditional loci of decision making on these issues (American Hospital Association, 1984, p. 1-2).

Some ethics committees exhibit change, evolving from evaluating research proposals to assessing the moral worthiness of direct medical care of particular patients. Richard H. Michaels and Thomas K. Oliver, Jr. at the Children's Hospital of Pittsburgh, Pennsylvania, originally created an HEC "to review clinical research proposals to assure that the rights of children are protected" (Michaels and Oliver, 1986, p. 566). As time passed, the committee defined itself more precisely. Their committee now serves at least three specific bioethical functions in the area of clinical care: (1) establishment of policy (e.g., guidelines for "brain death, cardiopulmonary resuscitation, innovative therapy, and human rights consultation"), (2) education, which includes "the development an annual hospital-based symposium on medical ethics with an invited national authority who lectures and presides at small group discussions" as well as the fostering self-education of HEC members by encouraging attendance at "regional and national meetings on medical ethics" and (3) optional, non-obligatory case consultation, which involves reviewing individual patients' records, clarifying issues to health care staff, patients, and family members (Michaels and Oliver, 1986, p. 556-557).

6.2. HEC Membership

Membership and composition of HECs varies widely and depends on the task and location of the committee. Committees formed within health care facilities tend to have a thick core of medical staff, with lay professionals (see below) while European committees, typically concentrating on research protocols and evaluation of proposals using patients or animals, have a thick core of scientists in the relevant fields; European committees also include a variety of lay professionals.

For instance, in dealing with ways of bettering physician cooperation in notifying authorities of involvement with euthanizing a patient, the Dutch government recently inaugurated five multi-disciplinary regional committees, each consisting of a lawyer, an ethicist, and a physician, who would in turn release their analysis to the prosecutor.

Though both North American and European committees utilize this multi-disciplinary approach, committees still lean heavily toward scientific/clinical representation in the UK (Neuberger, 1992) and Europe, though less so in the US. The American Academy of Pediatrics specifies at least eight members consisting of a nurse, two physicians, an ethicist or member of the clergy, a hospital administrator, a lawyer, someone familiar with disabilities, and a lay community leader (Fleischman, 1987, p. 384). The Judicial Council of the 1984 Annual Meeting of the American Medical Association recommends that HECs be multidisciplinary (Judicial Council, 1985). Michaels iterates the advocacy of a multidisciplinary committee:

> The membership should include both physicians and non-physicians with expert knowledge in relevant fields, and appropriate representation of a wide variety of medical and non-medical viewpoints (Michaels and Oliver, 1986, p. 572).

Of those clinical experts, the majority are physicians of one speciality or another. Although the heavy representation of physicians over other clinicians is found in hospitals, it is otherwise in long-term care facilities (LTCF). In LTCF, there is a greater representation of nurses over doctors.[5]

The heavy presence of clinical experts is understandable in the medical setting.[6] Still, the presence of social workers, psychologists, ministers, priests, rabbis, psychologists, even philosophers (or ethicists), attests to the growing belief that an interdisciplinary approach used in solving morally charged cases is considered better than a purely clinical approach. As one writer put it, "The interdisciplinarity [of HECs] is necessary to reflect the fact that moral insight and responsibility are shared across disciplines" (Mahowald, 1989, p. 243).

7. PROS AND CONS OF HECs

There are many pros and cons concerning HECs and ethics consultation. Some are more serious than others. Some are mendable; some are not. Some are theoretical; others are practical.

7.1. Positive Aspects of HECs

There are several virtues of HECs. Here are some perceived strengths of this mechanism of resolving moral dilemmas within the health care setting:

1. Committees allow for diverse points of view that enhance the education of all involved. The multidisciplinary mix of participants helps rather than hinders moral decision-making. Each member stands to learn from the other. The committee offers an excellent place within the health care setting for education to take place. First, the education could be centred on self-education. Later, it could involve other hospital staff not directly involved

with the committee. As Dana E. Johnson, a neonatologist at the University of Minnesota Hospitals and Clinics, in Minneapolis, says,

> The more you talk about [defective babies] in a committee, the better you will be able to educate yourself, the better you are going to be able to educate the parents, and the better decision you will be able to make (Johnson and Thompson, 1984, p. 729-730).

According to Ronald E. Cranford, "One of the values of the committees is to tell health care providers what's happening in society and to educate them to major developments as they occur" (Cranford, 1986, p. 14). An HEC could act as an educational conduit for those in need of relevant information and those in need of giving it. The forum provided by the committee structure allows a time and place for this to happen.

HECs simply offer "a wider perspective than individuals" (Uhlmann, *et al.*, 1987, p. 602). Because of this wider view, their usefulness could be beneficial in surrogate decision-making. Surrogate decision-making is a major issue for HECs in long-term care facilities (LTCF), where many elderly patients have no family to claim them for their own. This exasperating problem of surrogacy might best fall to the HECs, whose variety of concerned views and institutional goal or mission is to keep the patient as the primary focal point.

As well, the multidisciplinary mix of clinical and non-clinical participants that is present in many HECs in the US would help to avoid one-sided views that consider *merely* the clinical information or *merely* the psychological information.

2. Committees allow for a case by case review, retrospective and prospective (Lo, 1987, p. 46). The casuistic approach, both in retrospective and prospective review, allows committee members to attend to the details of a case. Case review makes the details of a patient's situation important. This has the effect of making the patient the centre of attraction or of importance, which is often thought to be the correct focus for health care staff and for HECs.

Alan Fleischman suggests that both types of review, especially retrospective, provide opportunities for self-education. As well, the review process gives all involved needed experience in dealing with actual cases that staff face or have faced. The next time a similar case arises, those involved will be prepared for what might arise. Guidelines for future decisions also could be drawn from these types of review (Fleischman, 1987).

3. Committees offer an institutional alternative to taking a hard-line sanctity of life stance to every medical anomaly that confronts health care staff, judicial review on a massive scale, or the traditional doctor/patient/family triad. The HEC offers the institution and its staff flexibility in reacting to hard cases on an individual basis. Faced with an ever-changing and increasing

technical array of life-support systems, the committee could offer a way to facilitate discussion where earlier there were no means to do so.

4. Committees provide reinforcement of a social nature to those who find themselves making and carrying out those decisions. The concurring judgement of an HEC would help bolster the attending health care staff in their shared values with the larger community. The moral milieu in which decisions are made would be confirmed, and the presence and supportive collective voice of the group would bolster the individual voice of the attending physician.

5. Committees allow decisions to be made by a group of disinterested persons (Lyon, 1985). American ethicist Ruth Macklin is adamant about members' disinterestedness concerning given cases, meaning not having ulterior motives on the outcome of the case (Macklin, 1987). If certain committee members have ulterior motives, the attempt to broaden a decisional setting helps to nullify or balance any ulterior motive that would otherwise loom too largely.

6. Committees provide what has been called "ethical comfort" (Fleischman, 1987, p. 389) to health care staff and to relatives of the patient. The increasing difficulty of making 'techno-ethical' judgements in the clinical setting will only make future decisions more complex, more difficult for the individual clinician to bear the weight alone. Depending on the case, guilt, inner conflict, frustration, uncertainty, grief can accompany health care workers' decisions and actions. Robert H. Sweeney, president of the National Association of Children's Hospitals and Related Institutions, thinks that HECs could be "an instrument of inner peace in situations of great stress" (Sweeney, 1987, p. 184).

7. Committees would help to diffuse responsibility. The consensus of those in the same or other fields would reinforce the practitioners' actions concerning a case. Those who do make morally troubling clinical decisions in patient care would have the institutional and psychological reinforcement of a group of clinical and non-clinical professionals.

8. As a viable alternative to the traditional physician-patient-family triangle, the courts and the legislatures, HECs offer a forum for that patient who cannot speak or think, who is comatose or infirm. The traditional physician-patient-family triangle has disadvantages, as do the courts and legislative action. HECs would offer an alternative when any of or all the other avenues fail or are deemed inappropriate.

9. Committees create a forum for disparate opinions from both within and outside the medical community to be voiced in an atmosphere of mutual tolerance. From the literature reviewed, the endorsement of group dynamics in resolving ethical dilemmas is widespread. For instance, an observer of the activities of various institutional review boards of the National Institutes of Health is confident of the positive aspects of group work: "When people as a

group discuss a procedure, they illuminate the dark corners no single one of them would have thought of if asked to evaluate it alone" (Paris and Reardon, 1986, p. 42). This view is supported by philosopher Richard Roelofs, after participating in one of the first HECs in the US, at Montefiore Hospital, in New York City, 1976-78. Roelofs states that at Montefiore "there was need, as well, for some regular forum, open to the hospital community at large, in which significant questions of law and ethics in medicine might be discussed" (Roelofs, 1980, p. 35).

Another writer, concentrating on committees within nursing homes, suggests that "the primary purpose of an ethics committee is to foster an institutional milieu that is sensitive to ethical priorities" (Brown, *et al.*, 1987, p. 1032). Finally, according to Corrine Bayley and Ronald Cranford, "In the short run, committees continue to perform a critical function by virtue of their very existence: they are tangible evidence of an institution's concern with ethical issues in health care" (Bayley and Cranford, 1986, p. 199). Their very presence, controversial as they sometimes are, offer a place where the ethical questions are never far off or foreign.

10. Committees help health care staff identify and recognize ethical issues in their field (Perkins and Saathoff, 1988). Highly publicized ethical issues involving Do Not Resuscitate (DNR) orders, continuation of life support systems or adult competency often attain a high-profile in public debate. Still, there are issues in health care that are not readily recognized for their ethical component. Routine lab tests, such as blood testing of clearly terminally ill patients, limb restraints for terminally ill patients, and procurement of a caretaker for incompetent adult patients may be such issues that must first be identified to come within the realm of ethical discussion. Ethical issues abound at the policy-level as well. As one speaker said at the 1988, Geneva conference on Health Policy, Ethics and Human Values: European and North American Perspectives:

> Health policy is inherently a value-laden enterprise. Every health policy is driven by assumptions about what ends are worth seeking, what constraints must be respected, what means are justified, what is most important (Gorovitz, 1988, p. 187).

7.2. Objections to HECs

1. Working in groups can also have negative connotations, one of which is the influence a bullying committee member may have on less aggressive participants. A second, related objection concerns the well-proven dangers inherent in group dynamics. Groups behave differently than individuals. Individuals would then 'think' differently, at least overtly so.

'Group dynamics' do offer the potential advantages of having several people from various disciplines, where each is incapable of purveying the same situation from so many angles. The diversity of views would, given the diversity of the committee membership, proffer a rich spectrum of views. This spectrum would offer members views not otherwise considered, if left to themselves. The presence of a multi-disciplinary committee would help to weaken the ideological walls that tend to separate and isolate fields of knowledge, even weaken that great wall of mutual deception that separates the sciences from the humanities. Stephen Ayres, who makes the above observation, does so by referring to C.P. Snow's notion of the 'two cultures' where there is a strict and abiding division between science and the humanities and the people that study them. Ayres thinks committees serve a positive function because

> Physicians are not always sufficiently humble, and scholarly nonphysicians are not always sufficiently knowledgeable about medical matters ... Comprehensive consultation or consensus gathering among individuals with differing backgrounds can benefit the quest for the 'right' decision. The obvious existence of at least two intellectual cultures in the health care structure is an important reason for patient care committees (Ayres, 1984, p. 33).

In a more recent analysis, Finnish philosopher Heta Häyry suggests that use of lay members is intended to supplement one of two functions that most European ethics committees exhibit: "an understanding of the moral ideals which prevail in the society or community where they function" (Häyry, 1998, p. 57).[7]

Ethics committee members could just as well give in to the pressures of thinking in groups, to peer pressure (Lo, 1987, p. 48). This could manifest itself in the "domination by members of one profession, ideology, or religious belief" (Young, 1986, p. 73). Häyry also observes that supporting lay members "are likely to side with their biomedical, frequently senior, colleagues if they have to choose between the promotion of research and the safety of the subjects of human or animal experimentation" (Häyry, 1998, p. 57). If this did happen, then committee members might reach their 'morally autonomous' decision to avoidance dissension. Some members are simply more 'powerful' than others in terms of their position, discipline, and psychological nature. Although even one's private reflections are 'tainted' by the public world in which one lives, it is another matter to have alternate views embodied by other people at the opposite end of the table. The influence one member of a group has on another is not to be discounted.

2. The scope and extent of a given HEC's agenda, if not already limited by committee members' own decision to limit it, would be too ambitious, too presumptuous. Individual health care facilities can encounter a variety of cases needing the help of an HEC. One case might involve a DNR order, another might involve a question of transplantation recipient, and yet another might concern a financial consideration involving a patient whose hospital

needs would involve detracting from other, perhaps more routine hospital services. Mark Siegler, in a famous article, criticizes HECs' (he uses the term IEC - Institutional Ethics Committee, but he is referring to HECs) *presumptuousness* in thinking it possible to consider all sorts of cases, which span many disciplines within medicine:

> IECs, [HECs] ... often think they are capable of analyzing, adjudicating, and resolving the most delicate and complex clinical matters. I think they are wrong (Siegler, 1986, p. 22-24.)

A possible response to this criticism would be to specialize committee work according to the problem at hand. Robert Veatch supports the idea that committees "can plausibly be legitimated for only one ethical task at a time" (Veatch, 1984, p. 41).

This is a question of scientific, clinical and moral competence. One solution to the criticism is to limit the domain of discussion and consideration. Another would entail making committee membership fluid. Active participation would be contingent upon the specific issue at hand. The logistics of either solution could entail bureaucratic nightmares for staffers. Outside major metropolitan areas, where 'networking'[8] (van Allen, 1987, p. 79) between health care facilities and universities takes place, the logistical job of collecting professionals from the relevant specialties could be impossible.

One way to avoid spreading the committee's commitment too thinly is to agree from the moment of its inception to limit discussion to one function. For instance, the function of education, first self-education, then later the education of others, recommends itself for beginning committees.

Another voice in the debate, Sister Margaret A. Farley, of Yale Divinity School, suggests that ethics committees' activities go beyond the few precursors outlined originally by the President's Report of 1983. There are many forces within the hospital that contribute to the betterment of the patient. If we assume that the ethics committee's main goal is to benefit the patient, then an increase in the variety and dimensions of committee topics recommends itself. Questions of social justice loom large in allocation questions, in the degree of employee participation health care workers have in the institutional framework, in how the health care facility runs, from advertising and hiring to financing and purchasing (Farley, 1984).

Although Farley acknowledges that many of these issues noted above really should be dealt with at the legislative level, since "Many social justice issues in health care simply cannot be solved at the institutional level" (Farley, 1984, p. 32), she is very positive about the development of ethics committees. Ethics committees may create the forum for such issues to be voiced by the various interested parties represented by individuals. She contends that

> Perhaps the major need for IECs (Institutional Ethics Committees) is to help find a new 'hermeneutic,' a new vantage point for interpreting health care delivery and our relations with one another in that enterprise (Farley, 1984, p. 35).

3. 'Many hands' make for no truly responsible hands. The term 'many hands' refers to the sheer number of participants - the number of participants varies widely - that would cause a "diffusion of responsibility" (McCormick, 1984, p. 154). McCormick rightly notes that the negative effects of an absolution of individual responsibility *vis-a-vis* the patients and regarding patient care could erode the often assumed primacy given the patient in current committees. Robert M. Veatch concurs with McCormick. Veatch contends that if a committee decides, recommends, educates, or writes policies for health care staff and others, the sheer number of members on the committee allows for the possibility of a bureaucratic escape clause for each individual committee member (Veatch, 1984, p. 36). This form of escape is possible in other collectives. Rosemarie Tong, author of *Ethics in Policy Analysis*, makes this point clear when she states that

> What is frightening about contemporary policy making is the telling that the decisions most likely to affect society for better or worse are made not by identifiable persons but by an amorphous collectivity whose willy-nilly process cannot be stopped once they are started (Tong, 1986, p. 68).

More specifically, Veatch states that "the committee might leave no one with the sense that he or she is responsible for the way a patient dies" (Veatch, 1984, p. 36). He is correct in claiming that one of the dangers of HECs include what he calls a "false sense of closure" (Veatch, 1984, p. 35) about the supposed consensus of moral values in a community. This would occur when the HEC is viewed by members of society (and the medical institution where it operates) as an adequate means by which moral dilemmas can be resolved by what could be an mixture of moral beliefs that happen to be near the discussion.

4. Vested interests among strong committee members, especially those firmly attached to the hospital administration, would tend to protect the institution's needs over others', for instance, patients'. Such protectionism would entail using the health care ethics committee as a shield against potential medical malpractice suits filed against the institution or its doctors.

Could HECs become servants of hospitals and their interests? (Kliegman, *et al.*, 1986). This is what Richard McCormick calls "in-house protectionism" (McCormick, 1984, p. 154). Could HECs

> Be transformed into devices designed to protect physicians and hospitals rather than the best interests of the patient or, worse, allow ... them to evolve into administrative tribunals or legal forums in which the law and due

> process, not ethics and the interests of the patient, are the paramount concerns ? (Paris and Reardon, 1986, p. 120).

The use of committee review to reduce or avoid legal liability is not new in the United States. According to George J. Annas, review committees for abortion and dialysis patient selection were used for just this purpose (Annas, 1984). Yet if this were part of HECs' underlying function, what has that to do with ethics? Just as importantly, what does the avoidance of legal liability have to do with what is best for the patient? Even if there are times when the avoidance of legal liability helps the patient, there is no logical necessity to this.

If a committee's function was partly to avoid liability, then the much vaunted autonomy of patients and the primacy of their care would be jeopardized. The individual patient already feels threatened by large scale medical operations. The sense of loss of control and uneven odds between patient and physician and patient and hospital could increase, if measures were not taken by the HEC to preclude ulterior motivation getting out of hand.

5. The doctor knows what is best. The very presence of a committee would be an affront to physicians with this view. Physicians holding this view must consider committees yet another invasion of their professional territory, "what they regard as the 'intrusion' of ethics on their turf" (McCormick, 1984, p. 153). Simply put, the doctor does not and cannot always know what is best, but often needs contrasting views, other opinions from colleagues. The question of what is best has moral implications as well as clinical ones. Physicians may be competent in dealing with clinical issues. However, they are not specially suited to determine the moral implications of a case, at least not without some help from others. Further, it is asking too much of physicians that they should know what to do in all cases. It would be a mistake on a doctor's part to assume such over-arching authority. The distribution of responsibility from the physician's viewpoint is not necessarily negative. There is such a variety of differing opinions that physicians have on moral themes. Allowing them to follow their conscience on moral issues is tantamount to allowing almost anything.

This fifth criticism rests on two *false* assumptions: (1) that there is no essential difference between technical and moral decisions and (2) that physicians, because of superior technical knowledge, are more adept at making the moral decisions concerning a given case, within the traditional confines of professional practice. Both assumptions are unfounded. There is an essential difference between technical and moral decision making: assuming one does not automatically accrue the other. Superior technical knowledge does not guarantee a 'good' moral decision. The paternalism of physicians and other health care staff in this discussion is unfounded. Health care staff may legitimately feel threatened by yet another institutional force invading their territory. Still, they must admit that their training does not qualify them to make moral decisions. Medical practice is not a private affair; its influence reaches practically all people at least some time in their lives. The practice of medicine is not a private affair, as McCormick correctly asserts. The very position which health care staff, specifically physicians, occupy is veined with many ulterior motives. Physicians can act as conduits for patient-hungry hospitals. As McCormick points out, "hospitals do not want to alienate their feeders" (McCormick, 1984, p. 153). This relationship would change or at least be monitored in those cases that are voluntarily brought to the attention of an HEC. The question that arises is the aptitude of the committee to discuss matters moral. This would be yet another criticism of HECs.

6. Call an HEC by any other name. Providing a universally accepted name to these committees has been problematic from the start. The variety of names also reflects a possible and very understandable misconception of observers. If we call a committee an *ethics* committee, then it *seems* as if this is the centre of ethical discussion in a society or a hospital. It also may seem to the public that *this* committee *is* the source of moral doctrine or moral truth. The public's perception of bio-ethics generally is often different from the perception of those within the activity itself. According to John W. Glaser, the public's perception of HECs is not on the mark, perhaps because of the name(s) we have given the committee:

> We have the impression, as a society, that ethical decisions are few, dramatic, far between and happen with a solemn awareness that 'this is an ethical issue'. HECs run the risk ... of reinforcing these distortions of ethical reality. Since no other forum gets the title 'ethics' we can be deceived into thinking that ethics happens - only, best, exclusively - here where it is announced in the title (Glaser, 1989, p. 275-276).

This misconception is an internal *and* external problem for participants and observers, respectively.[9] If the misconception occurs internally or externally, either wilfully or not, the HECs would represent *the* forum for ethical issues, *the* place where such issues can be clarified, understood, and analyzed in a professional manner by professionals. This issue is linked to the notion of expertise in ethics.

7. The non-clinical participants are ill-equipped for an intelligent discussion that ought or ought not to occur with a patient whose medical problems are beyond the understanding of the non-clinician's ability to grasp them. Participants' dearth of clinical knowledge would harm rather than benefit the patient. This is an important criticism, but its strength depends on what a *clinician* is. If a clinician is a physician, then there are many ethicists/philosophers actively working either full or part-time in the medical

setting as members of HECs or as ethics consultants who are *not* clinicians. If *clinician* is defined by another criterium other than membership in the group called physicians, then the strength of the objection would weaken.

Both Mark Siegler (1986) and John La Puma (1989) represent those who strongly support the view that these positions should be filled by people with a strong medical background, simply because so much depends on being able to assess a patient's situation concerning treatment procedures. Siegler, who is highly critical of HECs on several counts - many are noted in this section - favors instead the "formation of many small advisory groups possessing great clinical expertise in their own speciality and composed primarily of involved clinicians but with occasional representation of other experts" (Siegler, 1986, p. 23). He does, though, support "occasional representation of other experts" (1986, p. 23). Siegler makes a few medical analogies that show the severe limitations one medical expert in a given field has. Though their basic medical knowledge overlaps, the way they *apply* what they have learned to one or another speciality or sub-speciality differs greatly. If severe limitations exist here, then one can just imagine the insurmountable limitations a non-clinician would entertain in reviewing the technical subtleties of a given patient's care. Further, he suggests that a similar difficulty in applying "ethical principles such as beneficence, truth telling and autonomy" (1986, p. 24) exists in the areas of philosophy and theology (Siegler uses the term 'religious groups'). Siegler's solution to the problem of application in the ethical sphere is to limit the use of HECs as training grounds for clinical experts.

> The goal of such committees should be to develop clinicians from each of the clinical disciplines who have both the cognitive knowledge in ethics and law and the clinical experience to assist their colleagues in reaching sound clinical-ethical decisions (Siegler, 1986, p. 24).

Here the committee would act as teacher to medical specialists who would be in positions to help other health care specialists in the field. In his views on teaching medical ethics, Siegler is consistent in his view in leaning heavily toward clinicians in the leading role as teachers of ethics at the bedside.

La Puma supports a member having a strong background in medical technologies, terminologies, and "who understands the technical and personal details of a particular patient's care" (La Puma, 1989, p. 1110). In a letter to the editor that responds critically to Bernard Lo's positive assessment of HECs in a consulting capacity, La Puma suggests that a 'clinician-consultant' should be "competent in medicine and ethics" (La Puma, 1987, p. 1418). He does not explicitly say that a medical degree ought to be required for participation in an HEC or in doing work as an ethicist consultant. He does come very close though.[10] "Clinical credibility," for La Puma, along with "demonstrated subspeciality expertise," are two essential characteristics to the acceptance of a medical consultant (La Puma, 1987, p. 1418). Given the non-clinical backgrounds of committee members, clinical specialists' acceptance of non-clinical recommendations, thoughts, and efforts would not come easily.

This problem relates to any interdisciplinary issue. Similar arguments could be constructed to show that only experts in the field of nuclear energy know what is best in the construction of nuclear power plants. Only bio-chemists could judge issues relating to nerve warfare agents. Surely, this is not so. The moral dimension cuts through most fields, precisely because the moral dimension is concerned with the results of human action generally.

How can the deficiencies of expertise in medical/non-medical participants be overcome? No one can be expert in every field. Therefore, we rely on cooperation among professionals. Still, there is a requisite amount of knowledge of others' fields of expertise to talk across disciplines on a related issue. How much? Given this recurring problem, should there be a combined degree program for physician/philosophers or physician lawyers or for physician/philosopher/lawyers?

Others are not nearly so critical of non-clinical help in tough decisions. Jean-Pierre de Chadarévian, of Montreal Children's Hospital, readily accepted outside help in moral decision-making. In the initial formation of their first committee, de Chadarévian and medical colleagues found it difficult to assess the application of their committee mission to particular cases.[11] De Chadarévian reports that

> It became clear that most of us lacked the expertise and methods of approach that an ethicist devoting his or her time to the study of this complex subject could offer and teach. The feeling was shared by physicians, nurses, and other health professionals, and we decided to have an ethicist on the hospital staff. To us, he was seen as an expert adviser and not a 'medical consultant' (de Chadarévian, 1985, p. 188).

Alan Fleischman argues *for* mutual participation in various disciplines as a prerequisite for adequate moral discussion of clinical ethics. While arguing for a more case-oriented teaching program that would allow residents to acquire mastery in "moral diagnosis" and an awareness of the legal implications of newly developed technologies in perinatology, Fleischman hopes that

> Physician educators and their philosopher lawyer colleagues can work together to fruitfully integrate philosophic and legal theory with clinical case material. The philosophers can elucidate the ethical underpinnings of the physician-patient relationship, the lawyer can detail the legal consequences of physician actions, and the physicians can add their perceptions of the art of medicine and their intuitive responses to complex cases (Fleischman and Rhoden, 1988, p. 794-795).

The presence of a committee with a diversity of backgrounds can help to remind members of their professional limitations and at once remind each of the 'moral thread' that has strung them together in the medical setting.

8. CONCLUSION

HECs have established a presence in North American health care facilities and are experiencing some interest in European nations as a viable multi-disciplinary approach to moral resolution of difficult cases. The wide latitude of committee formation and function allows for flexibility of application to differing health care and cultural settings. Though not without inherent and situational difficulties, committees of the varieties described can offer a forum within the health care setting that fosters tolerance and acceptance of diverse opinion and professional background, where both the needs of the individual patient and the needs of the community can be heard.

NOTES

1. The 'best interest' standard is often found in deliberations of patient's treatment. Given the possibilities, what course of treatment would be in the patient's best interests, all other things considered?
2. National Society of Patient Representatives Survey on Institutional Ethics Committees, 1983: According to the survey, conducted in 1983, the following names for HECs were offered from respondents: Medical Ethics Committee; Medical-Moral Committee; Ethics Committee; Bioethics Committee; Biomedical Ethics Committee; Prognosis Committee; Critical Care Committee; Administrative Committee; Patient Care Committee; Human Values Steering Committee; Life Support Committee; Patient Care Evaluation Committee; Morals and Ethics Committee; Medical Bioethics Committee; Quality of Life Committee; Professional Activities Committee; Health Care Ethics Committee; Extraordinary Care Committee; Ethics Study Group; Medical-Moral and Social Justice Committee; Clinical Ethics Committee; Ethical Review Committee; Care of Terminally Ill Committee; Ethical Issues Committee; Medical Board; Ethics and Grievance Committee; Human Rights; Institutional Review Board.
3. Letter received from E.H.W. Kluge, July 17, 1989.
4. (1) HECs provide a consultative support for health care staff and families. Such a consultative role does not mean making decisions;
(2) HECs provide an institutional means by which particular ethical principles can be related to specific situations that one would find in the health care setting;
(3) HECs allow relevant members of the health care staff, in unison with members in related fields, the institutional possibility to address morally charged issues in a setting that would help to remind all of the seriousness of the issues;
(4) HECs would act as center for the ongoing education process that would be necessary for those charged with discussing issues within a variety of (usually) technical topics. Further, HECs would serve as catalysts which would provoke thought beyond specific types of moral

dilemmas such as the question of competency to "a consideration of more general bioethical issues" (Larcher *et al.*, 1997, p. 163).

5. Barbara Brown says that "the prominence of nurses in the formation and membership of LTCF ethics committee in initiating committee consultations is a striking, if not unexpected difference between nursing homes and the physician-centered hospital committees ..." (Brown *et al.*, 1987, p. 1032).
6. Mary B. Mahowald, a philosopher who has developed numerous HECs in the US, suggests having a greater number of physicians as members in order "to insure that a core of them are actually present at meetings" (Mahowald, 1989, p. 244).
7. The other expectation of European HECs, according to Häyry, is "to display ... professional scientific competence ..."
8. In fact, there are at least two networks established for that purpose: The Minnesota Network for Institutional Ethics Committees and the Delaware Valley Ethics Committee Network, both of which attempt to foster HEC committee development and to establish ties between existing committees, serving as clearinghouses for information and people with relevant skills.
9. An example of an internal misconception that results from the negative connotations that some physicians would automatically derive from the term 'committee' would be the following from Norman Fost: "The word *committee* is offensive to many physicians, suggesting an unwelcome bureaucratization of what should be personal and private decisions" (Fost, 1985, p. 2687).
10. La Puma defines 'clinician' as follows: "A clinician is one who understands particular patient's history, personal situation, and medical illness sufficiently well to help in managing the illness. A clinician is an experienced professional who attends and becomes involved with individual patients regularly ... If nonphysicians wish to be clinicians and work as ethics consultants, they need to acquire a similar core fund of clinical experience and expertise ... Becoming a clinician involves both didactic knowledge and everyday practical wisdom, and cannot easily be condensed into a primer for nonclinicians, as the following hypothetical case illustrates" (La Puma, 1987, p. 1110).
11. De Chadarévian briefly outlines the mission of the committee as being to reflect society's expectations, that is, to conform to them. Although a critique of the various philosophical assumptions made in the literature dealing with HECs comes further on, suffice it to say that this mission deserves serious scrutiny and review (de Chadarévian, 1985).

REFERENCES

Allen, C.L.: 1990, In medicine, ethics by committee. *Insight/Washington Times*, July 23; rpt. in *Newsbank*.

Allen, E. van: 1987, Ethics committees in Minnesota hospitals. *Minnesota Medicine* 70: 79.

American Hospital Association: 1984, *Guidelines: Hospital committees on biomedical ethics*. American Hospital Association, Chicago.

Annas, G.J.: 1984, Ethics committees in neonatal care: Substantive protection or procedural diversion? *American Journal of Public Health* 74(8): 843-845.

Ayres, S.M.: 1984, When values compete: Ethics committees and consensus. *Health Progress* 351 (9098): 32-34.

Bayley, C. and Cranford, R.E.: 1986, Ethics committees: What we all have learned. Reprinted in: *Making choices: Ethics issues for health care professional*. American Hospital Publishing, Inc., Chicago, IL., p. 193-199.

Bouton, K.: 1990, Painful decisions: The role of the medical ethicist. *The New York Times Magazine*, August 5.

Brown, B., Brown, A., Miles, S.H. and Aroskar, M.A.: 1987, The prevalence and design of ethics committees in nursing homes. *Journal of the American Geriatric Society* 35(11): 1028-1033.

Chadarévian, J.P. de: 1985, Letter to the editor. *New England Journal of Medicine* 312(3): 188-189.

Cranford, R.E.: 1986, quoted in: Hosford, B.: *Bioethics committees: The health care providers' guide*. Aspen Publication, Rockville, Maryland.

Farley, M.A.: 1984, Institutional ethics committees as social justice advocates. *Health Progress*, October: 32 ff.

Fleischman, A.: 1987, Bioethical committees in perinatology. *Clinics in Perinatology* 14(2): 379-393.

Fleischman, A.R. and Roden, N.K.: 1988: Perinatal law and ethics rounds. *Obstetrics & Gynecology* 71(5): 790-795.

Fost, N.: 1985, Hospital ethics committees: Administrative aspects. *JAMA* 253(18): 2687-2692.

Gambia Government/Medical Research Council Joint Ethical Committee: 1998, Ethical issues facing medical research in developing countries. *The Lancet*, January 24: 286.

Gerber, P.: 1988: Ethics and drug trials, editorial. *The Medical Journal of Australia* 179: 229-230.

Glaser, J.W.: 1989, Hospital ethics committees: One of many centers of responsibility. *Theoretical Medicine* 10(4): 275-288.

Gorovitz, S.: 1988, The viewpoint of the ethicist. In: Bankowski, Z. and Bryant, J.H. (eds), *Health policy, ethics and human values: European and North American perspectives, conference highlights, papers and conclusions*. World Health Organisation, Geneva, p. 187-190.

Have, H. ten and Janssens, R.: 1997, Regulating euthanasia in The Netherlands. *HEC Forum* 9(4): 393-399.

Häyry, H.: 1998, Should the decisions of ethics committees be based on community values? *Medicine, Health Care and Philosophy: A European Journal* 1(1): 57-60.

Johnson, D.E. and Thompson, T.R.: 1984, The 'Baby Doe' rule: Is it all bad? *Pediatrics* 73: 523-529.

Judicial Council: 1985, Guidelines for ethics committees in health care institutions. *JAMA* 253(18): 2698-2699.

Kliegman, R., Mahowald, M.B., *et al.*: 1986, In our best interests: Experience and the workings of an ethics review committee. *The Journal of Pediatrics* 108(2): 178-188.

Kurp, P.: 1988, For 'God Squads', a matter of life and death. Albany, New York *Knickerbocker News*, March 29; rpt. in *Newsbank*.

Larcher, V.F., *et al.*: 1997, Pediatrics at the cutting edge: Do we need clinical ethics committees. *Journal of Medical Ethics* 23(4): 245-250.

Lo, B.: 1987, Behind closed doors: Promises and pitfalls of ethics committees. *New England Journal of Medicine* 317(1).

Lyon, J.: 1985, *Playing God in the nursery*. W.W. Norton & Company, New York.

Macklin, R.: 1987, *Mortal choices: Bioethics in today's world*. Pantheon Books, New York.

Mahowald, M.B.: 1989, Hospital ethics committees: Diverse and problematic. *HEC Forum* 1(5): 237-246.

McCormick, R.A.: 1984, Ethics committees: Promise or peril? *Law, Medicine & Health Care* 12(4): 150-155.

Michaels, R. and Oliver, T.K., Jr: 1986, Human right consultation: A 12-year experience of a pediatric bioethics committee. *Pediatrics* 78(4): 566-572.

Murase, T.: 1986, Concerning the establishment of bioethics forum. *The Journal of the Japan Medical Association* 96(5): 989-92, quoted in: Sakai, A.: 1989, The present status of ethics committees in Japan's 80 medical schools. *HEC Forum* 1(4): 221-228.

National Society of Patient Representatives Survey on Institutional Ethics Committees: August 1983.

Neuberger, J.: 1992, Ethics and health care: the role of research ethics committees in the United Kingdom. *Research report no 12*. Kings Fund, London.

Nicholson, R.H.: 1998, Seeking harmony in discord. *Hastings Center Report* 28(1): 7.

Paris, J.J. and Reardon, F.E.: 1986, Ethics committees in critical care. *Critical Care Clinics* 2(1): 111-121.

Perkins, H.S. and Saathoff, B.S.: 1988, Impact of medical ethics consultation on physicians: An exploratory study. *The American Journal of Medicine* 85: 761-765.

President's Commission for the Study of Ethical Problems in Medicine and Biomedical and Behavioral Research: 1983, *Deciding to forego life-sustaining treatment: Ethical, medical and legal issues in treatment decisions*. Government Printing Office, Washington, DC.

Puma, J. La: 1987, Ethics by committee? Letter to the editor. *New England Journal of Medicine* 317(22): 1418.

Puma, J. La and Toulmin, S.: 1989, Ethics consultants and ethics committees. *Archives of Internal Medicine* 149: 1109-1112.

Robertson, J.A.: 1984, Committees as decision makers: Alternative structures and responsibilities. In: Cranford, R. and Doudera, A.E. (eds.), *Institutional ethics committees and health care decision making*. Health Administration Press, Ann Arbor, Michigan, p. 85-95.

Roelofs, R.: 1980, Reflections of a philosophical bureaucrat. In: Ruddick, W. (ed.), *Philosophers in medical centers*. The Society for Philosophy and Public Affairs, New York, p. 25-49.

Sherman, S.R.: 1984, Ethical implications of clinical judgments: The role of the Hospital Bioethic Committee (editorial). *Connecticut Medicine* 48(2): 131-132.

Siegler, M.: 1986, Ethics committees: Decisions by bureaucracy. *Hastings Center Report* 16(3): 22-24.

Stalder, G.: 1981, Ethical committees in a pediatric hospital. *European Journal of Pediatrics* 136: 119-122.

Sweeney, R.H.: 1987, Past, present, and future of hospital ethics committees. *Delaware Medical Journal* 59(3): 181-184.

Thornton, J.G. and Lilford, R.J.: 1995, Clinical ethics committee. *British Medical Journal* 311(7006): 667-670.

Tong, R.: 1986, *Ethics in policy analysis*. Prentice-Hall, Inc., Englewood Cliffs, N.J.

Toulmin, S.: 1988, Medical ethics in its American context: An historical survey. *Annals of the New York Academy of Sciences* 530: 7-15.

Uhlmann, R.F., *et al.*: 1987, Letter to the editor. *Annals of Internal Medicine* 107(4): 602.

Veatch, R.M.: 1984, The ethics of institutional ethics committees. In: Cranford, R. and Doudera, A.E., *Institutional ethics committees and health care decision making*. Health Administration Press, Ann Arbor, Michigan.

Waterston, T. and Sanders, D.: 1984, The right to die tangles between medicine and the law. Round the World sect., *The Lancet*, February 18: 387-388.

Yeo, M.: 1989, Primum non nocere: The ethics of teaching medical ethics. *Teaching Philosophy* 12(1): 23-34.

Young, M.: 1986, Hospital ethics committees. *Texas Medicine* 82: 72-74.

Youngner, S.J., Jackson, D.L., Coulton, D.L., *et al.*: 1983, Survey in president's commission for the study of ethical problems in medicine and biomedical and behavioral research. *Deciding to forego life-sustaining treatment*. Government Printing Office, Washington, D.C., Appendix F, p. 443-457.

BELA BLASSZAUER

ETHICS AND NATIONAL BORDERS

1. INTRODUCTION

Historical and economic differences appear in various ways in the ethics and the morals of a nation. 'Ethics' is used here as the systematic study of morals, while 'morals' refer to social practices of responsibility and rules for human behaviour. Although post-communist countries have significant religious and cultural diversity, as well as different historical and economic development, all seem to have at least one common feature, namely, they suffer from a serious moral crisis. It may be due to the transition from the so-called socialism to capitalism, or it may be due to the long-lasting system of oppression, where the individual good was sacrificed to an abstract common good, or perhaps to the utopian ideal of communism. Borders have not separated nations, but ideologies. Ethics has not influenced politics, but politics determined ethics. Consequently, there were no talks about rights, thus the reference to patient rights, for example, now comes as a shock to many. Nevertheless, in the world of communism one single right was declared: the right to free health care for everybody. It was, indeed, a great political and moral achievement, even though its realisation was far from perfect. Practically it was the only thing the communists could boost about and claim - on the basis of it - that Socialism was superior to capitalism. However, since the salaries of health professionals had been fixed on a very low level (for being employed in an 'unproductive' area), patients tried and still keep trying to obtain decent medical and humane treatment or any treatment at all by motivating physicians and/or nurses with envelopes padded with money. Thus tipping, an under-the-counter payment, has spread widely in the system, and it still exists as a primary obstacle to any real effort to reform health care. Besides these general similarities, there were others as well: for example, economic failures, lack of democracy, and the non-existing institutions of a democratically functioning system. However, each country within the former communist camp, has also had its own and somewhat unique history, culture and religious tradition.

H.A.M.J. ten Have and B. Gordijn (eds.), Bioethics in a European Perspective, 261-270.

2. THE EXAMPLE OF HUNGARY

On the crossroad of Europe, Hungary, for example, has had a history of wars for independence, many years of Turkish, Habsburg, German and Russian occupation, migrations and ethnic cleansing, and last but not least, many centuries of severe oppression by domestic landlords and foreign powers. Freedom was always considered a high value, while citizens have never experienced real democracy, thus could not really crave for it. The feudalistic system lasted too long, while industrialisation began rather late. In our times, while shifting toward capitalism, there is a high rate of unemployment, inflation, crime, alcoholism and poverty. In a country of ten million people, almost half a million are suffering from hunger, among them 150,000 children. Several millions, especially the old, the unemployed, the uneducated, and the low salaried people live a life of everyday struggle for survival. Despite the existence of many political parties, periodical elections, freedom of speech, free travel, and abundance of goods, all the ills of early capitalism are present. Corruption is widespread, the morals are further deteriorating, and ethics in general, medical ethics in particular, seem to have no place neither in academic circles nor in everyday life. The salaries, especially of those employed by the state, including the great majority of health care professionals, are extremely low, so is their self-esteem. While inflation and immorally low salaries create tremendous hardship for many health care providers, state subsidies disappear and prices of goods increase to Western levels. The health care system, which the communists were most proud of, is in economic and moral crisis. The fragmented reform efforts are leading nowhere, while the much needed drastic changes cannot come about due to the resistance of the medical-pharmaceutical establishment. The morbidity and mortality statistics clearly show that the nation is sick, but no solution seems to be in sight. The country is, indeed, in a critical period of transition, and the health care system with its annual 160 million patient/physician encounters is somewhere in the centre of attention and concern. In this context medical ethics could have a significant role in finding ways out of this moral crisis, and as a consequence, re-humanize medicine and improve the moral atmosphere both in the health care institutes and in society as a whole. In a country which has lacked democracy for so long, and which has a long way to go to develop it, it would be very necessary to implement at least those well-proven general principles that have helped to enhance significantly the patient/physician relationship in the developed countries over the last 30 years.

Since one of the most serious complaints about our health care system is the lack of information that patients experience when in clinics or hospitals, informed consent is one of those principles that should be implemented. Despite the Health Act of 1972 that compels physicians to inform patients about their health status, prognosis, planned therapy, risks and benefits, and despite the modification of this Act in 1990, which actually outlawed the therapeutic privilege, physicians are still very much reluctant to follow the legal norms and inform patients. There are, for example, considerable numbers of patients who do not even know what they are going to be operated on or what they have been operated for. The one time practice of having a patient sign a blank sheet of paper at hospital admittance, is still not a very uncommon solution for obtaining consent from the patient. Essentially, this 'blank sheet' states that the patient agrees with everything that becomes 'necessary' during his/her stay in the hospital or medical centre. Neither the patient, nor the physician knows at the time of signing the paper, what is going to happen, what intervention might be needed. Even if it is a kind of a consent, it is certainly not based on information, thus it is neither legally nor morally acceptable. The paternalistic Hungarian physician would like the patient to trust him unconditionally, follow his 'orders' and not to ask too many questions. Therefore it is striking, but hardly a coincidence, that such a basic principle as autonomy or the right to self-determination is painstakingly avoided in the Hungarian medical vocabulary, in fact, such words are practically never mentioned or written down by physicians. Nevertheless, due to the fact that malpractice suits are beginning to appear in Hungary too, physicians, though still reluctant, feel compelled to provide information to patients and seek their consent. So providing information takes the form of a 'ritual', which is usually no more than supplying a long document, many pages of written information about the planned intervention, supplemented with tables, figures, and statistical data. After reading it 'carefully', the patient is supposed to sign it, after which the physician may feel that legally nothing bad could happen to him. Since in Hungary nurses can only give information to patients if they are authorized by the patient's doctor, the communication between patients and health care providers is very much missing.

According to a survey carried out in three surgical institutes, 64% of the responding physicians inform patients according to their own judgement, while 18 % claim that informing patients is not the task of physicians. As a conclusion, the author of this survey states that the information physicians provide to patients is inadequate for making decisions (Jenei, 1996).

Patient rights are quite new in a country where even basic human rights have been ignored for so long. These rights came to the limelight when the new Health Act was debated (and in the winter of 1997 passed by the parliament). The most heated debate developed around the chapter on patient rights. Among them, of course, was the right to informed consent, including the refusal of treatment. This chapter on patient rights was vehemently criticized by physicians who claimed that the Act was against patients, against the medical profession, against physicians and the Medical

Chamber. It was considered by the Hungarian Medical Chamber a slap in the face, something that undermines patient's trust in physicians, forces physicians to resort to defensive medicine. According to the Chamber, the right to refusal, for example, might lead physicians to lose their professional autonomy, which in this part of the world seems to have been a synonym for unlimited power.

The claim that it was against patients was grounded on the argument that the right to refuse treatment was an introduction of passive euthanasia; the Hungarian Medical Chamber rejects it together with its active form, and instead favours the so called 'terminal palliative medicine'. The medical establishment has also warned against the development of defensive medicine in fear of malpractice suits in connection with the legal demand for informed consent. One physician has even called the new Health Act as sick, as such a legal norm would undermine the trust that patients have in physicians (Karloczai, 1997).

A member of the Hungarian Parliament, E. Pusztai, the former state secretary of the Ministry of Health, attacked the proposal on the ground that it forces a physician to violate his/her Hippocratic oath, more exactly the ancient norm of 'do no harm'. According to her, this principle must curtail the patient's rights to information (Pusztai, 1997).

The Hungarian Medical Chamber's ethical guideline on informed consent says the following:

> The information of the patient must foremost be in the interest and the benefit of the patient. Physicians while giving information must take into consideration the patient's tolerance, his/her personal circumstances, his/her level of intelligence and the type of his/her disease (Guideline, 1995).

These two sentences leave several doors open to physicians to control the quantity and quality of information to be disclosed. This instruction seems to empower the physician to do whatever he wants, to rely on his subjective judgement, his taste, his momentary mood as to how much information to give, or to give any at all.

In regard to informed consent, the media are confused as to where the approximate truth may be. Since the media professionals are not educated in bioethics, they cannot raise adequate questions, and cannot refer to sound counter-arguments. Therefore, many times the citizens are provided with one-sided information. Consequently, bioethics has a vital role in 'filling the gap', in providing comprehensive information to the public, and in educating people for the better understanding of moral issues affecting their lives.

To talk about rights in a country where some health care institutions are run similarly as a prison or a military compound, where there are rules, regulations, orders and permits for patients to follow, where patients can lawfully be forced to become teaching objects, it is indeed something of a revolutionary phenomenon. Democracy in post-communist countries develops slowly. A multi-party system, free election and the like are far from being indisputable proof of its existence. A real democracy should be experienced by the citizens and seen everyday in every area of life. At present people are still busy with their daily survival, which demands a great deal of conformity, subservience and skills in avoiding or solving conflicts.

The highly paternalistic practice of informing patients in Hungary, or rather the long traditional practice of silence, will not change from one day to the other. Nonetheless, it is imperative to improve communication radically between patients and physicians. The implementation of the principle of informed consent - as the most vital aspect of communication - could very much improve the morality of the health care system. The legal introduction of informed consent - among many obvious benefits - will result in the recognition of individual responsibility for one's own health as well.

Speaking about openness and free flow of information, the situation is not better in gaining any kind of other valuable information in the country. Although it is claimed that democracy exists, one can hardly obtain any fact about anything going on in the country. Let it be an extreme degree of corruption, an embezzlement of billions, or just the simple allocation of a university budget. Everyone knows that knowledge is power. In this part of the world, as far as vital information goes, only a few enjoy that privilege. The man on the street - including the author of this paper - knows very little about what is going on in big politics, or even in a state owned university where he happens to work, since truth is very much obscured, and information released by officials usually contradict each other, while deception, distortion, exaggeration and double talk are rather common.

Just to mention a simple example from the area of medicine: seven IVF centres compete with each other in the country. They claim various success rates. One is said to have achieved a 35 per cent success, announcing to be the best in Hungary (Forgacs, 1997). Another, a private one, claims that its own results are so good that the other six state run institutes cannot even come near to this, while admitting that there are no national statistics on the basis of which any kind of success rate could be assessed (Kaali, 1997).

Next is a case study which attempts to show the prevailing problems of patient rights in general, informed consent, second opinion and the rigid hierarchical system in particular.

3. CASE: A CHILD WITH CANCER

Annually there are about 300 new cases of children cancer in Hungary. A few years back R.B. (3,5 years old) was diagnosed with tumour. The father, a forensic photographer, took the child to one of the most prestigious pediatric oncology institutes in the country. He was given very limited information about his son's disease, and most of the information was communicated in Latin. The young patient underwent numerous examinations, among them spinal taps, which were done without anaesthesia, without pain relief. When the father asked why, the answer was that it had always been done that way. Once when paralysis of the child's face was diagnosed, the father expressed his profound doubt, and he himself voluntarily underwent the same examination, and it turned out that he too was suffering from the same paralysis. He had the instrument checked whether it was reliable. It turned out that it did not function properly at all. At one point the boy's blood sample had been mixed with another child's sample. This was also discovered by the father who did everything to oversee every medical step involving his son's treatment. He went as far as no parent has ever dared to go. He even forbade the cleaning person of the ward to enter his son's room, fearing that she would bring in all kinds of infectious diseases. Instead, he asked his wife to do the cleaning in his son's hospital room. While he ran from one place to another, having checked and rechecked various diagnostic results, consulted with top experts both in Hungary and abroad, he got into a 2 million forint debt. He was looked upon by the hospital staff, especially by the treating physicians, as a trouble-maker. The father, on the other hand, has very much wondered about what rights patients and parents have when sickness strikes. After, for example, blood was taken about 15 times from the boy, the father learnt that there was such thing as a kind of membrane, a protocatheter that could considerably lessen the pain that accompanies the drawing of blood. The father demanded such a catheter for his son, but according to the father, first he was informed that no such membrane was available in the institute, then by the time it turned out that there were indeed such catheters, it was not possible to use them, because their guarantee had expired without ever having being used.

The cancer therapy has affected the boy in such a way that he was loosing his hair. The father, in order to save the boy from a possible shock, had his own hair shaved off, and played with his son as a commando who looks more frightful if he is bald. In every respect the father has displayed a great courage in controlling the system, overseeing every intervention, seeking advices from all sources that he could seek out and afford, and kept notes of everything that happened to his son. For the physicians in the hospital his behaviour was not only strange but also highly disturbing. They have not been accustomed to it. Most of the parents of sick children accept medical authority silently, without questioning anything. R.B's father's most frequent experience was to see many parents with great fear of doing anything that would upset the physicians. They usually do not want any conflict with the providers, thus they close their eyes and see nothing wrong. The father has organized counselling without asking permission from his son's treating physicians or from the leaders of the hospital. That was seen by the physicians as a grave sin. The long struggle ended in success. The boy got well, at present he is in good condition, and seems to have been cured. He is alive, while many children with similarly dreadful diseases, as the father claims, have suffered a terrible death (Balogh, 1997).

4. CASE ANALYSIS

The father has decided to write a book about his experiences during his son's disease. He has actually written it, but before that he gave interviews to newspapers and magazines, telling about his and his son's experiences with the health care system and with the fight to overcome a serious disease. The medical profession, especially the top physicians in the institute where R.B. was treated, have come out with many statements denying everything, and trying everything to defend the institute and the involved health professionals. Two presidents of two pediatric societies have claimed that the work of the institute is on an international level and this institute is internationally recognized. They said that parents must agree with the proposed therapy, and in this case, R.B. was simply one child of the many children they were treating. The father's turn to the media, and making his damaging opinion known was - according to them - clearly unethical. They have seriously raised such questions as: Does a parent have a right to interfere with the physician's work? Could he ask for counselling without informing the treating physicians? Could a parent disobey the orders given by the physicians? Should physicians, and/or a professor tolerate the counselling with other doctors, the continuous control over physicians' work? To some of their own questions, they have given immediate answers as well. For example, yes, parents should know about their child's disease and they 'have to' consent to various treatments. The parents have no right to question the professional competence of the physician. Also, a parent has no right to go to the public and embarrass the hospital; he has to let the physician rely on his professional competence and consciousness and to work according to his oath. They regard this case as an example where the providers have made an extraordinary effort to do their best, and then there comes a lay person who interferes, asking for a second and third opinion, checking and double checking doctors, going to the media, and thus, undermining the trust in the medical profession (Pinter and Olah, 1998).

The following are some general statements from prominent persons which show we have a long way to go to achieve openness and democracy in our health care system. It is hoped that these randomly selected opinions shed some light on ethical relativism that is many times believed to be rooted in cultural and/or religious traditions. The president of the Hungarian Medical Chamber stated while talking about Hungarian health care in general, that "for a profession, whose object of work is a living person, it is necessary to have strict and rigid hierarchy, which is the precondition to decision-making, effective control and accountability" (Horvath, 1997). The truth is, however, that there is no control and no accountability whatsoever. Neither professional, nor societal. In the new Health Act there is a minor attempt to establish a little control, by way of the Ombudsman institution which is scheduled to be introduced in Hungary from the beginning of the year 2000. However, the Hungarian Medical Chamber was very much against even this kind of patient representation from the very beginning. The president of the Chamber said, for example,

> [The Act] wishes to introduce something which is a perfect example of the imbalance between the rights and duties of health professionals, namely it wants to have an ombudsman in every hospital. This has not really been elaborated on, and it is not based on international experience. This is unacceptable for the Hungarian Medical Chamber, mainly because there is such thing as medical confidentiality and the protection of patient/physician relationship (Gogl, 1997).

An economist, in his analysis, has called the hospital management in general a holy cow, whose primary aim is to uphold the status quo, and to preserve its interests. According to him, though 'scientific' argument never referred to this vested interest, but rather, in a hypocritical way, the management keeps talking about how much it worries about the health of the citizens. The Hungarian health care system has for some years now drifted to the Bermuda-Triangle of moral, managerial and economic crisis. Ethical norms are violated daily, while moral superficiality results in self-justification. Chaotic, more or less anarchic conditions become dominant, and within these conditions emerge those who take advantage of this situation selfishly and mercilessly (Lengyel, 1997).

A chief country doctor and hospital director states openly that

> Within the health care system the referral of patients depends largely on tipping, on the-under-the counter payments. One of the main problems of health care is the lack of professional control and reliable quality control. ... Considerable percentages of the retired people are unable to pay for the steadily increasing prices of drugs, and due to the lack of regular intake of their medicines, their health status becomes so serious that sooner or later they end up in hospitals, ... while some of them die because they reach a health care institution too late. For some of them a more expensive treatment is necessary since they could not follow the home therapy, which could have been successful, had they been able to purchase the necessary medicine (Kiss, 1998).

Sandor Nagy, a socialist representative in Parliament, has stated bluntly that

> Corruption has reached such extent in the country that it causes unbelievable moral and economic damages. It is just shocking that more and more people disobey the laws and norms. Soon the person will be surprised to see that

> someone honours some kind of rule. The violation of written and unwritten laws is going on openly (Nagy, 1997).

Writing about the plight of tumour patients, E. Scipiades describes the situation in the majority of the hospitals as feudalistic: "A number of our physicians and professors, just like the actors in Greek theatre, walking on Koturnus, and as half-Gods looking down from this height on patients and their relatives" (Scipiades, 1997).

The father, in our case discussion, has answered the two pediatricians' statement in the same media, where their criticism appeared. Thus, the debate, which was essentially ethical in nature, went on publicly. He has defended his close supervision of his son's treatment on the ground that a parent has this right. He was charged with suspiciously exercising control, and making unfounded accusations. He stated that one who is really concerned about the welfare of his child just cannot leave everything to the physician, who might learn things from his own errors. He believes that "some of the medical associations [are] longing for the old times, when a patient had no rights whatsoever to question the physician's work; when no physician was accountable when due to his/her error a patient died; when the medical society was a state within the state, with its own laws and ethics, and, above all, with a strong protecting shield" (Balogh, 1998).

5. CONCLUSION

As a summary, it may be concluded that although ethics and the moral level of a country may vary according to national borders, the economic as well as the political development exert the greatest influence on both. Political development is considered by the author to be reflected in visible democratic social institutions and mechanisms that can ensure the involvement of citizens in policy-making and exercising social control. There might be some truth in the claim that citizens have not grown up yet to be involved in deciding matters of such serious issues, as for example, how the health care system should be run. However, if it is true, then only those can be blamed for the 'ignorance' of the citizens who are in power and have the means to enlighten people, and help them to achieve the full (adult) status of citizens.

REFERENCES

Balogh, I.: 1997, Imadott gyermekem rakos (My beloved child has cancer). *Nepszabadsag* September 8: 8.

Balogh, I.: 1998, Az orvosok segelykialtasa (The physicians' call for help). *Nepszabadsag* February 6: 7.

Forgacs, V.: 1997, Megbizhato biralatot (Let's have reliable criticism). *Nepszabadsag* July 31: 12.
Gogl, A.: 1997, Egyenlo eselyt az egeszsegugyben (Let's have equal chance in the health care system!). *Nepszabadsag* April 23: 7.
Guideline: 1995, Ethical guideline on informed consent. *Magyar Orvos* 3 (November): 18.
Horvath, J.: 1997, Az egeszsegugyben tulzott a presztizs szerepe (The role of prestige is overemphasized in health care). *Magyar Hirlap* November 10: 4.
Jenei, I.: 1996, Betegtajekoztatasi szokasok harom sebeszeti intezetben (Practice of informing patients in three surgical institutes). *Lege Artis Medicinae* 6(9-10): 584-591.
Kaali, S.: 1997, Megbizhato tajekoztatast (Let's have reliable information). *Nepszabadsag* August 12: 12.
Karloczai, K.: 1997, Beteg jog a beteg jog (The proposed health act is sick). *Magyar Nemzet* September 12: 15.
Kiss, A.: 1998, Latlelet az egeszsegugyrol (Diagnosis of Health Care). *Nepszabadsag* January 24: 19.
Lengyel, L.A.: 1997, Korhazi mandarinok and kulik (Hospital mandarines and toilers). *Nepszabadsag* April 28: 11.
Nagy, S.: 1997, Leplezetlenul kijatsszak a torvenyeket (Open violations of the laws). *Nepszabadsag* September 1: 9.
Pinter, S. and Olah, E.: 1998, Az orvosok megertest es segitseget varnak (Physicians expect understanding and help). *Nepszabadsag* January 20: 9.
Pusztai, E.: 1997, A beteg erdekeben? (In the interest of the patient?). *Nepszabadsag* October 3: 12.
Scipiades, E.: 1997, A harmincszazalekos beteg eselytelensege (The 30% chance of the patient). *Magyar Hirlap* November 11: 7.

HENK A.M.J. TEN HAVE

ALLOCATION OF RESOURCES AND PERSONAL RESPONSIBILITY

1. INTRODUCTION

In many European countries, the health care resource allocation debate is characterized by a social context in which two values are generally accepted as fundamental, i.e. solidarity and equity. The guiding principles of most post-war governments, conservative or progressive, have been those of equality of access to health care and solidarity in sharing the financial burden proportionate to income. These principles are reflected in the health care systems.

However, because of the impact of financial and demographic determinants, it is very likely that future pressures on the health care system will further increase. Given this scenario, three options are available: (1) allocate more money for health care, (2) become more efficient, (3) make explicit choices about care. The first option is problematic because it will require at least in future higher taxation and insurance rates; it will also jeopardize other social goals, such as education and environment, competing for collective resources. The second option, increasing efficiency, has received much attention over the last decade. It was not considered ethical to make choices in health care and deny some patients necessary care as long as money was wasted with inefficient care. Many projects have started to deliver care efficiently, to make more efficient use of diagnostic tests and treatment schedules; and, indeed much more can be done to reduce wastage of resources. However, it is unlikely that even maximum efficiency will lead to a substantial reduction of the costs of important health care services. That implies that increasing efficiency can delay the need to make choices but it cannot prevent the necessity for choice in the longer run. Therefore, the third option - making explicit choices in health care - is the most realistic one. But then the question is: How should such choices be made? Several considerations and strategies for making choices have been discussed in an earlier chapter (ten Have, 1999).

H.A.M.J. ten Have and B. Gordijn (eds.), Bioethics in a European Perspective, 271-280.

2. PATIENT SELECTION

However, it is obvious that this approach to the problem of scarce resources requires a priority setting process at the macro-level of health policy that is deficient in everyday practical politics. As long as macro-level decisions are not made, the moral issue of making choices will be referred or transposed to the micro-level of doctor-patient interactions. Individual physicians have to select individual patients as candidates for scarce resources. This problem of patient selection is addressed in the ethical literature with the proposal, and subsequent analysis, of various selection criteria. A basic distinction is made between (1) medical criteria, such as medical benefit, imminent death, likelihood of benefit, length of benefit, and quality of benefit and (2) non-medical criteria, such as age, social value, special responsibility, ability to pay (Kilner, 1990).

Particular attention had recently been given to personal responsibility as a relevant item in the resource allocation debate. Many citizens live unhealthy or have risky lifestyles: they work too hard, eat too much, use alcohol and drugs, smoke, only move by car, and have dangerous hobbies. If solidarity is a basic principle in health care, everyone has an equal right to health care, regardless of their risky lifestyle. But precisely this solidarity can be undermined through the unhealthy behaviour of citizens. It is therefore suggested that equal access to health care requires an obligation to live as healthy as possible:

> Is it reasonable to require in return an obligation to be careful of one's health? Could such an obligation be enforced? Is it reasonable that those who live a healthy life must show absolute solidarity with those who do not? (Government Committee, 1992, p. 62)

3. CASE: SHORTAGE OF ICU BEDS

An Intensive Care Unit of a major regional hospital has inadequate capacity, because of a limited number of beds and specialized nurses. Arrangements have been made to take over from the Cardiology Department a 65-year-old patient with severe cardiac problems. She has recently been hospitalized in a bad condition and is in urgent need of major bypass surgery. The night before her operation, two drunk young men met with a car accident. One of them is dead, the other is severely wounded and admitted to the hospital. He needs medical care in the ICU.

The attending physician realizes that in this case the young driver will take the bed reserved for the older patient who, then, cannot have surgery, and will subsequently die. However, if the driver is not admitted to the ICU, he will die because of inadequate care. In this situation, the attendant calls the hospital's ethicist. She argues that in this particular case it is evident that the driver is to blame for his irresponsible behaviour. There have been many health education campaigns, including a large-scale recent one, warning the population against alcohol abuse and drunken driving. Therefore, the man would have known the risks of his behaviour and he should be held responsible for the consequences. The older patient should not be the victim of other people's immoral conduct.

4. THE POLITICS OF *HOMO SANUS*

The following argument is frequently used in many health policy documents. Since the beginning of this century we have witnessed astounding progress in medicine. This is reflected in the improved health status of the population of many western countries. However, in the last two decades we are confronted with the law of diminishing returns. The amelioration of the health status is slowing down and ever stronger efforts are required to yield additional positive effects. More expenditures for health care today neither reduce mortality nor eliminate the inequalities in health status between different parts of the population. At the same time a growing proportion of the population is ageing, and the very old people will consume most of the health care resources. The conclusion of this argument is that we can expect a health crisis early in the next millennium. At that moment, the health status of the population of western countries will start to deteriorate notwithstanding the enormous financial resources allocated to health care.

The prediction of such a crisis is an incentive to develop new health policy strategies and alternative scenarios that could reverse the downward trend. One of the responses is the construction of a new health model that re-connects health, economics and lifestyle in an effective way.

New health policy scenarios usually presuppose that resources will remain scarce and that medical progress legitimately will generate more and more claims to care and treatment. They focus therefore on other variables. The assumption is that future crisis is unavoidable as long as:

1. within the physical and social environment, factors harmful to human health continue to accumulate; and
2. people in modern society persist in dangerous and unhealthy ways of life.

Although both conditions can in principle be modified, attention is usually focused on the second one only. This selectivity is due to the fact that the identification of determinants of health is not an end in itself but a means towards an economically determined and more efficient social order. As an influential Dutch policy document, issued by the Ministry of Health, points out: "The challenge ... is to promote health within the predetermined framework of allocated resources" (Nota 2000, 1986, p. 10). The reorientation of health policy aims at using health promotion as a tool for distributing scarce resources efficiently, rather than at promoting health per se.

The precondition of such a policy is a model of health that is optimally geared to economic goals, and precisely such a model is introduced in the policy document. Health is defined in this document as:

> A situation of balance determined by the circumstances in which people find themselves and their ability to resist actual disturbances either alone or with the help of others (Nota 2000, 1986, p. 10).

The health model implied in this definition is very similar to the Health Field Concept (Lalonde, 1974). According to Lalonde, health is the result of four different determinants: human biology, environment, lifestyle and health care system.

The Health Field Concept is attractive as a policy instrument since it stipulates that health is not simply the result of the care system and since it facilitates a quantitative comparison of the relative impact of each determinant upon a specific health problem. Nevertheless, problems are unavoidable:

a. Health is determined by four factors, but what precisely is being determined? Is health an outcome, a product of all determinants togethers? Evidently, the determining factors are not themselves components of health. So what is it, on which these factors have such a decisive influence? As long as this has not been specified, the model offers us a formal rather than a material definition of health.

b. The model presents a specific relation between individual and collective responsibility for health. Within the set of determinants, lifestyle is granted an exceptional status since the other determinants are considered either as beyond control or as under state-control only. By assuming a proportionate but equally important role for all determinants, 'lifestyle' is installed as a significant factor in health, and with that the role of the individual is receiving special attention.

Lifestyle is defined as any behaviour influencing health. Moreover it is postulated that with such behaviour the individual agent is free and rational. The agent should have at least some ability to understand the situations in which he or she may be placed, to perceive and consider alternative courses of action in those situations, to appreciate what is to be said for or against the alternatives, to make a choice or decision, and to act correspondingly. This demonstrates that the concept of lifestyle, as applied in this health model, has obtained a specific meaning. It is not at all clear, however, that lifestyle is primarily an individual affair nor that each individual is *rational* in the minimal sense of having certain capacities of understanding and deliberation, as well as *free* in the sense that he is able to choose and to act in accordance with his decisions.

Despite these theoretical difficulties, the above-mentioned approach to health is influential in health education. Many authors, for instance Rouwenhorst (1981), who was the first professor of health education in the Netherlands, do not define health but refer to common sense. We all know, she argues, that health is not static but dynamic, *viz.* a process, a positive quality of life, a mode of being in the world. Such appeal to common knowledge keeps in the dark what health really is. But we seem right in refraining from theoretical disquisitions since we can use our pre-reflective experience for conceiving a scientific concept, *viz.* 'health behaviour'. Instead of speculating on the concept of health, we should rephrase and specify our common sense notions in order to obtain this operationally defined concept of health behaviour, that facilitates a scientific perspective and instigates socially acceptable educational practices.

The absence of conceptual analysis is striking in methodologies for educational interventions, such as those described by Green, *et al.* (1980) and Kok (1986). These experts present health education as a scientific activity following a cyclic pattern: problem analysis, behavioral determinants, behaviour modification, and evaluation. But it is significant that in health education practice little attention is paid to the starting-point of the theoretical framework. What the problem is, is often taken for granted. Therefore, the problem analysis is either simply passed over or exclusively dominated by epidemiological and medical data.

If so, it is clear that health itself, the very object of health education, is put in a medical perspective. Illustrative is a reference to research data showing that the most frequent causes of death are associated with behavioral patterns such as heavy drinking and smoking. From these data it is inferred that such behaviour is causally related to health (Kok, 1986). It is obvious that in this conclusion health is understood as the medically defined negative concept, *viz.* 'absence of disease' and 'absence of premature death' (since suicide and accident are among the listed causes of death). For health educators health is what it usually is in medicine. Like health policy makers, they seem to regard health as the formally defined output of a black box system. Doctors know what is in it; health educators are concerned with the relationship between input and output, for example the relation between specific eating habits and the incidence of a specific disease. 'Lifestyle' as a characteristic pattern of potentially harmful behaviour is at the input side of the black box, while behaviour-related disease and (premature) death are located at the output side, classified as 'health problems'.

5. PERSONAL RESPONSIBILITY

This brief analysis of health policy and health education literature is perhaps sufficiently indicative of the tendency to circumvent careful reflection on the concept of health. Without much theorizing, substitute concepts are introduced: 'healthy lifestyle' and 'health problem'. The problem of reaching consensus on a material definition of health can be avoided, since for practical purposes it is adequate to define health formally in the context of a black box system with its input and output variables. Doing so, health as a substantial value has been transformed into an instrumental value. This transformation is attractive in a situation of scarce resources for two reasons:
1. There is no need for an analysis of health as a goal in itself nor for a comparison with other goals of social policy. Health policy seems to be subordinated to the overall economic goal of limiting budgets and reducing the expenditures of the welfare state.
2. Health policy is an effective means of implementing this general economic policy. Because health is only formally defined, its meaning is a matter of individual preference. Whether health is a material or substantial value, how much weight is given to it, what qualities are thought to be its components, all these are issues for personal decision. According to this formal model of health, policy makers and health educators can only present empirical data concerning the association between a specific lifestyle-aspect and a certain problem but it is up to the individual whether or not to use the information.

Meanwhile, however, by the introduction in recent health sciences literature of the notion 'personal responsibility', a powerful moral instrument has become available to influence individual choices. If a particular problem is undesirable from a health policy perspective and if health education research shows the problem to be associated with a particular lifestyle, then health policy can attribute responsibility to individuals with that lifestyle, especially since lifestyle is supposedly the free choice of rational individuals.

Appealing to personal responsibility for health generates a series of moral arguments aimed at changing a hazardous lifestyle in view of future health problems. The notion of responsibility can therefore establish an efficacious feedback between 'health problems' and 'lifestyle'.

The concept of responsibility is a complicated one. It involves manifold issues and has different meanings, such as accountability, liability, rationality and absence of negligence (Agich, 1982). Attention should also be paid to the various ways in which it is functioning in the context of everyday practice and common language. Usually, responsibility is attributed by one agent to another agent. William Frankena (1973) makes a distinction between three attributive contexts:

a. Responsibility is attributed to a person with a praiseworthy character. We tend to assume that a responsible man will do what he has promised to do or has been asked to do, just because he has certain character traits.
b. We use the proposition 'X is responsible for Y' when Y is an act already performed. We hold X responsible for having done Y.
c. We use the proposition 'X is responsible for Y' when Y still has to be carried out. We hold X responsible for having to do Y in the future.

Ascription of responsibility in the first context can be considered a special case of the third context. If X is a responsible person we can be sure that he will perform Y.
Frankena's distinction between the second and third contexts points out an important difference between the practical uses of the concept of responsibility. In practice, 'responsibility' apparently has a prospective or retrospective force.
1) Prospectively, responsibility is assigned to the individual for his health, with the intention to influence his future behaviour. To attribute this kind of responsibility to someone, is equivalent to saying that he has an obligation to preserve his health. Through this ascription, we are attempting to guide and change the individual's behaviour. This use of the notion of responsibility has primarily an educational and motivating function. The moral appeal to maintain or adopt a healthy lifestyle presupposes that lifestyle is the result of a free and rational choice.
2) In the second context, ascription of responsibility has clearly a retrospective force. It implies a particular valuation of what has happened, combining causality with culpability. If an individual has a health problem, he is held causally responsible because of his unhealthy lifestyle in the past. And since he himself is the cause of his present problem, he is also answerable for the consequences of his prior behaviour. Ascription of responsibility in this retrospective sense is therefore retributive: it means disapproval and blame.

6. HEALTH PROMOTION ETHICS

In response to the prognosticated health crisis, a particular health model has been constructed in which health is the result of individual behaviour and the object of personal responsibility. It will now be clear that this model can serve two policy purposes:
1. It makes it easier to connect the prospective and retrospective senses of 'responsibility'. At least this is what seems to happen in practice. Attribution of responsibility may easily lead to retribution, when health promotion is not an end as such but an instrument of a social policy with primarily economic objectives. If it is urgent to reduce the costs of health care, and if it is known

that major expenditures are associated with certain patterns of behaviour, it is tempting to create an obligation to be healthy, accompanied by a system of sanctions.

2. As soon as the discourse of sanctions is introduced into health care, ethics and law become intermingled. Mere moral sanctions are limited and not often effective. If there is on the one hand a moral obligation to preserve one's health and on the other hand, in many countries, a legal right to health care, then it seems natural to assume that this right is suspended by neglect of the obligation. Health is capital entrusted to the individual and when it is not properly taken care of, the individual can no longer rightfully claim assistance. People are normally free to do as they choose; their choices and actions normally have reasons. But when their choices turn out badly and remain uninfluenced by moral appeals, legal and financial instruments are apparently thought justified from the policy perspective of averting the catastrophe scenario.

It is hardly realized how much the current health model is at variance with the medical model of disease. In medicine, patients are usually *not* held responsible for the genesis and evolution of their illnesses. Designating or labelling a condition as illness and trying to explain it in terms of disease, introduces excusability. To interpret a person's condition as illness in a medical sense implies at least three claims:

a. that the ill person is a patient or a victim, and that he therefore cannot be blamed for his condition. This medical judgement is contrary to everyday moralizing, but it certainly is a normative judgement interpreting this particular condition in terms of 'need' and 'vulnerability' rather than using traditional moralizing notions such as 'sin' and 'fault';

b. that professional care and scientific treatment are available for such conditions in general; whether they are actually effective is less important than the fact that they are intended for use for this category of problems;

c. that care and treatment are appropriate and morally desirable for this particular case since both are available and indicated, while the patient is suffering or going to suffer if no medical assistance is given.

The health model currently promoted in health policy and health education introduces a new dimension of normativity into health care that could regenerate traditional moralizing attitudes (de Beaufort, 1991). It is a major task for contemporary ethics to criticize the current tendency to connect the prospective and retrospective uses of the notion of responsibility and thus introducing retributive applications. This task is analogous to that of the past. The benefit of the medical model of disease was the introduction of excusability, and thus protection from society's moralizing judgements. For a long time it was taken for granted that the acceptance of this model should be paid for by a reduction of the patient's autonomy. More recently, however, it has repeatedly been argued in ethical literature, that reduced autonomy is not necessarily the price of excusability.

A similar task is waiting to be carried out by contemporary ethics: to argue that ascription of responsibility is not necessarily linked to a rhetoric of sanction and blame (ten Have, 1987). Although appealing to personal responsibility in a prospective sense is since long common practice in medicine, it is the political context of economic scarcity which has recently introduced the retrospective use of 'responsibility'. This is unfair as long as we cannot be sure in how far today's individual health problems are in fact the result of voluntarily chosen lifestyles (ten Have, ter Meulen, van Leeuwen, 1998).

7. CONCLUSION

As the presented case shows, it can be appealing in times of scarce resources, to use personal responsibility as a criterion for selecting patients. Often, lack of responsibility or evidence of irresponsible behaviour related to disease or injury, may be taken as a clue to assign patients a lower priority when scarce resources, like ICU beds, have to be distributed. Because of his irresponsible behaviour the patient has lost his claims to equal treatment; at least, it is unfair towards other persons who have meticulously cared for their health. Nonetheless, this argument of personal responsibility, although attractive, is morally problematic. Getting ill is not always preventable; it is most often not caused by individual failures of health behaviour. Even when it is possible to identify individual failures, such as drunken driving, it will be possible to discover individual failures in the other person too; cardiac problems are also associated with risky behaviour, though in the longer run. Retrospective association of health problems and unhealthy lifestyles is feasible at a categorical level (i.e., for a particular class or type of patients); retrospective attribution of personal responsibility for an existing illness at an individual level is always uncertain and problematic. The point also is that as soon as 'personal responsibility' is introduced as a criterion for patient selection, health professionals should engage in a moral surveillance of individual patients' lives. Even if it is possible to relate illness and lifestyle in a particular case, within the medical setting moral judgements about the origin and evolution of the illness should be irrelevant when an individual is in need of care and treatment.

REFERENCES

Agich, G.J.: 1982, The concept of responsibility in medicine. In: Agich, G.J. (ed.): *Responsibility in health care.* Reidel Publishing Comp., Dordrecht, p. 53-73.

Beaufort, I.D. de: 1991, Eigen schuld, dikke bult: een pijnlijke kwestie? In: *Kiezen en delen*. Achtergrondstudies in opdracht van de commissie Keuzen in de zorg. DOP, Den Haag, 6: 1-28.

Frankena, W.K.: 1973, *Ethics*. Englewood Cliffs NJ, Prentice-Hall.

Government Committee: 1992, *Choices in health care*. Ministry of Welfare, Health and Cultural Affairs, Rijswijk.

Green, L.W., Kreuter, M.W., Deeds, S.G. and Partridge, K.B.: 1980, *Health education planning, a diagnostic approach*. Mayfield, Palo Alto, California.

Have, H.A.M.J. ten: 1987, Ethische aspecten van GVO. In: Damoiseaux, V., Gerards, F.M., Kok, G.J., Nijhuis, F. (eds.): *Gezondheidsvoorlichting en -opvoeding. Van analyse tot effecten*. Van Gorcum, Assen/Maastricht, p. 183-195.

Have, H.A.M.J. ten, Meulen, R.H.J. ter, and Leeuwen, E. van: 1998, *Medische ethiek*. Bohn Stafleu Van Loghum, Houten/Diegem.

Have, H.A.M.J. ten: 2001, Choices in health care: Waiting list, rationing and priorities. This volume.

Kilner, J.F.: 1990, *Who lives? Who dies? Ethical criteria in patient selection*. Yale University Press, New Haven & London.

Kok, G.J.: 1986, *Gezondheidsmotivering. GVO als wetenschapsgebied*. Rede Rijksuniversiteit Limburg, Maastricht.

Lalonde, M.: 1974, *A new perspective on the health of Canadians*. Government of Canada, Ottawa.

Nota 2000. Over de ontwikkeling van gezondheidsbeleid: feiten, beschouwingen en beleidsvoornemens: 1986. Staatsuitgeverij, Den Haag.

Rouwenhorst, W.: 1981, *Om onze gezondheid: de rol van gezondheidsvoorlichting en -opvoeding*. Samson, Alphen aan den Rijn.

HUB ZWART

CONSENSUS FORMATION AS A BASIC STRATEGY IN ETHICS

1. INTRODUCTION

Consensus formation is an ethical strategy for reaching viable, tenable judgements or decisions in dealing with problematic cases - one strategy among others. In some cases, we may prefer to use other (perhaps more reliable) strategies, such as testing the logical coherence and empirical validity of judgements, or their concordance with canonical documents and authoritative statements. Whenever these strategies are not available, however, or whenever they fail to produce convincing results, opting for consensus formation becomes the most reasonable alternative.

The term *consensus* first emerged in the history of philosophy as the *consensus gentium*-argument (Suhr, 1971). Notably, it was used as an argument (one among others) to ascertain the existence of God. It was argued that, since throughout the world human communities and cultures seem to believe in God, one way or another, there must be some truth to this idea, even if incontestable proof of His existence (either of a metaphysical or of an empirical fashion) is beyond the limited capacities of human reason (under mundane circumstances). In the medieval epoch, the consensus argument was largely supplanted by strategies of sound reasoning and concordance with canonical documents or authoritative statements (such as the Bible or papal Bulls). In the 16th century, however, as traditional authorities found themselves fundamentally contested, the concept and strategy of consensus made its reappearance. The Protestant churches arranged a series of historical consensus meetings (Consensus Tigurinus (1549), Consensus Pastorum Genevensium (1551), Consensus Sendomiriensis (1570)) in order to contain the centrifugal tendencies, inherent to the protestant movement. Finally, in the present, the concept of consensus formation has proven its significance once again. The term 'consensus' is now used to refer to meetings of experts (such as physicians or medical ethicists) aimed at establishing a common standard of good professional practice. The *Appleton Consensus*, for example, refers to a meeting of bioethicists who tried

H.A.M.J. ten Have and B. Gordijn (eds.), Bioethics in a European Perspective, 281-288.

to reach consensus on a series of biomedical issues. *Consensus* now has become one of the standard terms (and standard techniques) of contemporary ethics (Zwart, 1998a).

Furthermore, the term consensus *a priori* contains a logic of its own. Whenever the word consensus is used, certain prepositions are already set to work. As soon as the word 'consensus' turns up, the moral discourse is bound to take a certain direction, determined by the tacit, inherent logic of the term. What does this logic look like?

To begin with, the term 'consensus' *a priori* refers to a middle condition, situated between two extremes - the one being a situation of complete agreement, the other a situation of complete disagreement. A completely closed community (fictitious no doubt) that unequivocally relies on some canonical document or other, interpreted by an authoritative readership, in accordance with a fixed set of interpretative rules, has no need whatsoever for a consensus strategy. All participants in a debate can be made to agree to a certain statement by more reliable and effective means, such as quoting (and adequately applying) the canonical documents at hand.

The other extreme (the situation of complete disagreement) is fictitious as well. If moral agreement is completely absent, consensus formation makes no sense at all and the parties involved will have recourse either to warfare or to arbitrary regulation. In short, consensus formation requires a considerable measure of agreement, while a limited number of (albeit important) issues is left open to debate. Logically speaking, moreover, the term *consensus* indicates that it is not something which can be *enforced* on others. Eventually, the consensus statement requires the free and deliberate consent of all parties involved. Also, the term *consensus* indicates that its outcome (the consensus statement) has not solely been adopted on rational grounds, but entails something like *sense* or *sensibility* as well.

Several strategies for reaching consensus have been developed. All these strategies will rely on some protocol or other. This protocol may be rudimentary (in which case the consensus process will be fairly open) or rather elaborate (in which case the consensus process will be fairly standardized). Yet, some kind of protocol will always be involved, at least in outline. Moreover, in every consensus procedure, there is a tendency towards more complete, more detailed protocolisation. Finally, it must be stressed that the aim of a moral protocol is not to ignore or disqualify the personal responsibility or sensibility of those involved, or to *force* them into some kind of final statement, but to indicate (as precisely as possible) where instances of serious disagreement are likely to occur and what the reasonable options are.

By focusing on case studies, I will now further clarify the logic of consensus formation in health care ethics, limiting myself to two cases: the case of animal experimentation (or more precisely, the role of the animal ethics committee in a university hospital) and the case of do-not-resuscitate-decisions. The analysis will not involve concrete, real-life cases, but rather focus on *typical* cases, that is, on the basic moral *scripts* bound to emerge in the real-life cases we may encounter.

2. CASE 1: ANIMAL EXPERIMENTATION FOR MEDICAL RESEARCH

In all university hospitals, animal experimentation takes place, often on a considerable scale. Without experimentation on animals, the practice and development of modern medicine would be impossible.

Now imagine a sensitive human person who, for the first time in his life, enters a university hospital laboratory where experimentation on animals is actually being performed.[1] *Animals like rats, mice and goats are subjected to experimental trials and bound (at least in the majority of cases) to experience suffering or some level of 'discomfort'. Finding himself confronted with instances of animal suffering, the sensitive individual is likely to experience feelings of uneasiness or even disgust. We start, that is, from a dual situation: on the one hand the moral image of the suffering animal, and on the other the (more or less spontaneous and immediate) experience of sensitive moral subject. Responding to the situation in an impulsive manner, the sensitive person might for example flee the laboratory, or try to free the suffering animals from their cages.*

The mere feeling that something is not right here, however, does not suffice as a well-considered moral judgement or as a legitimate ground for action. It is, for example, somewhat unlikely for the professional care-takers or researchers working in this same laboratory to experience similar feelings. Quite on the contrary, they will no doubt be eager to indicate that there are certain moral reasons (apparently convincing to them) for participating in this kind of animal practice. That is, they are able to account for and justify what they are doing. Thus, the consensus process is initiated as soon as the sensitive person articulates his reasons for criticizing the practice at hand, while the professional articulates his reasons for regarding it as admissible or even necessary. The sensitive visitor is called upon to explain *why* he experiences uneasiness or even disgust in the face of animal discomfort. His condemnation of the situation *as a whole* gives way to an effort to *analyze* it as precisely and meticulously as possible. Likewise, the professional care-taker will be called upon to explain why he regards the situation as basically legitimate, why under these circumstances cruelty towards animals is not regarded as immoral, but as morally justified, etcetera.

Should we study a series of discussions like the one described above, between the sensitive lay-person and the conscientious professional, a certain monotony becomes noticeable, a certain basic script is bound to emerge, and the speech acts of the participants turn out to be more and more predictable. At least implicitly, that is, they all adhere to a basic protocol. And we may try to articulate and formalize it, in order for the process of consensus formation to be furthered.

This is what happens, for example, during the meetings of an Animals Ethics Committee (AEC). On the one hand, it will be acknowledged that inducing discomfort to animals is problematic in itself. On the other hand, it will be acknowledged that a certain level of discomfort may be morally acceptable *if* the intentions behind it are legitimate ones - in terms of the scientific or social relevance of the experimental trial involved. A considerable number of participants in a consensus process is likely to agree, for instance, that it is morally permissible to sacrifice a limited number of rats in order to test an experimental cure for a disease from which a considerable number of human individuals are suffering. It then becomes the task of the AEC to determine at what point *precisely* a particular experiment on animals is regarded as morally justified, if both the animal *discomfort* and social or scientific *relevance* are taken into consideration.

It is possible now to discern a basic scheme to which the consensus formation process adheres. A series of stages and a number of transitions can be distinguished. The point of departure is the *image* of a situation - in this case the suffering animal, an image triggering uneasiness or even abhorrence, voiced in phrases like "Something here is not right!", or "This is not good!". It is the initial stage of the consensus formation process - a moral response triggered by an image of the situation as a whole.

Different people, however, will respond differently to different situations. The inherent logic of the consensus process now forces them to verbalize *why* they regard the situation as problematic, even objectionable. And they are able to do so by relying on certain basic *standard terms*, borrowed from the vocabulary of ethics, terms such as *discomfort* and *relevance*. Thus, the participants in the consensus formation process are provided with a limited set of items, a basic moral vocabulary that allows them to discern the basic structure of the situation and to really set the consensus formation process going.

Finally, as the consensus formation process proceeds further and further, a point will be reached where efforts towards quantification become relevant. For example, a scale will be introduced in order to determine the *precise level* of discomfort - that is, to *measure* it. A similar scale might be developed for relevance as well. Thus, it becomes possible to balance off discomfort and relevance in a less intuitive, more precise manner. Indeed, the consensus formation process is found to be guided by an inherent tendency to proceed from *images* (concrete situations) to *standard terms* (basic aspects or structures), and from standard terms to *numbers* (measurable variables).[2]

At a certain point, however, the consensus procedure runs the risk of becoming a routine and the sense of uneasiness is bound to recur - "Something is not right here!", "This is not good!", etcetera. The AEC and its procedures may now be regarded as biased and partial because of the fact that a fair majority of its members are themselves involved in research and will regard *any* level of discomfort acceptable as soon as *some* scientific relevance is to be expected. Moreover, the philosophical question might be raised whether a quantification in terms of physical discomfort really is an adequate way of capturing the moral phenomenon involved. At this point, we may feel that the standard moral vocabulary stands in need of broadening. A term like *intrinsic value*, for example, may serve to indicate that there is something of a problem in the instrumentalisation of animals *as such*, even in the absence of suffering. The introduction of such a term may encourage us to become more keen on limiting the number of experiments, for example. And finally, if taken to its logical conclusion, a proposal might be forwarded to quantify the concept of intrinsic value in some way - for instance by adding a fixed numeral to the score for *discomfort* by way of standard procedure.[3]

3. CASE 2: THE DECISION NOT TO RESUSCITATE

I will now turn to the analysis of a second *typical* case. Imagine the following situation.

A friendly and compassionate physician [A] pays a visit to an elderly patient and finds him in a deplorable state, which she describes as loss of decorum, loss of dignity, or grace. It is a terrible scene: the aspect of a patient whose physical state has suddenly deteriorated. In view of his physical condition, she decides not to resuscitate her patient, but allows him to die. Is there something wrong with this?

I think there is. Why this is so becomes clear as soon as another physician [B] pays a visit to this same patient and, finding him in the same deplorable state, describes it as a state of *emergency*. Instead of allowing him to die, she will immediately take a series of initiatives with the explicit objective of saving the patient's life. What we are faced with here is clearly a lack of consensus. Still, it is possible to initiate a process of consensus formation - on behalf of (more or less similar) future cases.

Let us analyze the case. To begin with we may note that both physicians are responding to a *scene*, an *image* of the situation as a *whole* [First Stage]. This situation is subsequently diagnosed in *moral* terms [Second Stage], borrowed from the basic vocabulary of ethics, such as *dignity* [physician A] or *emergency* [physician B]. Physician A basically claims that it is problematic, or even objectionable, to prolong the life of a patient in case of loss of dignity. Physician B, on the other hand, may well claim that, in cases of emergency, it is the physician's duty to save the patient's life, whatever the circumstances. At this point, the transition from image to standard term, from immediate response to a verbal analysis of structural aspects and items, has already occurred. The participants in the debate will now find that moral standard terms like *dignity* and *emergency* have an inherent logic of their own. As soon as the word 'loss of dignity' is introduced, some options will seem more plausible than others, and some trains of thought will seem more convincing than others, due to the moral presuppositions inherent in using this particular term. Phrases like *loss of dignity* allow us to make sense of the initial image or situation, of our immediate response to the situation as a whole, but may imply a bias as well and may even limit our possibilities for action. By elaborating the conflict between 'dignity' and 'emergence', we may be able to analyze the case at hand more carefully.

Thus, a consensus formation process is initiated. In the course of this process, a series of aspects is likely to be acknowledged as highly relevant.

1. It is the duty of the physician to save the patient's life. Whether or not this must be regarded as categorical and unconditional, is open to debate. Should physician B maintain that this principle remains valid at all times, regardless of the circumstances, the level of disagreement is such that we must have recourse to alternative strategies, such as arbitrary regulation - for instance by taking votes - instead of consensus formation proper. The consensus formation process presupposes that the range of moral principles is limited, and it tries to determine (as precisely as possible) where and on what grounds reasonable limits to the physician's tendency to intervene can be set.
2. The phrase loss of dignity indicates that there *is* such a limit. Moreover, a scale may be used to predict health prospects in terms of the patient's future quality of life.[4] Thus, the transition phase from standard terms to numerals is reached. On the other hand, we may have doubts whether this particular score, this particular form of quantification really covers what the physician initially experienced as she first uttered the term 'loss of dignity'. Quantification may well help us to increase the precision of our moral diagnosis, but may also cause us to neglect important aspects that were conveyed by our initial phrases and responses.
3. Finally, the will of the patient will by many be acknowledged as a decisive aspect. Others, however, will question whether patients are really able to anticipate future preferences.

These aspects (and other, additional ones) can be built-in into a moral protocol or 'consensus statement'. Such a statement will not solve all our problems, nor will it predetermine all our decisions (as an authoritative judgement in a closed community would), but it will help us to distinguish the clear cases from the difficult ones, and to indicate (as clearly as possible) where and how basic disagreements are likely to occur.

4. CONCLUSION

The consensus formation process adheres to a script that basically consists of three stages: *visualisation* (recognizing and responding to certain images, such as the image of the suffering animal or the image of the suffering elderly patient), *verbalisation* (articulating and justifying moral responses by means of a limited set of standard terms that allow us to discern the basic logical structure of the situation), and *quantification* (the effort to balance off relevant items in a more precise manner by introducing parameters, numerals and weights). A consensus statement, moreover, always remains open to contestation. Continuous questioning of the statement reached, of the provisional moral protocol, allows us to improve our efforts at visualisation, verbalisation and quantification. Thus, the consensus formation process may well be regarded as a practice of moral experimentation, indispensable for furthering the scientific quality and reliability of ethics.

NOTES

1. Cf. Zwart, 1998b.
2. This tendency to proceed from concrete, visual phenomena to verbal structures, and from verbal structures to numerical relationships, is of course a tendency which ethics has in common with any other science.
3. In fact, in a recent report on animal experimentation new model for balancing human and animal interests is presented. Items such as relevance and discomfort are to be carefully measured and scored in terms of a 10-point scale. Subsequently, the authors suggest that a fixed score of 2 points must be added in favor of animal interests in order to take their 'intrinsic value' into account (Vorstenbosch, *et al.*, 1997, p. 76).
4. The term quality of life is, of course, yet another standard term with a logic and persuasiveness of its own.

REFERENCES

Suhr, M.: 1971, Consensus gentium. In: Ritter, J. and Gruender, K. (eds.), *Historisches Wörterbuch der Philosophie*. Wissenschaftliche Buchgesellschaft, Darmstadt.

Vorstenbosch, J., Joles, J., Stafleu, F., and Tramper, R.: 1997, *Ethiek, dierproeven en de afweging van menselijke tegen dierlijke belangen*. Centrum voor Bio-ethiek en Gezondheidsrecht, Utrecht.

Zwart, H.: 1998a, Moral deliberation and moral warfare: Consensus formation in a pluralistic society [1998]. In: Have, H.A.M.J. ten, and Sass, H.M. (eds.), *Consensus formation in healthcare ethics*. Series: Philosophy and Medicine, Vol. 58; European Studies in Philosophy of Medicine, Vol. 2. Kluwer Academic Publishers, Dordrecht, p. 73-92.

Zwart, H.: 1998b, Moral legislation in the Netherlands on euthanasia and animal experimentation. *Gegenworte: Zeitschrift für den Disput über Wissen* 1: 24-30.

PART 4

PROCREATION AND REPRODUCTION

HENK A.M.J. TEN HAVE

PROCREATION AND REPRODUCTION

Introductory Comments

The Biological Time Bomb, published in 1968 by the British journalist Gordon Rattray Taylor, was an instant bestseller. The book presents an update of new and potential developments in reproductive medicine, such as test-tube babies and cloning. Fantastic scenarios are projected in the future with the prediction that many of them are almost reality: children born decades after the death of their fathers, asexual procreation of identical human beings, prolongation of human life over 150 years, mood and behaviour control by sophisticated new drugs, creation of cyborgs. Taylor warns against this future. Scientific tinkering with human life will lead to the destruction of mankind. Genetic technology and artificial reproduction will have similar deleterious effects as the explosion of the atomic bomb. The powers of technology cannot be trusted in the hands of scientists.

Taylor's warnings reflect a genre of popular books and science fiction, like *Faust, Frankenstein*, and *Brave New World*. The reproductive and genetic revolutions in medicine and biology always seem to elicit an ambivalent response: admiration and fascination as well as suspicion and awe. Since the 1960s the 'new biology' has also been an intricate stimulus to bioethical discourse. Scientists 'playing God' and 'remaking Eden' are critically interpellated for their moral commitments. In this area it seems that ethics can only follow the scientific and technological developments. Significant innovations and discoveries have been successively introduced into medical practice, often without the review procedures and stringent testing that is usual before the introduction of new medicines. Practical applications are also used to focus on the well-being of the individual, whereas particularly in this area of procreation and reproduction the qualities of community life are at stake. Popular representations emphasize precisely the social consequences and long-term political effects of the scientific manipulations at the beginning of human life. Genetics and reproductive medicine are a continuous site of contestation precisely because they are associated with human relationships and future ideals, but also power and control.

H.A.M.J. ten Have and B. Gordijn (eds.), Bioethics in a European Perspective, 291-293.

The new reproductive technologies have been rejected by the magisterial teaching of the Catholic Church. Nonetheless, in vitro fertilisation is practised in Catholic University Hospitals. Against this background, Paul Schotsmans from Belgium evaluates the moral debate concerning in vitro fertilisation. His starting-point is the notion of the human being as a person, not an unencumbered autonomous self but always in relationships, dialogue and communion with other beings. This notion of the person has motivated the development of the philosophical school of personalism, particularly influential in Continental Europe midway the twentieth century. The ethical considerations based on this personalist approach focus on three moral issues in regard to in vitro fertilisation: the quality of the relationship of the couple wishing a child, the respect for the human embryo, and the social responsibility of health care institutions applying reproductive technologies.

In the next chapter, Diego Gracia concentrates on the advancement of genetics. He describes different ethical approaches that are used to analyze the moral implications of the new genetics: the naturalistic, deontological, consequentialist, and responsibility model. Discussing these models in an assessment of genetic therapy, Gracia shows how only a re-structuring of the ethics of responsibility can lead to a balanced moral evaluation of genetic interventions, because it centres on the fundamental bioethical question: What are wise and rational goals of human life?

Genetic knowledge and technologies are interesting from a moral point of view because they have implications for the self-understanding of human beings. It is often argued that genetic interventions not only change the genetic make-up of the person but also the person himself of herself. The relationships between genetics and personal identity is discussed in the chapter of Ruth Chadwick from England. Critically reviewing various views that consider genes the essence of personhood, Chadwick analyzes the idea that gene therapy may have changed personal identity. The identity of individuals, however, is intimately connected with issues concerning the proper goals of medicine (as pointed out by Gracia) and with issues regarding the relationship between individual and community.

This last set of issues is the topic of the final chapter in this part of the book. Ten Have explores the interactions between culture and genetics. Analogous to the concept of medicalisation, 'geneticisation' has been introduced in the scholarly literature to signify the growing role of genetic vocabulary and models in present-day medicine and culture. This concept can be used as a heuristic tool to broaden the scope of bioethical debate beyond issues of individual autonomy, the right to know or not to know, and informed consent.

The case analyses in this part discuss a representative selection of moral problems in the area of reproductive medicine and genetics. First, Paul Schotsmans presents the problem of prenatal testing for a serious hereditary illness. He also analyzes the case from the perspectives of principlism and personalism, respectively. One of the more traditional controversies in bioethics is the issue of abortion. Due to new developments in genetics and reproductive medicine, abortion continues to be a major topic. In the second case analysis Bert Gordijn reviews the main ethical points of view regarding abortion. Comparable moral controversies arise with the issue of embryo experimentation, discussed in the third case by William Ellos from the U.S.A.

PAUL T. SCHOTSMANS

IN VITRO FERTILISATION AND ETHICS

1. INTRODUCTION

Louisa Brown, the first test tube baby, celebrated her twentieth birthday in the summer of 1998. Using the natural cycle, three births occurred in 1978 and 1979. The pace of development accelerated in 1980, e.g. the first Belgian test tube baby was born in 1981. Detailed follow-up studies revealed that IVF babies had virtually the same degree of anomalies as those conceived in vivo. As Edwards describes:

> Practitioners in assisted human reproduction should be very proud of their achievements for their patients. They have placed human conception firmly within the province of medicine, acted within acceptable ethical guidelines, dealt firmly with their own colleagues who transgressed acceptable limits, and kept the patients and the public fully informed of their work (Edwards, 1998, p. 17).

This sounds as a winning bulletin, although many ethical issues remain open, and this counts certainly for a Roman-Catholic moral theologian, who has to take into account the radical rejection of IVF by the Magisterium. The Vatican *Instruction on respect for human life in its origin and on the dignity of procreation* (1987) stated, indeed, that the church remains opposed from the moral point of view to homologous In Vitro Fertilisation: such fertilisation was called in itself illicit and in opposition to the dignity of procreation and of the conjugal union, even when everything is done to avoid the death of the human embryo. The application of the IVF technique became therefore 'illicit', even more when this Instruction was repeated later in the encyclical *Evangelium vitae*, a document with a higher moral authority than the previous *Instruction* (John Paul II, 1995). Four Catholic Universities (Lille in France, Nijmegen in Holland, Louvain-la-Neuve and Leuven in Belgium) reacted, however, immediately by the confirmation that "the efforts to integrate the new reproductive technologies within a humane context must not be suspended" (Borghgraef and Schotsmans, 1993, p. 82). What had been called 'illicit', was clinically further developed in their hospital policies, as "a neces-

H.A.M.J. ten Have and B. Gordijn (eds.), Bioethics in a European Perspective, 295-308.

sary ... medical alternative for the remediation of infertility". These Universities are at the moment still developing their IVF-practice.

This chapter will focus on an analysis of the ethical argumentation to justify the ethical integration of IVF, particularly in a Roman-Catholic environment. Indeed, there exists also a Roman-Catholic personalist approach that permits a different conclusion from the one reached by the Vatican's moral reasoning. I will therefore first clarify the way how this 'integrated personalism' handles the moral evaluation of the human act, as a guideline for the moral evaluation of medical developments. This will be presented in line with our moral theological basic insights. I will then further apply this structure of moral reflection to the technique of IVF, and, finally, end with a commentary about the so-called 'ethics of illicitness' and comment on the debate in the Catholic Church.

2. PERSONALIST ETHICS

The way in which ethicists have functioned in the recent decades, has been marked by extreme divergencies, leading to a strong confusion about what could be classified as 'morally adequate'. There is confusion in the use and presentation of ethical models. Generally we speak about utilitarianism, deontology and teleology as ethical models. The confusion is, however, sometimes so radical that bioethics or ethics in general has become something totally idiosyncratic: every ethicist seems to have his own system, his own way of talking, which certainly does not help much when realizing a clarification of the ethical dilemmas. Even worse, some types of ethical approaches function essentially as a possibility for rationalizing all kinds of medical research, developments of medical technology, application of new possibilities in human reproduction and so on. There is confusion in the presentation of value systems as rationale for these applications of some new techniques in human reproduction. We refer among others to the discussion in France about the right to have children - as if it is possible to speak about a right to have some human being at your disposal (cf. the debate on slavery). We refer also to the radical interpretation of the principle of autonomy in the Anglo-American ethical tradition, as if there is no longer a therapeutic relationship between physician and patient possible. To complete this picture, we can even refer to the confusion in the magisterial teaching of the Catholic Church: the ethical reasoning in the *Declaration on Euthanasia* (Congregation for the Doctrine of Faith, 1980) and the *Instruction on Respect for Human Life in Its Origin and on the Dignity of Procreation* (Congregation for the Doctrine of Faith, 1987) are indeed radically different.

This may clarify why personalism presents itself as 'an ethical frame of reference', so that those who are working in the context of personalism, can develop an identical ethical evaluation and structure their human practice in the light of this evaluation and their understanding of the implications in concrete human realities. For some outsiders, it may of course add to the already existent confusion. For those who practice this line of reasoning, it gives coherence and it offers an integration of several traditions in moral theology and moral philosophy.

Personalism suggests that it is essentially important to clarify the development of some anthropological options, which can then be used for the formation and function of the human conscience. By this, it should become clear that the reality of the human condition is much broader than what some may consider the absolute reign of autonomy: humans are situated, incarnated beings. By their very being, they are involved, fully intersubjective and social. To absolutize their freedom would mean to neglect fundamental dimensions of the human person: openness unto oneself, to others and the Other.

It seems therefore correct to state that in the search for the core of our being human, we are best served by recalling part of the history of anthropological reflection. What we find striking about the philosophical reflection in the twentieth century is that one can draw a virtually straight line from the source of this systematic reflection on humanity, which - as a systematic theory - more or less coincides with the beginning of the twentieth century, up to the present day, from the recognition of the openness of being human to the importance of social responsibility. This observation may help us to avoid falling into onesidedness. What we shall try to do here is to suggest a complementary vision on humanity and on being human. For it appears to us to be a dead end simply to stress one or another aspect of what it means to be human. According to this vision, to be human is both to exhibit and to participate in the wonder of a rich many-sidedness: to be human is to be rich in unicity and originality, but at the same time, originality is an empty concept if it does not include openness toward the other and if it does not involve cooperation with others for the expansion of a community in solidarity.

Another task we give ourselves is to continuously relate this anthropological approach to theological insights: this personalism flourishes indeed in the context of a Catholic community and it is absolutely necessary to make the foundational links to the basic insights of the Catholic tradition.

2.1. The Human Person: Unique and Original: Created as an Image of God

There was a period in which little if any attention was given to the mystery of the unique human being. Attention was focused mainly on human knowledge or, more broadly, human consciousness. However, the radical experience of the First World War opened the eyes of many European philosophers

(Strasser, 1963). Suddenly, one discovered the unique and concrete human person with his or her own life-project, his or her fears and expectations. It was discovered that the individual can only become a self by searching for his or her own identity. This is not egotism: the care for one's own being preserves the individual from being completely absorbed in the world of the 'they' in which one is obliged to sacrifice the unicity of being human to the whole of societal conventions and customs.

The being-human of each unique person - so the insight was clarified further on - is an attempt to realize oneself in freedom. The individual constructs the self with a view toward meaningful existence. With the knowledge of what one is and what one can become, one puts everything into the service of one's own project of life (Macquarrie, 1980). Here lies the contemporary anthropological foundation of human labour: the individual makes the self a project and puts everything in motion to realize this project. Furthermore, the initial insight clearly was that every human being is different, a unique and proper I, a person with unique talents, capacities, feelings and possibilities. In interaction with social-cultural surroundings, everything can be integrated towards the development of a unique, original personality with an individual character.

From a theological and religious point of view, this insight coincides with the Catholic tradition, stressing strongly the uniqueness of every created human being. The narrative of creation has mainly been understood as the creation of 'myself': a unique incarnation of God's love in reality.

The discovery of the unicity of being human was so profound that in the beginning it was formulated absolutely and radically. This might explain how things have gone awry, moving toward an egotistical fixation on the concept of autonomy. This could not be possible for a Catholic reflection: the reflection on what it means to be human never did stop with the experience of uniqueness. It also led to the appearance of a counter-movement in secular ethics. In medical ethics, and essentially in Anglo-American bioethics, the insight has broken through that it would be pernicious to base ethical reflection solely on the right to selfdetermination, as this concept of the human person is frequently translated.

This onesidedness was never present in the Catholic tradition: the reflection on what it means to be human was always related to the creational interpretation of every human person as in relation with his Creator, with God. It may explain why Catholic ethicists reacted very openly on the fundamental insight of existentialism: the unicity and originality of each personal experience of being human. They were indeed aware of the necessity to search for a complementary approach, one which is not limited to the unique, but isolated, experience of being an individual. This insight also became apparent in the secular tradition of ethical reflection.

2.2. *The Human Person: Relational and Inter-subjective: Created as a Response to God in Dialogue*

To grasp the whole of this mystery, we must consider as well the openness of each human being toward his fellow humans. As early as 1923, the Jewish philosopher Martin Buber wrote his pioneering work *I and Thou* (1923). With this most valuable contribution, a new insight broke through in the secular world, although this insight since long existed in the Judeo-Christian tradition: one can never be a human being on his own. As humans, we essentially stand in an open relation, involved with the reality in which we live, with other humans to whom we owe our existence and who continue to surround us, and ultimately with God. His insights were radicalized and re-translated by the French philosopher (with also a Jewish background) Emmanuel Levinas, stressing even stronger the radical alterity of the other (heteronomy) (Levinas, 1962). The ethical relationship is interpreted as a relationship in which the other offers him- or herself to me as radically different. Thus the other is no longer subjected to a meaning relative to my attitude to them. That most proper to the countenance of the other is that it appeals to me (Burggraeve, 1981), hence the ethical appeal: to what extent do I permit myself to be claimed by the other? In case of IVF, the other can be for the couple each other as the other partner, as also the growing embryo; for the physician, it is the couple as 'infertile' patients who have entrusted themselves into his care.

The limitation of this dialogical philosophy mainly lies in the fact that it overstresses the small-scale and the intimate encounter. However important this may be, this dialogical philosophy is only valuable if it is supplemented by an analysis of the uniqueness of every human being (cf. supra) and of one's solidary responsibility for a just society (cfr. infra). Concerning medical technology in human reproduction this approach represents, however, a clear indication of the basic criterion for ethical integration: the relational character of human sexuality and human reproduction.

2.3. *The Human Person: Communication and Solidarity: Created for the Realisation of the Kingdom of God*

The insight into the fullness of being human is again enlarged by the notion of participation in the community. We refer here to the phenomenon of living in a particular, concrete society as such and the ethical assignment that accompanies social living for realizing the good life. The investigation of the mystery of our being-human here encounters a new, fascinating discovery. For a long time, social commitment remained outside the scope of the majority of anthropological reflection, while it was a cornerstone of the Christian tradition: the interest for the most vulnerable persons and the preferential option for the poor were basic attitudes of the social participation of many Christian communities in society. This insight, however, remained passive until the breakthrough of the idea in secular reflection.

It is not surprising that this insight broke radically through because of the contrasting experiences of human beings confronting a social order that was largely inhuman. The excesses of the Industrial Revolution (in the preceding century) and of the dog-eat-dog relations between totalitarian states (in this century) functioned like a scream, heard by those who were committed to humanity. Therefore, it was primarily the victims of social desperation who gave voice to the task of solidary responsibility (cfr. the Frankfurter Schule with Th.W. Adorno and M. Horkheimer).

This might explain why those who accept such an approach go about designing an ideal image of a just society (the 'Kingdom of God' for the Christian believer). Yet, no matter how one presents it, it will never be possible to realize a perfect society. The most one can hope for is a movement toward the approximation of an ideal of justice, to come as close to it as possible. It will always remain possible to point out imperfections. Or, to express it in the terms of Emmanuel Levinas, even in the best welfare state the public administrator does not see the tears of the individual. And so we come to the implied concept of permanent revolution which is so particular to this approach: again and again what has already been achieved, or the already existent situation, continuously needs to be questioned and reoriented toward the more humane. The biblical condemnation of building images here takes on a new significance: the truth about human kind and society can only be understood as a permanent process of critical transcendence of the already existent and the already accomplished: *societas semper reformanda*.

Theologically, it has been mainly political theology that has made this shift in anthropological reasoning. I would here like to draw attention to a model that in my opinion has proven to encompass a greater ethical range and is therefore more adequate, namely the model of *communicative ethics* of the German philosopher Karl-Otto Apel (1973). From what was said above, it has already become clear that for the first time in the history of anthropological reflection, human beings have been explicitly confronted with the task of assuming solidary responsibility. Individual and relational ethics are no longer sufficient.

But still, the complementary construction of our approach should guard us against premature conclusions. Thus the central question remains: how can the personal decisions of conscience of a unique person express and integrate this connection with a solidary responsibility for society? With this model of communicative ethics, Apel is more or less trying to reconcile these two poles. He therefore reflects on the various ways that people use to communicate with each other. In speaking, taking a position, or defending a line of argumentation, one always does this in respect to others. If not, one's argumentation makes no sense. Even one who attempts to go about reflection entirely on one's own cannot escape from this. One who reflects can only articulate and verify thoughts in the form of a dialogue, even if it is a dialogue of somebody with himself. All creatures that can speak therefore have to be introduced into the rules of conversation and communication, - the ultimate justification of thinking can exclude neither a partner nor any potential contributions from participants to the discussion.

The coexistence of humankind is thus always an existence in communication. The ethical norm behind all this demands that not only assertions but also the claims of people over and against people must always be justified in dialogue. Here one arrives at the foundational demand that "not a single, limited individual interest of a human being may be sacrificed" and the ethical principle that "all needs of people - as virtual claims - must become the concern of society, at least insofar as they can be brought into accord with the needs of all the others" (Apel, 1973, p. 425). Therefore, someone who is presenting an argument, thinking, speaking, always postulates two things at the same time: the fact that there exists a society around him or her of which he or she is a member, and the hope and the expectation that everyone else can understand what is brought to the fore. At the same time, this offers us two fundamental principles for life in society: first, social life must concern us, in everything we do or omit, to secure the survival of humankind in society. It is clear that the image of a 'blessed future', of which we can only dream, is finally determinative of the way in which we are willing to let people today exercise their rights. To say it in another way: the second principle (the image of a perfect society) determines the content of the first (concrete solidarity with the whole of humankind).

2.4. The Personalist Criterion

These three fundamental value-orientations should normally need more clarification and philosophical foundation. It may, however, be clear that personalism tries to describe systematically the basic dimensions and relations of being a human person.

With these three fundamental value-orientations in mind we can form a moral criterion, with a personalist meaning: we say that an act is morally good if it serves the *humanum*, that is, if it in truth is beneficial to the human person adequately considered in these three basic value-orientations (Janssens, 1980).

In virtue of the historicity of the human person this criterion requests that we again and again reconsider which possibilities we have at our disposal at this point in history to serve the promotion of the human person. This is a demand of a dynamic ethics which summons us to the imperative of realizing what is better or more human. It implies also that we must recognize and respect the uniqueness of every human being: as a conscientious choice every unique human person must balance the values and non-values in such a way that he promotes the realisation of as many values as possible in order to realize the most humanly possible. In conjunction with this we must, in our acts, respect the originality of all as much as possible.

This criterion is experienced as an instrument for the responsible functioning of our conscience, ultimately the basis of the justification of our action. And as clearly understood in the Christian tradition: finally, every unique human being has to respond for his life by taking up full responsibility before himself, the other, the society and God.

3. A PERSONALIST APPROACH TO IN VITRO FERTILISATION

The interwovenness of the three characteristics of the human person becomes very clear: in order that there would be a true possibility for the child to grow, man and woman must have developed the quality of their relation in such a way that they can really be parents for their children. At the same time, our social responsibility requires that no single, individual human being may be sacrificed for the needs of others. The dilemmas concerning IVF can be grouped around three poles: the couple that cannot realize a legitimate wish for a child (quality of relation), the embryo itself (uniqueness of every human being), and the solidary responsibility of the society concerning the qualified application and organisation of this technique in modern societies.

3.1. The Stable Heterosexual Couple

First of all, the couple: in light of recent developments in sexual ethics, personalists do not see IVF as a substitution for sexual intimacy, but as a kind of prolongation of it. Catholic personalists refer to Vatican Council II, where not the "intention of nature inscribed in organs and their functions" was proposed as the ethical criterion, but "the person integrally and adequately considered". Concerning sexual ethics, the criterion of relation functions as a creative principle for ordering all the dimensions of human sexuality: the recreative, procreative and institutional dimensions. For all these reasons, the ethical requirement of a stable, heterosexual couple becomes an urgent clinical necessity. This implies that the personalist would react rather negatively on requests for the application of this technique for lesbian or homosexual couples, for single women and, of course, for surrogacy motherhood.

3.2. Respect for the Human Embryo

Secondly, the human embryo: every specific human life must be given a specific respect. The human zygote can never be treated as mere human biological material. Our anthropological frame of reference here offers an important perspective: the reason why we give so much importance to the respect for the human embryo lies in the fact that as humans we can never dispose of our equals. One of the fundamental characteristics of the other is indeed that he or she has been given to me, which is the basic idea of relational anthropology (Buber, 1923; Levinas, 1962). The fact that we cannot dispose of zygotes as we please is precisely connected to the fact that in dealing with human zygotes we touch upon a part of our being human from which one or even many poles of reciprocity may spring, just as each one of us is also a pole of reciprocity. The willingness to dispose of the human embryo is equivalent to giving in to the temptation to decide who may be our equals. This would mean that we ourselves are willing to dispose of the limits of a whole to which we belong by our own disposition. It would at the same time mean the denial of the existence of the other. Even more, with this we would disclaim that the existence of the other is finally the condition for the possibility of our own existence.

The human zygote thus claims from us a specific respect. If we do not respect human life in the first moments of its existence, then we would deny recognizing that the human beings we are ourselves are only there thanks to the existence of other human beings that we have not chosen ourselves. To eliminate those of our equals that do not suit us, or to dispose of them as we please, would - because of the ontological solidarity of the human entity - imply that we deny ourselves in our essence, or in other words, that we would disclaim the most essential character of our human condition. It would mean that we refuse to recognize that we ourselves had our origin thanks to others, organically through the genetic patrimonium that has been given to us by our begetters, psychologically through the network of relations in which our parents have raised us, symbolically through the mother tongue, the culture and the tradition that we inherited from them (Boné and Malherbe, 1985). With regard to the couple, the application of this technique can be justified if they can offer a human environment to the child. We would therefore suggest that only a stable couple who provides the gametes should be considered. Children indeed need parents, not just procreators. Every child should be born in a qualitative relational environment. A good father is in the first place a good husband, who can be a good partner to his beloved wife. A good mother is first of all a good partner to her beloved husband. Only when man and wife have become good partners, they will be able to be good parents. Clinically, this criterion may be expressed in the organisation of adequate counselling in the medical fertility team.

It has become general practice in secular settings that therapeutic experiments are performed on zygotes, embryos, foetuses or neonates. This creates a very important problem for the personalist approach. Personalism holds to the general principle of all medical experimentation on humans: as with all medical interventions on patients indeed, one must uphold as licit procedures carried out on the human embryo which respect the life and integrity of the embryo and do not involve disproportionate risks, but are directed towards its healing, the improvement of its condition of health, or its individual survival. No single human being may be sacrificed for the needs of others. A personalist approach expresses therefore a specific respect for the zygote and the human embryo. We can never treat them merely as objects, but must always value them as potential subjects. Researchers who consider experiments on embryos, therefore, must submit to the same code that applies to the whole of medical experimentation on humans. It would thus be irresponsible to conduct experiments when it is known beforehand that they cannot be of any benefit for the concerned human being.

3.3. Societal Control of the Quality of Application

On the social level, our concern is to reconcile the application and organisation of IVF with the demands of an adequate functioning of social responsibility in the society. This implies that hospitals and other health care institutions, practising IVF, should develop quality criteria for the application of the technique. It requires even more that couples should be fully informed about the procedure and should be given the possibility to give their consent and to withdraw the treatment at every moment they judge to be necessary. It requires also that Western democratic societies should develop guidelines concerning the qualitative organisation of fertility centres in their health care system. As we mentioned earlier, the coexistence of humankind is always an existence in communication. It may therefore be immoral that some Western states left the development of these techniques to the arbitrary course of events (as for example is the case in Belgium).

4. ETHICS OF ILLICITNESS? A DEBATE IN THE CATHOLIC CHURCH

The reader will have noticed that on the level of the ethical evaluation of this technique, this approach leads to an eventual integration of IVF in health care. This represents a different conclusion than the position taken by the Roman Congregation for the Doctrine of Faith in its 1987 Instruction. It is our opinion that reproductive technologies may be ethically integrated, at least on the conditions that they are performed with a therapeutic intention, inside marriage, using the gametes of the spouses and with a specific respect for the human embryos. The Congregation rejects this point of view (cf. supra). In light of the fact that critical reflection on the presuppositions of our own position is challenged by the Roman Instruction, we would like to develop three objections concerning the Roman position.

The Instruction uses a rather reductionist model for describing the unity of the spouses. A great deal of attention is given to the 'act of marriage'. This reflects the teaching of Paul VI in the encyclical *Humanae Vitae* (1968). Marriage is, however, a multidimensional reality: it includes bodily unity, but also the unity of affection and tenderness. Marriage can therefore not be reduced to bodily union alone. The Instruction considers further every act that does not reach the fullness of its meaning to be illicit. This a rather strange premise, that, humanly speaking, can almost be called untenable. It would lead to a fiasco with respect to moral conscience which, in more than one situation, has the duty to make prudential decisions. And finally, the Instruction does not take account of the rather frequent situations in which, even within marriage, a conflict can occur between the unitive and procreative dimensions. Every human being is inevitably confronted with the ambiguity of reality. For this, the Instruction is not blind, as is evident in some sections, but it does not enlarge this vision to encompass the ethical evaluation of reproductive technology. The Instruction offers the opinion that

> [Homologous in vitro fertilisation] is in itself illicit and in opposition to the dignity of procreation and of the conjugal union, even when everything is done to avoid the death of the human embryo (Instruction, 1987, p. 707).

This affirmation is clear, but is does not touch upon the essential question, namely to ascertain whether the 'ends of marriage' may also be realized by a specific, therapeutic intervention which precisely has as its goal to undo a *de facto* dissociation of these ends. For example, in the case of blocked fallopian tubes, the dissociation between the unitive and the procreative dimension is a result of a specific pathological reality that can be remedied by the technique of IVF. One can understand the Instruction only if it takes this position as a result of two disputable theories: a reductionist view on marriage and the failure to take into consideration the difference between acts which have a therapeutic intention and acts with no such therapeutic meaning. It would therefore be much more logical to conclude from the premises of the Instruction that therapeutic acts like IVF are a blessing, because they allow for the realisation of the fruitfulness of marriage.

In summary, IVF may be integrated from a comprehensive and dynamic anthropological viewpoint. Self-realisation, the quality of human relations and solidary responsibility are the perspectives from which we can evaluate the application of this technique. Within the framework of a stable couple and the normal prospects promising a balanced upbringing of the child, this technique offers the possibility of a valid answer to the sincere question of an infertile couple to have a child as a face from their own faces. Here a mentality that is anti-life is clearly broken through, and when the technical possibilities available to us are geared to the worthiness of being human, an open and generous 'yes' for the child in a stable family is possible.

5. A CHALLENGE FOR FUTURE REFLECTION

The Hospital (and Medical Faculty) Ethics Committee of the K.U. Leuven issued in its Renewed and Updated Recommendations on IVF (1989) a positive recommendation for the cryopreservation of human embryos (Borghgraef and Schotsmans, 1993). One of the justifications was a so-called "Ethics of the imperfect or the best possible (*minus malum*)":

> These recommendations can be fit in with the need to make a choice between two imperfect situations: conjugal infertility on the one hand and embryo manipulation on the other. The Committee would like to stress more particularly the specific character of medicine often faced with delicate situations calling for a decision. Most of the time, the physician is expected to carefully balance values and non-values (positive and negative indications). More than often he has to take the responsibility; he is not always able to predict the outcome of his decisions with certainty. As to IVF this means that the responsible physician, based on an attitude of respect for beginning human life, has to aim for the option which he considers to be the most humanly possible. The human embryo is human life in development and has to be treated with all due respect ... So far, the loss of a certain rate of embryos due to (cryopreservation) has been inevitable, thus creating an ethical problem. Here however, a major ethical criterion is the intention of the action, the fundamental disposition the action is based on. Therefore, everything should be done to maintain a fundamental attitude of respect for human life. Temporary freezing should only be considered as a means of increasing the chances of developing beginning life. Based on this fundamental attitude, fertilized oocytes should be frozen at the stage of their development, optimal for their chances of survival and for later successful implantation ... (Borghgraef and Schotsmans, 1993, p. 83-84).

The technique of IVF created not only the problem of cryopreservation, but also - as a consequence - the problem of the spare embryos. Personalism in confrontation with IVF is, again, deeply challenged on its translation of the basic principles in a concrete moral evaluation. The moral status of the human embryo represents, indeed, one of the most crucial points of debates concerning the ethical evaluation of IVF. It has also become more and more the concern of the Catholic Magisterium (cf. *Evangelium vitae*).

I would simply like to draw attention to the cautious approach of Catholic moralists who distance themselves from what Dunstan calls the 'practical absolutism' of the Roman position (Dunstan, 1988). It is also my opinion that the value-orientations of personalism can help to bring clarification in the extremely difficult application of how to express our respect for every human person. It would even be a symbol of continuity with the Catholic tradition that - more than until now has been done - the dimension of evolution and historicity would be integrated, to come to a renewed updating of personalist reflections concerning the human embryo.

6. CONCLUSION

IVF was and is a challenging issue presented by medical technology in human reproduction. It may also be illustrative of how personalism can deal with these new possibilities in medicine. This approach provides an ethical as well as a pedagogical frame of reference. From this perspective, the responsibility of every human being for the meaningful integration of these dimensions in his or her life-project can be taken up. This is the ultimate task of ethical reflection: to offer the opportunity to each person to contemplate his or her own situation in a sufficiently rational way, to become conscious of all the values and non-values that are present to them as human persons, and to open perspectives on the humanly desirable. At the same time, this approach contains an alternative to possible abuses and one-sided options. The quality of human relations and the respect for the human embryo are more important than a single-minded striving for scientific success.

In light of the Christian tradition such a conclusion reflects an openminded creational attitude: man as being created as an image of God, receiving the responsibility to make conscientious choices and to stand for these options. The openmindedness of the Christian tradition urges every Christian to respect the dynamism of human creation and to promote as fully as possible the human person in all his dimensions and relationships. It leaves space for an ethical *and* Christian integration of IVF as a therapeutic intervention for couples faced with the drama of infertility.

REFERENCES

Apel, K.O.: 1973, Das Apriori des Kommunikationsgemeinschaft und die Grundlagen der Ethik. In: Apel, K.O., *Transformation der Philosophie* II. Suhrkamp, Frankfurt am Main, p. 358-436.

Boné, E. and Malherbe, J.F.: 1985, *Engendrés par la science; enjeux éthiques des manipulations de la procréation*. Cerf, Paris.

Borghgraef, R. and Schotsmans, P. (eds.): 1993, *The technological advances in health sciences and the moral theological implications*. University Press, Leuven.

Buber, M.: 1923, *Ich und Du*. Im Insel Verlag, Leipzig.

Burggraeve, R., Levinas, E.: 1981, *The ethical basis for a humane society*. Center for Metaphysics and Philosophy of God, Leuven.

Congregation for the Doctrine of Faith: 1980, *Declaration on euthanasia*. Cittè del Vaticano.
Congregation for the Doctrine of Faith: 1987, Instruction on respect for human life in its origin and on the dignity of procreation. *Origins* 16: 698-711.
Dunstan, G.: 1988, The moral status of the human embryo. In: Bromham, D.R., Dalton, M.E., Jackson, J.C. (eds), *Philosophical ethics in reproductive medicine*. Proceedings of the first international conference on philosophical ethics in reproductive medicine. University of Leeds, 18-22 April 1988. Manchester University Press, Manchester, p. 3-9.
Edwards, R.G.: 1998, Introduction and development of IVF and its ethical regulation. In: Hildt, E. and Mieth, D. (eds), *In vitro fertilisation in the 1990s. Towards a medical, social and ethical evaluation*. Ashgate, Aldershot, p. 3-18.
Janssens, L.: 1980, Artificial insemination: Ethical considerations. *Louvain Studies* 8: 3-29.
John Paul II: 1995, *Evangelium vitae on the value and inviolability of human life*. Cittè del Vaticano.
Levinas, E.: 1962, Transcendance et hauteur. *Bulletin de la Société Française de Philosophie* 56: 89-113.
Macquarrie, J.: 1980, *Existentialism*. Penguin Books, Harmondsworth, p. 69-76.
Paul VI: 1968, *Encyclical letter on the regulation of births*. Cittè del Vaticano.
Strasser, S.: 1963, Fenomenologieën en psychologieën. *Algemeen Tijdschrift voor Wijsbegeerte en Psychologie* 56: 1-21.

DIEGO GRACIA

ETHICS, GENETICS, AND HUMAN GENE THERAPY

1. INTRODUCTION

Life is a miracle between two mysteries, the mystery of its beginning and the mystery of its end. The Greek word for beginning is *génesis*. This genesis has always been, and still is nowadays to some extent 'mysterious'. *Mystérion is* another Greek word; it means mysterious, but at the same time sacred. The beginning of life has something of the mysterious, and also of the sacred. This permits understanding why it has a religious dimension, which in our Western culture is paradigmatically represented by the traditions about the origins stated in the biblical book of *Genesis*. Every religion has said something about the origins in general, and the origin of life, in particular. But genesis has other dimensions, different from the religious one. For instance, there is a philosophical or metaphysical dimension, present in the history of philosophy since the first Presocratic thinkers up to now. And there is also another dimension, the moral one. The ethical question is always the same one: what should we do, and what ought not to be done? Is the manipulation of the beginning of human life compatible with its sacredness? And if the answer is positive: to what extent?

The ethical problems related to the beginning of life have always been important, but they are specially cogent nowadays, due to the discovery of another dimension of the genesis problem, the scientific one. It all began a century ago, when the science of genetics was founded. But the capability of manipulation of the beginning of life increased drastically during the fifties and sixties of our century, when molecular biology and molecular genetics appeared, and specially in the seventies, when the recombinant DNA technology became available. This technique permitted for the first time in the history of mankind the manipulation of genes and therefore the modification of the genetic codes. During the seventies this technique was used preferentially in micro-organisms; in the eighties it began to be applied to mammalians, giving place to new and unknown beings, now called transgenic animals; and during the nineties its application to human beings

H.A.M.J. ten Have and B. Gordijn (eds.), Bioethics in a European Perspective, 309-338.

began. Ethical problems have progressively turned higher and deeper. It is certain that what is technically possible is not always ethically permissible. Which are the limits, the moral limits in the use of these techniques?

Answering these questions, or at least bringing them into debate, is the goal of the new discipline called bioethics. The original idea of its founder, Van Rensselaer Potter, was promoting an harmonic development between biology and ethics (Potter, 1977). In fact, the new developments in biology, and especially in genetics, should and must be confronted with the classical ethical theories. Some authors think that the consequence of this contrast is completely frustrating, and that a new ethical paradigm is needed. For instance, David Heyd has written that the classic term 'ethics' must be reserved for the theories which deal with the normal problems of the normal people, and that the neologism 'genethics' must be introduced to design the new ethical problems raised by the new possibility of creating or transforming people. "Ethics is the theory of moral conduct in the world. Genetics is the science of creating new biological worlds", he says (Heyd, 1992, p. 23).

In any case, it is obvious that new genetics, and more specifically gene therapy, is a big challenge to the classic ethical theories. In the following I would like to analyze first, what genes are; second, the main lines of gene therapy; third, the different ethical approaches that can be used in analyzing its moral implications; fourth, the need for a new model of ethics of responsibility; and finally, the moral evaluation of gene therapy from the point of view of an ethics of responsibility.

2. GENES IN CONTEXT

As is well known, genetics as a science began in the 19th century, when Gregor Mendel presented the results of his experiments to the Natural Sciences Society of Brünn at two meetings in February and March of 1865. In the later description of his work, Mendel gave simple mathematical rules for heredity, which are still found, almost unchanged, in today's textbooks of genetics. The monograph was published in the transactions of the Brünn Society, under the title *Versuche über Pflanzen-Hybriden* ('Experiments in Plant Hybridisation') (Cookson, 1988, p. 24).

Mendel discovered some units of information which are transmitted through generations and are responsible for the similarities between ancestors and descendants, and in general inside families. Mendel did not define the way by which these units are produced. As a Catholic priest, it is certain he had a creationist idea of genes as units of information produced by God. The problem is whether they were created directly, as they actually are, or indirectly, through some mediations. And in this last case, the question is what kind of mediations are at stake. A contemporary of Mendel was the

British naturalist Charles Darwin, whose treatise *The Origin of Species by Means of Natural Selection* was published some years before the studies of Mendel, in 1859. Darwin thought that the information responsible for the biological characters is only transmitted to descendants when it permits a perfect adaptation to the environment. Some genetic variations will confer advantages and others disadvantages. Survival operates through a process of selection for advantageous characteristics and this process is called 'natural selection'. Under a given set of environmental conditions, the genetic make-up of an individual results in a particular phenotype, which must be adapted to the environment to allow the survival of the individual. It is therefore the environment that defines the biological information as right or wrong, successful or unsuccessful. Biological information is not right or wrong per se, but only when accepted by the environment. The environment and the adaptation to the environment are the framework and the criterion to judge some biological information as successful or unsuccessful.

Darwin could not interpret his data in genetic terms. But when genetics was rediscovered in 1900, and the term 'gene' appeared in 1909, genetic Darwinism emerged, being called Neodarwinism. Its main thesis was that genetic information is a consequence of the process of trial and error. Therefore, evolution is governed by a stochastic process of natural selection, and not by a deterministic one. Only nature can define what information is successful and must be transmitted to the following generations. Genetic information is changing by chance or at random, due to the influence of a lot of physical and chemical agents, which produce in it informational variations called mutations. And mutations can only be successful when nature accepts them and permits their transmission to the following generations. In other words, the information contained in genes is only a consequence, and therefore must be understood as the consequence of the stochastic or random game of nature.

This view is important in order to understand the real statute of the genetic information. Genes are not substantive entities with an autonomous dynamism and an internal teleology. This was the dream of the ancient biology: that natural things in general, and living organisms in particular, were substances with a perfectly determined internal finality. Today it is very difficult to defend either determinism, or teleology in nature. The universe as a whole certainly has an internal finality, governed by the so-called natural laws. But those laws are, at least to some extent stochastic, and therefore a perfect teleology cannot be attributed to every natural thing. For instance, if genetic information is the consequence of a complex and repeated process of trial and error, it is obvious that it is meaningless to talk of teleology when the consequence has been an error. The organism that cannot succeed in its adaptation to the environment, has not been teleologically or deterministically oriented. There is some global teleology, but compatible with indeterminism and with the possibility of errors. That is what has been called by Jacques Monod 'teleonomy' (Monod, 1970). Genes are the consequence of a stochastic teleology, which can be called teleonomy. Teleonomy is compatible with random variations and hazardous mutations. In other words, teleonomy is compatible with errors, precisely because it is the consequence of the process of trial and error. For instance the so-called congenital metabolic errors or errors of metabolism, can only be understood and interpreted this way. Genetic engineering and genetic therapy had no meaning in a perfectly teleological world; only teleonomy made them possible.

If so, then genes cannot be seen as substantive and autonomous entities, but as moments of the complex and dynamic process of nature. Genes must be seen from the point of view of nature, and not on the contrary. Nature is the only entity with complete substantivity, and genes are only partial and dependent realities. From this point of view, a living organism is not the consequence of the lineal expression of its genetic information, but the result of a complex interaction between two kinds of natural information, one collected by genes and the other offered by nature throughout its process of genesis and development. With only the genetic information there is no organism. The interaction between this specific information and the other information which is given by the environment, such as water, temperature, minerals, vitamins, hormones is required. Without those inductors the genetic information cannot be expressed. This means that the only substantive reality is the consequence of the interactions between those two kinds of information, and that genes have neither substantivity nor finality. This is important in order to define its ethical statute.

Generally, genetic diseases are defined as abnormalities at the DNA level, which result in an abnormal or absent corresponding protein (Frossard, 1991, p. 31-2). But this is not correct, because there is no abnormal protein in itself or per se. We consider a protein as abnormal when it is not compatible with an adequate adaptation of the organism with the environment.

Therefore, we can now define what is a genetic error. There is no correct or right genetic information per se. Genetic information is right or wrong only in view of the possibility of a successful adaptation to the natural environment, and therefore in view of the possibility of developing a plentiful life. What we call genetic errors or genetic diseases, are only negative possibilities of life.

There are negative possibilities of biological life. This is the case with the purely biological diseases, for instance the animal diseases, like tumours, infections, etc. Those diseases are also present in human beings, and when they are genetically determined, as for instance in the case of sickle cell anaemia or Huntington's disease. But in the case of human beings, negative possibilities of life are not only defined biologically but also biographically;

not only by facts but also by values. This is the second and most difficult part of the problem. Human beings are not completely adapted to the environment by nature, and must use a very peculiar biological function, intelligence, mind, in order to make its own process of adaptation. As the Spanish philosopher Xavier Zubiri has said, in human beings the adjustment to reality is not completely natural, like in animals, but unnatural or moral (Zubiri, 1986, p. 346 f). Animals have adapted themselves or disappeared. Human beings, on the contrary, must make their own process of adaptation. The human adaptation is not a natural property but a moral character. That is why human beings have the capability and also the duty to create new possibilities. The creation of possibilities is a moral enterprise. This means that genetic engineering can and must be used not only in curing the so-called genetic errors or genetic diseases, but also in enhancing human nature. This kind of genetic engineering is called positive, and in general is considered as morally unjustifiable. But its complete ban is simply impossible. If genetic engineering could really enhance human nature, and this enhancement could be compatible with freedom and equity, it is certain that it could be morally acceptable.

3. GENETIC THERAPY

There are two ideal types of molecular interventions in order to control genetic diseases. The first is to provide the natural protein which is absent in an organism, due to a genetic defect. This is the goal of the so-called recombinant DNA drugs. The second way is replacing the defective gene with a normal, healthy one. This method corresponds to the gene therapy. Therefore, genetic therapy in general has two different possibilities, one the recombinant production of pharmaceutical proteins, and the other the therapeutic intervention in genes. In the following I would like to analyze both procedures, assuming the descriptions made by Frossard (Frossard, 1991).

The industrial production of recombinant drugs began in 1982, when the FDA approved recombinant insulin. But the methodology used started in the early 1970s when Stanley Cohen and Herb Boyer inserted frog genes into plasmids of the bacterial species *Escherichia coli*, wondering whether they would be expressed. The experiment was a success. The bacteria did produce mRNAs coding for proteins, which until then had been produced only in frogs. Henceforward, humans could introduce any gene from any organism into cells grown in the laboratory and direct the production of the corresponding proteins. These cells act as factories that can be made to synthesise continuously the protein coded by the cloned gene. Since 1982, when recombinant insulin was approved and commercialised, several other recombinant DNA drugs have reached or are about to reach the market (Frossard, 1991, p. 32; Rollin, 1995).

The first kind of intervention we can make when a genetic error produces the lack of a protein, a diminished concentration of it or an alteration of its structure and function, is its recombinant DNA production and the use of it as a drug. In this case, we amend the way of nature, providing the natural protein to the organism. But there is another, more drastic possibility: the replacement of the defective gene with a normal, healthy one. This is the goal of the so-called gene therapy (Anderson and Friedmann, 1995). This is the technique that has started the biggest debates. What at first sight is a fantastic way of curing illnesses has turned out to be a complex ethical issue: to correct human genetic material is to correct what God - or Nature - has created.

Due to the ethical problems raised by gene therapy, genetic counselling has been more developed, as an alternative and less aggressive way. Genetic counselling is the attempt to reduce diseases through the genetic education of those families affected with genetic diseases and who may transmit them to the descendants (Biesecker, 1995). Genetic counselling will undoubtedly contribute to decreasing the number of births of individuals who are severely affected with genetic disorders - we have been witnessing such regression in the case of hereditary forms of anaemia called beta-thalassaemia. We know, however, that we will not eradicate genetic disease with public education alone. Genetic therapy constitutes the ultimate weapon towards that end (Frossard, 1991, p. 33).

There are two types of gene therapy. The first involves the manipulation of an egg or an early embryo, which means the correction of the genetic material will then occur in all the cells of the subsequent organism and also be transmitted to the offspring. This is the so-called germ cells therapy or germinal therapy. The second consists of manipulating the genome of somatic (non-sexual) cells, the aim being to correct the genetic make-up of a given cell line of an individual; here the modification introduced is not transmitted to the next generations (Frossard, 1991, p. 33; Wachbroit, 1995). This is called somatic cells therapy.

Gene therapy puts different ethical problems on the table, problems related with the technical means used to introduce genes into the host cells, and problems due to the negative or positive character of the therapy.

The first kind of ethical problems are created by the technical means used by gene therapy. To insert gene-containing DNA sequences into a host cell's DNA, scientists use preferentially retroviruses (these particular viruses integrate their own genetic material into the genome of the cells that they infect). Engineering a retroviral vector involves replacing the portion of DNA coding for the viral proteins, which represents up to 80 per cent of the virus's genetic material, with the foreign DNA sequence to be inserted. With the help of retroviral vectors, researchers should be able to introduce any gene or DNA fragment into human cells (Frossard, 1991, p. 34).

Until now, researchers have mostly carried out their experiments on rapidly dividing cells such as bone marrow cells - cells of the haematopoietic (blood precursor) system - and fibroblasts (skin cells). They have succeeded in transfecting bone marrow cells from mice, primates and humans with retroviral vectors containing several different genes. However, they are not yet technically able to replace precisely one gene with another one at a given genomic location. The success of integration of foreign genes in a cell's genome follows the laws of statistics; furthermore, these genes are, at best, randomly integrated. To make matters worse, the yield of integration is extremely low. Besides, some cells express foreign genes while other do not, and we still do not know the reason for that (Frossard, 1991, p. 34).

Gene therapy shows ethical problems due to the methods used to manipulate genes. But there are others that depend on its negative or positive character. At present, genetic manipulation on human embryos is banned, while that on somatic cells is permitted only under stringent conditions. My opinion is that a complete ban is unreasonable and impossible. If we accept the rule that genetic engineering or genetic therapy is morally correct when it increments the possibilities of human beings to be naturally and morally better, then there is no possibility of a global ban of this kind of techniques. Genetic engineering in general, and genetic therapy in particular, are not 'intrinsically' wrong, as many thinkers continuously say. There is neither a possibility of a global ban, nor are there reasons for a global acceptance. In every case it should be analyzed if they are used in order to improve the natural and the moral condition of the human beings, or on the contrary. Positive gene therapy can only be banned today because nobody has been able until now to demonstrate the reasonability of this technique.

The DNA recombinant technique has been used in order to correct, between others, two devastating childhood diseases: the lack of the adenosyndeaminase (ADA) enzyme, and cystic fibrosis (CF). Both are good examples of gene therapy in somatic cells, which implies less cogent ethical problems. There is also possible gene therapy in germ cells. All therapy will probably be in the future genetic therapy, or related with genetics. Genetics will be in the future the explanatory pathway of all diseases, from infections to tumours.

Molecular biology is becoming the final pathway of all diseases. From infectious diseases, as microbial or bacterial processes, to degenerative ones, like cancer, they all seem to have the same mechanism. Physiopathology must be understood today as the consequence of the molecular interchange between living beings and their environment. When this interchange becomes dangerous for a good process of adaptation to the environment, then disease becomes present.

The molecular interchange is always made between two different entities, called genome and environment. They do not have the same origin. Genome is an environmental product. But we can talk of living beings only when genome exists, and therefore when it has some autonomy or specificity. Living beings are defined by their genome, and their life consists in the process of interaction between the project of life prefigured in the genome and the possibilities of development offered by the environment. Life is the molecular interchange between genetic material and environment. Therefore, disease is always a molecular process of maladjustment, due to a molecular problem in the genomic level, or in the environment, or in the interchange between both.

This means that today, and especially in the near future, physiopathology will be reduced to molecular biology. Therefore, we can consider molecular biology as the final pathway of all human diseases. There is no other possibility. This means that there are different fields of application of molecular biology. One is the molecular biology of genetic information. But there is also a molecular biology of the interaction between genes and environment. This last biology also has different levels. The first is the interaction between genes and environment during the embryogenic process, generally known as developmental biology. And the other is the interaction after the embryonic period, during the whole life of the organism. Immunity is, for instance, an interesting example of the exchange between genes and environment during a lifetime.

Today, it is impossible to understand the disease process from outside the molecular biology. From infection to cancer, everything has the same explanation. There are substances, like X-ray, or some chemical products, which produce genetic mutations, and consequently diseases. In other cases, disease is due to the movement of genes from one position to another in the same cell (called 'transposons'), or to the exchange of genetic material between different bacteria ('plasmids'), or between animals (viruses). All of them alter chromosomal structure, making deletions, inversions, and other rearrangements. It is becoming clear that such changes are a critical feature of chromosome evolution. As Laurie Garrett has written: "at the bacterial level, genetics, far from being the rigid blueprint envisioned less than a decade earlier, was more akin to a game of Scrabble in which each organism came into existence with a finite set of letter tiles, or genes, but jumbled those tiles around according to a set of rules creating a vast variety of different words" (Garrett, 1995, p. 225).

This Scrabble game has its own logic, which produces a great amount of suffering to animals and human beings. Nature is enhancing its structure this way. Can we help nature introducing intelligence as a new factor of change and evolution? And if the answer is positive, with which conditions? That is the ethical problem that we must analyze now.

4. ETHICS AND GENETICS

Ethics is a complex behaviour of human beings, which has at least two different levels, one intuitive and the other reflective. Intuition is the first and immediate way of acting morally. Every situation produces an ethical reaction of approval or rejection. Those reactions are not only emotional but also intellectual. Nevertheless, they are not reflective. This is why a rational ethics needs a second step, in which emotions and intuitions are submitted to a rational and critical analysis, in search of its reflective consistency.

Emotions and intuitions are not of a great help in genetic questions. On the one hand, nature produces a general feeling of admiration and respect. But on the other hand, human beings consider that things can be manipulated looking for the personal perfection and happiness. This antagonism leads to a counterintuitive or paradoxical situation, that is to say, to an ethical conflict. The resolution of this conflict can only be made by reason. As Mark J. Hanson has said talking about cloning, this "is one of those many issues in which moral intuitions may not seem a solid enough resting place" (Hanson, 1997, cover). Therefore, a more accurate analysis of the ethical implications of genetic manipulation and genetic therapy is necessary, in search of a convincing foundation of our moral decisions (Murray, 1995; Juengst and Walters, 1995).

From my point of view, there are at least four different ethical models to approach these problems: the naturalistic, deontological, consequentialist and responsibility model. In the following, I would like to analyze the internal logic of each model, and the pros and cons of every one of them. Only at the end of this analysis a rational moral position can be defended.

4.1. The Naturalistic Model

The core problem of ethics is not the question whether there are principles, but the nature and characteristics of the principles. Without some a priori principles, judgements in general and ethical judgements in particular will be impossible.

But the determination of the logical status of moral problems has been a very difficult task, only accomplished through centenary debates. One important debate turned around 'naturalism', understood as the possibility of translating moral or prescriptive propositions in other propositions of a descriptive or non-moral character. Classical ethics was always naturalistic. It was criticized by Moore at the beginning of this century because of the 'naturalistic fallacy'.

The naturalistic approach considers nature as the foundation of moral order, and states that the manipulation of the natural order is always wrong and bad. This 'naive naturalism' is present in the popular criticism of genetic manipulation. This is, for instance, the philosophical background of the publications of Jeremy Rifkin, entitled *Who Should Play God?* (Rifkin and Howard, 1977a, 1977b), and *Algeny* (Rifkin and Perlas, 1983). Its own author explains the meaning of this neologism in the following terms:

> Algeny is about to give definition and purpose to the age of biotechnology. Dr. Joshua Lederberg, the Nobel laureate biologist who now serves as president of Rockefeller University, first coined the term. Algeny means to change the essence of a living thing by transforming it from one state to another; more specifically, the upgrading of existing organisms and the design of wholly new ones with the intent of 'perfecting' their performance. But algeny is much more. It is humanity's attempt to give metaphysical meaning to its emerging technological relationship with nature. Algeny is a way of thinking about nature, and it is this new way of thinking that sets the frame for the unfolding of the next great epoch in history ... Algeny is both philosophy and process. It is a way of perceiving nature and a way of acting on nature at the same time. It is a revolution in thought commensurate in scale to the revolution in technology that is emerging. We are moving from the alchemic metaphor to the algenic metaphor (Rifkin and Perlas, 1983, p. 17f).

There is, therefore, a concept of nature we can call algenic, which affirms and accepts the power of manipulation of nature by human beings. Man is not a slave but a master of nature. Anti-algenic mentality thinks exactly the opposite. Nature has an internal order we must respect. As Rifkin says:

> Plato, St. Thomas Aquinas, Charles Darwin ... these were not evil men. Their cosmologies were not the product of intrigue. These learned gentlemen were merely trying to express, as best they could, the workings of nature. They truly believed that their formulations were an act of discovery, an unmasking of the universal scheme of things. They sought the truth and, to a man, believed that it existed somewhere outside themselves. They were convinced that their cosmologies were an accurate description of the way the world was.
>
> Some futurists think this is all about to change. They point to the emergence of a whole new vocabulary of words and terms as proof of sorts that the self-deception that has guided our cosmologies over the millennia is about to be expurgated once and for all. For example, they point out that the idea of an 'objective' reality is giving way to the idea of a 'perspective' reality. The idea that future states are subject to ironclad laws of causality is giving way to the idea that the future is a trajectory of 'creative possibilities'. The idea of 'deterministic outcomes' is being replaced with the idea of 'likely scenarios'. The idea of 'permanent truths' is being replaced with the idea of 'useful models'. Many philosophers and scientists are convinced that this abrupt

change in vocabulary signals a departure from the long-existing hubris by which humanity has cast itself as 'the measure of all things'. Quite the contrary; the new language is not an expression of humility ... It is not humility that animates the new cosmological jargon but bravado. When we take a closer look, the new vocabulary suddenly takes on an entirely new appearance, at once menacing and intoxicating. Perspectives, scenarios, models, creative possibilities. These are the words of authorship, the words of a creator, an architect, a designer. Humanity is abandoning the idea that the universe operates by ironclad truths because it no longer feels the need to be constrained by such fetters. Nature is being made anew, this time by human beings. We no longer feel ourselves to be guests in someone else's home and therefore obliged to make our behaviour conform with a set of pre-existing cosmic rules. It is our creation now. We make the rules. We establish the parameters of reality. We create the world, and because we do, we no longer feel beholden to outside forces. We no longer have to justify our behaviour, for we are now the architects of the universe. We are responsible to nothing outside ourselves, for we are the kingdom, the power, and the glory forever and ever (Rifkin and Perlas, 1983, p. 242ff).

The basic idea of naturalism is always the same, that nature has an internal order, and that this order is sacred and must be respected. It is not casual that Rifkin could write at the end of his book:

What is the purpose of life? Why are we here? When confronted with our own existence, two choices present themselves. To accept life as a gift to be enjoyed or as an obstacle to overcome. If we experience life as a gift, we give thanks. Giving thanks means sharing our good fortune by helping coextend the gift of life to the rest of posterity. Indeed, if wisdom exists at all, it resides in the knowledge that life can be truly enjoyed only if it is generously shared and extended. If, however, we experience life as an obstacle to overcome, then we will be relentless in our search for ways to defeat its most essential attribute, its temporary nature, its limited duration. We will devour the life around us in order to extend our own. We will exhaust the very reservoirs of life from which the future is secured, all in an effort to secure our own future in perpetuity (Rifkin and Perlas, 1983, p. 254f).

Naturalism has its own language and metaphors. In relation with genetic manipulation, its favourite metaphor is 'playing God'. There are many books and articles with this same title (Baker and Mason, 1976; Lubow, 1977; Rifkin and Howard, 1977a, 1977b; Goodfield, 1977; Chase, 1987; Seydel, 1990; Peters, 1995). As J.V.C. Glover has pointed out:

The first objection people often produce when asked about this kind of interventions is that it is wrong for us to play God. That's a phrase one often hears in the debate. But there's a real question about what exactly this objection comes to. ... Religious believers sometimes think that there's a plan that God has and that any intervention of this kind messes it up. ... To believers who says this, the point worth making is that they've got to explain why ordinary medical intervention to save lives is not messing up God's

> plans, whereas this kind of genetic intervention would be (Glover, 1989, p. 196).

The influence of naturalism diminished from the end of the Middle Ages, when the idea of an ordered and perfect nature, directly created by God, came to be questioned. Teleology of natural things was denied, and therefore the shift from descriptive judgements to prescriptive ones. Morality, therefore, could no more be founded on the contingent natural order but on the absolute order of reason. From naturalism to rationalism: this was the shift of ethics at the end of the Middle Ages and at the beginning of the Modern Age.

4.2. The Deontological Model

The main philosophical question has always been the same: can the content of moral principles be affirmed as absolute and exceptionless, or not? Is moral reason capable of formulating absolute material principles or not?

Ancient philosophy resolved this question appealing to the natural order, a sort of internal legality in nature, which is morally binding for all human beings. This was the so-called 'natural law', which had at the same time a formal and a material character (Finnis, 1980). But in modern times, when metaphysical naturalism began to be criticised, philosophers started thinking that only reason is capable of defining absolute and exceptionless imperatives. The new rationalistic approach substituted the old naturalistic ethics. Its main characteristic was the defence of a hard deontologism, understood as the possibility of defining moral absolutes by the power of human reason alone.

It was in modern times when the distinction between analytical and synthetical judgements was made. The typical examples of the first type of judgements were the specific of logic and mathematics; and of the second, those of all empirical sciences, like physics or medicine. Rationalists in general thought that moral principles were of the first kind, and that only the application of these principles to concrete and empirical situations had the variability and contingency specific of the second. Kant introduced important changes in this theory, affirming the synthetic character of all ethical judgements. But in opposition to the empiricists, he continued stating the capability of the human mind to establish absolute and exceptionless moral principles, not only canonical and formal, like the categorical imperative, but also material and deontological principles (as in Kantian ethics is the case with all perfect duties; see Kant, 1995, p. 1080, note 12).

Deontologism is applied to all those ethical theories that affirm the capability of human reason to establish absolute and exceptionless moral precepts. It was Broad who distinguished in 1930 moral theories in two opposite sets, one including those ethical methods which believe in the existence of absolute and exceptionless deontological principles, and a second one which included all those theories which defend that moral reason is incapable of affirming absolutely any normative proposition. Broad called the first kind of methods or systems 'deontological', and the second 'teleological' (Broad, 1930, p. 162). Deontology, therefore, is equivalent to 'unconditional obligation', and teleology to 'conditional obligation'. As Broad said:

> Deontological theories hold that there are ethical propositions of the form: 'Such and such a kind of action would always be right (or wrong) in such and such circumstances, no matter what its consequences might be' (Broad, 1930, p. 206).

Therefore, teleological theories must be those which do not admit the possibility of affirming that some type of action will always be right or wrong, and which hence think that consequences are important in defining the rightness or wrongness of an action. This way, the ethical systems of Descartes, Spinoza, Leibniz, and Kant are typical models of deontology, whilst those of Hume, Bentham, Stuart Mill and Moore are teleological. In general, it can be said that old ethical theories were deontological, whilst the new ones are preferentially teleological.

The ethical judgements about genetic engineering and gene therapy made from the point of view of the modern deontological theories are highly different from those obtained from the old naturalistic models. In the present time, nature is no longer considered a perfect and normative order, but an incomplete work that must be perfectioned by human beings. Nature is a moral enterprise, which has no other limits than the absolute respect of human beings, the only one moral agent (or using the words of Kant: the only end in itself) with dignity and no price.

When man is considered the only one moral legislator or moral end, to which all other things should be submitted as means, then a new theological metaphor begins to appear, that of man as a *petit Dieu*, as Leibniz said, and therefore as co-operator of God in the process of creation of nature. Man is similar to God especially by freedom and reason, and therefore human beings must use freely their reason in order to complete the creative process of God. Creation is unfinished and must be completed by human beings. Horace Freeland Judson has written a book entitled *The Eighth Day of Creation* (Judson, 1979). Creation does not last seven days but eight, and man is the author of the latest. This is why Albert Rosenfeld has entitled another book *The Second Genesis* (Rosenfeld, 1975). Religion has been interpreted in Western cultures frequently in naturalistic categories, but not necessarily. Also antinaturalism can be compatible with religious beliefs.

4.3. The Consequentialist Model

The third rational approach to ethics in general and to gen-ethics in particular is consequentialism, and more concretely, utilitarianism. This approach argues that genetic engineering and genetic therapy are morally correct when useful for all, or at least for the greatest number of human beings.

This approach deals more and better with means than with ends. Utilitarianism thinks that human beings are ends in themselves, but considers the moral obligation to be to reach the greatest happiness of the greatest number of people, not necessarily of all of them. The idea of end has not, therefore, an absolute character, and the first moral duty is not the respect of all human beings but the maximisation of good consequences for the highest possible number of people.

There are different types of utilitarianisms, at least two, called *act utilitarianism* and *rule utilitarianism*. The first defines the moral duty as the maximisation of utility in every concrete act, whilst in the second maximisation is applied to rules and not to acts, or also to the full set of acts which are made with the same rule. The first and more classic type is act utilitarianism, which is present in the founders of the school, Jeremy Bentham and John Stuart Mill. The utilitarianism of our century is rule utilitarianism. This rule utilitarianism is completed in some authors with other novelties, at least two, non-naturalism and universalisability.

Non-naturalism was introduced at the beginning of this century by Moore, criticising the idea of Bentham and other utilitarians that the moral predicate 'good' can be identified with other natural or non-moral predicates, like 'pleasure' and 'well-being'. Moore thought that good is a primary and simple quality, which cannot be reduced to another quality different from itself. Good is good, and we cannot say more. The only thing we know is that the moral obligation of every human being is the maximisation of good.

The other great novelty introduced in the utilitarian theory during the twentieth century has been universalisability and universalisation. The main representative author of this theory has been Richard Hare. The old sentence of Bentham, "the greatest happiness of the greatest number is the measure of right and wrong" (Bentham, 1988, p. 393), is now transformed in this other: "the greatest happiness [or the greatest satisfaction of preferences] of all is the measure of right and wrong". The difference is important, because here the principle of utility is combined with the Kantian principle of universalisation.

But the usual and popular utilitarianism ignores in general these subtleties introduced by authors like Moore and Hare, and continues the old naturalistic and non-universalistic line, which emphasises the importance of maximising utilities and consequences, whilst some or many people could be hurt in their physical or moral life. In this sense, utilitarianism uses some kind of strategic reason, which considers that individuals can be sacrificed in favour of or subordinated to the total amount of happiness of the community. This mentality is very frequent among economists and politicians, and permits understanding why some kind of utilitarianism is so widespread in democratic countries.

Utilitarianism has been frequently used to judge the ethical problems raised by genetic engineering. If deontologists were primarily interested in preserving human integrity and dignity, utilitarians pay special attention to the general amount of consequences. If the consequences are good, like in the so-called negative therapy, that which tries to correct genetic diseases, genetic engineering is not only morally permitted but a moral duty. And positive therapy, that which looks for the enhancement of human nature, must be generally banned because the consequences cannot be considered good neither for individuals nor for the community. Only in the case in which the enhancement of a human trait could be thought of as individual and socially good, it could be permitted. If we analyze the literature written in the last decades about the ethics of genetic engineering, and specially the reports and recommendations of the national and international commissions, we could realise that the criteria used to justify or to ban those techniques are generally prudential, in the sense of a ponderation of the consequences which the use of these techniques can have for individuals and specially for society (Toulmin, 1975; 1982; 1987; Jonsen, 1991). Talking about the pioneer National Commission, Jonsen has written: "The commissioners began to do public ethics almost by an American instinct that was inherited from James and Dewey: try to get the facts as fully as possible, talk with well-informed persons, invite all interested persons to have their say, argue in public about what you have learned, and then try to find where each member agrees and disagrees. Formal ethical theories and principles were not conspicuous, although sharp thinking by educated ethicists, working their way through the arguments, was indispensable" (Jonsen, 1998, p. 115 ff).

I would like to stress the prudential mentality with which these judgements have been made. There is not a substantive ethics, but only a prudential and strategic use of the new means offered by science and technology. Substantive ethics is more interested in ends than in means. It assumes that an act is right if it subordinates the means to the ends. Strategic ethics, on the contrary, is more interested in means than in ends. The first moral duty is the prudential and wise use of means. The major amount of work about ethics and the new genetics has been made with this mentality, looking for a prudential and strategic use of the new scientific and technical means, more than for a substantive analysis of the ends, and the way of subordinating the new means with the respect and dignity of human beings, the only one moral end.

The great tragedy of our situation is, perhaps, that substantive ethics has been made with naturalistic and deontological categories, today untenable, and that the only one alternative to those approaches has been a merely strategic ethics, only preoccupied with the prudential and strategic management of consequences. Our problem is, therefore, how to articulate in a new form principles and consequences in the moral discourse, avoiding at the same time a rigid and ideal deontologism and an extreme and partial teleologism. This is the new attempt of the so-called ethics of responsibility.

4.4. The Responsibility Model

All ethical theories must affirm some principle as absolute. The problem is not whether there is an absolute principle, but what kind of principles can be defined as such. Deontological theories are those which consider material or normative principles as absolute. Teleological theories also affirm some principles as absolute, but only in its logical, formal or structural character, not in its content. There is no moral theory that does not assume some canonical principle as imperative and categorical. The moral problem is not whether categorical imperatives exist, but how they can be formulated, and whether they have at the same time a deontological character.

According to deontological theories, categorical imperatives are not only canonical but also deontological, whilst teleological theories argue that they are only canonical, and that all deontological norms must by definition only be hypothetical imperatives. Therefore, there are two types of deontologism, *hard deontologism*, which affirms normative principles as absolute and exceptionless, and *soft deontologism*, which considers normative principles as non-absolute and with exceptions. Today hard deontologism is hardly defensible. As Sidgwick said, all ethical systems have some intuitive principles, apparently deontological and absolute, derived from what he called 'common-sense morality', but as soon as we bring the principles of common-sense morality face to face with difficult and unusual situations, its apparent self-evidence vanishes, and either they are purely tautological, or they cease to be self-evident (Sidgwick, 1981, p. 215 ff; Broad, 1930, p. 151). In the same way, Beauchamp and Childress think that absolute principles "are rare and rarely play a role in moral controversy" (Beauchamp and Childress, 1994, p. 32). They play no role in moral controversy at all, because, due to its absolute character, discussion is by definition impossible. Absolute principles are only structural conditions of morality. And when formulating material and normative principles, the absoluteness disappears. Certainly, deontological principles also have a universal and absolute form. But this form is more a regulatory ideal, in the Kantian meaning, than an empirical rule. As Beauchamp and Childress say, "in light of the enormous range of possibilities for contingent conflicts among rules, absolute rules are best construed as ideals than finished products" (Beauchamp and Childress, 1994, p. 33).

If this is so, we can then conclude that there are canonical principles, which include, at least, the principle of universalisability of respect for human beings. This principle is, as Kant thought, categorical and imperative, but purely formal, because universalisability is practised with a formal or structural character, *viz.* that human beings are moral agents and therefore must be respected. It defines 'what' must be done, but not 'how', 'when', 'to whom', etc. For instance, it defines 'which' human beings must be respected, but not 'who' is a human being, or 'when' one begins to be human, etc. 'What' the principle says is that human beings are 'ends in themselves' and not merely 'means'. And because they are ends in themselves, they are the ends of the ends they choose in their lives. Humans are the ends of the ends. In Kantian terms: "Persons are, therefore, not merely subjective ends, whose existence as an effect of our actions has a value for us; but such beings are objective ends, i.e., exist as ends in themselves. Such an end is one for which there can be substituted no other end to which such beings should serve merely as means, for otherwise nothing at all of absolute value would be found anywhere" (Kant, 1995, p. 1085). This is the reason why they have 'dignity' and not only 'price'. Human beings, all human beings, therefore, must be absolutely respected, as the objective ends they are. This is the 'form' of morality.

All moral judgements must not only have a formal structure, or canonical form, but also some content, which must be assumed deontologically, as a duty. This is the place of the deontological principles, like the four principles of bioethics (see Part I of this volume). Here the universalisability is not made over the formal or structural moral condition of human beings, but only over the specific conditions of the moral action; not over 'what' must be done, but over 'how' it must be done, 'who', 'when', 'where', etc. These conditions have by definition material content, and the universalisation therefore goes always beyond the empirical data. Here the inductive process is necessarily incomplete and imperfect, and the universal propositions suffer from a defect of empirical base. Hence, the contingency and fallibility is unavoidable. Deontological principles must be necessarily soft.

Soft deontological principles are right or wrong according to their conformity with the hard canon. Therefore, when a deontological principle does not respect the canon of morality in a concrete situation, it must be modified in that specific situation. This means that the application of the deontological principles must always take into account the circumstances and consequences of each case, and that therefore there is another moment in the process of moral decision-making, that of evaluation of circumstances and consequences. It can be concluded, then, that the decision-making process has two material moments: one universal and deontological and the other particular and teleological. The first says what must be done in general. The second defines the exceptions to the universal norm depending on the specific circumstances and consequences. And the reason for making exceptions is the same as the reason with which we justify the norm, the formal canon of morality. For instance, the canonical principle of respect for human beings leads to the formulation of the duty of truth telling as a deontological principle. But in some exceptional cases, in view of the circumstances and consequences proper of a specific situation, telling the truth can be considered a violation of the duty of respect for human beings. In this case, it is evident that lying can be permitted as an exception, as the only way of respecting a certain human being here and now.

Consequently, moral reasoning has necessarily three moments: one formal and canonical, *viz.* the imperative of respect for all human beings, and two other material, one deontological (softly deontological) and the other teleological. Deontology and teleology are not exactly two different ethical theories, but two unavoidable moments of moral reasoning (Beauchamp and Childress, 1994, p. 109ff). The unavoidable inadequacy of deontological principles to the canon of morality makes it necessary to analyze and evaluate the circumstances and consequences of every act, in order to accommodate every act as much as possible to the canon of morality.

If this is so, then the criterion to order deontological principles in a concrete situation is simply the evaluation of circumstances and consequences, rather than the conditions described by Beauchamp and Childress in the fourth edition of their book (Beauchamp and Childress, 1994, p. 34). At the same time, there is no evidence that deontological principles should all be *prima facie* of equal importance, or of the same level, as they, following Ross, say. On the contrary, there is important historical evidence that some deontological principles have priority over others. For instance, it has always been thought that some public and common goods have moral precedence over private ones. This can be called the rule of hierarchy of the public duties over the private ones. There is another rule, the genetic rule, by which public norms are constructed by private individuals, throughout general or democratic consensus. Genetically, therefore, private rules or principles have priority over public rules or principles, but hierarchically, on the contrary, public principles are prior to private ones.

According to the two rules defined before, the four principles of bioethics can and must be organised on two different levels. Here I basically disagree with the point of view of Beauchamp and Childress. To my mind, they do not correctly define the content of the four principles, and consequently do not order them adequately. In my view, autonomy and beneficence should be defined as principles of private character, primarily related with private goods, whilst nonmaleficence and justice should be understood as public principles, and defined as the rational agreements of societies on the contents of the common good. This means that autonomy (the capacity of acting as a moral agent) and beneficence (the set of ends and means everyone considers good for him or herself and for his or her moral friends, in order to achieve his or her ideals of perfection and happiness) are genetically prior to the other two principles, non-maleficence (the agreements reached by society over the mutual respect for the physical and spiritual life of all individuals, friends or strangers) and justice (the public agreements reached over the mutual respect for human beings, either friends or strangers, in social life). Everyone must be nonmaleficent and just in his or her relationship with the rest of human beings, friends or strangers, and must prioritise this to the prosecution of ones' own particular or private goals of perfection and happiness. But because non-maleficence and justice are public principles, they must not only have a particular or private expression (what my moral friends and I consider nonmaleficent or just), but also a public one. This public expression must be reached by public consensus or democratic social agreement, and be defined by laws (Bayertz, 1994). A social group can consider capital punishment a maleficent act, while another dies not. Of course, a particular person can also think that capital punishment is always maleficent and immoral, but in this case his or her moral duty is, first of all, not to involve him or herself in this kind of acts, and second, to persuade society of the immorality of these acts, making everything possible to democratically change the law. Legality and morality are different, but intrinsically related, because law is always an expression of the moral values of a society.

This leads us to the last issue, the distinction between principles and norms. Deontological principles are intellectual criteria for defining moral duties. The four principles of bioethics have unavoidably this character. Norms, on the contrary, are concrete rules of action. As Clouser and Gert have pointed out (Clouser and Gert, 1990; 1994, p. 251-266; Clouser, 1995; Gert, Culver and Clouser, 1997), the four principles are not, properly speaking, norms of action. Going further, not all the bioethical principles necessarily generate norms. The content of a norm can be maleficent or non-maleficent, just or unjust, beneficent or not, but it makes no sense to say it is or is not autonomous. If a norm has been made or imposed by coercion, it must be considered maleficent, and not non-autonomous. Therefore, we can conclude that there are four bioethical principles, but the norms according to their content can only be classified in three categories: nonmaleficent, just and beneficent.

The general consequence of this analysis is that some principles are unavoidable in ethics. But not all principles are defensible. There are, in my opinion, two types of untenable principles. On the one hand, those which affirm the absoluteness of deontological principles, for example the theories of the natural law. On the other hand, those which assert the completely conventional character of moral principles, including the formal and canonical. None of them are rationally defensible, and therefore responsible. Only an intermediate way is logically possible and ethically responsible.

This moderate or responsible principlism affirms: (1) The absoluteness of the formal principle of respect for all human beings, but (2) the relative character of all material and deontological principles. (3) Those principles are internally structured on two levels, one composed by the public principles of non-maleficence and justice, and the other for the private principles of autonomy and beneficence. When public principles are generated through participative and democratic ways, they then have priority over the private. For instance, non-maleficence and justice always have priority over beneficence. (4) Norms derive directly from deontological principles, being universal by definition, but also non-absolute, and therefore with exceptions. (5) Exceptions are always made according to the circumstances and consequences proper to specific cases. Exceptions are justified by the strict inadequacy between the formal principle of respect for all human beings, and any possible system of specific norms. Therefore, we can conclude that the form of the duty is always absolute, but its content must be by definition contingent and relative. Kant explained this idea, when at the beginning of the *Grounding for the Metaphysics of Morals* he wrote: "There is no possibility of thinking of anything at all in the world, or even out of it, which can be regarded as good without qualification, except a good will" (Kant, 1995, p. 1062).

5. RESPONSIBILITY AND BIOETHICS

When we remember the amount of work made by bioethicists during the last twenty-five years, we realise that it has dealt more with means than with ends. The great amount of new technical means produced in the last decades, and the new capability of manipulating human life they have introduced in medicine, have obliged us to an accurate analysis of the ethical implications of its use. This is why bioethics has dealt almost exclusively with the ethical implications of the technical means of the new medicine.

But an ethics that deals only with means is merely strategic and prudential. Ethics deals not only deals with means but also and specially with ends. Ethics is the critical and rational analysis of the ends of human life.

If this is so, we must then conclude that ethics, medical ethics, and bioethics, have been working in the wrong way, or, at least, that they have been incapable until now of dealing with real, fundamental and core problems. During its twenty-five years of life, bioethics has been dealing fundamentally with the rational, wise and prudent management of new means provided by the rapid development of technology. But the debate about ends has failed completely.

Bioethics must promote the rational discussion and the adoption of responsible and wise ends in our societies. This must be its main objective. Bioethics must abandon the idea of being only a practical way of resolving the conflicts reached by the new means proper of modern medicine. It must also go out of the hospitals, and begin a new era of debating the ends assumed by our societies about life and death, not only in the health arena but also in all other social places of debate. From means to ends; from hospital to society: these are the main changes necessary to introduce immediately in bioethics.

My opinion is that the miscarried goals of our medicine are a consequence among others of a more general problem suffered by our societies, due to the uncompleted liberal revolution they have had, and the defective democracy they are living in.

In the eighteenth century, when liberal revolutions took place, Western countries began to assume pluralism in two specific fields, the religious and the political, and in the latter field they built up the parliamentary system, in order to manage political disagreements throughout the so-called 'representative democracy'. But every day it becomes clearer that representative democracy is a necessary condition of a pluralistic and liberal society, but not a sufficient one; in other words, the liberal revolution has not finished, and only by completing this revolution can liberal democracy be morally defensible. If the first part of the liberal revolution assumed the sovereignty of people in defining their political values, the second part must reach another undeniable goal, the sovereignty of society in defining its moral values and goals. In the liberal societies of the nineteenth and twentieth centuries, moral values were not advocated by society and individuals but by churches, ideologists, mass media communicators, politicians, and in some cases also philosophers. This is especially evident in some specific fields, like the ethics of the human body, which until the birth of bioethics at the end of the sixties and the beginning of the seventies, remained in the hands of churches and priests. Bioethics is part of this general revolution which began in the sixties with the Civil Rights movement, and which is giving back to civil society the power and the capacity of discussing, debating and reaching agreements on all kinds of value questions, with the security that only in this way society can be 'demoralised', and therefore become more human, free and moral. This second liberal revolution, the 'moral liberal revolution' (instead of the 'political liberal revolution' of the eighteenth and nineteenth centuries), needs new ways of social work, which in contrast with 'representative' ways built-up throughout the liberal revolutions of the last centuries, as political parties, parliaments, etc. are generally called 'participative' ways. The biggest moral problem we have in our societies is how to organise social participation in the debate about value questions, convinced that only through this pathway, the same representative democracy can be legitimised.

This new way was openéd in the fifties, with the Civil Rights movement, but it has been continued by many other civil and social movements, as the feminists' movement, the consumers' associations, the gay and lesbian groups, the patients' rights teams, the antipsychiatric movement, and so on. In my opinion this is one of the reasons for the birth of *communitarianism* as a movement. If we leave the antiliberal, conservative and reactionary forms of communitarianism aside, it seems evident that the right meaning of this movement is a need for improving communitarian participation in defining and deciding all kind of value questions.

The lack of participation and debate about the ends of human life has been probably due to the misconception of a peculiar premise to the liberal way of life, that of neutrality. Liberalism assumes that the ends of human life must be determined privately by individuals, according to everyone's substantive idea of good. Nobody can intervene in these matters. They are absolutely private and personal. And the state must be strictly neutral in these value questions and substantive ideas of good life.

But neutrality has at least two completely different meanings. One is the neutrality *from* coercion, and the other the neutrality *to* debate or to discuss. As it is well known, traditionally a difference has been established between two different types of freedom, *freedom-from* and *freedom-to*. Neutrality-from is the commitment not to force or compel others in order to impose on them a particular conception of good life. Neutrality-from is the practical commitment of respect for all conceptions of good life. But this practical neutrality cannot be confused with the ideological neutrality, the idea that there is no possibility of arguing rationally about values and ends, therefore, about the substantive ideas of good. Liberalism has frequently confused those two types of neutrality. Liberalism has thought that it must be neutral in both senses of the word, practically and theoretically. That is why the debate about ends has been so weak. It is necessary to react, and bioethics can be interpreted as a reaction, which must combine the neutrality from coercion with a belligerent attitude in all value questions related to the goals of human life and the substantive conceptions of good. The most important mistake of liberalism has been the lack of a profound debate about ends and goals in life. As the report of the Project on *Goals of Medicine*, lead by the Hastings Center, states:

> It is not unreasonable to say that as society goes, so goes medicine. A transformation of medicine ideally requires a transformation of society; they can no longer be kept separate. To rethink the goals of medicine requires, at the same time, rethinking the goals and values of society; and of the cultural substrate of society (Callahan, 1996, p. 6).

Here neutrality is not only impossible but also incorrect and immoral. This has been, as is well known, the critique made against liberalism by communitarian thinkers. It is not a coincidence that a communitarian bioethicist, Ezekiel J. Emanuel, has been the author of a book entitled *The Ends of Human Life* (Emanuel, 1991). The thesis of this book is that medical ethics has been so irresolute because it has centred its interest only on the ethical analysis of medical technology, that is, on the evaluation of means, of the new means, and not on the ends. I agree with Emanuel that the true ethical problems of medicine are not these but the right analysis of ends. And the debate about ends must be made not only by individuals but also by society as a whole. Therefore, medical ethics must be placed, as in Aristotelian times, in the framework of political philosophy. Emanuel understands political philosophy as:

> Concerned with the collection of communal ends in a polity, articulated, elaborated, and organised into a coherent framework to guide the ethical reasoning of citizens as members of that community in issues of public concern. Political philosophy reflects and embodies shared values, or what is sometimes referred to as the common morality (Emanuel, 1991, p. 22f).

In my opinion the critique made by Emanuel in regard to bioethics is completely pertinent. But it is necessary to go further, affirming that ethics consists not only in the management of the shared values of a community, but also in its rational and critical analysis. Ethics is a part of philosophy, understood as the critical analysis of the ends of human life. The political management of shared values is necessary as well as their critical analysis, and that is the field proper of ethics. Ethics is the part of philosophy that deals with the ends of human life and the means needed to reach them. Medical ethics has dealt traditionally with means and in the micro-level proper to private and personal decisions. Now it is necessary to change from the analysis of means to the study of ends, and from the micro- to the macro-level. Interest in ends forces us towards a rigorous change of perspective.

The need for this change has been recognised by many important liberal thinkers, like John Rawls, who in his last book, *Political Liberalism*, deals with this problem, and acknowledges the need for integrating the communitarian perspectives in the theory and practice of political liberalism (Rawls, 1993).

This is, too, the core problem of the present debates in the field of health care. As argued above, the bioethical movement must be understood within this perspective, and one of the most important practical consequences accomplished by this movement, *viz.* Ethics Committees, are deprived of meaning out of this framework. Ethics Committees must be understood as ways of improving liberal democratic participation in the value conflicts arising in the process of health care. This permits understanding why in the last years many proposals have been made in this direction. For instance, papers of Leonard Fleck (Fleck, 1974; 1979; 1986; 1987; 1990; 1991; 1992; 1994a; 1994b; 1994c; 1994d) and Amy Gutmann (Gutmann, 1987; Gutmann and Thompson, 1996, 1997). Gutmann's thesis is that in a liberal society disagreements must be deliberated, and that human nature has four different causes of disagreements, and therefore four purposes of deliberation: first, scarcity of resources (not always, but frequently, and not only in health care, but in any other aspect of human life); second, our limited generosity; third, incompatible moral values; and fourth, incomplete understanding. Those four are inescapable human conditions in a liberal society, which can only be managed correctly improving social participation and value deliberation. A true re-moralisation of our society is not possible without the improvement of two important changes, 'participation' and 'universalisation'. Only in this way can the goals of every human being and the goals of our communities be correctly established. And bioethics must be an important way of promotion and discussion of these kind of questions.

To summarize: the success of bioethics has been due to civil society feeling the need of a reflective and responsible analysis of values. In previous centuries, Western society thought that law could be a correct way of managing values. Today, on the contrary, a big gap between state and civil society has been established, and people in general think that law is only the minimal common denominator of values defended by individuals and social groups. The social debate on values is one of the great goals of our societies from the sixties to the present, and bioethics is one of the ways open to achieve this goal. Therefore, I think bioethics must assume this responsibility of being a forum of debate and education in value questions related to body and to life. And specifically, medical bioethics must be understood as a place of debate for the value questions related with the management of the human body and the human life. A big change is necessary when defining the goals of the management of body and life not only in our societies, but also in health care professionals. And medical bioethics has this important role to play. The goal of medical bioethics is training health professionals and health consumers in value questions, in order to increase the correctness and the quality of their decisions. Only through understanding bioethics as a responsible way of changing the present goals of medicine, this change will succeed.

6. RESPONSIBLE GENETICS

Gene therapy cannot be correctly understood and judged either from the naturalistic point of view, from the hard deontologism, or from the pure consequentialism. Extremes like dogmatism and strategic rationality are here untenable. Gene therapy cannot be absolutely banned or absolutely accepted. Between these two extremes it is necessary to construct an intermediate and responsible way.

In general, people accept that genetic engineering should be permitted to correct the genetic defects or genetic diseases, and banned in those other cases in which its goal is only the enhancement of human nature. This is the intuitive criterion used by normal people when discussing these questions, and it is also the recommendation made by experts in reports (Gracia, 1993).

But the problem, as we have analyzed before, is that disease is not an absolutely clear concept. Biological dysfunction or defect is a necessary condition of disease, but not a sufficient one. Disease is not only a biological problem but also a social and moral construction. In fact, it is dependent on values in the same way, or even more than on facts. It can be said that disease is a biological dysfunction in the context of a specific biography, or that it is at the same time a biological and a biographical question.

Biography is the result of articulating the specific resources everyone has (biological, economical, geographical, etc.) with their own goals, values, aspirations, and preferences. This is the point in which health and disease are moral constructions. They depend on the goals of perfection and happiness assumed by societies in general and by human beings in particular. If the human ideal of perfection is being always taller and stronger, genetic engineering will be used, with no doubt, to enhance these characters. Human beings can deny their contingency, dreaming irrationally of the possibility of a world in which man will be deprived of negative contingencies, like suffering, diseases, frustration and mortality. This irrational idea of a plentiful human life leads to thinking that a completely beautiful, good, wealthy and healthy mankind, living in a perfect and unlimited well-being, is possible. But this is a great mistake. This is an irrational goal, which produces necessarily frustration and unhappiness. An irrational idea of happiness leads paradoxically to unhappiness. And the question is whether we will use genetic engineering in a rational and responsible way, or, on the contrary, we will be looking for a non-existing and frustrating heaven. As Philippe Frossard has written:

> The very idea of selecting for desired traits is not only theoretically impossible, because of DNA mutation and recombination events; it is also absurd. Where would we stop? Any physical and behavioral trait can become a subject of disagreement. Some members of our society would certainly want to go for blue-eyed, blond-haired types - there is a historical precedent for that. And what about our conception of intelligence? Should we seek to raise systematically all IQs? The mean value in the population would still be 100. How about height? ... What can possibly be an ideal height? If everybody was 7 feet high, those who wanted to feel taller would have to go for 8 feet. Along the same lines, why not take into consideration the size of certain external organs? This would certainly introduce an even stronger bias

> toward assortative mating. ... Both the idea of moulding humans according to predetermined quality criteria and the means of doing so have been put into application in several instances. The list of a few precedents - including the Nazi experience, sterility campaigns for the 'mentally abnormal', choice of children's gender (male in 98 per cent of cases), storage of sperm from Nobel Prize recipients as a marketable commodity - remind us that these situations have led to bitter objections at best and atrocities at worst (Frossard, 1991, p. 233-5).

It all depends on the set of values assumed as goals of human life by individuals and by societies. And this is the real scenery for ethics. The ethical question is not whether negative genetic engineering must be permitted and promoted, or whether positive genetic engineering should be banned. This is a merely strategic approach to the problem. The real question is which are our individual and collective goals in human life? And this question does not have a unique and definitive answer. In fact, everyone is changing little by little the values and goals of his or her life, and societies are also modifying continuously their organic structure of values. The aim of ethics is not the interference with these processes, but the rational and critical analysis of their content. Ethics is the critical discussion or the rational deliberation about the ends of human life. No more and no less.

Bioethics must be an important and privileged place of rational and participative deliberation about the wise use of the body and life. From Socrates up to today, this has always been the real place of philosophy. There is not a strategic discussion about means, but the rational debate about ends. And this is the great problem in our society. All of us are convinced of well-being as the only end of human life. In fact, our canonical definition of health says that it is a perfect physical, psychological and social well-being, and not only the absence of disease. Is it a wise goal for human life? Is it a rational goal for medicine? What are the goals to which genetic engineering should be submitted? These are the questions that everyone, individually and collectively, must deliberate and answer. And bioethics should join and help each one of us in this way.

7. SOME CONCLUSIONS

There is no possibility of establishing a clear-cut definition of health and disease using only biological data. Health and disease are not only biological concepts but also and principally cultural and social concepts. Health can only be defined as the capacity of achieving the personal goals each one has assumed in his or her life, and disease as the incapability of doing that, or the impossibility of achieving his or her own vital goals.

Therapy is also a cultural, axiological and moral concept, and not merely a biological one. Therapy is not only the cure for biological diseases, but also the enhancement of human nature, and therefore the reversion of negative possibilities of life into positive ones. This is the reason why there is no possibility of establishing a clear cut between genetic engineering as a method of curing some diseases, and genetic engineering as a way for enhancing human nature. In this point, genetic engineering is not different from other kind of therapies. For instance, Viagra can be seen as a drug that cures some symptoms, or as a means for the enhancement of human nature. The same with Prozac.

The main question is not, therefore, the biological definition of the concepts of health, disease, medicine, and therapy, but what the goals of human life are, because the other terms can only be understood in an instrumental way, as means, technical means in order to achieve the goals everybody has assumed in his or her life. This is the real moral question, the definition of human ends and the adequacy of means to ends.

From this point of view, genetic engineering must be understood only as a means, with a morality that depends on its relationship with the ends. Because genetic engineering is only a 'mean', it receives its moral qualification, principally, from the 'ends' to which it is ordered. If the ends are correct or morally good, genetic engineering, either negative or positive, either performed in somatic or in germ line cells, should not be banned. The other way, the solution of the conflicts between means, is merely strategic. An ethical analysis of the problem leads to the discussion of the ends involved, and the correlation between ends and means.

The ends of human life are first of all private. Everyone has his or her own ends, and the human right to freedom of conscience protects everyone in defining and achieving his or her own ends. The main private goal of human beings is always the same, the search for happiness. The problem is that some ideals of happiness are inhuman, because they do not respect all human beings as ends in themselves, and not merely as means; and others because they are untenable, impossible, and therefore frustrating. This is a very paradoxical consequence that the search for happiness can lead to unhappiness. In general, it can be said that the irrational goals can not be achieved, and as a consequence they always generate frustration, and unhappiness. This is the case of thinking that science will be capable of reaching a completely happy world, in which disease and suffering will not be present. This belief, due to its irrational character, generates necessarily frustration and unhappiness. Looking irrationally for happiness leads necessarily to unhappiness.

There are human goals other than the private ones. These are the so-called public goals. Because for all of those who live in society, there are some ends which must be common, assumed and respected equally by all.

These collective and public ends, which are compelling for all, must also be established by all, throughout a process of deliberative participation. Only in this way can prohibitions be morally and legally binding for all.

This is the only right way in order to establish some limits to genetic engineering. There is no other manner than to reach some common agreements about what is inconvenient to all human beings and what is not. The method must therefore support participative deliberation. This is the actual way of reaching true moral and prudential agreements about the collective ends of human life and about the set of prohibitions we should all respect.

Bioethics can not be understood as the prudential resolution of conflicts between means, but as the discourse and debate about the ends of human life. Which are the wise goals of human life? Which are the rational goals of medicine? Which are the goals to which genetic engineering should be submitted? These are the questions that everyone, individually and collectively, must deliberate and answer. And bioethics should join and help each one of us in this way.

REFERENCES

Anderson, W.F. and Friedmann, T.: 1995, Strategies for gene therapy. In: Reich, W.T. (ed.), *Encyclopedia of Bioethics*. MacMillan, New York, Vol. 2, p. 907-914.

Baker, K. and Mason, S.: 1976, The limits of 'playing God'. *Keeping Posted* 22(2): 19-23.

Bayertz, K. (ed.): 1994, *The concept of moral consensus*. Kluwer Academic Publishers, Dordrecht.

Beauchamp, T.L. and Childress, J.F.: 1994, *Principles of biomedical ethics*. Oxford University Press, New York.

Bentham, J.: 1988, *A fragment on government*. In: Burns, J.H., Dinwiddy, J.R., Rosen, F. (eds.), *The collected works of Jeremy Bentham*. Oxford University Press, London and Oxford.

Biesecker, B.B.: 1995, Practice of genetic counselling. In: Reich, W.T. (ed.), *Encyclopedia of bioethics*. MacMillan, New York, Vol. 2, p. 923-927.

Broad, C.D.: 1930, *Five types of ethical theory*. Routledge & Kegan Paul, London.

Callahan, D. (ed.): 1996, The goals of medicine: Setting new priorities. *Hastings Center Report* Special Supplement, November-December.

Chase, A.: 1987, *Playing God in Yellowstone*. A Harvest/HBJ Book, Harcourt Brace Jovanovich Publishers, San Diego.

Clouser, K.D.: 1995, Common morality as an alternative to principlism. *Kennedy Institute of Ethics Journal* 5: 219-236.

Clouser, K.D. and Gert, B.: 1990, A critique of principlism. *Journal of Medicine and Philosophy* 15: 219-236.

Clouser, K.D. and Gert, B.: 1994, Morality vs. Principlism. In: Gillon, R. (ed.), *Principles of health care ethics*. John Wiley & Sons, Chichester, p. 251-266.

Cookson, W.: 1988, *The gene hunters: Adventures in the genome jungle*. Aurum Press, London.

Emanuel, E.J.: 1991, *The ends of human life. Medical ethics in a liberal polity*. Harvard University Press, Cambridge, Mass.

Finnis, J.: 1980, *Natural law and natural rights*. Clarendon Press, Oxford.

Fleck, L.M.: 1974, Justice, age rationing, and the problem of identifiable lives. In: Hackler, Ch. (ed.), *Health care for an aging population*. State University of New York, Albany, NY, p. 93-105.

Fleck, L.M.: 1979, Abortion, deformed foetuses, and the Omega pill. *Philosophical Studies* 36(3): 271-283.

Fleck, L.M.: 1986, Confidentiality: Moral obligation or outmoded conept? *Health Progress* 67(4): 17-20, 61.

Fleck, L.M.: 1987, DRGs: Justice and the invisible rationing of health care resources. *Journal of Medicine and Philosophy* 12(2): 65-196.

Fleck, L.M.: 1990, The Oregon medicaid experiment: Is it just enough? *Business and Professional Ethics Journal* 9(3 and 4): 201-217.

Fleck, L.M.: 1991, How just must we be? In: Humber, J.M. and Almeder, R.F. (eds.), *Biomedical ethics reviews 1990*. Humana Press, Clifton, NJ, p. 131-188.

Fleck, L.M.: 1992, Just health care rationing: A democratic decisionmaking approach. *University of Pennsylvania Law Review* 140(5): 1597-1636.

Fleck, L.M.: 1994a, Just genetics: a problem agenda. In: Murphy, T.F. and Lappe, M.A. (eds.), *Justice and the human genome project*. University of California Press, Berkeley, CA, p. 133-152.

Fleck, L.M.: 1994b, Just caring: Lessons from Oregon and Canada, in: Hackler, Ch. (ed.), *Health care for an aging population*. State University of New York Press, Albany, NY, p. 177-198.

Fleck, L.M.: 1994c, Just caring: Health reform and health care rationing. *Journal of Medicine and Philosophy* 19(5): 435-443.

Fleck, L.M.: 1994d, Just caring: Oregon, health care rationing, and informed democratic deliberation. *Journal of Medicine and Philosophy* 19(4): 367-388.

Frosssard, P.: 1991, *The lottery of life: The new genetics and the future of mankind*. Bantam Press, London.

Garrett, L.: 1995, *The coming plague: Newly emerging diseases in a world out of balance*. Penguin Books, New York.

Gert, B., Culver, C.M. and Clouser, K.D.: 1997, *Bioethics: a return to fundamentals*. Oxford University Press, New York-Oxford.

Goodfield, J.: 1977, *Playing God: Genetic engineering and the manipulation of life*. Random House, New York.

Glover, J.V.C.: 1989, Human genetic engineering - how far should we go? In: Weatherall, D. and Shelley, J.H. (eds), *Social consequences of genetic engineering*. Elsevier Science Publisher, New York, NY, p. 193-202.

Gracia, D.: 1993, Ethical and social aspects of the human genome analysis project. In: Haker, H., Hearn, R. and Steigleder, K. (eds), *Ethics of human genome analysis. European perspectives*. Attempto Verlag, Tübingen, p. 247-271.

Gutmann, A.: 1987, *Democratic education*. Princeton University Press, Princeton.

Gutmann, A. and Thompson, D.F.: 1996, *Democracy and disagreement*. Harvard University Press, Harvard, Mass.

Gutmann, A. and Thompson, D.F.: 1997, Deliberating about bioethics. *Hastings Center Report* 27(3): 38-41.

Hanson, M.J.: 1997, At the center. *Hastings Center Report* 27(3): cover.

Heyd, D.: 1992, *Genethics: Moral issues in the creation of people*. University of California Press, Berkeley.

Jonsen, A.R.: 1991, Casuistry as methodology in clinical ethics. *Theoretical Medicine* 12(4): 295-307.

Jonsen, A.R.: 1998, *The birth of bioethics*. Oxford University Press, New York.

Judson, H.F.: 1979, *The eighth day of creation. Makers of the revolution in biology*. Simon and Schuster, New York.

Juengst, E. and Walters, L.: 1995, Ethical and social issues. In: Reich, W.T. (ed.), *Encyclopedia of bioethics*. MacMillan, New York, Vol. 2, p. 914-922.

Kant, I.: 1995, *Grounding for the metaphysics of morals*. In: Cahn, S.M. (ed.), *Classics of Western philosophy*. Hackett Publishing Company, Indianapolis, Cambridge.

Lubow, A.: 1977, Playing God with DNA. *New Times* 8(1): 48f.

Monod, J.: 1970, *Le hasard et la nécessité*. Éditions du Seuil, Paris.

Murray, R.F.: 1995, Genetic counseling: Ethical issues. In: Reich, W.T. (ed.), *Encyclopedia of bioethics*. MacMillan, New York, Vol. 2, p. 927-932.

Peters, T.: 1995, Playing God and germline intervention. *Journal of Medicine and Philosophy* 20(4): 365-386.

Potter, V.R.: 1977, Evolving ethical concepts. *Bioscience* 27(4): 251-253.

Rawls, J.: 1993, *Political liberalism*. Columbia University Press, New York.

Rifkin, J. and Perlas, N.: 1983, *Algeny*. The Viking Press, New York.

Rifkin, J. and Howard, T.: 1997a, Who should play God? *Progressive* 41(12): 16-22.

Rifkin, J. and Howard, T.: 1977b, *Who should play God: The artificial creation of life and what it means for the future of the human race*. Dell Publishing Company, New York.

Rollin, B.E.: 1995, Genetic engineering: Animals and plants. In: Reich, W.T. (ed.), *Encyclopedia of bioethics*. MacMillan, New York, Vol. 2, p. 932-936.

Rosenfeld, A.: 1975, *The second genesis*. Vintage, New York.

Seydel, F.D.: 1990, Human gene therapy: Playing God? In: Dainiak, N. (ed.), *The biology of hematopoiesis*. Willey, New York, p. 339-346.

Sidgwick, H.: 1981, *The methods of ethics*. Hackett Publishing Company, Indianapolis, Indiana.

Toulmin, S.: 1975, Exploring the moderate consensus. *Hastings Center Report* 5(3): 31-40.

Toulmin, S.: 1982, How medicine saved the life of ethics. *Perspectives in Biology and Medicine* 25: 736-750.

Toulmin, S.: 1987, The national commission on human experimentation: procedures and outcomes. In: Engelhardt, H.T. and Caplan, A.L. (eds.): *Scientific controversies: Case studies in the resolution and closure of disputes in science and technology*. Cambridge University Press, New York, pp. 599-613.

Wachbroit, R.: 1995, Human genetic engineering. In: Reich, W.T. (ed.), *Encyclopedia of bioethics*. MacMillan, New York, Vol. 2, p. 936-940.

Zubiri, X.: 1986, *Sobre el hombre*. Alianza Editorial, Madrid.

RUTH F. CHADWICK

GENETIC INTERVENTIONS AND PERSONAL IDENTITY

1. INTRODUCTION

A defining feature of gene therapy, in contrast to conventional medical treatment, is that it is explicitly designed to bring about changes at the genetic level, by, for example, introducing a functioning gene into a human being who lacks one. Questions arise as to whether the alteration of an individual's genetic make-up could bring about a change in who that person is, i.e. an identity change, and if so whether this is something about which there are grounds for concern from a moral point of view. In order to answer these questions it is necessary to examine what the connection is, if any, between genes and the person, and how that might be affected by gene therapy.

There might be several different ways in which genes and the person could be connected, genes as the 'essence' of the person, or genes as a criterion of identity over time. What some commentators appear to be interested in, however, is a person's sense of their own identity. It is important to distinguish these different levels; otherwise the debate about the acceptability of gene therapy is liable to confusion.

2. AM I MY GENES?

The debate as to what counts as essentially 'me' is longstanding - whether, for example, I am my brain; whether the person can be reduced to the combination of physical and mental attributes that make up a person as we experience them; whether there is something over and above this set of attributes. It would be beyond the scope of this discussion to rehearse the arguments for and against competing accounts of the person. Rather, it will be relevant to address the ways in which an account in terms of DNA relates to these debates.

H.A.M.J. ten Have and B. Gordijn (eds.), Bioethics in a European Perspective, 339-349.

First, a physical account of the person. A standard objection to any physical account of the person, such as that which identifies the person with his or her brain, is the possibility of division and the problem of how to account for that. The suggestion that the essence of the person might consist in their DNA is open to the objections to which a physical criterion is subject. In so far as it has a physical existence - it is material stuff - it can have bits added and removed. Also, whereas the philosophical puzzles generated by the possibility of dividing a brain and transplanting the two halves into different people are thought experiments, we know that the DNA in one fertilised egg can and does split into two when identical twins are formed. This has led to difficulties about establishing when a person's life story begins. Bernard Williams, for example, formulated a Zygotic Principle (ZP) to deal with this question, such that "A possibility in which a given human being, A, features is one that preserves the identity of the zygote from which A developed" (Williams, 1990, p. 16). Williams subsequently modified ZP in order to deal with the possibility of monozygotic twinning, so that "a story is about A it is about an individual who developed from the earliest item from which A in fact uniquely developed" (*ibid.*, p. 178).

The brain has had a central place in discussions of personal identity because of its role as the necessary condition of consciousness and mental life, and this has contributed to the importance of psychological criteria, rather than a physical substrate, for the essence of the person. One of the most influential accounts of what it means to be a person has been the view that a person is a being with self-awareness. The idea that we could hold up a disk containing our genetic profile and say "That's me", as has been envisaged by some genome scientists (cf. Gilbert, 1992, p. 83), does not take this into account. On the self-awareness account, it is the capacity to give voice to the statement "That's me" that is crucial, not the information itself. This objection however might not count against a view which gives importance to potentiality. One of the considerations supporting giving moral consideration to embryos and foetuses has been the view that all the genetic information is in place that, other things being equal, will lead to the development of a person, is in one place, as stated in the discussion of the physical criterion above. So in this sense the traditional argument for potentiality supports the view that the genetic information has a crucial role to play. This does not establish, however, that the genes are the person, but only that they constitute a necessary condition for the development of the person and their psychological states.

Perhaps, however, the attraction of DNA as the essence of the person is its candidature for the part of some fact over and above physical or psychological attributes, that constitutes the essence of the person. This seems to be what is suggested in accounts that put forward DNA as the modern secular equivalent of the soul. Nelkin and Lindee, for example, have drawn attention to the parallels between the Christian soul and DNA, which, they suggest, are "more than linguistic or metaphorical" (Nelkin and Lindee, 1995, p. 41):

> DNA has taken on the social and cultural functions of the soul. It is the essential entity - the location of the true self - in the narratives of biological determinism (*ibid.*, p. 41-2).

It has also taken on the soul's role as the guarantor of immortality. We find a degree of immortality in passing on our genes to our descendants. Cloning offers a vision of passing on all of them rather than only half, and thus may appear more attractive to those in search of continued existence. Unlike a soul, however, as we have already seen, DNA has a physical existence. What makes DNA attractive as a soul-like candidate, however, is its double aspect - as both physical stuff and as containing 'information'. Whereas we know that all the cells of the body are replaced over a period of years, in each new replacement cell the nucleus will contain the same set of genes. It is this that makes cloning by somatic cell nuclear transfer a possibility, and which has led to the metaphor of the 'blueprint'. Our DNA, it is said, contains all the information that influences our physical and mental characteristics. The genetic information gathered inside the nucleus governs the future development of an identifiable individual.

There are problems with this account. First, the blueprint metaphor has to be recognised for what it is - a metaphor. The implications of the adoption of this metaphor for our ways of seeing persons have to be acknowledged:

> The expansion of information technologies and communication systems ... have laid the foundations for the formation and validation of different forms of knowledge and new ways of seeing the body and the self (Stacey, 1997, p. 159).

And

> In genetics, for example, the body is conceptualised through informational metaphors, broken down into microscopic units which can be reprogrammed ... The body is seen as a potentially correctly programmed system which, through scientific intervention, might be improved by the replacement of abnormal genes with normal ones (Stacey, 1997, p. 159).

This way of seeing the person, as a complex information system, competes with other perspectives. Nelkin and Lindee link the genetic perspective with biological determinism, but environmental factors influence both physical characteristics such as stature; and mental attributes such as intellectual achievement. Even if we accept that all the information is contained in the DNA, laying down the potential attributes of the person, environmental factors will affect which parts of that potential are realised. So what does it mean to say that my genes are me, if large parts of my genome will never be activated?

A possible response to this might be that what is important is precisely that my genome endures throughout all environmental changes whether or not parts of it remain inactivated, just as the soul was thought to exist, intact, throughout all the vicissitudes of life. This interpretation could give support to the view that the genes are the essence of the person, which might manifest differently in the phenotype according to environment.

Against this view of genes as the 'essential encoder' Hugh Miller argues for a commonsense view of the person, traced to an Aristotelian account:

> Persons are physical (biological) entities with distinctively rational capacities, or modes of functioning. They are material bodies that can do certain distinctive things (Miller, 1998, p. 194).

On his view the identity of a person is determined by the character he or she develops as he passes through space and time. Miller rejects the idea of an immaterial self such as a Cartesian ego, while wanting to retain the idea of the essence of personhood being linked with both free will and moral responsibility, and with our biological nature. Psychological criteria are therefore central but cannot dispense with attachment to a material body located in space and time. The question then arises: to what extent do genes determine character? Miller argues that there are three possible answers to this, corresponding to different degrees of determinism. His conclusion is that genetically determined traits function as parameters within which a human being is free to develop a unique character which makes him or her the person he or she is (Miller, 1998). DNA is a necessary condition of personal identity but should not be conflated with it. The rival accounts of the essence of the person as consisting in the genetic blueprint or in the character are thus intricately intertwined with questions of determinism, free will and responsibility.

As Miller has argued, sameness of DNA cannot logically guarantee sameness of person, because otherwise identical twins would be one and the same person and this is not the case. In addition to travelling distinct spatio-temporal paths, they have different life stories and develop different characters, although identical twins, unlike clones, even share the same mitochondrial DNA. Miller further argues, however, that if two organisms of the same species have different DNA blueprints, then it follows from the logic of identity that they must be different persons. But this does not follow unless it is assumed that personhood does exist in the DNA, and Miller argues, as stated above, that while DNA is a necessary condition of personal identity, it should not be conflated with the essential encoder. If this is so it would be important for the consideration of identity over time.

3. IDENTITY OVER TIME

In accordance with different views on what constitutes the essence of a person, there are associated criteria of identity over time. Depending on the view that is taken on the first question, there will be differences of opinion on how much change can take place, for example in gene therapy, without implying the conclusion that an identity change has taken place.

The use of genetic fingerprinting as a means of identification, to show that the individual being accused at time t2 is one and the same individual as the person who committed the crime at time t1, does not show that DNA is the criterion of identity in a deep sense. Fingerprints, after all, were used as a means of identification, but without any suggestion that they constituted the essence of the person.

If personal identity consists in something other than the DNA, then it would in principle be possible to bring about some change at the DNA level, for example by gene therapy, which would not constitute a change in the identity of the person. There are different possibilities:

1. that any change in the DNA brings about an identity change;
2. that a change in a certain proportion brings about an identity change;
3. that a change in a key part brings about an identity change.

1. will only be the case if 'I' am identical with my complete set of genes. Given the facts, however, that human beings share over 99% of their genes, and doubts about whether a gene that expressed in only one organ can appropriately be called 'human' (cf. Pottage, 1998, p. 759), it seems implausible to accept that no change, even in one gene, would be permissible;
2. depends on a physical criterion of personal identity with all the associated problems;
3. depends on the prior question of which aspects of the person are thought to be the sine qua non, for example, which psychological attributes or character traits; and if that can be answered can we identify which genes are associated with the characteristics in question?

4. GENE THERAPY

Gene therapy might be supposed to pose a threat to personal identity in so far as the therapy introduces a identity-changing effect (see, e.g. Elliott, 1993). It is germ-line, rather than somatic therapy, however, that is normally considered to be susceptible to this problem. The conceptual distinction between somatic and germ-line therapy is typically drawn via the point that while somatic therapy affects the body cells of an individual, germ-line therapy also affects the reproductive cells, thereby having an impact on that individual's children and ultimately on the gene pool of the species (see, for example, Clothier, 1992). This attempt to distinguish between them conceptually has never, however, been entirely clear-cut. It has for some time been envisaged as a possibility that cells being introduced into the body in the course of somatic therapy could recombine with other viruses and infect the germ cells. *New Scientist* has reported evidence of such 'contamination' (14 March 1998). The proposals for in utero gene therapy have also given rise to speculation about inadvertent transfer into germ cells (Schneider and Coutelle, 1999, p. 256-7). More radically, the Dolly experience has undermined the distinction between types of cell. The implication that every somatic cell is now a potential embryo has led to the necessity of reexamining our definitions of concepts such as 'embryo'.

Somatic therapy carried out on an adult with informed consent is commonly considered to be no different in principle from an organ transplant such as the introduction of a donor kidney (which also contains extraneous genetic material). It is of course true that, especially in the early days, recipients of donor organs experienced some psychological problems connected with their sense of identity, but these have not been construed as challenges to identity in a deep sense. As Bernard Williams puts it "In general, 'same X which consists of parts' doesn't entail an X which consists of the same parts. There's no puzzle about replacement cells" (Williams, 1990), so analogously it would be difficult to sustain an argument for the view that inhaling a functioning gene in, for example a nasal spray, as has been tried in the case of cystic fibrosis, would pose a threat to any essential core. And adding a functioning gene is what is normally envisaged, although as replacement becomes more common it is conceivable that this might be thought to make a difference.

In germ-line therapy however, what is under consideration is therapy to change the genetic constitution of an individual at the embryonic stage. As such, any change would be global rather than local, i.e. it would not be targeting a particular part of the body such as the lungs or bone marrow. Whether such an intervention would have an identity-changing effect or not depends, as already indicated, on whether at the embryonic stage we do have a person - whether there is a full 'blueprint' of the individual in the embryo that does in fact constitute the essential identity of that person. Some would argue that what we have at this stage of development is not a person but a potential person. It could still be argued, however, that a genetic intervention at this stage will affect, if not determine, who that individual will be.

There is disagreement here both on the status of the embryo and on the relation between genes and personal identity. In this situation it might be helpful to try another approach and ask whether a future person could legitimately claim to have a grievance as a result of gene therapy performed at the embryonic stage.

5. WHAT GRIEVANCES MIGHT A FUTURE PERSON HAVE?

One way of addressing this question is to ask what grievances a future person might have. As I have argued elsewhere (Chadwick, 1998), in response to Bernard Williams' suggestion that there are only two, *viz.* "I should have had a nicer time" and "It would have been better if I had never existed" (Williams, 1990), there are other possibilities, including "I should not have had my genome altered", "I should have been someone else". "I should have been free of this genetic disorder", "I should have been given genetic immunisation", and "I should not have been brought into existence". I suggested that those grievances which were coherent presupposed sameness of identity; the grievance which deals explicitly with identity change, namely "I should have been someone else" is not coherent because, as Williams pointed out, I could not have been someone else; what could have been the case is that I did not exist and someone else did.

This analysis, however, is incomplete, because the grievance may lie elsewhere than in an identity issue of this sort: it may lie instead in the individual's sense of their own identity.

The person who says "I should not have had my genome altered" may have a coherent grievance but not one that lies in a personal identity issue. The statement of the grievance presupposes that identity has been preserved. Nevertheless it may be perceived as a personal identity issue.

There are at least two ways in which a person's perception of their identity may be affected. One possible grievance a future person might have is "I don't know who I am" meaning, that he or she does not know what his or her genetic origins were. This issue has been more commonly raised where paternity has been in doubt, but reproductive technology has changed that. In the light of the possibility of reproductive cloning, one question that may be asked is "Am I a copy?" Why is this important? Given genetic heritage, the individual nevertheless develops in a unique way and also has the potential for what is called self-creation. Concern for genetic origins is, however, still regarded as very important in our society. Marilyn Strathern has suggested that this may disappear:

> Perhaps the current interest in genetic origins will turn out to have been more of a radical ... break with the past, and with the old reproductive model, than it is an evolutionary development of what we already know ... I am not so certain that we shall in future need representations of downward inheritance or of relations embodied in relationships: all that we shall need is the programme ... Questions that the individual person once asked of him- or

> herself about origin and links need no longer be asked ... when they can be asked of the individual's genome (Strathern, 1992, p. 178).

This in itself, however, is presupposing that identity consists in the genes and not in other factors such as situatedness, which is precisely in dispute.

The second way that a person's perception of their identity may be involved is in so far as they identify themselves with their genes or particular characteristics which they take to be genetic. We see this, for example, in arguments put forward by disability rights organisations in opposition to the new genetics and its attempts to 'cure' conditions which are not regarded as disorders by their possessors. There are at least two separate strands to be disentangled here. The first is the concern that attempts to eliminate or cure genetic conditions will lead to a society less tolerant of disability. This is typically countered by an argument that what is the object of these attempts is not the people but the conditions. At this point the second strand comes into play, which is that it is not possible to distinguish these two elements, because their identity is dependent upon their genetic condition. Whereas an individual who suffers from a disease such as smallpox can conceptualise themselves without this disease, this is not possible in the case of certain genetic conditions.

The question arises as to whether this is true of all conditions that have a genetic component. In the case of a predisposition to a multifactorial disease, such as breast cancer, the individual concerned may not know of the predisposition, and so her identity has been constructed without this knowledge. It has been argued, therefore, that there may be a right not to know genetic information on the grounds that it may constitute a threat to the individual's sense of their identity (cf. Chadwick, *et al.*, 1997). The fact that this argument is put forward for the right not to know that one has a genetic predisposition on these grounds supports an argument for the view that one's knowledge about one's genes plays an important part in constructing one's identity in this sense. (Indeed, both origins and predispositions are important in this respect).

The point is, how seriously should we take this as an objection to gene therapy? It is easy enough to make sense of the person who has a grievance that they have been given information which has changed for the worse their perception of their self and their future; but what about the individual who objects to gene therapy to remove deafness on the grounds that it has turned them into a different person? What we have seen above is that if this was an identity change in the deep sense they would not be able to voice this as a coherent grievance. Let us look at what coherent grievances may be expressed:

> I am no longer the same person [because my condition played an essential part in constructing my identity]. An attempt to eliminate my condition

> constitutes an attack on what is essentially me. I am no longer able to regard myself in the same way (Strathern, 1992, p. 178).

The third is the most straightforward. It draws attention to psychological consequences of interventions which should be taken into account. In so far as the first and second are arguments about an individual's perception of themselves they can be construed as coherent grievances. How conclusive they are as objections to action is another question and will depend on the weight given to this argument in comparison with other arguments for and against gene therapy. It would be possible, for example, given appropriate technology, to construct an argument for gene therapy as a means to preserving an individual's sense of self, e.g., for an individual at risk of premature Alzheimer's. One point to note here is that this objection is no longer limited to being an objection to germ-line therapy; somatic therapy also would be subject to this objection - and indeed certain non-genetic interventions such as cochlear implants in the case of deafness. In so far as the latter is true, it highlights the possible irrelevance of genetics to issues of personal identity under this interpretation.

What these possible grievances should highlight is the inadequacy of giving our attention to identity issues only in the deep sense of identity as opposed to the individual's self-perception. MacIntyre has drawn attention to the inevitable failure of a concentration on issues of 'strict' identity:

> Possessing only the resources of psychological continuity, we have to be able to respond to the imputation of strict identity. I am forever whatever I have been at any time for others - and I may be at any time called upon to answer for it - no matter how changed I may be now (MacIntyre, 1981, p. 202).

The concept of a person is an abstraction: "the characters in a history are not a collection of persons, but the concept of a person is that of a character abstracted from a history" (*ibid.*).

This narrative conception of the self also has implications for gene therapy via the concept of collective identity. This is a further sense in which it might be said to be of concern that we interfere at the genetic level. First there is collective identity as a species. Nelkin and Lindee, having suggested that DNA is the secular equivalent of soul stuff, proceed to suggest that our genes are seen as defining us as a species - distinct from others - of those who share our DNA. This is yet another sense in which DNA may be seen as definitive of identity. More pertinent to the present discussion, however, is collective identity of those who share a particular genetic characteristic.

6. COLLECTIVE IDENTITY

MacIntyre has argued that "the self has to find its moral identity in and through its membership in communities such as those of the family, the neighbourhood, the city and the tribe" (MacIntyre, 1981, p. 203). This may be what is at stake in certain disability rights arguments, where people see their identity intimately associated with a particular condition and group - as in the deaf culture argument, for example. The promise of genetic cures for deafness is seen as a threat to deaf culture (cf. Grundfast and Rosen, 1992). A full treatment of this argument requires a consideration not only of personal identity issues but also of issues of justice and of the proper goals of medicine. The essential point for present purposes is that it is crucial to distinguish between arguments that are opposing genetic developments such as gene therapy on the grounds of individual identity, on the one hand, and those which oppose them on the grounds of collective identity, on the other. As already mentioned, however, this argument is not confined to genetics, because it has also been applied to cochlear implants.

7. CONCLUSION

What then can we learn from this? First, that it cannot be established that the essential 'me', my identity in the strict sense, lies in my complete set of genes. It seems implausible to suppose that there could be no change at the genetic level without producing an identity change. Other perspectives locate the essential me elsewhere, such as in psychological characteristics or in character, but the nature of the relationship between genes and these characteristics is controversial, depending as it does on different views about determinism.

Our genes may however play an essential part in constructing our sense of self, particularly in so far as we are aware of some essential characteristic. In that case it is the phenotype that becomes important and interventions of a nongenetic kind may be just as threatening as genetic ones.

How seriously are we to take the arguments about threats to personal identity as objections to genetic interventions? We have seen that no coherent grievance could be expressed to changes to identity in the strict sense. A coherent objection could be expressed to changes to one's perception of one's identity in so far as it can have serious psycho-social consequences and these have to be taken into account. They do not, however, form a conclusive argument against gene therapy.

Issues of collective identity cannot be resolved via the identity issue alone. They require an examination of issues of justice and the relationship between individual and community. As MacIntyre says, the location of the moral identity of the self in communities "does not entail that the self has to accept the moral limitations of the particularity of those forms of community" (MacIntyre, 1981, p. 205).

ACKNOWLEDGMENT

This text has been published as: Chadwick, R.: 2000, Gene therapy and personal identity. In: Becker, G.K. (ed.), *Moral status of persons*. Rodopi, Amsterdam, p. 183-194. Reproduced by permission from publisher.

REFERENCES

Chadwick, R.: 1998, Gene therapy. In: Singer, P. and Kuhse, H. (eds.), *A companion to bioethics*. Blackwell, Oxford.

Chadwick, R., Levitt, M.A. and Shickle, D. (eds.): 1997, *The right to know and the right not to know*. Avebury, Aldershot.

Clothier, C.: 1992, *Report of the committee on the ethics of gene therapy*. HMSO, London.

Elliott, R.: 1993, Identity and the ethics of gene therapy. *Bioethics* 7: 27-40.

Gilbert, W.: 1992, A vision of the grail. In: Kevles, D.J. and Hood, L. (eds.), *The code of codes: Scientific and social issues in the human genome project*. Harvard University Press, Cambridge (Mass.)

Grundfast, K. and Rosen, J: 1992, Ethical and cultural considerations in research on hereditary deafness. *Molecular Biology and Genetics* 25(5): 973.

MacIntyre, A.: 1981, *After virtue*. Duckworth, London.

Miller, H. III: 1998, DNA blueprints, personhood, and genetic privacy. *Health Matrix* 8(2): 179-221.

Nelkin, D. and Lindee, M.S.: 1995, *The DNA mystique: The gene as a cultural icon*. W.H. Freeman, New York.

Pottage, A.: 1998, The inscription of life in law: genes, patents and bio-politics. *Modern Law Review* 61(5): 740-65.

Schneider, H. and Coutelle, C.: 1999, In utero gene therapy. *Nature Medicine* 5(3): 256-7.

Stacey, J.: 1997, *Teratologies: a cultural study of cancer*. Routledge, London.

Strathern, M.: 1992, *Reproducing the future: Essays on anthropology, kinship and the new reproductive technologies*. Manchester University Press, Manchester.

Williams, B.: 1990, Who might I have been? In: Chadwick, D., *et al*. (eds.), *Human genetic information: Science, law and ethics*. Wiley, Chichester, p. 167-73.

HENK A.M.J. TEN HAVE

GENETICS AND CULTURE

1. INTRODUCTION

In present-day culture, genetical issues have a strong public representation. The media are frequently drawing our attention to genetic discoveries. These reports seem to follow a particular pattern: the gene mutation associated with a particular disease has been identified; an effective therapy for the disease is lacking, and it is not probable that the discovery itself will enhance the development of therapeutic possibilities. However, the discovery will lead to a more or less accurate prediction of the presence and prognosis of the disease when individuals are tested, even long before the onset of symptoms.

An example of such a reporting pattern occurred when the genetics research group from Nijmegen University announced that they had isolated the genetic structure responsible for Steinert's disease. This is one of the most prevalent muscular dystrophies, occurring in 1 out of 8000 persons. It is characterized by weakness of limb muscles and facial muscles. The onset is very variable, but for many patients it may be as late as the fourth decade. No cure is as yet available. The discovery was published in a scientific journal but also widely reported in the media. In newspaper interviews the researchers pointed out the practical implications. Although this scientific breakthrough will not lead to any therapy, the researchers argued in *De Telegraaf* (6 February 1992), - a national newspaper -, that "it makes it possible to detect with hundred percent certainty the disease long before the onset of the first symptoms". Also the severity of the disease can be prognosticated from the extent of the abnormalities of the DNA.

Public representations of genetic research, following the above pattern, generate a complex set of questions regarding the impact and value of genetic knowledge. The ethical debate however, tends to focus on the impact of genetics at the level of individuals. The emphasis on the proper management of information by individual citizen has a tendency within moral debate to neglect the social dimension of genetic information. Clarification of the right to know and the right not to know, although valuable in itself as a possible way to empower individual persons, also needs to elucidate the cultural

H.A.M.J. ten Have and B. Gordijn (eds.), Bioethics in a European Perspective, 351-368.

context within which genetic knowledge is promulgated, as well as the social processes involved in the dissemination of genetic technologies. The penetrating social impact of genetic technology, and its diversified cultural manifestations should lead, for example, to a critical attitude towards moral statements that individual persons are free to chose among available genetic options and will not be directed into unwanted scenarios.

The moral debate concerning genetic technologies can further be hampered by the immediate interest to translate the public fascination with new data, devices and discoveries into practical applications. Communication of new genetic discoveries reveals a paradoxic tension between knowledge and application. On the one hand, researchers publish the results of their projects because knowledge in itself is valuable. One of the rationales for the Human Genome Project is that it will lead to gains in basic knowledge. On the other hand, and at the same time it is stipulated that genetic research has potential for medical advancement. Publication of knowledge claims, especially in the public media, seems almost always to be accompanied by expositions of the potential practical implications and the relevancy to patient care. This immediate linking of knowledge and application creates particular difficulties for moral debate. The fact that knowledge is available should not in itself dictate its application. What is necessary is prior identification of the goals that we want to accomplish in using the knowledge, a careful balancing of the benefits and harms generated through the application of knowledge, and a delineation of the norms and values that should be respected. Multiplication of technological possibilities therefore calls for a concomitant development of the moral framework guiding and regulating potential and actual application of genetic technology. In order to promote human use of new technologies, ethical reflection will be unavoidable. At the same time, such reflection often is already more or less orientated on particular applications since these are pre-given and postulated together with the knowledge claims.

The intertwinement of knowledge and application claims also calls into question the responsibility of the human genetics community to communicate clearly and accurately about the nature and significance of genetic information. Sometimes, communication is overstated, for example when it is claimed that the Human Genome Project will provide the ultimate answers to the chemical underpinnings of human existence (Watson, 1990).

Moreover, such representation of genetic research can lead to particular public perceptions. It creates, for example, the impression that knowledge about many individual genes is knowledge about how the genome functions in people. It also leads to the fact, discussed by Fogle (1995), that genes are viewed by the public as entities, each of which controls one portion of the phenotype, rather than as integrated into a system.

2. ETHICS AND GENETICS

The current development of genetics is a challenge, particularly to societies, to reflect upon the future evolution of human life and social existence. It is often argued that genetic information is special and that it therefore requires special ethical treatment. Genetic knowledge is not private information, but necessarily implies relatives. Genetic information is also potentially valuable to third parties, such as insurance companies, employers, and prosecutors. Genetic technology can affect future generations. For these reasons, developing a framework of moral norms and regulations should involve all members of society. The purpose of the ethical debate is to develop guidelines and standards for the appropriate use of gene technology. Obtaining more information is not necessarily better, unless there is a clear perception of the benefits, goals, or uses which may be approached or realized with such knowledge. The fact that genetic information is available for practical use, does not imply that it is morally justified to actually use the knowledge. The morally relevant point is how to make the best possible use of the available genetic information. Various moral principles, rights and rules have been developed to delineate what is regarded as appropriate use. The main focus of the ethics literature is precisely here: reflecting upon and evaluating the rapid evolution of genetics, ethicists try to analyze the potential effects of genetic information and to determine the conditions for justified applications of gene technology.

However, it is also possible to approach genetical issues from another perspective. While not denying that significant moral questions may arise in using and applying genetic knowledge, ethics may also raise the question whether gene technology itself is morally neutral. A crucial concern is the moral value and meaning of genetic information. Let us assume, for the sake of argument, that the Human Genome Project has realized its claims: mapping of the human genome has been successful; all human genes have been located on the chromosomes. Diagnostic tests to identify all disease genes and predict any genetic dispositions and susceptibilities are flooding the health market. Assuming that the Genome Project has been ultimately and completely successful, we still have to concern us with questions about the moral value of predictive knowledge of future human existence.

3. GENETICISATION

The development of genetics as a science is more and more associated with a growing influence of genetic knowledge and technology in particular areas of society and culture. This influence manifests itself directly, through the application of genetic testing, for example in prenatal care and in various insurance arrangements, as well as indirectly through new imagery and concepts of health, disease, disorder (ten Have, 1997; Hoedemaekers and ten Have, 1999). In the 1980s and the 1990s, genetic explanations have become more attractive. From an analysis of film, television, news reports, comic books, ads and cartoons, it is shown that in popular culture, the gene is a very powerful image. It is considered not only as the unit of heredity, but as a cultural icon, as an entity crucial for understanding human identity, everyday behaviour, interpersonal relations and social problems. Nelkin and Lindee (1995) have related the growing impact of the genetic imagery in popular culture to 'genetic essentialism', the belief that human beings in all their complexity are products of a molecular text. The relation between genes and persons are examined in an earlier chapter of this volume (Chadwick, 2001).

Moreover, the expansion of the science of genetics, as well as the significance of genetics in the socio-cultural context of postmodern human beings also have repercussions for health care and medicine as well as science in general. Molecular biochemistry now has stronger claims to be the most fundamental science in medicine and the life sciences as ever before. There is also the general conviction that future genetics will drastically change medical diagnosis, treatment and prevention. In order to identify and analyze the various cultural processes related to the biomolecular life sciences, the concept of 'geneticisation' has been introduced in the scholarly debate.

3.1. The Thesis

Abby Lippman (1991), a social scientist of McGill University in Montreal, introduced the concept of 'geneticisation' to describe the various interlocking and imperceptible mechanisms of interaction between medicine, genetics, society and culture. Lippman postulates that Western culture currently is deeply involved in a process of geneticisation. This process implies a redefinition of individuals in terms of DNA codes, a new language to describe and interpret human life and behaviour in a genomic vocabulary of codes, blueprints, traits, dispositions, genetic mapping, and a gentechnological approach to disease, health and the body. Geneticisation is defined as

> The ongoing process by which priority is given to differences between individuals based on their DNA codes, with most disorders, behaviors and physiological variations ... structured as, at least in part, hereditary (Lippman, 1993, p. 178).

Introducing the concept of 'geneticisation', Lippman touches on a kind of awareness that seems to be more widely shared, although it often is not well articulated. A growing number of studies nowadays is aimed at exploring our culture's fascination with genetics (Koechlin and Ammann, 1997; Katz Rothman, 1998; Van Dijck, 1998; Glasner and Rothman, 1998). Genetic technology is not merely regarded as a new technology that is available for responsible use by autonomous consumers, but rather as a potential transformation of human understanding and existence. For example, the opening sentences in the recent book by Barbara Katz Rothman are as follows:

> Genetics isn't just a science. It's becoming more than that. It's a way of thinking, an ideology. We're coming to see life through a 'prism of heritability', a 'discourse of gene action', a genetics frame. Genetics is the single best explanation, the most comprehensive theory since God. Whatever the question is, genetics is the answer (Katz Rothman, 1998, p. 13).

Genetic thinking is considered a way of understanding the world, genetic practice is a way of imagining the future.

Through these more general implications, genetics has proliferated as a public issue. Despite the ubiquitous permeation of genetic thinking, and despite its apparent popularity, new genetic advances are not welcomed with total acceptance or univocal acclamation. Over the last fifty years, the advancement of genetics has always been controversial. The implication of this historical lesson is drawn by José van Dijck: "The dissemination of genetic knowledge is not uniquely contingent on the advancement of science and technology, but is equally dependent on the development of images and imaginations" (1998, p. 2). Therefore, it is necessary to study the popular representations of the new genetics, the various ways in which the public face of genetics is shaped. The popularisation of genetics is associated with contestation. The story of geneticisation therefore is not only the story of successes and breakthroughs, but also one of challenges, protests and criticism. In her study of the popular images of genetics, Van Dijck distinguishes four stages in the story of geneticisation: (1) the introduction of the 'new biology' in the 1950s and 1960s (with new images arising; attempts to dissociate genetics from former eugenics; disputes over 'biofears' and 'biofantasies'), (2) the DNA debate in the 1970s with political controversy over biohazards (politicisation of genetics; disputes over the safety of DNA research; increasing awareness of social and ethical implications), (3) the growth of the biotechnology business in the 1980s (industrialisation of genetics; 'biobucks' and 'biomania'), (4) the initiation and implementation of the Human Genome Project (medicalisation of genetics; creating new medical images for genetics; the 'biophoria' of genome mapping). These stages do not represent chronological phases; they rather signify epistemological and conceptual shifts.

3.2. The Critique

The thesis of geneticisation is apparently fruitful in generating new types of research and in directing scholarly attention to dimensions of genetic technology that are usually neglected in bioethical analyses. However, it also evokes sometimes vehement critique from geneticists and philosophers who reject it on various grounds. The critique on the thesis focuses on the following dimensions.

3.2.1. The Usefulness of the Concept

In response to a publication on geneticisation, Niermeijer, professor of clinical genetics in Rotterdam, argues that the thesis creates similar misunderstandings as the earlier debate on medicalisation (1998). The argument is twofold. First, the concept of geneticisation is useless since there already is widespread public debate focusing on the social consequences of genetics. Second, the concept creates misunderstanding by suggesting that genetic technology leads to new phenomena and situations, whereas the psychosocial and cultural effects of new genetic information are already known for a long time. In reaction to this critique, it is pointed out that geneticisation is not a new empirical phenomenon but a new theoretical concept that unfolds particular dimensions and brings in new perspectives in the debate on present-day genetics. Although there have been many types of debate and many aspects discussed up to now, the orientation on cultural and social implications of the new genetics is rather new. Dismissal of the concept of geneticisation as useless, therefore, removes those items from the agenda of public debate that it intends to bring into the discussion (van Zwieten and ten Have, 1998a, 1998b).

3.2.2. The Empirical Basis

From an analysis of literature Hedgecoe (1998) concludes that the ideas about geneticisation are not based on convincing empirical evidence but rather on theory-derived polemic. Close scrutiny of the thesis shows that it lacks "adequate grounding in empirical reality". For his conclusion, Hedgecoe refers to research data of Condit, showing that public perceptions of genetics are not necessarily deterministic. Hedgecoe also concludes from Condit's publications that there is no evidence that the use of genetic explanations in public discourse is more common now than in the past. This points to the fact, in Hedgecoe's opinion, that the thesis of geneticisation is no more than a sweeping claim. What is needed are small scale studies, focusing on individual elements of geneticisation. As a warning signal, Hedgecoe recapitulates the medicalisation debate of the 1970s; he points out that overstated and inconsistent claims have been made and that the empirical basis was not sound.

This type of critique raises questions concerning the status of the geneticisation thesis. Is it correct to construe the thesis primarily as an empirical claim which can then be falsified on the basis of empirical evidence? Is the thesis a sociological explanation of the facts of scientific and everyday-life reality? Or is the thesis, as I would defend, a philosophical interpretation of the self-understanding of today's human life and culture? Apart from the dispute about the status, there is also confusion about the methodology used by proponents of the thesis. Those who oppose the thesis seem to proceed from a positivistic point of view that explanatory theories should be deduced from a representative collection of empirical data, whereas geneticisation seems to be a theory based on understanding the interactions of science and society. The old dichotomy between *Erklären* and *Verstehen* is relevant here. Criticisms of the geneticisation thesis which is developed within the humanities, cultural sciences and philosophy, seem to presuppose the priority of the methodology prevailing in the natural sciences. Critique, such as Hedgecoe's, therefore is a symptom of the same phenomenon identified and criticized in the geneticisation thesis, *viz.* the uncritical predominance of mechanistic and reductionistic images; the only acceptable method of explanation and theory formation appears to be the model of the natural sciences, just as human existence is more and more explained in terms of molecular biology.

At the same time, the geneticisation thesis is not a fata morgana or ideal construct; it is about empirical reality. The connection between reality and theory, however, seems to be different as constructed by opponents of the thesis. In philosophical discourse only a few examples will suffice to make a specific point plausible. A few well-selected examples will lead to a new interpretation of the same reality, whether or not these examples are statistically representative of the majority of cases. For instance, on a global scale, ethnic cleansing is not a widespread phenomenon. Even in locations where it occurs it is very hard to convincingly prove its empirical reality. Nonetheless, the implications of the phenomenon for our philosophical self-understanding as human beings, for culture and politics are enormous. It should therefore be considered a category mistake when notions and explanations from philosophical discourse are tested with the instruments and methods from the empirical sciences. These category mistakes are common. For example, the studies of Michel Foucault on clinical medicine have been rejected by some medical historians with the claim that they do not take into account many relevant data from the history of medicine. The same mistaken approach is used in refutations of the work of Illich on medicalisation.

Finally, examples of the small scale studies advocated by critics are already available. Processes of geneticisation have been analyzed in the case of screening and counselling programmes for beta-thalassaemia in Cyprus (Hoedemaekers and ten Have, 1998).

3.2.3. The Ambiguity of the Concept

Another critique is that the concept of geneticisation is unclear (Hedgecoe, 1998). On the one hand it is difficult to demarcate it from related concepts, such as 'genetic essentialism'. On the other hand, a clear-cut definition of geneticisation is missing. Descriptions of the concept used to be comprehensive, wide-ranging, complex, and therefore ambiguous. Because of the conceptual unclarity geneticisation will probably undergo the same fate as the discussion about medicalisation; it will increasingly be regarded as unhelpful.

Indeed, it is correct that at the moment various concepts and variable descriptions of the same concept are used. There is a definite need of conceptual clarification. There is no a priori reason why such clarification is not possible. In fact, one of the newly introduced items on the agenda of bioethical research is precisely the interaction between socio-cultural influences and genetic technology. Articulating and specifying the concept of geneticisation will be a necessary condition for further development of this new research area. In this respect, the analogy with the medicalisation debate in philosophy of medicine can prove to be more helpful than assumed by the critics.

3.3. Geneticisation as Heuristic Tool in the Moral Debate

The medicalisation debate should indeed be regarded as a precursor to geneticisation. Lessons from this debate should be used to develop the recent debate on the socio-cultural impact of gene technology. These statements at the same time must be qualified. Medicalisation as well as geneticisation seem to be instantiations of more encompassing processes. Prima facie there is much similarity with Foucault's philosophy of normalisation: since early nineteenth century, medicine creates social order by its polarized distinction between 'ill' and 'health'. The theory of medicine (classification of diseases), the human body and society as a whole became closely interconnected. Concomitant with an epistemological shift towards the significance of knowing the interior of the body, the position and value of medical knowledge in society was elevated. Biopolitics transforms human beings into subjects. There is no escape from medical power; even the expressions of patients can be seen as an extension of medical power. Within society modes of power are developing that admit forms of individualisation while at the same time denying other forms; the same movement that empowers individuals and liberates them from some forms of oppression results in other forms of domination. This is also the Janus-face of medicalisation: at the same time as it provides certain benefits it also subjects patients to certain forms of discipline.

Arney and Bergen (1984) emphasize that medicine is not simply 'medicalizing'. Instead of using domination and control, the field of medical power has been reformulated. The locus of medical power is no longer the individual physician but is instead located in large, pervasive structures encompassing physician and patient alike. Medical power also is no longer exclusionary but has become incorporative; challenges from alternative health care, holism, bioethics, hospice movement are rapidly incorporated into 'orthodox' medical practice. The new field of medical power, therefore, is not so much dependent on domination and control as it is on monitoring and surveillance. Technologies of monitoring and surveillance incite discourse; they make the intimacies of the patient visible, they leave visible records. Everything must be noted, recorded, and objected to analysis.

The concept of geneticisation can be somewhat further explained by relating it to the concept of medicalisation. The process of medicalisation can occur on different levels: (1) conceptually, when a medical vocabulary is used to define a problem, (2) institutionally, when medical professionals confer legitimacy upon a problem, (3) at the level of the doctor-patient relationship, when the actual diagnosis and treatment of a problem takes place (Conrad and Schneider, 1980). Medicalisation is also associated with several consequences: it is a mechanism of social control through the expansion of professional power over wider spheres of life, it locates the source of trouble in the individual body, it implies a particular allocation of responsibility and blame, and it produces dependency on professional and technological intervention (Crawford, 1980; Zola, 1975; Illich, 1975).

In analogy, the concept of geneticisation can be studied on various levels:

1. conceptually, when a genetic terminology is used to define problems;
2. institutionally, when specific expertise is required to deal with problems;
3. culturally, when genetic knowledge and technology lead to changing individual and social attitudes towards reproduction, health care, prevention and control of disease;
4. philosophically, when genetic imagery produces particular views on human identity, interpersonal relationships and individual responsibility. In contradistinction with medicalisation, the concept of geneticisation seems to be broader because it also refers to developments and differences in the interaction between genetics and medicine; there is, for example, not simply an expansion of concepts of health and disease into everyday life, but a fundamental transformation of the concepts themselves. In medicine, there is also a tendency to use a genetic model of disease explanation, as well as a growing influence of genetic technologies in medical practice (Hoedemaekers and ten Have, 1999).

Using the concept of geneticisation also requires a critical analysis of theoretical developments following the introduction of the medicalisation thesis (Williams and Calnan, 1996). Particularly the perspective that patients are not passive 'docile bodies' under the control of medical power, but articulate consumers and autonomous decision-makers needs to be taken seriously, because the moral requirements of non-directiveness and respect for individual autonomy are strongly emphasized in present-day clinical genetics.

4. FUTURE SOCIETY

Prima facie, it seems unavoidable that the future will bring us a society within which all potentially useful genetic information is freely available and actually applied. In principle, every member of this society will be able to foretell his individual fate from reading his genes, and to adapt his personal lifeplan in accordance with such predictive knowledge.

In the opinion of 'geneticisation' authors as Lippman (1992), this future has already partly begun. Society is involved in a process of geneticisation. As an instantiation of the more encompassing process of medicalisation, this process involves a redefinition of individuals in terms of DNA codes. Postmodern society is using a new genomic language to communicate about human life. Disease, health and the body are explained in terms of molecular biology. Nelkin and Lindee (1995), in their book *The DNA Mystique*, examining popular sources such as television, radio talk shows, comic books and science fiction, show how popular images "convey a striking picture of the gene as powerful, deterministic, and central to an understanding of both everyday behaviour and the 'secret of life'" (1995, p. 2). It seems that the cultural meaning of DNA nowadays is remarkably similar to that of the immortal soul of Christian theology. The bio-information metaphor and cartographic metaphor, often used in the context of the genome project, are in fact reworkings of the mechanical metaphor that has been frequently used in the past in medical discourses on the body. These linguistic (and often also visual) representations of the body carry with them the importance of a technological approach: machinery is used to fix machinery. They represent the body as being comprised of "a multitude of tiny interchangeable parts, rendering the body amenable to objectification and technological tinkering in the interest of developing the 'perfect' human" (Lupton, 1994, p. 61).

How should this development towards a geneticized future be evaluated? It is at least important to try to identify what influences, and what determines this development. It seems that this development towards a geneticized future is possible because of the consensus regarding two ideals in current moral debate: the ideal of value neutrality of clinical genetics and the ideal of individual responsibility in health matters.

4.1. Non-directiveness

One of the prime tenets of genetic counselling is patient autonomy. Once genetic information is available, the basic rule is that patients or clients should be able to use the information according to their personal views. Geneticists or counsellors should not seek to tell patients or clients whether they should obtain particular information or what they should do with the information if they acquire it. In other words, the goal of genetic counselling or screening is to inform patients or clients about what is possible and what their options are (Collins, 1991). The leading principle of counselling and screening therefore is non-directiveness. Accurate information should be provided to the person concerned regarding the nature of potential genetic conditions, the prognosis, possible treatments and preventive strategies. The experts providing such information should not, in any respect, try to influence the decisions made by the persons who are counselled or screened.

The moral ideal underlying this practice of clinical genetics is value-neutrality. The genetic expert is withholding any normative judgement regarding the obtaining and application of genetic information; his aim is merely to provide information and to help the patients or clients to work through possible options. It is evident that this ideal in itself is a weak counterbalance to tendencies to make genetic tests more generally accessible. Patient values are to be decisive whenever choices have to be made on the basis of genetic information. When respect for individual autonomy is the basic norm guiding the use of genetic information, it is also reasonable to expect that predictive 'combi-tests' will eventually be on sale in the supermarket or drugstore (de Wert, 1994).

4.2. Individual Responsibility for Health

A second determinant that may further increase the significance of genetic information is the ideal of individual responsibility for personal health. Health policy and health education, especially in times of limited budgets and reduced expenditures, increasingly appeal to the notion of 'personal responsibility'. If health policy defines a particular problem as undesirable, and if health education research shows the problem to be associated with a particular life-style, then health policy can attribute responsibility to those individuals that exhibit that life-style, particularly since life-style is supposedly the free choice of rational individuals.

Traditionally, in health care the rhetorics of responsibility is used in a specific way (see also: ten Have, 1999). In the medical model of disease, patients are usually not held responsible for the genesis and evolution of their illnesses. Diagnosing a condition as disease introduces excusability. When a person's condition is interpreted as illness, the medical judgement implies that he cannot be blamed for his condition, and that treatment and care are appropriate and morally desirable. In this traditional model, the notion of responsibility is used with prospective force: it is equivalent to saying that a person has an obligation to preserve his health. Through assigning responsibility to the individual for his future health, an attempt is made to guide and change the individual's behaviour. Such practical use of the concept is different from the retrospective ascription of responsibility. The latter use implies an evaluation of what has happened. If an individual has a health problem, he is held causally responsible because of his unhealthy life-style or risky behaviour in the past. This use combines causality with culpability. Since the person himself is the cause of his present problem, he is also answerable for the consequences of his prior behaviour. Retrospective use of the concept of responsibility therefore is retributive; it implies disapproval and blame.

In present-day health policy there seems to be a development towards connecting the prospective and retrospective senses of 'responsibility' (ten Have and Loughlin, 1994; see this volume: ten Have, 2001). Usually, the line of argumentation is as follows. If there is an urgent need to reduce the costs of health care, and if at the same time it is scientifically argued that major expenditures are associated with certain patterns of behaviour, it is tempting to create an obligation to be healthy and to introduce some system of sanction for those who do not implement such obligation. In a liberal society, individuals are normally free to do as they choose. In this respect, caring for your health is not different from other dimensions of personal life. But when individual choices turn out badly, and when individuals remain uninfluenced by moral appeals of health educators, legal and financial sanctions may be thought justified.

Today, a similar argument is used concerning predictive information. It may be prudent to use genetic diagnosis to predict future disabilities, and therefore appeals to (prospective) responsibility may be justified; but this argument in practice is often linked with the argument that individuals who deliberately have not used diagnostic possibilities, should be (retrospectively) responsible for adverse consequences for themselves or their offspring. When, for example, a couple decides not to use prenatal diagnosis, or not to terminate pregnancy in case of diagnosed fetal disorders, it is argued that the couple then is responsible for the suffering of the child, when indeed a child with handicaps is born (Hilhorst, 1993). If suffering could have been avoided, and a choice is made not to use predictive opportunities, parents should bear the consequences of their irresponsible choice; they can no longer argue that suffering has befallen them; they have themselves to blame.

This line of argumentation, if indeed taken seriously, will be a significant stimulus for individuals to obtain genetic information as much as possible, particularly when there is a threat that governments, insurance companies and employers will work with a system of incentives and disincentives. When there is a cultural imagery that future diseases, disorders and disabilities can be foretold by examining the individual's genome, persons can no longer claim that they are victims, if they have deliberately decided not to use predictive diagnosis. It has been their voluntary choice not to know, and not to eliminate potential disadvantages to their health. Fate has been replaced by choice.

5. A GENETIC CIVILISATION STRATEGY?

The ideals of value-neutrality of clinical genetics and of personal responsibility for health, prevailing in current bioethical debate, may indeed generate a situation where the availability of genetic information in itself produces its wide-spread application. In this view, human beings in the next millennium will be dominated by predictive knowledge of their genome and driven by new norms in interpersonal behaviour.

Such assumption is not unrealistic since we have witnessed a similar change in normative behaviour patterns at the close of the last century (ten Have, 1990). With the rise of new knowledge about the origin and transmission of infectious diseases, in many countries philanthropic activities have been organized to civilize the public through inculcating the new hygienic norms. Philanthropists launched a large scale offensive to civilize the habits and life-styles of the masses. As enlightened men, they coupled assistance with moralisation. Norms of behaviour, such as cleanliness, domestic nursing, soberness were transmitted not by repression or coercion but by the subtle means of advice, persuasion and education. The result was the normalisation of individual behaviour. The new norms of a healthy, regular, and disciplined conduct passed into domestic life; the strategy succeeded in having the norms internalized. Hygienism thus produced a new behaviour pattern in the general population.

Why could a similar transformation of life-styles not occur today as a result of new genetic information? Though it is hard to forecast the future, two factors can be identified that may prevent, hinder or at least restrict this development towards geneticisation of future human existence.

The first factor is the need to make some delineation between disease and health, normality and abnormality, given the uncontrollable wealth of information that will in the end be available. In the current ethical debate, the above distinctions are increasingly problematic. It is apparently difficult to make use of the traditional distinctions in determining what conditions should be screened or not. Perhaps it is even thought impossible to apply them as normative criteria guiding potential genetic screening programmes. Nonetheless, the exponential growth of genetic data and resulting possibilities of detection, will inevitably lead to an urgent need of selection; without selective use and meaningful criteria to make distinctions of value among the immense data available, the usefulness of data will be questionable. The multiplication of possibilities for testing will at the same time increase the necessity to reach consensus regarding those conditions and predispositions that seriously restrict the functioning of human beings within community, and those that are within the bounds of reasonable variations of human functions and structures. Of course, at the moment it is unclear how such distinctions can be made and morally justified. But the acknowledgement that it will be an extremely difficult task should not lead to the conclusion that it is impossible. Right here is a major challenge to philosophical reflection. Many moral discussions about whether or not to apply genetic knowledge seem essentially to focus on this issue in particular (for example, the debate on the development and use of human growth hormone; Wilkie, 1993).

The second factor is the normativity of medicine. Medicine regards itself ultimately as a helping and caring profession, not merely as service institution. In such a self-conception, value-neutrality is not an appropriate position to guide medical activities. Physicians in this view adhere to professional norms that go beyond value-neutrality. Diagnosis, therapy, prevention are guided and motivated by specific values, *viz.* promotion of health, relief of suffering, elimination of disease. From this value perspective, respect for individual autonomy is only an instrumental value, necessary in order to accomplish the values intrinsic to medicine as helping and caring profession. The norm of non-directiveness in clinical human genetics, therefore, is inadequate from a medical point of view. It may have been prudent to introduce this norm against the background of historical misuse of genetic information. It may be desirable as a practical norm as long as genetic information is mainly related to genetic risks to the offspring. But it can be argued that in the present situation, where genetic testing is more and more concerned with detecting genetic risks for the future health of the individual person who is tested, the normative attitude of clinical geneticists should shift from neutrality to prescriptivity (de Wert, 1999). A similar point is made by Caplan: it is likely that a shift will occur from a normative stance of value-neutrality toward "an ethic in which the promotion of genetic health and the amelioration, prevention, and correction of genetic disease are the foundation of clinical and public health practice" (Caplan, 1992, p. 134). Decisions made on the basis of genetic information, should in this view aim at promoting health and alleviating disease. There is no reason to think that advocating these values in the realm of human genetics is inappropriate or unethical. Studies of the practice of clinical human genetics in fact indicate that those professionals who now offer genetic screening and testing services do not always act in conformity with their self-imposed ideal of value-neutrality (Fletcher and Wertz, 1988).

6. CONCLUSION

In postmodern society two determinants are at work that will probably lead to a future where individual existence will be to a large extent affected and permeated with predictive genetic information. First, we witness the current domination of the moral principle of respect for personal autonomy; the individual ought to choose among the potential of genetic tests those possibilities that fit his or her lifeplan. Second, society is moralizing individual responsibility in the sense that persons who do not use the opportunities to foresee and prevent future suffering, have to face the consequences. Both factors give a strong push to know as much as possible about our life in the near and distant future. In this perspective the collective destiny of human beings in Western societies will be deeply geneticized.

However, there are reasons to question the prediction of further geneticisation.

First, a clear opposition exists between the above determinants; the first emphasizes the interest of the individual, the second the community interest. It is not evident what interest will prevail; it is not obvious that one interest will definitely overrule the other.

Second, autonomous individuals will not at random use everything available; they will sooner or later start to wonder what may be the meaning and relevancy of all knowledge available and obtainable. Even within a fully free health market, individuals will not consume everything; they will attempt to make a distinction between appropriate and inappropriate, intelligible and unintelligible uses of genetic tests. This will instigate a public debate concerning the significance of genetic testing and genetic information, the more so since powerful parties such as insurance companies have an obvious interest in promoting testing.

Third, it is doubtful whether future medicine will depart so radically from its present-day value orientation, especially in the European setting. The autonomous request of individual patients will be a significant moral factor, but at the same time, medicine will also want to be guided by its own norms to make distinctions between disease and health, normality and abnormality. Beyond the individual demands and subjective complaints, medicine will continue to strive for a more rather than less objective determination of needs, signs and symptoms. Apparently, a full geneticisation of human existence in the future may only occur when we abandon the philosophical attempt to differentiate between 'healthy' and 'ill', 'normal' and 'abnormal'.

Finally, this analysis illustrates the advantage of the concept of 'geneticisation': it operates as heuristic tool, like the concept of 'medicalisation' in the medical-philosophical debates of the 1970s. It discloses particular areas for philosophical scrutiny, it re-directs and re-focuses moral discussion. In creating and facilitating different ethical perspectives, the concept of geneticisation particularly draws attention to (1) socio-ethical issues and (2) an interpretative ethical methodology.

The challenge of the current development of genetics for bioethics can be explored more fully when attention is given to its social and cultural implications. The concept of geneticisation can instigate us to change our perspective. While not denying that significant moral questions arise in the use and application of genetic knowledge, ethics may also address the question whether genetic information itself has any moral value and meaning. Regardless of the significance of personal autonomy, there is also the question of what the social and cultural consequences will be of new genetic knowledge. What does it mean for society and culture in general when every member of society will be able to foretell his individual fate from reading his genes, and to adapt his personal lifeplan in accordance with such predictive knowledge? What are the implications for our notions of life and illness when disease, health and the body are predominantly explained in terms of molecular biology? What will be the effect of a technological discourse about the human body?

The concept of geneticisation is re-orientating attention away from moral topics related to the current emphasis on individual autonomy, such as non-directiveness in clinical genetics and the ideal of individual responsibility in health matters. This re-orientation creates space for seriously questioning the dominant bioethical discourse with its emphasis on individual freedom to choose. In a liberal society, it is argued, individuals are normally free to do as they choose. In this respect, caring for your health is not different from other dimensions of personal life. But when individual choices turn out badly, and when individuals remain uninfluenced by moral appeals of health educators, legal and financial sanctions may be thought justified. If suffering is in practice avoidable, and individuals freely decide not to use predictive opportunities, they should bear the consequences of their choices. The logic of individual choice and responsibility necessarily includes the logic of blaming the victim.

The heuristic value of the concept of geneticisation is precisely here: it introduces into the bioethical debate moral issues and methods that tend to be 'forgotten', neglected or disregarded. Geneticisation, in the words of Van Dijck (1998, p. 29) is "a gradual expansion of loci of contestation where meanings of genetics are weighed". The concept therefore informs bioethics that biomedicine and bioscience should be associated with biocriticism. A central thesis in the epistemology of the French philosopher Gaston Bachelard (1938) is that the problem of the growth of science must be formulated in terms of obstacles. There is no history of science without shadows, without failures, dissensus, and conflicts.

REFERENCES

Arney, W.R. and Bergen, B.J.: 1984, *Medicine and the management of living. Taming the last great beast*. University of Chicago Press, Chicago/London.

Bachelard, G.: 1938, *La formation de l'esprit scientifique*. Vrin, Paris.

Caplan, A.L.: 1992, *If I were a rich man could I buy a pancreas? and other essays on the ethics of health care*. Indiana University Press, Bloomington and Indianapolis.

Chadwick, R.: 2001, Genetic interventions and personal identity. This volume.

Collins, F.S.: 1991, Medical and ethical consequences of the Human Genome Project. *Journal of Clinical Ethics* 2 (4): 260-267.

Conrad, P. and Schneider, J.: 1980, Looking at levels of medicalization: a comment on Strong's critique of the thesis of medical imperialism. *Social Science and Medicine* 14A: 75-79.

Crawford, R.: 1980, Healthism and the medicalization of everyday life. *International Journal of Health Services* 10(3): 365-388.

Dijck, J. van: 1998, *Imagenation. Popular images of genetics*. Macmillan Press, London.

Fletcher, J. and Wertz, D.: 1988, *Ethics and applied human genetics: A cross-cultural perspective*. Springer Verlag, Heidelberg.

Fogle, T.: 1995, Information metaphors and the human genome project. *Perspectives in Biology and Medicine* 38(4): 535-547.

Glasner, P. and Rothman, H. (eds.): 1998, *Genetic imaginations. Ethical, legal and social issues in human genome research*. Ashgate, Aldershot.

Have, H.A.M.J. ten: 1990, Knowledge and practice in European medicine: The case of infectious diseases. In: Have, H.A.M.J. ten, Kimsma, G.K. and Spicker, S.F. (eds.), *The growth of medical knowledge*. Kluwer Academic Publishers, Dordrecht/Boston/London, p. 15-40.

Have, H.A.M.J. ten and Loughlin, M.: 1994, Responsibilities and rationalities: Should the patient be blamed? *Health Care Analysis* 2(2): 119-127.

Have, H.A.M.J. ten: 1997, Living with the future: Genetic information and human existence. In: Chadwick, R., Levitt, M. and Shickle, D. (eds.): *The right to know and the right not to know*. Avebury, Aldershot, p. 87-95.

Have, H.A.M.J. ten: 2001, Allocation of resources and personal responsibility. This volume, case 7.

Hedgecoe, A.: 1998, Geneticization, medicalisation and polemics. *Medicine, Health Care and Philosophy* 1(3): 235-224.

Hilhorst, M.T.: 1993, Aangeboren en aangedane handicaps: maakt het moreel verschil? In: Beaufort, I.D. de, and Hilhorst, M.T. (eds.), *Kind, ziekte en ethiek*. Ambo, Baarn, p. 67-91.

Hoedemaekers, R. and Have, H.A.M.J. ten: 1998, Geneticization: The Cyprus paradigm. *Journal of Medicine and Philosophy* 23(4): 274-287.

Hoedemaekers, R. and Have, H.A.M.J. ten: 1999, Genetic health and genetic disease. In: Launis, V., Pietarinen, J. and Räikkä, J. (eds.), *Genes and morality. New essays*. Rodopi, Amsterdam/Atlanta, p. 121-143.

Illich, I.: 1975, The medicalization of life. *Journal of Medical Ethics* 1(1): 73-77.

Katz Rothman, B.: 1998, *Genetic maps and human imaginations. The limits of science in understanding who we are*. Norton, New York/London.

Koechlin, F. and Ammann, D. (eds.): 1997, *Mythos Gen*. Utzinger/Stemmle Verlag, Rieden bei Baden.

Lippman, A.: 1991, Prenatal genetic testing and screening: Constructing needs and reinforcing inequities. *American Journal of Law and Medicine* 17: 15-50.

Lippman, A.: 1992, Led (astray) by genetic maps: The cartography of the human genome and health care. *Social Science and Medicine* 35(12): 1469-1476.

Lippman, A.: 1993, Prenatal genetic testing and geneticization: Mother matters for all. *Fetal Diagn Ther* 8(suppl 1): 175-188.

Lupton, D.: 1994, *Medicine as culture. Illness, disease and the body in Western societies*. Sage Publications, London.
Nelkin, D. and Lindee, M.S.: 1995, *The DNA mystique. The gene as a cultural icon*. W.H. Freeman and Company, New York.
Niermeijer, M.F.: 1998, Geneticalisering. Misleidend onbegrip door onjuiste informatie. *Medisch Contact* 53: 641-642.
Watson, J.D.: 1990, The human genome project: Past, present, and future. *Science* 248: 44-48.
Wert, G. de: 1994, De oorlog tegen kanker, de jacht op kankergenen, en de speurtocht naar de ethiek. *Tijdschrift Kanker* 18(2): 41-55.
Wert, G. de: 1999, *Met het oog op de toekomst. Voortplantingstechnologie, erfelijkheidsonderzoek en ethiek*. Thesis Publishers, Amsterdam.
Wilkie, T.: 1993, *Perilous Knowledge. The Human Genome Project and its implications*. Faber and Faber, London/Boston.
Williams, S.J. and Calnan, M. (eds.): 1996, *Modern medicine. Lay perspectives and experiences*. UCL Press, London.
Zola, I.K.: 1975, In the name of health and illness: On some socio-political consequences of medical influence. *Social Science and Medicine* 9: 83-87.
Zwieten, M.C.B. van, and Have, H.A.M.J. ten: 1998, Geneticalisering, een nieuw concept. *Medisch Contact* 53: 398-400.
Zwieten, M.C.B. van, and Have, H.A.M.J. ten: 1998, Geneticalisering. Naschrift. *Medisch Contact* 53: 642.

PAUL T. SCHOTSMANS

PRENATAL TESTING FOR HUNTINGTON'S DISEASE

1. CASE

The case we present has been described in the literature (Higgins, *et al.*, 1990) in a period where the gene for Huntington's disease (HD) had not yet been discovered. The ethical dilemmas as we see them, are, however, still strongly present: the principle of autonomy versus responsibility for future generations is only one example in this case. And even more, some couples still require the exclusion test as an alternative for the direct testing of their unborn child.

It is also our intention not only to present the case and give an overview of the most important dilemmas, but we will meanwhile present a method for what we think to be an adequate ethical clarification.

A woman in her early 20s requested prenatal diagnosis but did not wish to have any alteration in her knowledge of her own risk for HD. In this family it was possible to clearly distinguish the maternal grandmother's marker from that of the affected maternal grandfather. The mother inherited the marker A from her mother and inherited marker B from her affected father. The foetus proved to have inherited marker B, implying that the foetus inherited a chromosome 4 from its affected grandfather. The likelihood that the foetus has inherited the gene for HD rose from 25% to close to 50%. The pregnancy was terminated at 13 weeks gestation (Higgins, et al., 1987, p. 9).

Subsequently, there has been considerable ongoing contact with other family members, with much discussion centring around whether prenatal exclusion testing should have been offered. The unaffected parent of our proband has expressed concern for the future implications of having terminated a pregnancy at 50% risk for a late-onset illness. She has requested that such testing be avoided.

H.A.M.J. ten Have and B. Gordijn (eds.), Bioethics in a European Perspective, 369-383.

2. A MODEL FOR ETHICAL ANALYSIS

In order to clarify our ethical approach it may be helpful to take advantage of the tri-level distinction made by the German philosopher Max Scheler: practical morality, ethos and ethics (Scheler, 1916, 1973). Scheler explains that the way people behave can be described as a kind of 'practical morality': it may be approached and verified with sociological and statistical tools of research. In medicine, this can be brought together under another heading, namely the scientific and/or clinical state of the art. To start every ethical reflection with this analysis is at the same time giving the priority to the medical experience, which of course is not only that being shared by physicians, but also by patients or clients, nurses and others belonging to the process. In the case of human genetics, this implies the description, understanding and awareness of the scientific and clinical evolutions. It is therefore that we will describe all the necessary historical, scientific and medical elements to understand the inherent dilemma of our case.

The second stage is a little bit more difficult to grasp: Scheler spoke about the 'ethos', which is more oriented to the moral sensibilities operating within a particular culture. Valuations, intuitions or subjective feelings and ideas about a particular ethical dilemma must indeed be clarified. When this more subjective side is not brought under the light of an ethical clarification, the unconscious presuppositions may remain hidden and block the ethical reflection. Human genetics is particularly interesting to try to develop an idea about the reactions of people to the rapid changes of paradigms. Some commentaries played with anxieties and emotional barriers to block an open dialogue over human genetics. It would however, be better simply to express all the emotions, intuitions and anxieties in order to have an idea about how to integrate them in the ethical process.

Finally, the third stage is the place of rational reflection on the practical morality and on the ethos. This stage incorporates both previous stages and tries to clarify the values and norms which are playing through the ethical debate. Some consider this stage as a radically neutral enterprise, whereby all references to normative theories must be banned or may only be indicated, without using them as a matrix for problem-solving. Others (including myself) are convinced that this is the normative dimension of the ethical process: by the integration of an ethical model, principle or hierarchy of principles or by referring to a specific anthropology, values and disvalues are brought in comparison in order to realize the best possible way to achieve the normative orientations. It may be clear that this third stage is at the same time the battle-field of the ethical debate: normative positions may be very different, also concerning medicine in general and human genetics in particular.

In order to make an ethical discussion possible, we will limit ourselves in the description of the third phase to the clarification of the most important values, norms and principles, as they are playing a role in the debate about the case (for a more normative approach about personalism in an analogous field see this volume: Schotsmans, 2001).

3. CASE ANALYSIS

3.1. Practical Morality: The Medical State of the Art

3.1.1. Some Basic Insights

To have a clear idea about the crucial ethical dilemmas requires an insight in the medical facts. In this case, we have to clarify the historical background and some recent developments concerning the research on the so-called Huntington's Disease.

HD is a disorder of the central nervous system and is thus rightly classified as a neurological disorder; yet it is in many ways a condition whose effects extend across many fields and which is encountered by clinicians in widely differing specialities. Research scientists in increasing numbers are also involved with HD as biochemists, neuropharmacologists and molecular biologists. In both clinical and basic science aspects this disorder serves as a model from which we can learn much about other progressive genetic disorders of the nervous system.

The description by the Long Island physician George Huntington in 1872 of the disease that has subsequently borne his name is one of the most remarkable in the history of medicine. It was not the first description of the disorder, but it stands out as the first full delineation of the condition as a specific disease entity, quite separate from other forms of chorea. Huntington's paper was given before the Meigs and Mason Academy of Medicine at Middleport, Ohio, on 15 February 1872 and published only 2 months later in the Philadelphia Journal, *The Medical and Surgical Reporter* (Harper, 1991, p. 3). All the cardinal features of HD can be recognized in this description: the adult onset, progressive course and eventually fatal outcome; the choreic movements combined with mental impairment, and risk of suicide.

Typically, HD begins in midlife, between the ages of 30 and 45, although onset may occur at any age (a juvenile variant of the disease affects those who develop symptoms before adulthood). Early symptoms can be mild enough to go unnoticed at first and may include depression, mood swings, forgetfulness, clumsiness, twitching, and lack of coordination. As the disease progresses, the severity of symptoms increases. Concentration and short-term memory are diminished and involuntary movements become more pronounced; walking and everyday activities become difficult, and speech and swallowing abilities deteriorate. While nothing has been proven to slow or stop the relentless course of HD, which may run for 10 to 15 years or more, medications and therapies are available that can relieve or control many of the symptoms of HD.

Facts at a Glance
Huntington's Disease Society of America Website

What is HD ?

An inherited, degenerative brain disease, which means that is a disease of both mind and body. Symptoms generally appear between 30 and 50 years of age, but have appeared as young as 2 and as old as 70. Each child of an HD-affected parent has a 50% chance of inheriting the disorder and is said to be 'at risk'. HD usually progresses over a 10 to 25 year period.

Characteristic symptoms
- Personality changes, depression, mood swings
- Unsteady gait, involuntary movements
- Slurred speech; Impaired judgement
- Difficulty in swallowing
- Intoxicated appearance

Research breakthrough
- After a 10 year search, scientists announced in March 1993 that they had found the gene which causes HD
- Researchers are now trying to determine what the gene's normal function is and how the flawed gene causes HD
- Many avenues of research are being investigated in the search for a treatment or cure for HD

Genetic testing
- The gene discovery has made possible a new predictive test for HD allowing those at risk to find out whether or not they will develop the disease
- The new test generally requires a blood sample only from the person being tested, unlike the old which required samples from several family members
- Some testing centres may request a sample from a parent for confirmation purposes
- Pre- and post-test counselling is a necessary and important part of the testing procedure

Affected population
- Affects all races and ethnic groups and both sexes
- HD does not skip generations ; if you do not inherit the gene, you cannot pass it on
- If you carry the gene, you will develop the disease if you live long enough, and you can pass it on

With the isolation of the HD gene in 1993, a direct gene test has been developed by which people at risk can learn with a high degree of certainty whether or not they will develop the disease at some point in the future (the test cannot predict when). While this presymptomatic test is a boon for some, others prefer not to know their fate, particularly in the absence of a cure for

HD. The gene for HD was discovered on the short arm of chromosome 4. It is actually a segment of DNA which contains a stretch of repeated 'nucleotides', which are the basic building blocks of DNA.

The decision about whether or not to be tested for HD is a very personal one which requires careful consideration. The results of this test may affect many aspects of one's life including personal well-being, relationships with family and friends, employment, insurance eligibility and other areas. Those who decide to be tested for HD often do so to assist them in preparing to make a major life decision (such as marriage or planning a family). Many people, on the other hand, consciously decide not to take the test, preferring to live with uncertainty rather than with the knowledge that they will one day develop HD.

3.1.2. What is Exclusion Testing for Huntington's Disease in Pregnancy?

Prenatal diagnosis has so far played a very small part in relation to DNA-based prediction in HD. This is not just because it is a late-onset disorder, allowing many years of healthy life before onset; the attitudes of those seeking presymptomatic testing have clearly shown that planning reproductive choices is a major reason for requesting this. Few pregnancies occur to those who are already affected, so that undertaking a specific prenatal diagnosis inevitably implies a presymptomatic test for the healthy person at risk. Few couples will wish to undergo such a double ordeal ; it is much more likely that most of those with an adverse presymptomatic test result will refrain from childbearing entirely, while those with a favourable result will go ahead without the need for prenatal tests.

A somewhat different approach to prenatal diagnosis is what has been termed 'prenatal exclusion testing' which requires some explanation. In 1984, Harper and Sarfarazi proposed prenatal exclusion testing as a method whereby a parent who is at 50% risk of carrying the gene for HD can elect to have children who are at low risk of being gene carriers by excluding in the foetus the parental allele which is at risk of being linked to the HD gene mutation (Harper and Sarfarazi, 1985). In 1987 Quarrell *et alii* described the possibilities of exclusion testing after the finding in 1983 of a DNA probe (G8) which was localised to the short arm of chromosome 4 and assigned the locus D4S10 (Quarrell, Meredith, Tyler, Youngman, Upadhyaya and Harper, 1987). It was noted earlier that only a quarter of individuals at risk for HD in South Wales had a pedigree structure appropriate for presymptomatic testing. This study also examined whether testing would be feasible for a pregnancy and found by contrast that for this group it would be possible in almost 90%, the same figure being the case for a series of pregnancies that had actually occurred in those on the HD register. There is however, a major difference between this approach and full prenatal diagnosis, giving rise to the term 'exclusion test'. If the marker coming from its non-HD grandparent is the one transmitted to the foetus, then its risk of HD will clearly be reduced greatly to around half the recombination rate (1-2%). However, if the marker inherited by the foetus is that from its HD grandparent, this does not mean that it will be affected by HD but simply that it will have the same risks as its parent, i.e. an increase from 25% to around 50%. This produces two problems: the first is a decision to terminate taken on the basis of a 50% risk. Secondly, the risk to the pregnancy is linked to that of the at-risk parent; should the pregnancy continue and the parent later develop HD, then that same high risk will apply to the child and an 'involuntary' presymptomatic test will have been done.

In proposing the exclusion test, which uses anonymous, polymorphic markers closely linked to the HD gene, Harper and Sarfarzi pointed out that it has two main advantages: firstly, since previous determination of the genetic linkage phase between the marker and disease gene in the parent at risk of transmitting the HD gene to the foetus is not required, the test is available to most families including those where the pedigree structure is unsuitable for conventional presymptomatic testing; secondly, a high risk test result for the foetus does not alter the risk status of the parent. However, implicit in the exclusion test is the assumption that the pregnancy will be terminated if the result indicates the foetus has 50% risk of carrying the HD gene. If the high risk pregnancy is not terminated, then a unique situation may arise, for if the parent develops the condition then the child will most likely (around 96% probability) become affected also (Tolmie, *et al.*, 1995).

Most studies of attitudes of individuals at risk for HD have concentrated on tests for the individual ; therefore it is not clear how many couples will wish to make use of exclusion tests for HD but the test is likely to benefit those who, without them, would not have had children and for whom predictive tests are either unacceptable or unavailable. Many ethical and counselling problems arose including lack of understanding of the test. It can be concluded from this experience of the Cardiff centre that prenatal exclusion testing is very different from presymptomatic testing but needs equally thorough counselling.

3.1.3. The Legal Background

Concerning our case: in the absence of criminal law restricting abortion, the legal question of whether abortion constitutes medical professional misconduct relates to whether it is ethically discussed and undertaken. Under the doctrine of informed consent, it is essential that the proband should be thoroughly advised about the limits of prenatal testing, as well as about all alternatives in predictive testing. In the majority of the legal systems, the rights of the pregnant woman take precedence as long as the foetus is not viable. Individuals are free to take action which they feel is appropriate to their circumstances. There is no legal impediment to exclusion testing.

3.2. The Ethos

3.2.1. The Ethos of the Counsellors

The predominant ethos among genetic counsellors may be observed in the influential 1983 report of the American President's Commission on Bioethics: they noted the high priority that postwar genetic counselling has placed on the respect for the autonomy of clients (Walters, 1993). This approach was based in part on a rejection of the coercive eugenic policies that had been employed in the United States and Nazi Germany earlier in the twentieth century. It was also indebted to Carl Rogers' notion of client-centred therapy.

Wertz and Fletcher have demonstrated the extent to which the principle of autonomy takes precedence over other ethical principles among medical geneticists (Wertz and Fletcher, 1989). In their responses to 14 clinical cases, medical geneticists from 18 countries cited the principle of autonomy as their first or second reason for their answers 59% of the time. The consideration cited next most often, the principle of non-maleficence, was cited only 20% of the time. The principle of beneficence ran a distant third at 11%, while the principle of justice was appealed to in only 5% of the responses.

There are two primary implications of the principle of respect for autonomy in the genetic counselling relationship. The first is that the genetic counsellors are generally committed to respecting their clients' freedom to make their own decisions. The second is that genetic counsellors have a strong commitment to protecting the privacy of their clients and the confidentiality of information about the lifestyles or genetic conditions of their clients.

3.2.2. The Ethos of the Population

HD is an interesting entrance into the exploration of the moral sensitivities of a large population and/or the affected families. Predictive testing for HD is considered to be a test case for predictive testing for other late onset diseases, monogenic as well as multifactorial disorders. In the hypothetical situation of having a 50% risk for developing HD, about half of a group of 169 women (aged 21-35 years), tested by a Leuven group, expressed interest in a predictive test. As to the question of giving results of predictive tests to third parties, the group would be very reluctant to inform the employer or the insurer, but not their own family. Prenatal testing for late onset diseases was considered acceptable by half of the women; only one quarter of the total group would terminate a pregnancy of a child that might develop a late onset disease (Decruyenaere, *et al.*, 1993).

Although the interest in predictive testing for HD is very large, many of those who express their interest, express their intention of postponing the final decision for various reasons. To have certainty about the future and to make arrangements for the future play a major part in the decision of the total group. Making decisions concerning children and to a larger extent informing children about their risk status are important factors in deciding in favour of the test (Evers-Kiebooms, *et al.*, 1989).

3.3. Ethics: A Description and Analysis of Ethical Values and Principles

3.3.1. The 'Principlist' versus the 'Personalist' Approach

It is well known that Beauchamp and Childress in their famous textbook have distinguished four ethical principles: beneficence, nonmaleficence, respect for autonomy and justice (Beauchamp and Childress, 1979, 1994). The first of these principles, beneficence, is also called well-being or welfare in the broad sense. The principle of beneficence focuses our attention on the benefits and harms of an action or policy. In other words, this principle asks us to look at the consequences, outcomes, effects, or utility of what we do. Many Anglo-American ethicists share the view that the four principles are helpful in analyzing ethical decisions or the policies adopted by professional groups or public policy makers. The four principles do not exhaust the categories that can be applied to decision making or policy making. However, the principles can serve as a kind of checklist, or points to consider, in ethical analysis.

Some philosophers have argued that in cases of conflict among the four principles, one of the four should always take precedence over the others, or even that there is an order of priority among the principles. Thus, some utilitarians argue that the net benefit of an action or policy should be our primary concern. In contrast, some egalitarians would argue that maximizing net benefit is less important than equalizing welfare - insofar as possible. Libertarians and other proponents of individual liberty assert that achieving net benefit or producing equality of welfare is less important than respecting the free choices of persons.

These differences of approach make clear that we need a moral theory or even better, an anthropological framework to clarify the balance of values and nonvalues in the discussion about the case. Elsewhere in this book (see chapter 8), we have clarified our personalist approach: in virtue of the historicity of the human person we must always reconsider which possibilities we have at our disposal at this point in history to serve the promotion of the human person. The possibility of a direct gene/DNA test since 1993 is therefore a very important event. If the diagnosis of HD has been confirmed within the family, one can directly look for the mutation itself in each individual of the family. Values as the uniqueness of every individual (also the unborn foetus), the relational structure of genetic counselling and the far-reaching social influences of prenatal diagnosis must be balanced against each other in order to realize as many values as possible and to avoid nonvalues as much as possible. A careful moral clarification cannot be made without deciding for a hierarchy: it is our moral duty to protect as much as possible the realisation of the wellbeing of all involved human beings, in the short and the long run. Personalism is therefore more explicit in clarifying a value hierarchy than principlism, which is not more than a kind of checklist of principles (Schotsmans, 1989, 1994).

3.3.2. A Crucial Dilemma in Genetic Counselling: A Right to Know, a Right Not to Know or a Duty to Know

Margery Shaw starts her comment in the *American Journal of Human Genetics* as follows: "Armed with today's technology, it is now possible to begin to eradicate the Huntington gene from our species. All of us would be happy to welcome the day when no one ever suffers the devastating effects of Huntington Chorea. But in order to approach this goal, which is attainable within the next two generations, it is necessary that those who possibly or certainly carry the gene take positive steps to prevent its transmission" (Shaw, 1987, p. 243). This is clearly a commentary which for some observers is transgressing the limits of genetic counselling. To understand this we must - probably a little bit more rationally - clarify the various tensions identified between rights to know and not to know. We will follow carefully Chadwick's description of the most important arguments (Chadwick, 1997).

3.3.2.a. Arguments in Favour of a Right Not to Know

The argument for a right not to know is typically made by or on behalf of an individual, e.g. when a third party wants access to information, which he or she does not wish to know about himself or herself - when a genetic relative wants it, or an insurer wants it, or an employer requires pre-employment screening. It is important to recognise however that it is not simply a case of X wanting information about Y. X may want information about X which will inevitably give Y information about Y which Y does not want. Suppose, for example, that of two genetically identical twins, one wants to be tested for a genetic predisposition and the other does not. Even if the one who is tested agrees not to disclose to the other the result of a test, the choices he or she makes thereafter may be revealing. What are the arguments for a right not to know ?

The negative approach: one strategy that might be adopted is the negative one of making the case that the arguments for a right to know are inadequate: do people actually use the genetic knowledge in reproductive decision-making or in long-term planning ? As far as the moral argument is concerned, however, whether or not people do actually use the information in particular ways cannot by itself determine whether or not they have the right to it. It would nevertheless be important information in carrying out a utilitarian calculus about the costs and benefits likely to follow the overriding of confidentiality. More positive arguments for a right not to know are of both non-consequentialist and consequentialist sorts.

1. The human condition is one of limited knowledge. There is an argument that since the human condition is one of limited knowledge, it does not make sense to say that we ought to know, or that there is a duty to know. The objection to this argument is that it does not follow from the fact that the human condition is one of limited knowledge and that it is not possible that we should know everything, that there is a right not to know every particular thing where that knowledge is available.
2. Consequentialist arguments. Knowledge causes distress: one argument for the right not to know is the harm that can result from knowledge. A distinction is commonly drawn however between knowledge and its use, the idea being that knowledge in itself is morally neutral. In the context under discussion the application of this point would be that it is our attitude to genetic knowledge rather than the knowledge itself which is significant. From a consequentialist point of view, what has to be considered is whether the overall benefits of knowledge outweigh the disadvantages. The lack of knowledge can also cause harm: decisions taken in ignorance, in reproductive matters for example, have the potential to lead to harm that could have been avoided. Knowledge helps us to avoid bad outcomes and choose good ones. Certain kinds of misery however should perhaps be given special weighting. It may not be justifiable to take away hope from a person by exposing them to knowledge they do not want.

Apart from the shock and unhappiness of coming to terms with an unpleasant diagnosis, genetic knowledge may have serious social consequences for the individual in terms of stigmatisation and discrimination. These however are strictly speaking arguments for restricting the access of others to the knowledge rather than for a right of the individual concerned not to know.
3. What of the argument from self-determination (cf. also Dierickx, 1998)? If the right to know is based on autonomy as expressed in a claim to self-determination, then perhaps the same argument can allow an individual to waive that right ? If we understand autonomy in a wider sense, however, as empowerment, then the argument sounds rather different. To waive knowledge is to waive empowerment - is this a right ? It might be argued that such a right conflicts with responsibility. On the other hand it might be argued that genetic knowledge is not empowering, at least not always, and that in some sense restriction of knowledge might be necessary to protect autonomy or an individual's sense of self.
4. Some of the most interesting arguments concern this sense of self and notions of integrity and privacy. The concept of privacy suggests a boundary around the self which should not be violated. It is not equivalent to self-determination - it is not saying that individuals should be able to do or have what they want, but making a point about the inviolability of the private sphere. A woman who has a genetic predisposition to develop breast cancer

in later life may have a self-image that is incompatible with this as a possible future. It might be argued that this example is unrealistic, because, given that breast cancer is multifactorial, every woman must acknowledge some risk, unless her image of her possible future is misguided. But some risks may be so remote as to seem virtually inconceivable. Genetics makes risks more or less probable. Is it justifiable to intrude on this woman's self-perception ? Is there more to this than the removal of hope ? To what extent should the desire of an individual to retain a particular self-image take priority over other considerations ?

5. There are arguments to suggest that the purported right not to know relies on an over-individualistic approach to the ethical issues. These arguments are to a considerable extent concerned with the notion of responsibility rather than rights. They are more developed by personalist than by principlist theories.

3.3.2.b. Arguments Against a Right Not to Know

1. The argument against such a right may be based on the value of solidarity. Making vital information about one's genetic disorder available to persons with whom one shares genes is a highly ethical act of solidarity with one's own group. In order to carry out this act of solidarity, however, one has to have the information to share, which in turn suggests a responsibility for others. These consequences may be of importance both to persons currently alive who are making reproductive decisions and to as yet unborn generations.

2. Also public health considerations could provide an argument against a right not to know. The individual's self-image should perhaps give way before this. But has society yet accepted the idea of duties of solidarity in genetic research ? Such duties must be first accepted by the individual: the duty to participate cannot simply be imposed.

3. The recent move towards a more communitarian approach to ethics and medical ethics in particular has a special significance in genetics with its emphasis on relatedness. It has given rise to the suggestion that emphasis on individual rights should be supplemented if not replaced by consideration of individual responsibilities. Then concern for individual confidentiality and privacy would be at the very least mediated by responsibility to share genetic knowledge, which in turn implies a responsibility to know it. It might be argued, however, that the responsibility one has to one's partner and family is of a different kind from one's responsibility to institutions and employers, just as their respective needs to know are different. This at least points the way forward, to consideration of the extent of the feasibility and desirability of solidarity in genetics.

4. And finally, some argue that the right not to know can only be justified from an ideological and obscurantistic point of view (Report of the Belgian

Federal Committee on Bioethics about the European Convention on Human Rights and Biomedicine, 1998). It must be counterbalanced by a duty to know.

3.3.2.c. A Duty to Know?
Shaw would argue that the at-risk individual has in certain circumstances an ethical duty to know whether or not he or she is a carrier. These situations arise when a third party might be harmed by a person's refusal to be tested: "Knowingly, capriciously, or negligently transmitting a defective gene that causes pain and suffering and an agonizing death to an offspring is certainly a moral wrong if not a legal harm. Thus, if reproduction is contemplated (or not consciously prevented) there is an ethical obligation not to harm the offspring and one's genotype should be determined so that appropriate steps can be taken to avert the disease in future generations. In addition to the duty to know one's own carrier status for the benefit of future offspring I would also argue that a spouse has the right to know the results of the test" (Shaw, 1987, p. 245).

3.4. Directive versus Non-directive Genetic Counselling

Genetic counselling is when an individual, a couple or a family put questions about a medical condition or disease that is, or may be, genetic in origin to a health professional (the genetic counsellor) (Clarke, 1994). As mentioned in the President's Commission Report, genetic counselling helps people with a potential or manifest genetic problem understand and, if possible, adjust to genetic information ; when necessary, it aids them in making decisions about what course to follow. It is an individualized process in which a specialist in medical genetics confers with an individual, or couple, or sometimes a group seeking additional information or assistance (1983). The President's Commission discerned five major categories of moral principles playing a role in genetic counselling: confidentiality, autonomy, knowledge, well-being and equity.

A more debated question concerns the neutrality of the counselling process. Those who do genetic counselling agree that it should always be done in a morally neutral manner. This is reflected in professional discussions of the goals of genetic counselling, in the norms that should govern the behaviour of clinical geneticists and counsellors, and in discussions of the techniques and methods that counsellors should use to attain their goals. The long dominant view of the goals, norms, and methods thought appropriate in genetic counselling can be accurately described as an ethos of value neutrality (Caplan, 1993). One of the reasons why it is possible to argue that genetic counselling is not morally neutral is that it is not exactly obvious, despite the frequency with which such claims are made, what an ethos of neutrality entails. The most frequent injunctions as to how to achieve an ethos of neutrality are exemplified in warnings that counselling must be nondirective. Moral neutrality is indeed often equated with nondirectiveness. It is used to describe the stance that the counsellor should adopt toward the counsellee, one of openness and a willingness to listen. Nondirectiveness has its roots in a theoretical position within psychiatry, social work, and psychoanalysis, that prescribes nondirectiveness as the best stance for eliciting information from a patient so that the patient may come to have an insight about his or her psychological problems.

Directive counselling would permit or require the counsellor to be active, willing to engage in challenge, argument, and confrontation with clients. Those who favour nondirective counselling among genetic counselling professionals are usually referring not to a neutral or indifferent moral outlook but, rather, to a passive role in which counsellors try to be responsive to client needs and questions and avoid challenges or confrontations in seeking to accomplish their educational goals. But the question remains if it will be possible in the future to keep this passive role as the only attitude of clinical geneticists. The relationship between the counsellor and the counsellee can in any case never be totally non-directive.

4. CONCLUSION

Because there is unanimous agreement that predictive testing should not be done for children, prenatal diagnosis is not offered for HD unless the couple agrees to terminate a pregnancy at increased risk. The eventual appeal to the prenatal exclusion test brings its own ethical dilemmas. Many questions remain and many times couples are confronted with a serious ethical dilemma. We have tried to present the most important data and clarifications in order to create the possibilities for a balanced ethical judgement. In any case, the ethical clarification is an instrument for a personal and conscientious decision. This is what ethics is about: to promote the quality of the ethical decision-making process of all those who are confronted with ethical dilemmas in their life.

REFERENCES

Beauchamp, T.L. and Childress, J.F.: 1994, *Principles of biomedical ethics*. 4th edition. Oxford University Press, New York.

Caplan, A.L.: 1993, Neutrality is not morality: the ethics of genetic counseling. In: Bartels, D.M., LeRoy, B.S. and Caplan, A.L. (eds.): *Prescribing our future. Ethical challenges in genetic counseling*. Aldine De Gruyter, New York, p. 149-165.

Chadwick, R. (ed.): 1987, *Ethics, reproduction and genetic control*. Routledge, London.

Chadwick, R.: 1997, The philosophy of the right to know and the right not to know. In: Chadwick, R., Levitt, M. and Shickle, D. (eds.), *The right to know and the right not to know*. Avebury, Aldershot, p. 13-22.

Clarke, A. (ed.): 1994, *Genetic counselling. Practice and principles*. Routledge, London.

Decruyenaere, M., Evers-Kiebooms, G. and Van den Berghe, H.: 1993, Perception of predictive testing for Huntington's disease by young women: preferring uncertainty to certainty? *Journal of Medical Genetics* 30: 557-561.

Dierickx, K.: 1998, Het menselijk genoom en het recht om niet te weten. Een verkenning van de morele gronden. In: Nys, H. (ed.): 1998, *De conventie mensenrechten en biogeneeskunde van de Raad van Europa. Inhoud en gevolgen voor patiënten en hulpverleners*. Intersentia Rechtswetenschappen, Antwerpen, p. 175-184.

Evers-Kiebooms, G., Swerts, A., Cassiman, J.J. and Berghe, H. van den: 1989, The motivation of at-risk individuals and their partners in deciding for or against predictive testing for Huntington's disease. *Clinical Genetics* 35: 29-40.

Harper, P.S. (ed.): 1991, *Huntington's disease. Major problems in neurology*. Vol. 22, W.B. Saunders Company Ltd., Toronto.

Harper, P.S. and Sarfarazi, M.: 1985, Genetic prediciton and family structure in Huntington's Chorea. *BMJ* 290: 1929-31.

Higgins, M., Bloch, M., Kanani, S., Quarrell O.W.J., Theilman J., Hedrick A., Dickens B., Lynch A. and Hayden, M.: 1990, Ethical and legal dilemmas arising during predictive testing for adult-onset disease: The experience of Huntington disease. *American Journal of Human Genetics* 47: 4-12.

President's Commission for the Study of Ethical Problems in Medicine and Biomedical and Behavioral Research: 1983, *Screening and counseling for genetic conditions. A report on the ethical, social and legal implications of genetic screening, counseling, and education programs*. United States Government Printing Office, Washington, D.C.

Quarrell, O.W.J., Meredith, A.L., Tyler A., Youngman, S., Upadhyaya, M. and Harper, P.S.: 1987, Exclusion testing for Huntington's disease in pregnancy with a closely linked DNA marker. *The Lancet*: 1281-1283.

Scheler, M.: 1916, *Der Formalismus in der Ethik und die materiale Wertethik*. Verlag Max Niemeyer, Halle. (cf. *Formalismus in ethics and non-formal ethics of values. A new attempt toward the foundation of an ethical personalism*: 1973, Northwestern University Press, Evanston).

Schotsmans, P.: 1989, Brave New World binnen handbereik? De genetische uitdaging aan de ethiek. *Concilium* 3: 248-258 (cf. English Translation in Concilium: Brave new world within reach? The genetic challenge to ethics).

Schotsmans, P.: 1994, Ethical considerations on the growing potential for human genetic control: Genome projects. *Bulletin of the Polish Academy of Sciences. Biological Sciences* 42(1): 1-23.

Schotsmans, P.: 2001, In vitro fertilization and ethics. This volume.

Shaw, M.W.: 1987, Testing for the Huntington gene: A right to know, a right not to know, or a duty to know. *American Journal of Medical Genetics* 26: 243-246.

Tibben, A.: 1993, *What is knowledge but grieving? On psychological effects of presymptomatic DNA-testing for Huntington's disease*. Koninklijke Bibliotheek, Den Haag.

Tolmie, J.L., *et al.*: 1995, The prenatal exclusion test for Huntington's disease: Experience in the West of Scotland. 1986-1993. *Journal of Medical Genetics* 32: 97-101.

Walters, L.R.,: 1993, Ethical obligations of genetic counselors. In: Bartels, D.M., LeRoy, B.S. and Caplan, A.L. (eds.): *Prescribing our future. Ethical challenges in genetic counseling*. Aldine De Gruyter, New York, p. 131-147.

Wert, G. de: 1992, Predictive testing for Huntington disease and the right not to know. Some ethical reflections. In: Evers-Kiebooms, G., *et al.*, *Psychosocial aspects of genetic counseling. March of Dimes Birth Defects Foundation* 28(1), Wiley-Liss, New York.

Wertz, C.D. and Fletcher, J.C. (eds.): 1989, *Ethics and human genetics. A cross-cultural perspective*. Springer Verlag, Berlin/Heidelberg.

BERT GORDIJN

ABORTION

1. INTRODUCTION

Most European countries have revised their abortion laws towards a more liberal approach during the last few decades.[1] Nevertheless, the question of when, if ever, abortion is morally permissible is still widely discussed and fiercely debated in academic circles as well as amongst a broader public. The character of these discussions is often very aggressive. Particularly in the USA the controversy on abortion is sometimes so violent that as a result more than once medical doctors working in abortion clinics were killed. Therefore, at the turn of the millennium abortion still is a major troubling public policy issue. The different views towards the moral assessment of abortion can be categorised into three general positions: (1) the pro-life view, (2) the pro-choice view and (3) a position in between.[2]

Advocates of the pro-life view believe that abortion is seldom if ever morally licit. According to them, the foetus has to be regarded as a person from the moment of conception. Consequently, abortion involves the killing of a person. Advocates of the pro-choice view, however, believe that abortion is always or almost always morally justifiable. They hold that there is a moral asymmetry between the moral status of the foetus and that of the pregnant woman. Although abortion involves the termination of foetal life, it does not involve the killing of a person. Those who uphold the intermediate view believe that abortion is sometimes morally permissible, and sometimes not. They differ widely in their normative and theoretical views. Most of them hold that the moral permissibility of abortion can depend on a lot of different factors such as the foetus' prospect for future health and welfare, the woman's reason for seeking abortion and the stage of foetal development.

It looks as if during decades of intense debate the defenders of these three different positions have not moved one inch. Therefore, abortion presents itself as one of the least tractable issues in bioethics. One important reason for the unchanging status quo in the discussions about the ethical aspects of abortion is a seemingly insoluble disagreement on a more theoretical level, namely with regard to the conceptual question of what is a

H.A.M.J. ten Have and B. Gordijn (eds.), Bioethics in a European Perspective, 385-397.

person. The advocates of all three positions seem to agree on the normative thesis that it is morally wrong to kill an innocent person. They differ, however, with regard to the question of whether, and if yes, from which phase of development an embryo or a foetus can be regarded as a person.

In this contribution, the three main positions towards abortion will be explained through a sketch of the different reactions to a case of unwanted pregnancy. Thereupon, the more fundamental conceptual theories on personhood at the early stages of human life will be explained.

2. CASE: AN UNWANTED PREGNANCY[3]

A 36-years-old divorced mother of two children (16 and 17 years old), has an unwanted pregnancy. For a number of months she has been having a new relationship with a man who is also divorced and who claimed to have been sterilised three years ago. Apparently the operation has not been totally effective; it is possible that a spontaneous recovery of the vas deferens has taken place.

According to the woman her social situation does not allow a pregnancy. After her divorce she began a university degree course, with which she is halfway by now. Also, with her children almost grown up her life has been organised in such a way that a baby does not fit in it anymore.

Her boyfriend is a lot more favourable to the pregnancy. He feels responsible for it, moreover he is opposed on principle to induced abortion. Much sooner than he had planned, he is prepared to let their relationship take a more definite shape and together with her he wants to bring up the child.

3. THREE MAIN POSITIONS TOWARDS ABORTION

3.1. The Pro-life View

As already mentioned above, the advocates of the pro-life view believe that abortion is rarely if ever morally allowable. The foetus has to be regarded as a person from the conception onwards. Consequently, abortion involves the killing of a person. Two different pro-life views can be distinguished. Defenders of the extreme pro-life view hold that abortion can never be morally justified even if the pregnant woman's life is endangered. The advocates of a more moderate pro-life view, however, argue that an exception has to be made when the pregnant woman's life can be saved through an abortion or when her health is seriously endangered through a continuation of the pregnancy.

The central norm in the pro-life view, obviously, is respect for human life. Advocates of the extreme pro-life view could argue that since human life is the necessary condition of all other values, it ought to be respected unconditionally. From a religious perspective it is often argued that since life is a gift from God, it cannot be rejected. Defenders of the more moderate pro-life view make an exception in situations in which prolongation of the pregnancy would kill the pregnant woman. Only in these situations they allow termination of the pregnancy. The grounds are similar to the reasons that allow for killing in order to save one's own life in situations of self-defence. Upholders of the pro-life view do not believe that human life and human personhood can be conceptually dissociated. Personhood is just one of the characteristics of human life. All human life, therefore, has the basic rights that come with personhood, e.g., the right to life and to have the integrity of one's body respected.

It is not difficult to imagine how an upholder of the pro-life view would respond to the case above. It would not matter whether he was an adherent of the moderate or the extreme pro-life view. He would certainly regard an abortion as morally impermissible in this case, because it would imply the killing of human life. Nothing can allow for killing human life except perhaps defending or saving your own life, which is obviously not the case in the described situation. According to the pro-life view, therefore, the pregnant woman in the case would have the moral duty to bring to term her pregnancy irrespective of all the future problems that could perhaps present themselves.

3.2. The Pro-choice View

Advocates of the extreme pro-choice view hold that abortion is ethically permissible as long as the women asking for it are making autonomous choices irrespective of any further circumstances. Defenders of a more moderate pro-choice view argue that abortions are ethically allowed only if the autonomous choices of the women asking for them are non-trivial. Contrary to what proponents of the pro-life view hold, adherents of both pro-choice views believe that there is a fundamental difference between the moral status of the foetus and that of a pregnant woman. Whereas the pregnant woman is considered as a full person, the foetus is not. Accordingly, abortion does not involve the killing of a full person, although it involves the termination of foetal life.

Women have the right to decide for themselves whether they want to bring their pregnancy to term or not. Interference with a woman's reproductive choices would imply a violation of her fundamental right to self-determination as autonomous person. Forbidding abortion would, therefore, not be consistent with respect for women as autonomous persons. Conformingly, banning abortion is morally impermissible.

Respect for autonomous choices, obviously, is the central norm in the pro-choice view. There is a lot of debate with regard to the exact defining properties of autonomy. Most authors, however, concur in holding that the behaviour and choices of an agent can only be regarded as autonomous, if the agent displays a certain independence from controlling influences (e.g., coercion, psychosis, alcoholic intoxication) and some kind of capacity for intentional action. Moreover, the choices must be made on the basis of true information and understanding of the relevant facts. Respect for autonomous choices demands no interference with the autonomous decisions of an agent as long as his actions do not negatively influence the autonomy of other agents. Furthermore, it also implies the positive duty to try to enhance or restore the autonomy of an agent when this is needed. For example, if a woman wants to terminate her pregnancy because she falsely believes that her foetus has a serious genetic disease, respect for autonomy demands that it is pointed out to her that the disease will only show up later in life and that its symptoms can be controlled easily. By giving her the true information concerning the health condition of her future child she can reconsider her decision in the light of a better and truer understanding of the relevant facts.

How would an upholder of the pro-choice view assess the situation of the 36-years-old divorced mother in the above case? Supposing that she is autonomous (there is nothing that contradicts this hypothesis) and after having considered the matter thoroughly she decides to have the foetus aborted, then this decision ought to be respected.

3.3. The Intermediate View

Adherents of the intermediate view believe that the ethical judgement of concrete cases of abortion depends on the specific circumstances. According to them terminating a pregnancy can be morally justified under a whole range of conditions, whereas under different circumstances they would disallow the abortion. Their opinions differ, however, with regard to the specific situations in which abortion is deemed to be ethically allowed. A lot of different factors can be relevant for an ethical assessment.

The moral permissibility of abortion can, for example, depend on the woman's reason for seeking abortion. These reasons have to display some degree of seriousness and gravity to be able to justify an abortion. Most adherents of the intermediate view would, for example, hold that an abortion is morally permitted when the woman who wants the abortion has been raped and fears serious emotional problems when bringing the pregnancy to term. Also, severe foetal deformity could count as an acceptable rationale for terminating a pregnancy. On the other hand, some reasons would simply be regarded as too whimsical to be considered as morally relevant reasons to stop a pregnancy. For example, when abortion is asked for because otherwise the pregnancy would negatively interfere with a planned holiday, most adherents of this view would probably regard such a motive as being too frivolous to count as a serious rationale for abortion.

The potential health or welfare prospects of the foetus can also be deemed relevant. From this perspective, for example, most proponents of the intermediate view would argue that in case of a pregnancy through incest resulting in an extreme rejection of the future child by the pregnant woman the unhappy future welfare prospects of the foetus would allow an abortion. When, on the contrary, abortion is wanted because ultrasonography showed that the foetus has only nine fingers instead of ten, the concomitant insubstantial limitation of future welfare and health would be too negligible to allow an abortion.

Also, the question of the extent to which the prospective parents are responsible for the prospective mother becoming pregnant can be deemed morally significant. If, for example, a woman is pregnant owing to rape or contraceptive failure she is evidently not responsible for the pregnancy. Under such circumstances most proponents of the intermediate view would allow abortions that in cases of full responsibility of the prospective parents they would disallow. From this point of view, therefore, the need for justifying abortion grows with the responsibility the prospective parents are deemed to bear for the pregnancy.

The stage of foetal development, finally, is another example of a morally relevant characteristic from an intermediate point of view. Most proponents defend the belief that the difficulty of justifying an abortion increases proportionally with the stage of development of the foetus. Obviously, this thesis can only be defended with some sort of gradualist view on the personhood of a foetus.[4]

The intermediate view on abortion displays a great variety. Therefore, the opinion from this perspective on the above case could take several forms. It could, for example, be argued that since the prospective parents do not seem to be responsible for the pregnancy in this case and the social situation of the pregnant woman appears not to be very favourable for fitting in a baby, an abortion is not morally impermissible.

On the other hand, an adherent of this view could also reason that the rationale brought forward by the woman is not substantial enough to allow an abortion, especially because her partner is favourable to the pregnancy. Considering that he is willing to let their relationship take a more definite shape and to bring up the child together with her, it would not seem to be impossible for the couple to come to terms with the new situation without the woman giving up her university education. Supposing, moreover, that the foetus appears to be fairly healthy and the pregnancy is already in an advanced stage, it could be argued that under these conditions an abortion can not be morally justified.

4. ABORTION AND PERSONHOOD

In the scholarly debate about abortion the concept of the person plays an important role. This role stems from the fact that many authors cherish the assumption that this concept is needed to analyze problems of moral status. They suppose that a person, as a matter of principle, possesses certain basic moral rights that must be protected.[5] In contrast to rights which depend upon particular circumstances, e.g., promises or legal contracts, basic moral rights are independent of any such special circumstances. They are usually held to include the rights to life, liberty, self-determination and freedom from infliction of physical harm.

Discussions about the morality of abortion on a more theoretical level have usually focused on the question of whether, in a given situation, a human being can already be regarded as a person. The argument goes as follows. Each and every person has a basic right to life. Accordingly, when in a given situation a human being is said to be a person, its life cannot be taken without violating this right to life. The life of a human being, however, that is not considered to be a person can conversely be ended without special moral concern.[6] Therefore, to decide on the moral permissibility of abortion with regard to a particular foetus, it is important to be able to determine whether that foetus can already be regarded as a person. Therefore, in the debate on the moral acceptability of abortion the answer to the question: "When does a person begin to exist?" decides whether, and, if yes, until which phase of development the abortion of an embryo or a foetus can be morally justified.

However, although the concept of the person plays an important role in ethical debates, the term 'person' is used with a wide variety of meanings.[7] It seems as though every author has his own particular concept of the person. Because of this enormous variety of concepts, discussions constantly arise about which entities it does and does not include.[8]

Generally, three different approaches with regard to personhood can be distinguished.[9] According to the conceptionalist view, personhood starts at conception and can be understood largely in biological or theological terms. Proponents of the conceptionalist view, correspondingly, consider abortion to be a violation of the fundamental right to life of the embryo or foetus.[10] Proponents of the pro-life view on abortions would evidently defend the conceptionalist theory of personhood.

Following another theory, personhood has to be understood as a matter of the development of the capacities that require some form of consciousness. According to the advocates of this view, the properties that constitute the necessary and sufficient conditions for personhood are acquired only in a relatively late stage in foetal development or even only in infancy. From this perspective, abortions can generally not be considered as violating any right to life since abortions usually take place before the embryo or the foetus becomes a person. Therefore, the advocates of the pro-choice view on abortion tend to point out that personhood is essentially dependent on capacities that require consciousness.

According to the gradualist view, finally, an embryo or a foetus gradually obtains more moral status as it develops. From this perspective, the difficulty of morally justifying an abortion increases proportionally with the stage of development of the foetus. That is why on the basis of this theory, abortions in a later stage of foetal development can be morally disapproved of whereas abortions in an earlier stage can be allowable. Most adherents of the intermediate view on abortion feel attracted to some kind of gradualist theory with regard to personhood.

4.1. The Conceptionalist Theory of Personhood

The conceptionalist theory of personhood states that personhood already comes into existence at the conception. Advocates of this theory regard personhood as being an all-or-nothing matter. There are two versions of the conceptionalist theory of personhood: the biological and the theological one.

Adherents of the biological version of the conceptionalist theory on personhood argue that from a biological point of view, as soon as the egg-cell has been fertilised a human being exists. The zygote carries all the information for the gradual development into a foetus who in his turn gradually becomes a child and later on a mature human being. Biological conceptionalists tend to point out that with regard to the process of development and growth of a human being there does not seem to be a moment in time that displays significant discontinuity. Development starts with cell division at the level of the fertilised egg-cell and goes on continuously. Therefore, there cannot be any moment at which a foetus who was not already a person suddenly turns into one. Accordingly, biological conceptionalists regard embryos and foetuses as being in full possession of all the basic rights of a person from the moment of fertilisation.

According to theological conceptionalists, God determines whether egg-cells are being fertilised or not (cf. Gen. 29,31; 30,22; 49,25; Ruth 4,13). If a zygote comes into being, this means that God wanted to create a new and unique human being. This new human being, as all other human beings, has a unique relationship with God. Therefore, the life of any human being has an intrinsic value, irrespective of any further circumstances. Accordingly, it is not ethically allowed to interfere aggressively in this relationship between human beings and God by willingly terminating a life. Theological conceptionalists hold that a zygote already deserves all the respect that we owe to a person.[11] Since one of the most important basic rights of a person is the right to life, it is not morally licit to terminate a pregnancy.

Both biological and theological conceptionalists concur in rejecting birth as a criterion of demarcation between persons and non-persons or not-yet persons. Birth is rather considered to be a morally indifferent event. From a biological perspective, for example, there is no relevant structural or developmental difference between an eight-month-old foetus who has just been born and an eight-month-old one who is still in utero. Therefore, there is no reason at all to consider the first as possessing basic rights which the second would lack. Both must have the same moral status.

Also, from a theological conceptionalist point of view birth lacks any moral significance. Since the unborn child is in possession of the whole set of basic rights from the moment of conception, birth makes no difference with regard to moral status.

4.2. Locke's Theory of Personhood and its Successors

Most pro-choice theorists look upon abortion as morally justified because they have fixed a person's beginning at some point relatively late in its foetal development or even in its infancy.[12] In contrast with the conceptionalists, they do not regard personhood as being determined solely by biological properties or divine arrangement but as having to do with characteristics of consciousness.

This view on personhood can be traced back to John Locke's (1632-1704) ideas on the person. From his point of view consciousness is an essential characteristic of a person which he regards as being:

> A thinking intelligent being, that has reason and reflection, and can consider itself as itself, the same thinking thing in different times and places; which it does only by that consciousness which is inseparable from thinking, and as it seems to me essential to it: it being impossible for anyone to perceive, without perceiving that he does perceive (Locke, 1963, p. 55).

According to Locke, the concept of the person and the one of the self are identical. That is why he uses these two terms interchangeably. In another definition the central role of consciousness as a necessary condition for personhood is expressed again:

> Self is that conscious thinking thing (whatever substance made up of, whether spiritual or material, simple or compounded, it matters not) which is sensible, or conscious of pleasure and pain, capable of happiness or misery, and so is concerned for itself, as far as that consciousness extends (Locke, 1963, p. 62).

Locke embraces the concept of thinking in a very broad sense: it seems to comprise all sensations and perceptions. On the other hand, with his concept of *consciousness* Locke appears to mean something that we would nowadays probably call 'self-consciousness': the capacity to recognise particular sensations and perceptions to be ours, i.e. a capacity which exclusively refers to one's own contents of consciousness. This capacity does not function with regard to the sensations and perceptions belonging to other persons. Locke continues his reflections as follows:

> Thus every one finds, that whilst comprehended under that consciousness, the little finger is as much a part of himself, as what is most so. Upon separation of this little finger, should this consciousness go along with the little finger, and leave the rest of the body, it is evident the little finger would be the person, the same person; and self then would have nothing to do with the rest of the body. As in this case it is the consciousness that goes along with the substance, when one part is separate from another, which makes the same person, and constitutes this inseparable self (Locke, 1963, p. 62-63).

Here, it becomes more than clear that Locke was not interested in biological properties as the basis for personhood. He rather considers them as being fairly irrelevant.

Locke's ideas on the person have been enormously influential in contemporary ethics. With Locke, many present authors regard consciousness as a *sine qua non* for personhood (cf. Lizza, 1993, p. 355 and Warren, 1997, p. 94). Locke's concept of consciousness, however, was not very clear. What is more, Locke failed to clarify whether he views consciousness only as a necessary condition for personhood among others or also as a sufficient one. Consequently, many different interpretations and modifications of Locke's idea of consciousness as a condition for personhood have been developed.[13]

In the contemporary ethical debate, authors influenced by Locke concur in viewing the person as an entity capable of at least some rudimentary form of consciousness. However, most of them vary in developing this idea beyond the mere possession of consciousness. So in specifying further properties or characteristics that go to make up a person opinions differ. A list (Tooley, 1983, p. 90-91) of the more important properties that have been proposed - solely or in combination with others - as being necessary conditions for personhood beyond the mere possession of consciousness gives an impression of the enormous variety of concepts of the person by the authors influenced by Locke:

> The capacity to experience pleasure and/or pain;
> The capacity to have desires;
> The capacity to remember past events;
> The capacity to have expectations with respect to future events;
> An awareness of the passage of time;
> The property of being a continuous, conscious self, or subject of mental states, construed in a minimal way, as nothing more than a construct of appropriately related mental states;

The property of being a continuous conscious self, construed as a pure ego, that is, as an entity that is distinct from the experiences and other mental states that it has;
The capacity for self-consciousness, that is to be aware of the fact that one is a continuing, conscious subject of mental states;
The property of having mental states that involve propositional attitudes, such as beliefs and desires;
The capacity to have thought episodes, that is, states of consciousness involving intentionality;
The capacity to reason;
The capacity to solve problems;
The property of being autonomous, that is of having the capacity to make decisions based upon an evaluation of relevant considerations;
The capacity to use language;
The ability to interact socially with others.

In contrast to the conceptionalists, authors within the Lockean tradition who beyond the mere possession of consciousness regard social and interactive aspects as being necessary conditions for personhood, may be willing to attribute personhood to a premature neonate while denying it to the nine-month foetus who is still in utero. For those authors who over and above think that properties that can only be acquired later in infancy like self-consciousness and rationality are essential to personhood, even normal neonates cannot be regarded as persons.

4.3. The Gradualist Theory of Personhood

Most adherents of the intermediate view on abortion feel attracted to some kind of gradualist theory with regard to personhood. According to this view, a foetus gradually obtains more moral status as it develops. A first premise of this theory is that moral status is not an all-or-nothing phenomenon but a phenomenon that admits of different degrees. Another supposition underlying the gradualist theory of personhood is that with regard to the process of development and growth of a human being there are moments in time or periods that display morally significant discontinuities. Therefore, there are moments or stages at which, for example a foetus who was a person in a lesser degree suddenly or gradually turns into one of a higher degree. Accordingly, adherents of the gradualist theory regard embryos and foetuses as being in possession of only a few of the basic rights of a full person. From the moment of fertilisation on there will be a gain in moral status according to the stage of embryonic or foetal development.

According to the advocates of the gradualist view, the moral status of unborn human life is somehow connected with a variety of morally relevant characteristics and properties which are acquired step by step during embryonic and foetal development. For example, nidation is often considered as a morally significant event in embryonic development because before nidation the embryo cannot quite be regarded as an individual: it sometimes happens that it splits up in two separate ones. Moreover, before nidation two separate embryos can join becoming one individual. After nidation, the individuality of the embryo is fixed. Also, the first primitive forms and the further development of sentience can be regarded as morally significant, since sentience is a necessary condition for feeling pain and pleasure. Furthermore, there are gradualists who hold that the end of the embryonic stage implies a gain in moral status, since after this process all the organ systems are present in the organism. Finally, birth is often considered to be a morally relevant event from a gradualist perspective, because after birth the child is involved in social and interactive processes.

5. CONCLUSION

The different views towards the moral assessment of abortion can be categorised into three general positions: the pro-life view, the pro-choice view and the intermediate view. According to the first theory, abortion is never or almost never morally allowable. Adherents of the second view hold that abortion is almost always licit. The third view is in between.

All participants in the debate seem to concur in holding that it is morally wrong to kill an innocent person. They differ, however, with regard to the question of whether, and if yes, from which phase of development an embryo or a foetus can be regarded as a person. Therefore, an important reason for the impasse in the discussions about the ethical aspects of abortion is the seemingly insoluble disagreement on a more theoretical level, namely with regard to the conceptual question of what is a person.

As to this question, there are three main theories. The adherents of the pro-life view on abortion defend the conceptionalist theory of the person according to which an embryo is a person from the moment of conception. Advocates of the pro-choice view argue that personhood has to be understood as a matter of the development of the capacities that require some form of consciousness. The upholders of the intermediate view, finally, defend the idea that an embryo or a foetus gradually obtains more moral status as it develops. The discussion on this underlying concept of the person seems to have reached an impasse.

Of course, this situation is intellectually unsatisfying. Therefore, it is necessary to try to find another theoretical approach to the problems of moral status of unborn human life. Would it perchance be possible to analyze ethical questions about the moral status of the foetus without the concept of the person? Perhaps this would be an interesting challenge in the debate on abortion. If we would try to analyze questions like the following without using the concept of the person, it might be easier to come to some sort of agreement: What is the moral significance of conception and nidation? How does the commencement of the nervous system influence the moral status of the foetus? Does the completion of the embryo-genesis or the ability to survive independently of the body of the mother change the set of moral attributes of the unborn? What is the moral meaning of birth?

NOTES

1. Abortion is the termination of foetal life *in utero*. In most modern countries more liberal laws making abortion legal under certain conditions were developed at the end of the sixties or the beginning of the seventies.
2. Cf. Davis (1992) who uses a very similar division of those who discuss ethical aspects of abortion. She distinguishes Restrictives, Permissives and Moderates.
3. This case is described by Essed (1992, p. 34-35).
4. See section IV. 3. for further discussion.
5. Cf., for example, Warren: "Whatever else we are, we are persons; and it seems likely that this fact will prove fundamental to the justification of the strong moral status that most of us want for ourselves and those we care about." (Warren, 1997, p. 90).
6. Cf., for example, Robertson, 1991, p. 295.
7. Cf. with respect to different definitions and opinions about the person within the debates of today: Dennet 1976; Doran, 1989; Teichman, 1985 and Vincent, 1989.
8. The amazing diversity of the concept of the person causes a lot of conceptual problems in ethical discussions. Sapontzis (1981), for example, holds that there are at least two sorts of concepts which are lumped under the label of 'person': descriptive and evaluative ones. Furthermore, he states that the relation between the two is seriously misunderstood. Cf. also Wiggins: "On occasion, almost everyone feels difficulties in holding in a single focus three different ideas: (a) the idea of a person as object of biological, anatomical, and neurophysiological inquiry; (b) the idea of the person as subject of consciousness; and (c) the idea of the person as locus of all sorts of moral attributes and the source or conceptual origin of all value." (Wiggins, 1987, p.56).
9. See for further discussion of the concept of the person, this volume: chapter 4.
10. See, for example, Kreeft, 1990.
11. See, for example, Donum Vitae (1987, I, 1).
12. See, for example, Singer, 1979 and Tooley, 1972.
13. See Gordijn, 1996, p.143-148 for further discussion.

REFERENCES

Congregatie voor de Geloofsleer: 1987, *Donum Vitae*. Osservatore Romano, Rome. Dutch translation in: *Archief van de Kerken* 42, 353-379.

Davis, N.A.; 1992, Abortion. In: Becker, L.C. and Becker, C.B. (eds.): *Encyclopedia of ethics*. Garland Publishing, Inc., New York, Vol. I, p. 2-6.

Dennet, D.C.: 1976, Conditions of personhood. In: Rorty, A. (ed.): *The identities of persons*. University of California Press, Berkeley, p. 175-196.

Doran, K.: 1989, *What is a person. The concept and the implications for ethics*. Lewiston, NY, Queenston, Ontario.
Essed, G.G.M.: 1992, Gynaecologie-obstetrie. Vakinhoudelijke aspecten. In: Have, H.A.M.J. ten, *et al.* (eds.), *Ethiek en recht in de gezondheidszorg*. Kluwer, Deventer, XI 1-41.
Gordijn, B.: 1996, *Die Person und die Unbestimmbarkeit ihrer Grenzen. Eine grundlegende Kritik an der Debatte über Personenidentität*. Peter Lang, Frankfurt am Main.
Kreeft P.: 1990, Human personhood begins at conception. *Journal of Biblical Ethics in Medicine* 4: 8-11.
Lizza, J.P.: 1993, Persons and death: What's metaphysically wrong with our current statutory definition of death? *Journal of Medicine and Philosophy* 18: 451-374.
Locke, J.: 1963, *The works of John Locke*, A new edition, corrected. 10 Vols. London 1823, Reprinted by Scientia Verlag, Aalen, Vol. II.
Robertson, J.A.: 1991, What we may do with pre-embryos: A response to Richard A. McCormick. *Kennedy Institute of Ethics Journal* 1(4): 293-302.
Sapontzis, S.F.: 1981, A critique of personhood. *Ethics* 91: 607-618.
Singer, P.: 1979, *Practical ethics*. Cambridge University Press, Cambridge.
Teichman, J.: 1985, The definition of person. *Philosophy* 60: 175-185.
Tooley, M.: 1972, Abortion and infanticide. *Philosophy and Public Affairs* 2: 37-65.
Tooley, M.: 1983, *Abortion and infanticide*. Oxford University Press, Oxford.
Vincent, A.: 1989, Can groups be persons? *The Review of Metaphysics* 42: 687-715.
Warren, M.A.: 1997, *Moral status. Obligations to persons and other living things*. Oxford University Press, Oxford.
Wiggins, D.: 1987, The person as object of science, as subject of experience, and as locus of value. In: Peacocke, A. and Gillett, G. (eds.), *Persons and personality: A contemporary inquiry*. Blackwell, Oxford, p. 56-74.

WILLIAM J. ELLOS, S.J.

EMBRYO EXPERIMENTATION

1. INTRODUCTION

Starting from modest beginnings some twenty years ago Saint Carealot Medical Research Centre has emerged as one of the truly world leaders not only in patient care but in the critical area of genetic research. But since Saint Carealot remains very much a patient oriented operation as well, the research facility makes its expertise available to actual patients in the Centre's now world renowned hospital by way of providing new and essential information on genetic matters. A number of physicians who are primarily employed at the research centre also practice medicine in the hospital's well developed reproductive fertility enhancement clinic. Given the reputation of the Centre many childless couples come not only for medically scientific information but also for personal guidance and help from the hospital's team of psychological and pastoral counsellors. In recent years the teams of pastoral counsellors have been greatly expanded to include both lay and clerical, male and female, representatives of a number of quite diverse religious denominations.

2. CASE

Otan and Era Tonnel enjoy a deeply committed marriage experience permeated by ever enduring sorrow, the inability to have children. Over seven years together they have been patient in effort and in prayer but with no result. They have been clients of the reproductive clinic for the past two years. The situation has been rather difficult for them since they are not citizens of the country in which Saint Carlot is located and, even though the train ride there is not very long from their home, they do feel themselves to be in something of an alien land. Even though he is not Catholic, Otan is extremely respectful of Era's quite devout practice of her religion and this has helped him to feel more at home in the foreign setting where Era appears to be comfortable. To this point they have been using, under

H.A.M.J. ten Have and B. Gordijn (eds.), Bioethics in a European Perspective, 399-404.

the supervision of the clinic, a number of fertility enhancing drugs but sadly to no avail. They are beginning to become desperate.

Since they are both university educated people they have shown a great deal of interest in the actual scientific details of the drug regimes and read even the most technical literature with a good deal of understanding. In the course of this reading it becomes clear to them that the next reasonable move would be to request some sort of in vitro fertilisation procedure. Era has religious reasons for having problems with this which Otan does not share and he continues to defer to her on the matter. But as time goes on a sense of strained urgency so strongly develops that they both realize that neither of them is in a psychological state conducive to the making of clear plans and decisions. They then decide to bring their concerns to the expert staff members of the clinic and to abide by their advice. But Era is especially concerned about the advice to be received as to the use of the extra embryos which will not be implanted in her uterus. Will there be experimentation done on them? Will they be implanted in another uterus? Will they be in some way preserved or will they be just discarded?

3. COMMENT

The founders of Saint Carealot were much concerned that they would have a sort of a model hospital and research centre in the Catholic tradition and so well serve both the religious and scientific community. In this matter they were especially concerned to follow the ancient tradition of natural law ethics. As time went on a number of problems arose as to the interpretation of this tradition. One version, often called *vitalism*, is greatly concerned with the preservation and enhancement of biological life. Actually this vitalistic concern was one of the main reasons for the development of the reproductive research centre. The discovery of the genetic code as the very map of life itself prompted the expectation that more knowledge of the structure of this code would lead probably soon to great enhancement of human life itself.

The other version of natural law stresses more a level of human responsibility known as *right reason*. With roots both in Aristotle and Aquinas this points out the essential need for a harmonious working and enhancement of the body-mind/soul relationship. But this right reason aspect of natural law also connects to vitalism in that it is presumed that a healthy body will be a strong partner of healthy mind. At any rate it is essential in the process of right reason to be both attentive to the natural rhythms of the mind and body and to act always as much as possible in harmony with them.

But problems arise in the areas of genetics and embryos. Are these minute biological units really bodies at all or more just the building blocks of what eventually become bodies? In the case of genes there is a clear realisation that no single gene alone could possibly be a body as it is but the essential coding element of a part of that body, sort of a blueprint or map. But it is quite a different kind of thing from a road map which is certainly not part of the countryside it portrays in the way that a gene is an internal living element of the object it maps. While in most cells these internal maps are complete, the sex cells might be considered to be but incomplete halves of a fully complete biological entity. Even though there will certainly be a fascination and focus on the gene itself at least until such time as the completion of the mapping of the total human genome, there is now a realisation of course that genes only live in the cellular environments constitutive of ever larger phenotypically biological units. The study of the smallest of these stable units, the human embryo, may be essential in learning of the complex interactions between phenotype and genotype which constitute the essential biological substrate for all the natural vitalistic rhythms of the mind and body interactions.

Now here we run historically into a specifically European problem as regards this relationship of mind to body. The philosophical foundation for this interaction is to be found in Aristotle who in fact took over from his teacher, Plato, a very inchoate and puzzling notion of the material, including the bodily world. Generally distrustful of it, he concentrated on an idealistically permanent world. Excellent biologist that he was and certainly concerned with the reality of change as well as permanence, Aristotle remained Platonic enough to speak of individual biological units as each possessing an idealistically permanent aspect and a changeably material aspect neither of which is strictly speaking in our contemporary sense either mind or matter. Aristotle's Greek words for these two principles, *psyche* and *hyle* were translated and used by Aquinas as the Latin *forma* and *materia* neither of them exactly form and matter in the contemporary sense of the terms. But *psyche* also becomes in Latin *anima* which has also the meaning of soul and so is generated the insolvable problem of the location of this physical soul which in its philosophical origins is not enough of an entity of itself to have any such kind of location.

René Descartes, while he set out to do something quite different, in fact sets the pattern for the modern consideration of the mind/body relationship. In the face of a climate of nearly universal philosophical scepticism Descartes looked for a way of establishing some basis for complete and total certitude. The famous move as outlined in his *Meditations* involves a process of methodical doubting. It is vital to realize that this is not a real but only a methodical doubt, for to truly doubt everything would be philosophical madness. Nonetheless as a method Descartes requires us to do just that until we reach a point in which it is impossible for us to further doubt. That point is generally positively presented in the famous dictum "I think, therefore I am". This would say that I cannot doubt even in the process of doubting that I am thinking. Contemporary understandings of Descartes which will be very important for our consideration of the ethics of embryo research realise that Descartes cannot fundamentally so much not doubt that he is thinking as that he cannot doubt that he is doubting. As the deconstructionists will show, certitude resides radically in the incertitude of doubt.

There were at least three major consequences of the Cartesian move with serious implications for questions of embryo research. First the concentration on thinking as the ultimate human experience led Descartes to consider human beings primarily as thinking things. This led immediately to serious questions about the relation of the mind to the body which Descartes was never able to really solve. Certainly the more componential presentation of soul as a factor always relative in some inchoate way to the material body which is central to the Aristotle/Aquinas synthesis is now transformed into a situation of a severe bifurcation of soul and body. This has profound consequences for the ethics of embryo research. An Aristotle/Aquinas approach would consider the embryo to be clearly an integral unit of both soul and body. Hence the tampering with one would automatically entail tampering with the other as in deepest reality there really are not one and another, soul and body, but only one holistic entity. Caution and right reason would demand extreme hesitancy and care in any intervention. But if soul and body à la Descartes are really quite separate, then manipulation of the material embryo becomes much more a tolerable kind of operation.

The second consequence of the Cartesian move was a move which went on to permeate all the modern period into a concern and fascination with the inner workings of the mind. Along with this there developed strong tendencies to ever more and more identify and isolate different aspects of mental experience for categorisation and analysis. It is one of the many reasons why it is only with the advent of the contemporary period roughly at the turn of this now fast waning last century that breakthrough progress could be made in the physical and biological sciences. But the influence of the modern period remains strong in the ever present analytic tendencies to differentiate, categorize and dissect. This is especially clear in the debate about embryo research where a good deal of the discussion turns on the question as to a distinction between an embryo and a zygote. Analytically speaking a zygote is a fertilized ovum which is not yet implanted in a uterine wall. Zygote biological units are notoriously unstable with a very high percentage of them self-destructing and washing out in the normal menstrual flow. Implanted embryos are dramatically stable with a majority in the high ninetieth percentile going on to become at birth perfectly healthy babies.

What do we have in the petri dish, zygotes or embryos? Even employing vitalist natural law principles a case might be made that, since the vast majority of zygotes self destruct anyway, what is wrong with our experimenting on them and then discarding them?

But the older Aristotle/Aquinas view would look much more for the continuities rather than the disruptions of natural processes, and so would not admit the ethical validity of the analytic zygote/embryo distinction. And this is just what the Vatican has done in forbidding either zygote or embryo experimentation.

The third consequence of the Cartesian revolution has to do with the way in which we have come to conceive and construe our very notion of what is a human person. There is a very strong tendency to think of a person in terms of mind. And so in the related abortion debate and now also in discussions about termination of life it is the level of mental development, awareness and abilities which determine whether a person should live or die. The Cartesian concentration on mental factors as being primarily cognitive rather than emotive has led us, even in the face of contemporary psychological attempts to redress the balance, to consider persons to be primarily thinking units with emotional appendages somehow connected to and instigated by bodily functions. If the zygote/embryo shows no evidence of thinking, should we seriously take into consideration any bodily emotive factors it may feel?

A final aspect of the debate highlights again from a different perspective the influence of the analytic bias. This has to do with an ever stronger growing consideration of the human person as being defined by the rights which that individual may achieve or possess. Rooted in often quite different or even antagonistic legal and cultural systems there has arisen a world-wide consideration for fundamental human rights. While the language of rights may lead us to think of them uncritically as things, they are, of course, actually a type of human relationship, namely a claim relationship. So that if one person makes a rights claim on another person or group, then that person or group has a right also to consider appropriate response. Some rights are considered so fundamental that the only proper response should be the honouring of the rights claim. But even at this level which certainly entails rights to life and bodily integrity, rights are rooted in individualistic factors which tend to separate rather than unite people. They must always be fought for and maintained especially by rational and intelligent people. But what of an embryonic humanoid entity totally incapable of this kind of activity? Does it have any rights?

While our contemporary century certainly has in many ways moved toward a more integral view of the human mind/body, person/society condition, the influence of the modern period has precluded a creative connection with older holistic traditions. Issues such as embryo experimentation, cloning and genetics will force us in the next century to do so.

4. CONCLUSION

Two aspects of the ancient natural law tradition serve to highlight the need for a holistic approach in all matters in the area of embryo experimentation. One aspect, *vitalism*, demands always the preservation and enhancement of every kind of biological life. Another aspect of this tradition rooted in Aristotle but brilliantly developed by Aquinas points out in right reason the essential need for each and every human being to act so in rhythm with the basic harmonies of both mind and matter, soul and body as to further the health and well-being of all. Genes are living maps, embryos already formed bodies.

Two factors in the Cartesian philosophical revolution argue for bifurcations. The split between mind and body in prioritizing the former allow more for the manipulation of the latter. A fascination with categorisation and analysis facilitates the bracketing of both mental and biological units for separate consideration and care. Since a person is basically mind even emotions are of less importance and subservient. Since obviously an embryo cannot think, it is not a human person and so can be dealt with in a somewhat cavalier way.

PART 5

SUFFERING AND DEATH

HENK A.M.J. TEN HAVE

SUFFERING AND DEATH

Introductory Comments

Francisco de Quevedo published his famous *Sueños* in 1627. As a contemporary of Cervantes and Calderón, he is one of the giants of Spanish literature. The fifth dream of this fictional story is the Dream of Death. The author reports how he is taken by Death herself to her court-room. Here, deceased persons are judged before (more often than not) going to hell. The first group he encounters in this dream are the physicians. They are blamed because their ultimate goal is to make healthy persons ill and to take care that the ill will never recover. Death herself explains to the narrator that all people eventually die from the effects of medical treatment. The correct answer to the question why someone has expired, is that he died of doctor X or Y who treated him, rather than from typhoid or fever. In short, physicians are simply 'graduated poison'.

Quevedo's stories point to an important fact. For a long time during its history, medicine was not only powerless in the face of death, but also it could make the patient's situation worse. Patients did not trust their doctors, not only because of lack of knowledge and incompetence, but also because the drugs that they control could as easily be beneficial as toxic and killing. Hence for a long time it was conventional wisdom: when you are ill, never call a physician because then you will get another illness.

This ancient situation dramatically changed since the second part of the nineteenth century. Based on the models and methods of the natural sciences, medicine transformed into a powerful, scientific discipline. The promise of Francis Bacon and René Descartes, contemporaries of Quevedo, that medical science in future can cure diseases and prolongate human life, was gradually materialized.

However, the success of modern medicine is double-edged. Since medicine can cure disease and prevent illness, it has grown into one of the major determinants of improved health, longer life and better quality of existence. But at the same time, medicine can sustain human life when there

H.A.M.J. ten Have and B. Gordijn (eds.), Bioethics in a European Perspective, 407-409.

is no longer any chance of recovery. It can postpone death beyond the point where existence is worthwhile for the patient.

Issues of death and dying have been on the agenda of bioethics from its earliest days. The most significant moral aspects will be addressed in this part of the book.

In the first chapter Wim Dekkers focuses on images of death and dying. In historical perspective, current medical approaches and attitudes towards death are new. Uncovering philosophical and theological thinking about death does not provide an archaeology of ideas, but may help us to better understand our present situation. Dekkers analyses various attitudes towards death in Western culture. He questions whether today we have really learnt to deal with the perennial fears and perplexities of death. We are in need of a new *ars moriendi*. But we also need more abilities to care for the dying in a humane way.

Zbigniew Szawarski from Poland, specifies in the second chapter the fundamental ethical issues in terminal care. He differentiates between several kinds of death, and shows how each kind has its specific implications for terminal care and moral evaluation. Although death has always been the most basic and natural event in human life, we have competing views on what is the good for dying persons. Szawarski distinguishes two medical strategies in the face of death. On the one hand, we want to affirm life and do everything to preserve it; on the other hand, we want to affirm death.

Medicine usually chooses the first strategy, trying to preserve human life as long and vigorously as possible. But, one of the major issues in bioethics concerns situations where the focus on prolongation of life is morally problematic. Medical interventions may create problems because continuation of treatment is no longer desirable or has become futile. Franz Josef Illhardt from Germany addresses in the third chapter the question of limits to medical treatment, even when withholding or withdrawing treatment implies that the patient dies. He analyzes the moral criteria and arguments as well as the decision-making process; terminating treatment is in fact an essential component of moral medical practice.

The strategy of affirming death is one of the most poignant controversies in today's bioethics. Is it ethically justifiable that physicians actively bring about the death of a patient in particular circumstances? In the final chapter of this part of the book, ten Have discusses the topic of euthanasia. He specifically focuses on the Netherlands, until now the only country where euthanasia is not only widely practised, but also to a certain extent legally regulated. In the debate on euthanasia in general, two moral arguments play a major role: the voluntary request of the patient, and the suffering of the patient. The interplay of these arguments is cause for moral concern since it leads to a gradual expansion of the practice of medically assisted death to various categories of incompetent patients.

The case analyses in this part illustrate the various moral problems discussed in the chapters. Marcel Verweij from the Netherlands comments upon a case of a do-not-resuscitate order, analyzing various criteria of foregoing life-sustaining treatment. Rien Janssens from the Netherlands concentrates on the issue of pain management. Improved palliative care is usually regarded as a way to prevent requests for euthanasia. Whether or not this will be successful, depends very much on the quality of the support and care for our dying fellow human beings. The case of a patient in persistent vegetative state is finally presented and discussed by Roberto Mordacci from Italy. It is a final occasion to review the basic notions of this area of bioethics: the wishes of the patient, sanctity of life, quality of life, and proportionate treatment.

WIM J.M. DEKKERS

IMAGES OF DEATH AND DYING

1. INTRODUCTION: WHOLE DEATH

Though death is usually easy to recognize, it is difficult to define. Perhaps the most significant of all changes which have occurred in the understanding of death during the past decades is the alteration in the definition of death itself. The modern debate about death and dying is mostly about how to define death, how to develop death-criteria, and how to develop tests which are appropriate to meet those criteria (Dekkers, 1995). Nowadays, more than ever before we have to face the saying: "mors certa, hora incerta": we can be sure of death, but not of the precise moment of death. The need to assess the exact moment of death, for example for transplantation purposes, is greater than ever before. The discussion is centred around distinctions between biological, personal and social death, between death of the organism as a whole and death of the whole organism (including all tissues and cells), between natural death and non-natural death, and between whole brain death, neocortical death and brainstem death. Although in most Western countries the whole brain death definition has become generally accepted and is invariably included in the legislation of those countries where organ transplantation is practised, the question whether brain dead people are 'really' dead is still heavily discussed.[1]

A second characteristic of the modern debate about death is the preponderance of legal and ethical discussions about decisions to make at the end of life. A huge amount of literature exists about legal and ethical aspects of euthanasia, physician-assisted suicide, withholding or withdrawing treatment, and the ending of life not on request. According to Callahan, Western society is more comfortable with a legal than with a philosophical or religious discourse, and more at ease with moral language focused on the making of decisions than with the wisdom of those decisions. He writes:

> The debate has mainly been about law, regulation, moral rules, and medical practice, and about making legal, or ethical, or medical choices about dying. It has not been about death itself, about how we should think it through in our lives (Callahan, 1993, p. 13).

H.A.M.J. ten Have and B. Gordijn (eds.), Bioethics in a European Perspective, 411-431.

Like Callahan, I believe that we need to reflect on death itself, trying to find meaning in it, if indeed there is meaning to be found at all. Although the term 'death itself' is problematic because it is begging the question of what death itself is, this problem will be left aside in this chapter. The starting point for my reflections on death and dying is our intuitive knowledge of death: death as a total phenomenon which we all know from daily experience, death which has been a puzzling phenomenon or even a mystery for every human being and which has been a central issue in religion, philosophy and theology throughout the ages. If one focuses on death as a total phenomenon, it is difficult to distinguish clearly between thoughts about the process of dying on the one hand and about what comes after death (be it a hell, purgatory, heaven or 'nothing') on the other hand. These two issues are closely related. I prefer to call death which is intuitively known as a total phenomenon 'whole death'. Death in this broad sense of the word is recognised as universal and inevitable. Death is the individual and collective fate of human beings and of all living creatures. Questions which arise in this context are: What is the value of death? What does it mean? Is death final? Is death a good thing or a bad thing? Can life be meaningful if it ends in death? (Momeyer, 1995). Death in this broader context brings us to the field of theology and philosophy, but it is especially in religion that death is conceptualized. However, it is important to note that the answers to the questions just mentioned are a matter of belief and conjecture. There is no way to test any religious concept of death. No religion can blame another one for not complying with the so-called facts (Kuitert, 1998).

The aim of this chapter is to shed some light on our current view of death and our attitude toward it, especially with regard to the way in which death is approached in the practice and theory of medicine and health care. From a historical point of view, the way we deal with death in our modern society is quite new and unprecedented. First, technological developments in modern medicine have provided us with the means to keep seriously ill people alive and to postpone death. Secondly, the influence of so-called 'Western activism' has led to the tendency not only to postpone death, but also to hasten death in situations where natural death comes, supposedly, too late. Thus, the argument is that we can learn a lot from history about forgotten, yet meaningful attitudes toward death. Therefore, it is necessary to analyze some general-philosophical concepts of death (section 2) and some cultural-historically defined attitudes toward death (section 3). Then, a cultural-philosophical diagnosis of the way in which we approach death at the end of the twentieth century is presented (section 4). We are in need of a new *ars moriendi* and in this respect we may learn a lot from history. Finally, the metaphor of 'coming home' will be discussed, which may be fruitful in developing such a new *ars moriendi* (section 5).

2. PHILOSOPHICAL CONCEPTS OF WHOLE DEATH

Since time immemorial, the meaning of death and dying has been the subject of philosophical and theological scrutiny in every culture. From a philosophical perspective, recognizing the inevitability of death is different from supposing death is final. At a general level philosophical reflections on death divide those who deny the finality of death, and suppose there is some form of life or consciousness after death, from those who regard (physical) death as final, as the total termination of biological and mental life. Probably the single most persistent theme in Western philosophical reflection on death is the view that death is not the termination of the self, but its transformation into another form of existence. The conviction that individual human beings survive death, perhaps eternally, has been very differently elaborated in the history of philosophy, but in some form has persisted and frequently dominated throughout Western civilisation (Momeyer, 1995).

Throughout history philosophers have dealt with the themes of death and dying in many different ways. If we take a bird's-eye view of the rich history of philosophical thinking about death, we can distinguish between four general philosophical attitudes that can be taken with respect to death (Douma, 1998): (1) death is repressed or denied, (2) death is experienced as a liberation, (3) death is viewed as the source of life and (4) death is taken under one's own control.

2.1. Death Repressed or Denied

It is possible to repress death by claiming that it has no significance. The Greek philosopher Epicurus (341-270 BC) famously argued that death cannot be regarded as a bad thing. He saw no reason to fear death, believing that in death the soul, composed of the finest atoms, simply dissipated, so that there was nothing left to have experiences. In his letter to Menoeceus he writes:

> Become accustomed to the belief that death is nothing to us. For all good and evil consists in sensation, but death is deprivation of sensation. And therefore a right understanding that death is nothing to us makes the mortality of life enjoyable, not because it adds to it an infinite span of time, but because it takes away the craving for immortality. ... So death, the most terrifying of ills, is nothing to us, since so long as we exist, death is not with us; but when death comes, then we do not exist. It does not then concern either the living or the dead, since for the former it is not, and the latter are no more (Epicurus, 1926, p. 85).

Tolstoy's Ivan Ilyich could not accept that he was a mortal being and consequently had to die. Realizing that he had only a few weeks to live, he posed the following questions, which may be considered as a late echo of the saying of Epicurus: "I shall be no more, then what will there be? There will be nothing. Then where shall I be when I am no more?" (Tolstoy, 1960, p. 135). In the depths of his heart Ivan Ilyich knew he was dying, but he simply did not and could not grasp it. The syllogism which he had once learned "Caius is a man, men are mortal, therefore Caius is mortal" (Tolstoy, 1960, p. 137) had seemed to him all his life true as applied to Caius but certainly not with respect to himself. That Caius - man in the abstract - was mortal, was perfectly correct. But he was not Caius, nor man in the abstract. It is right for Caius to die,

> But for me, little Vanya, Ivan Ilyich, with all my thoughts and emotions - it's a different matter altogether. It cannot be that I ought to die. That would be too terrible (Tolstoy, 1960, p. 137).

Ivan Ilyich tried to drive the false, erroneous, morbid thought of his own mortality away and supplant it with other proper, wholesome thoughts. He spent most of his time in attempts to restore the mental balance which had death kept out of sight. But in the end he failed in doing so. The idea of being mortal and as it were the reality itself kept coming back again and confronting him. Even worse than his physical sufferings were his mental sufferings which may illustrate that anyone who is guided by the Epicurus' sophism deceives himself.

2.2. *Death as Liberation*

Death can be experienced as a liberation, that is, as the liberation of the soul released from the body. In Plato's dialogue *Phaedo* death is described as a separation of body and soul and as a purification of the soul. The soul is imprisoned in the body during earthly life and is liberated at death. The significance of human existence lies in the 'excarnation'. Corporeality hinders man and therefore being freed from the body is a liberation. Thus, Plato's view of death is inseparable from his doctrine of the soul, and his identification of the soul with personhood. The soul is immortal and is independent from the human body. It returns after physical death to where it originally comes from, the world of ideas. In Plato's judgements in *Phaedo*, intellectual pursuits are the most noble, but these are consistently hindered by bodily appetites and bodily limitations of sensory experience. Hence, Plato argues, the true philosopher aspires to death, and lives to die, in the expectation that only the soul's liberation from embodiment will make possible the fullest attainment of knowledge. A philosopher is not so much concerned with the body, but keeps his attention as much as he can away from it and toward the soul. Socrates assures Simmias,

> That true philosophers make dying their profession, and that to them of all men death is least alarming (Plato, 1978, p. 50: *Phaedo*, 67e).[2]

On the one hand, the Christian tradition has been strongly influenced by Platonic dualism. The depreciation of physical life as lower compared to the higher and immortal life of the soul has had enormous consequences within Christianity (Douma, 1998). On the other hand, however, within Christianity there has always been an impediment to a full application of platonic dualism. If the Bible, for instance, speaks of the resurrection of the dead, this means that body and soul can be separated, but not placed over each other. It is not the 'excarnation', but the 're-incarnation' confirmed by the resurrection of Jesus Christ, that is central in Christian faith. Thus from a Christian perspective, death may be considered a liberation. It is, however, not a liberation from the physical, but from the sinful and consequently temporal life of man.

2.3. Death as the Driving Force of Life

In the third place, death may be experienced as the driving force of life. In this context, the German philosopher Heidegger must be mentioned. Central in his conception of death is a particular attitude toward life. In *Being and Time* he employs death to characterize the dynamic wholeness of human existence (*Dasein*) (Heidegger, 1962, sections 46-53). Death is a necessary element of human existence. Human existence is essentially characterized by the way that it is 'being-toward-death' (*Sein zum Tode*). As long as death has not occurred, human existence has not yet reached his end, since there is a 'not yet' still to come. Death brings life to an end and makes man's existence into a dynamic whole. Death represents that point in life at which the existence as a whole can be investigated (Macann, 1993, p. 96-97). Heidegger criticizes the notion of death as something totally distinct from life. In his view, death is not just an event or the end of all events, it is also an attitude toward life. He considers death as an anticipated possibility at every moment of life. Real human life is only possible if and insofar as human beings are aware of this final possibility of life. Death is a task that every human being has to fulfil, him or herself.[3]

However, not to think about death is the most common attitude that man adopts toward the end of his life. People continually try to escape from the death-linked 'real' existence into a 'lower' form of existence and therefore do not have the courage to be as they 'really' are. This 'lower' form of existence refers to being human, when it has lost sight of what it is to be itself, so much that it is capable of doing and thinking only what 'they' do or think or what 'one' does or thinks.[4] In such a situation the self has lost its selfhood, has ceased to be itself and has become what others want it to be. People who do have this courage may perceive what human life really is. In this sense, death may be considered as the source of life. Death fully belongs to life. It is an indispensable factor in being able to live our lives.

There is a certain analogy between Heidegger's ideas and the Christian faith. In both at least, death is taken seriously, but it occurs in a very different way. With Heidegger death is bound up in the ontological structure of being human. According to the Christian faith, however, death has intruded as an enemy and therefore does not ontologically belong to human life. From a Christian standpoint, one can hardly make death the motor of life. It is perfectly human to loathe death. From a Christian perspective death is not an indispensable power, making life human (Douma, 1998).

2.4. Death Taken under one's Own Control

In the fourth place death may be considered a phenomenon that man ought to take personal control of. A well-known example of this attitude towards death is the death of Socrates which is reported by Plato (1978) in the dialogue *Phaedo*. As we have seen, Socrates regards death as a transfer of the soul from one stage to another. In the dialogue *Apology* Socrates is tentative in his assertions about death. There he says:

> For let me tell you, gentlemen, that to be afraid of death is only another form of thinking that one is wise when one is not; it is to think that one knows what one does not know. No one knows with regard to death whether it is not really the greatest blessing that can happen to a man, but people dread it as though they were certain that it is the greatest evil, and this ignorance, which thinks that it knows what it does not, must surely be ignorance most culpable (Plato, 1978, p. 15; *Apology* 29a-b).

Later, having been sentenced to death, Socrates argues that death is either dreamless sleep from which we do not awaken or transport to a place where we might ever after commune with those who preceded us in death. In his view, the first situation (the termination of the self) is not fearsome, the second one (in which one has contact with other people) is to be joyfully celebrated.

> Death is one of two things. Either it is annihilation, and the dead have no consciousness of anything, or, as we are told, it is really a change - a migration of the soul from this place to another. Now if there is no consciousness but only a dreamless sleep, death must be a marvellous gain. ... If on the other hand death is a removal from here to some other place, and if what we are told is true, that all the dead are there, what greater blessing could there be than this, gentlemen? (Plato, 1978, p. 25; *Apology* 40d-e).

At the end of his defense when he has expressed his readiness to die, Socrates uttered his deepest and most influential conviction about life and death as follows:

> You too, gentlemen of the jury, must look forward to death with confidence, and fix your minds on this one belief, which is certain - that nothing can harm a good man either in life or after death, and his fortunes are not a matter of indifference to the gods (Plato, 1978, p. 25; *Apology* 41c-d).

The consequence of Socrates' persisting in his refusal to give up his principles was that he was condemned to death and was forced to drink the poisoned cup. Thus, he deliberately chose death. In this sense one can say that he had control over his own death. Socrates' attitude during his trial and execution was a model and inspiration for Stoics of all eras. According to Seneca (ca. 3-65 A.D.) and Epictetus (ca. 60-100 A.D.) man must take control over death. Man chooses for life, but man can also choose, if desirable, for death. In their view, suicide may be seen as the ultimate consequence of a person's autonomy and autarchy.

3. CULTURAL ATTITUDES TOWARD DEATH IN HISTORY

If one tries to take a bird's-eye view of man's attitudes toward death from a cultural-historical perspective, one may distinguish between three 'ideal-typical' attitudes (Beerling, 1976): the fatalistic attitude, the activistic attitude, and the indifferent attitude. The fatalistic attitude toward death, which supposedly predominated in the Middle Ages, accepts death as an inevitable phenomenon. One dies when one's time has come. The activistic attitude tends to try to prolong life and postpone death using whatever means are at one's disposal. This is a not unfamiliar approach to Western citizens at the end of the twentieth century. An indifferent or neutral attitude tends to deny death and its significance for human existence. Here Epicurus' view should be mentioned.

If one wants to explore the Western image of death at the end of the twentieth century, one cannot ignore the work of the French historian Philippe Ariès (1914-1984). He is one of the most important authors in the last two or three decades on this subject. In *Western Attitudes toward Death* Ariès (1974) distinguishes between four images of death: Tamed Death, One's Own Death, Thy Death and Forbidden Death. In a later study, *The Hour of Our Death* (1982) he speaks about five models of death: The Tame Death, The Death of the Self, Remote and Imminent Death, The Death of the Other, and The Invisible Death. By and large, however, the overall argument of his analyses has not changed. It is his intention to cast an eye over a thousand-year landscape like an astronaut looking down at the distant earth. His hypothesis is that there is a relationship between "man's attitude toward death and his awareness of self, of his degree of existence, or simply of his individuality" (Ariès, 1982, p. 602). Notwithstanding the fact that Aries' approach raises considerable methodological and historiographical questions, his analyses provide a great deal of insight.[5]

3.1. Tamed Death

The image of death as a tamed death covers a period of approximately a thousand years. Central to this image is a familiarity with death which implies a sort of acceptance of the order of nature and of the collective destiny of mankind. This attitude can be summarized in the phrase *memento mori*. Another important item and a recurrent motif in medieval art is the personification of death. Often death is represented as a skeleton with an hourglass or with a scythe. Sometimes death is an enemy who violently takes the living from their life. Sometimes he is a friend who confers with the living about the right time in which to go.

Ariès mentions the following characteristics of 'tamed death'. First, man is usually forewarned. Medieval man does not die without having had time to realize that he is going to die. In the literature, one can often find sentences like: "he knew that his death was near" or "he felt that his time had come". The warning comes through natural signs or through an inner conviction. It is a sort of spontaneous realisation. Knowing that the end is near, the dying person prepares for death. One is not afraid of death, but afraid of not being forewarned and of dying alone.[6] Secondly, one awaits death lying down. This ritual position was stipulated by the thirteenth century Christian liturgists. The dying man must lie on his back so that his face is always turned toward heaven. Thirdly, death is a ritual organized and presided over by the dying person himself. The ritual of the dying is carried out in a ceremonial manner. Dying is a public ceremony. The dying man's bedchamber becomes a public place to be entered freely. It is essential that parents, friends, and neighbours are present. A fourth important aspect of familiarity with death is the coexistence of the living and the dead. This is a new phenomenon, unknown in pagan Antiquity and early Christianity. Despite their familiarity with death, the ancients feared being near the dead and kept them at a distance. The world of the living had to be kept separate from that of the dead. From the sixth century onward there is a change into what Ariès calls "the promiscuity between the dead and the living". The coexistence of the living and the dead finds its expression in the art-historical theme of the 'dance of death'.

3.2. One's Own Death

In the period of 'One's own death' beginning in the eleventh century, the traditional familiarity with death does not disappear but alters partially. Gradually a personal meaning for man's traditional familiarity with death emerges. There is an increasingly close relationship between death and individual life. In the mirror of his own death each man discovers the secret of his individual self. Ariès mentions three characteristics.

First, the idea of the Last Judgement is that Christ will return at the end of the world. The dead who entrust their bodies to the care of the Church go to sleep and are at rest until the day of the great return, the Second Coming of Christ. On the Last Day there will be a resurrection of the dead. In the twelfth century the scene changes. The apocalyptic inspiration and the evocation of the Second Coming of Christ are almost blotted out, and room is made for the idea of individual judgement. In early Christian times no place was given to individual responsibility, to counting good and bad deeds. From the twelfth century on, however, each man is to be judged according to the balance sheet of his own life. Moreover, a deep-rooted refusal to link the decay of the body with the end of physical being began to exist. In this period, belief in physical life after physical death predominated. The physical existence after death was not believed to continue for eternity, but only to the end of the world. From that period on, the idea of the Last Judgement has been linked to that of individual biography. This biography ends not at the hour of death, but on the Last Day. The book of life is closed not at the moment of death, but on the last day of the world, at the end of time.

A second phenomenon from this period is that the time between death and the end of the world gradually becomes shorter. The individual judgement is no longer situated at the Second Coming of Christ, but in the bedchamber, around the deathbed. This results in books on the proper manner of dying. The *ars moriendi* (the art of dying) identifies an important subgenre of medieval literature of the fifteenth and sixteenth centuries. These little manuals on dying are meant to guide the reader's conduct in his hour of death.

The third phenomenon regards the appearance of the putrefied cadaver in art and literature. The horror of physical death and decomposition is a familiar theme in fifteenth and sixteenth century poetry. Decomposition is a sign of man's failure. Man of the late Middle Ages is very conscious that he has merely been granted a short stay, that the delay between birth and death can be a brief one, and that death is ever-present.

3.3. Thy Death

From the eighteenth century onward, Western society has tended to give death a new meaning. Until this period the emphasis was on familiarity with death and with the dead. Death was most of all a concern of the dying person and of him alone. From the eighteenth century on, man is more concerned with the death of others than with his own death. The dying person still retains the initiative in the ceremonies surrounding his death, but the attitude of those present changes. Although the dying person retains the leading role, the bystanders are no longer the passive, prayerful observers of the past. A new intolerance of separation gives rise to the expression of sorrow. People are troubled not only by being at the bedside of the dying or by the memory of the deceased, but also by the very idea of death.

The memory of a much-loved person and the loss of that person inspires a new cult of mourning, tombs, cemeteries and a romantic approach to death in the nineteenth and twentieth century. Ariès calls the nineteenth century the era of mourning. Survivors accept the death of another person less easily than they did in the past. The death which is feared is no longer so much the death of oneself as the death of another person. These feelings also lie at the origin of a cult of memory. The tombs of the dead begin to serve as a symbol of their presence after death. The common desire is to keep the dead at home by burying them on the family property, or else to be able to visit them in a public cemetery. People go to visit the tomb of a dear one as one would go to a relative's home.

3.4. Forbidden Death

The phase 'forbidden death' emerges in the twentieth century. From this time on death becomes wild and, although omnipresent in the past, disappears. Death becomes shameful and forbidden, an idea which evolves rapidly between about 1930 and 1950. According to Ariès, this is primarily due to a change in the place of death. One no longer dies at home in the bosom of one's family, but in the hospital. The hospital has become the place to receive care which can no longer be given at home. Death in the hospital is no longer a ritual or ceremony, over which, amidst his assembled relatives and friends, the dying person presides.

The funeral rites are modified as well. One tries to reduce the unavoidable operations needed to dispose of the body to a minimum. That death has occurred should be as invisible as possible to neighbours, friends, colleagues and children. Dark clothes are no longer worn. In the case of cremation, the decision is more than just a desire to break with Christian tradition. (Re-)introducing cremation can be seen as a manifestation of enlightenment, of modernity. According to Ariès, however, the deepest motivation to cremate is that cremation is the most radical means of getting rid of the body.

This flight from death cannot be attributed to an indifference toward the dead person. In reality the contrary is true. The forbidding of public manifestations of sorrow, the obligation to suffer secretly and alone aggravates the trauma stemming from the loss of a loved one. The English sociologist Gorer (1965) speaks about 'a conspiracy of silence' about death, as though it were a pornographic subject. According to him death has become a taboo and has in the twentieth century replaced sex as the principal forbidden subject. The more liberated from the Victorian constraints concerning sex society is, the more it rejects things having to do with death. There is an almost total suppression of everything reminding us of death.

4. DEATH AT THE END OF THE TWENTIETH CENTURY

In the same vein as Ariès, many other authors have criticized the attitude toward death in the second and third quarter of the twentieth century. Although Elias strongly criticizes Ariès' historiographic methodology (see note 5), his own reflections share a great deal with those of Ariès. According to Elias (1985), the isolation of the ageing and dying seen in modern societies is precisely one of the weaknesses of those societies. Elias mentions four special features of these modern societies which shape the image of death and what he calls the 'social repression of death' (Elias, 1985, p. 55). First, the length of individual life has increased considerably. In a society where the average life expectancy is seventy-five years, death is more remote for a person of twenty or even thirty than in a society where the average life expectancy is forty. It is understandable that the longer the life expectancy becomes in a society, the easier it is to avoid thinking about death during one's own life. Secondly, death is seen as more of a final stage of an artificial process than as a natural consequence. This way of experiencing death has gained significance through progress in medical science and technology. Thirdly, modern developed societies show a relatively high degree of so-called 'internal pacification'. When people in these societies imagine the dying process, they probably think, first of all, of a peaceful death in bed resulting from illness and old age. As a fourth and most important feature of developed societies affecting this attitude toward death, Elias mentions their high degree of individualisation. In these societies, people generally see themselves as fundamentally independent and individual beings, as isolated subjects. Elias calls this specific way of experiencing oneself the self-image of the *homo clausus*. The idea that one dies alone emphasizes the feeling that one lives alone. The image of one's own death is closely related to the image of one's own life.

The sociologist Prior (1989) has analyzed the 'construction' of death at the end of the nineteenth and beginning of the twentieth century in Belfast, in daily life, medical sciences, medical practices, and funeral practices. He argues that death in modern Western society has been hidden, isolated, privatized, bureaucratized, medicalized, hospitalized and 'dehumanized' (*ibid.*, p. 15). According to him death is hidden in the folds of everyday consciousness. This does not mean that death is totally absent. On the contrary, death is an important object of scientific research and, as such, is neither hidden nor forbidden in areas of studies such as demography, pathology and sociology. Thus, instead of arguing that death is forbidden - as Ariès does - we can better say that death is nowadays visible only through an objective and scientific language.

What can we now learn from the thoughts of authors like Ariès, Elias and Prior? Do their analyses and insights also apply to the present situation? What is our modern view of death at the end of the twentieth century? With respect to questions like these, I confine myself to two points. First, the ongoing medicalisation of death. If it is true that death is still medicalized, this would imply that we are still in the fourth phase described by Ariès, that is, the phase of 'forbidden death'. Secondly, however, there is the question of whether death is still as much of a taboo as it was some decades ago and whether we may possibly add a fifth phase to the four stages described by Ariès.

4.1. Medicalisation of Death

In most western countries and urban centres of the world, death has become highly institutionalized by medical, social and associated volunteer and professional organisations. Many authors point to an increasing medicalisation of death, which signifies that death, first of all, has become the responsibility of the medical profession.[7] The medicalisation of death appears in - among other things - the change in the place where people die. The most common place of death in the United States is the hospital or nursing home (Corr, *et al.*, 1994, p. 19). In the United Kingdom 54% of deaths occur in a hospital, 23% at home, 13% in a nursing home, 4% in a hospice and 6% somewhere else (Field and James, 1993). In the Netherlands these figures are only slightly different: 40% die in a hospital, 26% at home, 13% in a nursing home, 17% in a rest home or home for elderly people, and 4% somewhere else (van den Akker, 1994). Only a few people die without ever being diagnosed and treated for a particular disease. According to the Dutch Remmelink Report, in 30% of all cases of death in the Netherlands (40.000 out of 130.000 people), some medical decision at the end of life has influenced the course of the disease process and the occurrence of death (van der Maas, *et al.*, 1992). These figures suggest that death is medicalized and increasingly under the influence of the expansion of medical power.

Due to developments in medical technology, death has become a technical process, a process which is controllable and manageable. Technology is a constitutive factor in modern medicine and health care. It is part of our Western activism. We tend to control not only the birth of our children, the diseases by which we will be affected, the form of our body, but also the time and 'nature' of our death. There is a quest for total control of death. Phenomena and notions such as 'desired death', 'death by choice' and 'a right to die' have gained a special meaning. This meaning can only be understood by referring to this atmosphere of 'managing death'. Up until the nineteenth century, medicine hardly played a role in the constitution of the image of death. In those times, the doctor was usually absent at the hour of death. Nowadays, medicine has amply made up the difference. This development had started already in the beginning of the twentieth century, but nowadays death is seen from a medical perspective more than ever before. Dying and death are medicalized (Burgess, 1993). In the course of time medicine has developed from a rather marginal societal phenomenon into a most important cultural achievement. From an art to heal the sick and wounded, medicine has grown into a technologically oriented science, a science which increasingly penetrates our daily lives.

Illich has argued that the detrimental effects of modern medicine constitute an epidemic of clinical, social and cultural iatrogenesis. Cultural iatrogenesis means that professionally organized medicine undermines "the ability of individuals to face their reality, to express their own values, and to accept inevitable and often irremediable pain and impairment, decline, and death" (Illich, 1976, p. 133). Cultural iatrogenesis encompasses the dependence of people in modern industrial societies upon medical care to solve all their problems. Also death has come to be seen less as an inevitable part of life and more as a failure of treatment. Many physicians believe that a patient is dying not because of the disease he or she is suffering from, but because there are no further medical or technological strategies available to keep the patient alive. Callahan writes:

> Death is not construed as an inevitable biological denouement but as a medical failure ... Death has been moved out of nature into the realm of human responsibility (Callahan, 1993, p. 64).

We have made mortality itself our fault, so to speak, our responsibility. No death is 'natural' any longer.

Callahan is right when he argues that although death is omnipresent in the practice of medicine, it has no well-understood place in medical theory (Callahan, 1993, p. 14). Describing the image of death in modern technologically oriented medicine, he explicitly deals with Ariès' notion of tame death. According to Callahan, Ariès' story about the past is incomplete (Callahan, 1993, p. 52), but in his view, generally speaking, Ariès account of tame death is plausible and believable (Callahan, 1993, p. 28). Callahan introduces the notion of a 'wild death'. Wild death is not yet tamed. Wild death is "not only a technological death, but a hidden, dirty death, one that is shunned, feared, and denied" (Callahan, 1993, p. 30).

4.2. Is Death Still a Taboo?

Despite this ongoing process of medicalisation and professionalisation, there is another, wider cultural development which deserves our attention, that is, a gradual change in our attitude toward death. Although for many people death still remains a taboo (Walter, 1993), there are signs that this taboo is crumbling away. A counter-movement against the medicalisation and professionalisation of death has emerged.

For about two or three decades, attempts have been made to face death in a more proper manner than before. Dying patients are no longer hidden away in hospitals. Death is finding an increasingly central place in society and in our personal lives. We are becoming more and more aware of the fact that death is an undeniable, inevitable event from which we cannot hide and that facing death may ease our own dying and enrich our lives. In this context the hospice-movement has to be mentioned. This movement was set up by Dame Cicely Saunders at the end of the sixties with the intention of providing dying patients a more human way of dying than is possible in a highly technologically oriented medicine and to give dying patients their place back in society. "Neither to hasten death, nor to postpone death" is one of the central guidelines of the hospice movement (Saunders, 1993). At the moment hospices and other practices of palliative care are established in many Western countries to take care of terminal and dying patients.

Funeral rites are gradually changing as well. The interest in rituals surrounding death has grown and the public is increasingly interested in funeral practices. Funeral exhibitions and shops are shooting up like mushrooms. There is a growing desire to lay out the body of a deceased loved one oneself and to place the dead body on a bier at home. The management of funerals is slowly becoming a more 'human' praxis. An increasing number of people prefer an individual and personal burial. In particular, patients with AIDS have contributed to these developments. Often they express clear wishes about the way the burial or cremation of their own dead body should take place. Though by now cremation is widely accepted in different religious traditions (while for many centuries cremation was considered a pagan custom), the percentage of cremations has not increased appreciably in the last four decades. The interest in rituals surrounding death has grown and thus the interest in burials as well. It is often felt that burials meet the needs of a ritual better than cremations do. Furthermore, it seems as if some elements of Ariès' third phase (the phase of the death of the other) are coming back again. "The Netherlands gradually abandons its closed cult of mourning", was the headline of a recent article in a Dutch newspaper (*NRC Handelsblad*, July 23, 1996). Nowadays it is accepted that neighbours, friends, colleagues and children may notice that a death has occurred. Public mourning seems to be accepted again.

Callahan argues that, if Ariès were alive today, he would not find much reason for a revised judgement about the denial of death (Callahan, 1993, p. 34). However, the above considerations may lead to the conclusion that death is no longer clearly 'forbidden' as described by Ariès and no longer 'wild' as described by Callahan. There are signs that for one or two decades now a fifth phase has been emerging, which includes various elements of the first three phases which were described by Ariès. We are at the beginning of a new period concerning our attitude toward death and that we are as a consequence in need of a renewed art of dying (*ars moriendi*).

5. TOWARD A NEW *ARS MORIENDI*

The gradual disappearance of the taboo regarding death is a favourable development, but it is just the beginning. We have the assignment of mastering death in a new and adequate way. A new art of dying is called for here (Post, 1993). Medicine and health care may play an important role in looking for such a new understanding of death. As Callahan (1993) argues, medicine should promote both a good life and - what he calls - 'a peaceful death'. He distinguishes between the older tame death and a peaceful death. The difference between a tame death, as described by Ariès, and a peaceful death, as advocated by Callahan, resides in the fact that technological advances allow for the management of the moment and circumstances of death. A peaceful death combines the advantage of a tame death with the contemporary possibilities provided by palliative care to soften and control death.

With regard to a new *ars moriendi*, however, it is important to widen the scope from medicine and palliative care to culture. There is much truth in the argument of authors like Callahan that Western culture places too much emphasis on the ideals of autonomy, independence and self-sufficiency. What is meant here is the neoliberal form of autonomy which has been emphasized so much in modern health care and health care ethics. However, autonomy may also stand for being aware of the fragility of human existence and for accepting one's own mortality. This notion of autonomy can be opposed to the predominant, neoliberal idea of individual autonomy which is considered to be strongly connected with being independent.

In this respect it is worthwhile to rethink the notion of autonomy which is taken from classic Greek-Roman philosophy, in particular as it is described by Seneca in his views on human finiteness. Seneca was very much influenced by Socrates' thoughts on death and his notion that "to philosophize is to know how to die". According to Seneca (1979), human finiteness derives from man's corporeal existence. Corporeal beings are mortal by nature. Becoming human means learning to live consciously as finite and mortal beings. It is everyone's task to find an adequate attitude towards the transient and changeable dimensions of human life. Independent and autonomous is the individual who can give meaning to his or her own mortality and who knows how to incorporate the finiteness of his or her existence into daily life. Dependent is the individual who does not recognize or denies the limitations of his or her corporeality.

Referring to Seneca's view of autonomy, Manschot (1992) suggests 'Socratic autonomy', an autonomy which has been described in Plato's dialogue *Alcibiades*.[8] Manschot considers 'Socratic autonomy' or 'real' independence the conscious and mature designing of man's finiteness and mortality. In the Socratic view, people who seem to be dependent because they need care from others can be preeminently independent and autonomous because they are thoroughly confronted with the fragility of human existence.

We are currently faced with multiple representations of what makes a death a 'good' or 'bad' death (Bradbury, 1993). Yet the notions of 'Socratic autonomy' and 'peaceful death' form an excellent starting-point to develop an adequate image of death and a new way of dying. However, despite what Stoic philosophers think, the notion of 'Socratic autonomy' does not necessarily imply the moral acceptability of suicide, assisted suicide or euthanasia. Furthermore, one can have a peaceful death only if one is also taken care of by others in the dying process. Thus, the right balance between care for one's self and care for the other is at stake here. In the next section a metaphor will be presented which may be helpful in the search for that balance and in bridging the gap between an (abstract) image of death and the practice of medicine and health care, in particular of palliative care.

5.1. Coming Home

Metaphors are among the most important tools for trying to comprehend partially what cannot be comprehended totally. Metaphors may help us to construe a conceptual image of matters, affairs and situations which are difficult to describe in a more exact, rational way. The essence of a metaphor is understanding and experiencing one kind of thing in terms of another (Lakoff and Johnson, 1980, p. 5). Metaphors are pervasive not only in everyday language and thought, but also in action and everyday activities. No metaphor can ever be comprehended or even adequately represented independently of its experiential basis (*ibid.*, p. 19). Thus, metaphors are capable of giving us a new understanding of our experience. Metaphorical thinking may be described as 'imaginative rationality' (*ibid.*, p. 235). In particular, new metaphors are capable of creating new understandings and, therefore, new realities and practices.

A metaphor which, on the one hand, may ease our understanding of the phenomena of death and dying and, on the other hand, may lay the foundations of a new way of dying is 'coming home'. The goals of medicine and health care, in particular those of terminal and palliative care, can be adequately described in terms of 'coming home' or 'being at home'. The goal of palliative care, then, should be 'to bring the patient home safely'. In the following this metaphor will be elucidated by recalling five possible meanings of the expression 'coming home'. The presupposition here is that care for the dying should be an integral part of the care of our fellow human beings, in particular of palliative care. It is an integral part of the medical ethos to assist people in dying and to ensure a humane death with dignity. It belongs to the goals of medicine and health care not only to promote a good life but also a good death.

First, we may take the expression 'coming home' literally. If the patient is admitted to a hospital, nursing home or another health care institution, caregivers should strive to send the patient home, that is, to his own house, as soon as possible. At home is, so to speak, the 'natural' place to live and also to end one's life.

Second, many people do not live in their own house, but in a nursing home, a home for the elderly, a psychiatric institution, or an institution for mentally handicapped or chronically ill people. Care given to them must focus on providing them with a home in the sense of a homelike environment. They must feel at home in a 'surrogate home'.

Homeless people do not have a house of their own, nor a surrogate home. Still one can say they have a home, that is, their own body. This means that a third meaning of being home exists, that is, feeling comfortable in one's own body. In this context, the body is to be seen as a place where a person should feel at home. This means that this place should be maintained as much as possible and repaired if necessary. The person is the most important steward of his own body, but physicians can also be seen as stewards of the patient's body. From time immemorial bodily well-being has been an important goal of medicine.

Being at home can further be viewed on a fourth level. Ideally, patients feel at home in a non-material environment too. For caregivers this implies paying attention to the patient's psychic condition, his preferences, and his personal norms and values. For many patients in a psychogeriatric nursing home, for example, 'home' appears to be a key word. Demented people living in a nursing home often say they have to go home because, for example, their children are coming home from school. Moreover, often they are diligently searching for a safe non-material home in the sense of spiritual familiarity and safety.

The fifth meaning of coming home is related to the (religious) conviction that man after death returns to his place of origin. On the one hand, dying means leaving home in the four meanings of the word mentioned above. On the other hand, dying also means coming home in this fifth sense. Monotheistic religions such as Judaism, Christianity and Islam have rationalized death as one's return to the Creator. People from these religions often consider their dying as a coming home, as a reunion not only with their Creator, but also with relatives and beloved ones who preceded them in death.

5.2. Dying

The metaphors 'being home' and 'coming home' may potentially create new understandings of death and dying and new practices concerning care for dying people. These metaphors refer rather to the moral than to the medical-technical or care aspects of health care. They may provide new ways of thinking about what role caregivers could play in the life of patients. The leading question for caregivers should be: what can I contribute to the safe coming home of the patient, in terms of the five meanings just mentioned? In addition, these metaphors refer to the idea of a good life and a good death, which have always been considered the object of ethics. By taking care of a patient's safe coming home in the several meanings of the word, the caregiver can help his patient to live or to complete a good life.

From a religious perspective one can argue that it does not matter where and how one dies, assuming that one has a safe 'coming home' in the fifth meaning. Still, the place where and the manner in which one dies influence the quality of 'coming home' and, therefore, can contribute to dying in dignity. As has been indicated above, the culture of dying, mourning, and funeral rites is slowly changing. Increasingly people wish to die at home and to make the dying process a meaningful experience for the dying person himself as well as for the relatives. It is worthwhile to support this increasing desire to die at home in the midst of relatives and friends. In connection with this, it is worth mentioning the goal of the Foundation Dutch Hospice Movement: "to support already existing possibilities for terminally ill people to complete their life in their own environment" (Sluis, 1997). It is telling that the Dutch hospices are sometimes called 'Nearly-at-Home Home' (*Bijna Thuis Huis*). The term 'nearly at home' can be understood in two ways: (1) as a home that approaches as much as possible one's own, ideal home and (2) as an environment in which one is almost at home in the religious sense of the word.

6. EPILOGUE

For about three decades the term 'bioethics' has commonly been used to cover the field of medical ethics and health care ethics. 'Bioethics' literally means 'ethics of life'. Though contemporary bioethics also incorporates issues of death and dying, the time is ripe to develop a 'thanato-ethics'. This discipline should focus on ethical aspects of death and dying and of care for the dying, the central point being that we cannot think of life without considering death and the other way around. In this respect we can learn a lot from history. The Socratic (and Heideggerian) principle "to philosophize is to know how to die", the medieval familiarity with death, and modern insights regarding care, supported by medical technology if necessary, are leading principles. Although it may be desirable in palliative care not to focus attention totally on (future) death, we cannot escape from (re)thinking death and trying to integrate it in our life. Of course, we cannot literally integrate death into our lives. We must integrate death in the sense that we must be able to give a meaning to death - our own death and the death of others as well - in the story of our own lives or in the context of something which transcends our personal lives.

NOTES

1. Apart from philosophical-theological ideas and convictions, there is also empirical evidence for the argument that whole brain death is not the death of the whole organism. From this perspective brain dead patients are dying, they are beyond a certain point of no return, but they are not yet 'really' dead. Modern debates concerning the definition of death have culminated in the question of whether we need a concept of whole-brain death at all. Veatch (1993) argues that no one really believes that literally all functions of the entire brain must be lost for an individual to be dead. In his view, a better definition of death involves a higher brain orientation. Truog (1997) argues that it would be best to abandon the concept of brain death altogether because it is theoretically incoherent, internally inconsistent and confusing in practice. Bernat (1998), however, argues that whole brain death should remain the standard for determining death in much of the Western world and that this concept best maps onto our everyday conception of death.
2. Many philosophers since Socrates and Plato have considered death in the same vein. Cicero (106-63 BC) said that philosophizing is nothing else than getting ready to die: "Tota enim philosophorum vita, ut ait idem, commentatio mortis est" (Cicero, 1957, p. 97: 30,74-31). Michel de Montaigne (1533-1592) wrote: "To philosophize is to learn how to die" (de Montaigne, 1991, p. 89).
3. For a criticism of Heidegger's notion of 'being-toward-death', particularly with regard to Sartre's view, see Cebik (1980).
4. 'They' and 'one' are translations of the German expression 'das Man'. This term sounds the same as both the English noun 'man' (German: 'Mann') and the personal pronoun 'one' (German: 'man') (Macann, 1993, p. 84).

5. According to Verhoeven (1978), Ariès has made a typical literary mistake in interchanging 'narrated death' for 'real death'. Sentences like "he felt that his death was near" in almost all cases referred to by Ariès, were derived from literary stories. What Ariès considers a ritual of dying is, in fact, a narrative technique. 'Tamed death' may be nothing else than narrated death, a literary fiction. Elias (1985) is even more radical in his criticism. According to him, Ariès' selection of facts is based on a preconceived opinion. He fails to point out that the medieval epics are idealizations of knightly life, selective wishful images that often throw more light on thoughts of the poet and his audience than on the way one really died in the Middle Ages.
6. Also in the Bible, death is viewed as an inevitable event that must be accepted. Those who reached an old age, such as Abraham, Isaac, and Jacob, died "satisfied with life". There is something peaceful about the way the Bible describes the death of these biblical figures. The dying person knows that he must leave this life. He does not die alone , but is surrounded by his family and speaks words of admonishment and comfort. Patriarchs and others who had enjoyed a long life could peacefully close their eyes (Douma, 1998).
7. However, as Wortmann (1998) rightly argues, the term 'professionalization of death' is preferable, to include all other professional disciplines concerned with death.
8. See Plato (1980). It is debatable, however, whether the dialogue *Alcibiades* was written by Plato himself or by one of his pupils.

REFERENCES

Akker, P.A.M. van den: 1994, *Leven met de dood.* Bohn Stafleu Van Loghum, Houten.

Ariès, Ph.: 1974, *Western attitudes toward death. From the Middle Ages to the present.* The John Hopkins University Press, Baltimore and London.

Ariès, Ph.: 1982, *The hour of our death.* Vintage Books, New York.

Beerling, R.F., *et al.*: 1976, *De dood.* Meulenhoff, Amsterdam.

Bernat, J.L: 1998, A defense of the whole-brain concept of death. *Hastings Center Report* 28(2): 14-23.

Bradbury, M.: 1993, Contemporary representations of 'good' and 'bad' death. In: Dickenson, D. and Johnson, M. (eds.), *Death, dying & bereavement.* SAGE Publications, London, p. 68-71.

Burgess, M.M.: 1993, The medicalization of dying. *Journal of Medicine and Philosophy* 20: 269-280.

Callahan, D.: 1993, *The troubled dream of life. Living with mortality.* Simon & Schuster, New York.

Cebik, L.B.: 1980, The significance of death for the living. *Metamedicine* 1: 67-83.

Cicero: 1957, *Ciceronis Tusculanarum Disputationum Libri V.* Mit Benützung von Otto Heines Ausgabe erklärt von Max Pohlenz. Erstes Heft: Libri I et II. B.G. Teubner, Stuttgart.

Corr, C.A., Nabe, C.M. and Corr, D.M.: 1994, *Death and dying, life and living.* Brooks Cole Publishing Company, Pacific Grove.

Dekkers, W.J.M.: 1995: What do we call 'death'? Some reflections on the end of life in Western culture. *Ethical Perspectives* 2: 188-198.

Douma, J.: 1998, Death in Christianity. In: Karim, A.B.M.F., Kuitert, H.M., Newling, D.W.W. and Wortman, V. (eds.), *Death. Medical, spiritual and social care of the dying.* VU University Press, Amsterdam, p. 143-156.

Epicurus: 1926, *The extant remains.* With short critical apparatus, translation and notes by C. Bailey. Clarendon Press, Oxford.

Elias, N.: 1985, *The loneliness of the dying.* Blackwell, Oxford.

Field, D. and James, N.: 1993, Where and how people die. In: Clarke, D. (ed.), *The future for palliative care. Issues of policy and practice.* The Open University Press, Buckingham, p. 6-29.

Gorer, G.: 1965, *Death, grief, and mourning in contemporary Britain.* Cresset Press, London.

Heidegger, M.: 1962, *Being and time.* Translated by J. Macquarrie and E. Robinson. Harper & Row, New York.

Illich, I.: 1976, *Limits to medicine. Medical nemesis: The expropriation of health.* Penguin books, Harmondsworth.

Lakoff, G. and Johnson, M.: 1980, *Metaphors we live by.* The University of Chicago Press, Chicago and London.

Kuitert, H.M.: 1998, Editorial comments on death from a religious perspective. In: Karim, A.B.M.F., Kuitert, H.M., Newling, D.W.W. and Wortman, V. (eds.), *Death. Medical, spiritual and social care of the dying.* VU University Press, Amsterdam, p. 113-118.

Maas, P.J. van der, Delden, J.J.M. van, Pijnenborg, L.: 1992, *Euthanasia and other medical decisions concerning the end of life.* Elsevier, Amsterdam.

Macann, Chr.: 1993, *Four phenomenological philosophers. Husserl, Heidegger, Sartre, Merleau-Ponty.* Routledge, London and New York.

Manschot, H.: 1992, *Levenskunst of lijfsbehoud? Een humanistische kritiek op het beginsel van autonomie in de gezondheidszorg.* Universiteit voor Humanistiek, Utrecht.

Momeyer, R.W.: 1995, Death. Western philosophical thought. In: Reich, W.Th. (ed.), *Encyclopedia of bioethics.* Volume I. Simon & Schuster Macmillan, New York, p. 498-505.

Montaigne, M. de: 1991, *The complete essays.* Translated and edited with an introduction and notes by M.A. Screech. Penguin Books, London.

Plato: 1978, *The collected dialogues of Plato.* Edited by E. Hamilton and H. Cairns. Princeton University Press, Princeton.

Plato: 1980, Alcibiades I. In: *Verzameld werk.* Part III. De Nederlandsche boekhandel/Ambo, Antwerpen/Baarn, p. 8-68.

Post, S.G.: 1993, *Inquiries in bioethics.* Georgetown University Press, Washington D.C., Chapter 5.

Prior, L.: 1989, *The social organization of death. Medical discourse and social practices in Belfast.* St. Martin's Press, New York.

Saunders, C.: 1993, Introduction - 'History and challenge'. In: Saunders, C. and Sykes, N. (eds.), *The management of terminal malignant disease.* Edward Arnold, London, p. 1-14.

Seneca: 1979, *Het gelukkige leven.* Ambo, Baarn.

Sluis, P.: 1997, Stichting Nederlandse Hospice Beweging 10 jaar oud. *Medisch Contact* 52: 171-172.

Tolstoy, L.: 1960, *The death of Ivan Ilyich and other stories.* Penguin Books, London.

Truog, R.D.: 1997: Is it time to abandon brain death? *Hastings Center Report* 27(1): 29-37.

Veatch, R.M.: 1993, The impending collapse of the whole-brain definition of death. *Hastings Center Report* 23(4): 18-24.

Verhoeven, C.W.M.: 1978, Doodsproblematiek en levensfilosofie. In: Verhoeven, C.W.M., *et al.*, *Dood en stervensbegeleiding.* Callenbach, Nijkerk, p. 11-28.

Walter, T.: 1993, Modern death: Taboo or not taboo? In: Dickenson, D., and Johnson, M. (eds.), *Death, dying and bereavement.* SAGE Publications, London, p. 33-44.

Wortmann, V.: 1998, Death in the East and the West. In: Karim, A.B.M.F., Kuitert, H.M., Newling, D.W.W. and Wortman, V. (eds.), *Death. Medical, spiritual and social care of the dying.* VU University Press, Amsterdam, p. 219-236.

ZBIGNIEW SZAWARSKI

TERMINAL CARE AND ETHICS

1. INTRODUCTION

Human life is a process. As with any process it implies movement and change over time. And as any process it has its beginning and end. What constitutes the beginning of human life, what is its end, and what happens between the beginning and the end of our life is a perennial challenge for philosophy, arts, and sciences. Some things are, however, absolutely certain - all human individuals are conceived and born by human parents, all human individuals are mortal, and all human individuals have to reach a certain stage of maturity before they can conceive, bear and care for their progeny. Growth, maturity, and then decline is the universal cycle of all living beings.

The history of people is something more than simply the history of their growing and declining bodies. We are acting and thinking bodies. We have needs and desires, we are aware of our past and try to shape our future. We live in the world of things and values. Sharing our biological nature with animals and going through the same cycle of birth, development, and decline, we differ from them in one fundamental aspect, - we are human persons.

Philosophers sometimes make a useful distinction between a human being and a person. *X* is a human being if it belongs to the species of *Homo sapiens* and this is a judgement of fact. *X* is a person if it has a special moral status and this is a judgement of value. *X* is a human person if it belongs to the human species and has a special moral status. Although it seems to be evident that not all persons are humans (e.g., God, angels or extraterrestrials, if they exist) it is not certain if membership of the human species implies logically being a human person. One can arguably hold that although a human zygote or a comatose patient belong to the human species, nevertheless they are not human persons. And that is precisely the point where we are faced with our first philosophical difficulty. Becoming and ceasing to be a human being is a process which has its particular, well described, stages. How and when do we become human persons? What is the human entity that goes on through the successive stages of biological development and degeneration? A fully grown adult differs from the small

H.A.M.J. ten Have and B. Gordijn (eds.), Bioethics in a European Perspective, 433-451.

child he has been many years ago, but it still seems to be the same person. A dying septuagenarian in no way resembles the unborn baby he used to be at one stage of his existence. It is certainly the same biological organism although in a different stage of its development. But is it the same person? It seems that there is only one alternative: either it is or it is not the same person. But making such a decision at this stage would simply be begging the question. Until we know precisely how to recognize a human person, until we define exactly the sufficient and necessary conditions for being a human person it is impossible to decide when a being is or is not a human person.

I am not going to discuss here the extremely difficult question concerning the adequate definition of a human person. Nevertheless as my topic is terminal care for human persons, I am bound to explain my usage of that term. I propose to define 'person' as a *moral agent.* So if somebody is a person it means that he is a moral agent and *vice versa*: if he is a moral agent, then he is a person. To be a moral agent is simply to be able to act morally or, in other words, to be capable of moral responsibility. However, such a definition of persons seems to be apparently false and counter-intuitive; it excludes from the class of persons too many categories of people. We would like to grant a moral status to human embryos, infants, children, the mentally retarded and the mentally ill, as well as seriously demented or PVS patients. Although their moral agency is sometimes indeed more than questionable, they nevertheless remain in a specific relation to us. They may be our children, parents, relatives, friends, patients and how we treat them is not a matter of moral indifference.

Probably the simplest way to cope with that objection is to introduce the concept of *moral patient.* The moral patient is the direct opposite of the moral agent; it is a human being who is not capable of moral action, because he has not reached yet a proper stage of moral development or because, having achieved the proper stage of moral development, he has lost the disposition to be morally responsible. A newborn baby is certainly not capable of acting morally but one day when it reaches a proper stage of moral development it will become a moral agent and a moral person. Before this happens it is only *a future person.* A profoundly demented patient who has totally lost all capacity of moral decision making is in that sense *a former person.* So according to their possibility of moral action and responsibility we have three categories of persons: *future, present,* and *former persons.* Although these classifications embrace the whole human life span, from the moment of conception until the moment of becoming a human corpse nevertheless they ignore one particular class of adult human beings. These are the human beings who look and behave as if they were human persons yet they are essentially not able to act as *moral* agents. They are psychopaths. I propose to call them *defective persons.* The important thing is that all these categories of persons (including psychopaths) belong to *the moral community* and hence they are objects of moral concern. The moral community consists of moral agents and moral patients. And it is of moral relevance to its members how we treat our fellow beings whatever their status as human agents. They all belong to the human community. They are all involved in the moral web of values, rights and obligations.

A history of my life is thus the history of continuous change. It is the change of states of my body. It is the change of states of my mind. And it is the change in my status as a social being. At the beginning of my life I am simply a vulnerable object of care by my parents. When I reach full maturity I begin to care for myself, my children, my parents or my partner. And when I enter into the final stage of my life, I become again vulnerable and the object of care. As a moral agent I am both the subject and the object of care. As a moral patient I can be only the object of care.

As our individual life is a continuous process it is virtually impossible to make a clear cut-off between the different stages of our life. It is quite evident that a small baby is not yet a moral agent and that a teenager is, but it is impossible to say when precisely the change occurred. And the same process takes place at the end stage of our life: the transition from the living body to the human corpse is continuous; one state emerges from the other in an unbroken sequence and that is why we cannot agree to one universal definition or one set of criteria of death. But it is still the same individual human life in the same continuity of changes. Although we shall never be able to discover when precisely a future person becomes a present person and when a present person changes into a former one, yet we can be absolutely certain that it is the continuity of changes that unifies particular, infinitesimal sequences in our life. My personal history unfolds over time through a trajectory from its beginning to its end, whatever the nature of these boundaries. Each of us has been born in some place at a certain time (i.e. was a future person), each of us lives his own life (i.e. is a present person) and each of us will die one day thus becoming a former person. The problem is when and how do we become former persons. How and when do we begin to die and what is the right way of caring for dying people?

2. HOW DO WE DIE?

We know how people die. Sometimes they die at the moment of, or soon after, birth. Sometimes they die in natural disasters or car accidents. They die of hunger, malnutrition or dehydration. They die in wars, revolutions, and ethnic cleansing. They are killed by madmen, tortured to death, or executed. They die at sea, in the jungle, and in the mountains. And most often they die in hospitals, nursing homes or their own beds. Whatever is the time, place and the way of one's death, its essence is always the same: it is a process of disintegration of the human mind and the body which transforms the living being into a corpse, the former person. What is important is the pace of that process. Sometimes people die suddenly - they die without even being aware of their death. Sometimes they are dying for weeks, months, and even years. There are philosophers who claim that life is the art of dying or that we all live towards death. However the theoretical awareness of my own mortality and finitude is not identical with the reality of the process of dying. I can now genuinely enjoy my life just because I know that I will not live for ever. But struggling with pain, suffering, indignity, humiliation, despair, and hopelessness - and that is what people usually feel when they are dying - is something dramatically different from the abstract, philosophical awareness of our mortality. To know that people must die one day, and even to see other people die is not the same as dying. Although I cannot experience my own death, I see how other people die and I can envisage what it is like to be dying.

Not all kinds of death involve terminal care. There is no need of terminal care for people who die suddenly, or who suffer an injury or a disease that is treatable. Only when medicine is powerless, when it cannot stop the progress of disease and degeneration and bring the patient back to normal life, only then the problem of palliative and terminal care appears. We realize then that there is nothing we can do to save the patient's life but we still care about the way he is dying; we think that we can *help*; that we can ameliorate his pain and suffering, that we can ease his death. Thus, terminal care is simply being with and caring for the patient at the time of their death.

But, as I said before, there are different kinds of death and not each kind of dying involves special problems for the carers and the family of the patient not to mention the patient himself. Here are a few examples of people's dying.

1. Mr. J.W. (86), a retired clergyman, died whilst he slept. A day before he eagerly worked in his garden. At first, it was a shock for his wife, but on the other hand, she was not too surprised. Mr. J.W. died and that was what people of his age usually do. He died peacefully after a long and worthwhile life. Some people would say, it was a perfect example of natural death.

2. Miss C.Q. (19), a student of the University of Wales Cardiff, died 23 hours after her admission to hospital. She died from meningitis and was hardly aware she was dying. The medical staff did everything to save her life. Unfortunately, the antibiotics were administered too late to stop the progress of disease.

3. Mrs. R.S. (80) a widow, living alone in her tiny and cold flat, fell down and broke her femur and several ribs. She was admitted to the local hospital and during the surgery it was discovered that she had a cancerous mass in the bone. The patient, although fully alert and competent has not been told the diagnosis. Her leg was repaired and she is being sedated with heavy pain

killers. No other diagnostic procedures were carried out. The family lives quite far away and is not able to look after the old lady.

4. Mr. P.K. (38), a bachelor and a medical doctor, with diagnosis of AIDS, was admitted to a hospice for AIDS patients. Although his body is incredibly thin, weak, and frail, his mind is still lively and acute. He needs assistance almost in all everyday activities but desperately does not want to die 'without dignity', i.e., as he sees it, in the state of total mental degeneration. He would like to have a friend who will stay with him and help him to die when he decides that the proper time has come.

5. Mr. E.C. (74), is suffering from advanced Alzheimer's disease, and is a resident in a small nursing home in Spain. Typically, he spends all his days constrained to his chair or bed. He has no visitors.

6. Tony Bland (19), a victim of the Hillsborough stadium disaster in 1989, has spent 4 years in the state of persistent vegetative state. In 1993 his parents won the right from the House of Lords to withdraw his life support.

7. Andrew Devine (30), diagnosed originally as a PVS victim of the same Hillsborough disaster, needed 5 years to recover sufficiently to communicate some simple ideas by using a touch-sensitive buzzer switch. His state is defined now as 'locked-in-syndrome'. He is conscious and aware of his surroundings, but being totally paralysed he can usually respond only by eye movements.

Although all these people have become or are becoming former persons, there are some relevant differences in the way they are dying. This is evident if one considers where they are dying, who is caring for them, how long they are dying, and what is the mode of their death.

The first two questions are closely related. In modern societies people typically die in hospitals where they are provided with expert, professional care. If they do not die in a hospital, they die either in a hospice, or in a long-term care institution, or at home. Each place has its own specific clinical and moral problems and implies particular consequences for terminal care. So the place where the patient dies is not irrelevant for him and his family.

2.1. Hospital Death[1]

People do not go normally to hospital to die. They go to hospital because they desperately want to live. They believe in the power of medical knowledge and technology. They believe in the magic of modern science. They know also that in the hospital they will meet a highly qualified, expert team of specialists who will competently diagnose their condition and apply the best possible treatment. However, life is never so simple. Because, in spite of the fact that each patient is a unique individual human person, with a unique life history and a system of individual values, paradoxically, the structure of the hospital drama is almost always the same for all involved parties. Here is a short synopsis of that drama.

2.1.1. The Stage

A large, modern building with a maze of corridors and thousands of rooms and compartments. Sometimes you need a map to find your way around. If you are lucky or rich, you can find yourself in a comfortable single apartment with your own TV and a bunch of flowers on the table. But typically patients are gathered together in enormous wards with 6, 12, or even more beds. This is a white, sterile, emotionally cold, and impersonal place where a number of people in white coats and uniforms walk swiftly around with their thoughtfully concentrated faces, and another considerable larger number lie bedridden, surrounded by mysterious and glittering medical equipment. The moment at which you enter the medical ward signifies the end of your private life. You are now a hospital patient and to be a hospital patient is simply to be a cog in the hospital machine, totally subjected to the rules and everyday routine of that complex, bureaucratic, and thoroughly authoritarian institution.

2.1.2. Dramatis Personae

The main protagonists of the hospital drama are the patient and the hospital staff. The patient is vulnerable, ill, suffering and uncertain of the gravity of his condition and his chances for the future. The staff are the highly qualified, expert professionals who know their station and its duties. If the patient is incompetent, the family plays a principal role as a legal guardian and representative of his interests. Sometimes one can see, backstage, a passing figure of a clergyman, a psychologist, or a lawyer. They do not, however, normally play any key roles. The three main protagonists of the drama are the patient, the professional, and the family or its legal representative.

2.1.3. The Drama

The essence of the hospital drama is an inevitable conflict of interests and aspirations of the main protagonists. If the patient is sick, and the professional can easily treat his ailments (e.g., a trivial appendicitis or tonsillitis) everybody is happy and there is no drama. All roles and expectations are perfectly fulfilled. Yet, if the condition is really serious and irreversible (like advanced cancer, multiple sclerosis, or AIDS), and if all known and applied treatments are failing, then you have a potential for a real and powerful moral drama, because nobody knows with absolute certainty what the good of the patient is, what the obligations of the doctor are, and how to proceed in a situation of conflicting wishes, aspirations, or principles. One may say, that if the patient is competent, he and only he has an ultimate and paramount say. If he wants to be left alone, he must be left alone. This is the gospel of the patient's autonomy. But what shall we do if the patient wants to die and his family and doctors want to keep him alive as long as possible? What shall we do if the patient is clearly incompetent (e.g., in irreversible coma or persistent vegetative state) or his competence is seriously diminished by age or the progress of disease? What shall we do if the doctors advise and initiate aggressive treatment without any consent of the patient and his family? Should cardiopulmonary resuscitation (CPR) be a standard treatment for all dying patients? How long should we keep people on respirators? How do we know when and how we should stop the treatment? And last but not least, what constitutes a treatment in terminal illness? Is feeding and hydration a form of treatment or terminal care only? Should we treat accidental and reversible illness in terminally ill patients? How about living wills, assisted suicide, and euthanasia? Probably, every protagonist of the hospital drama will agree that we have a right to die with dignity, but what does this mean in the hospital milieu?

2.1.4. Conclusion

The philosophy of modern hospital medicine is to cure and discharge the patient to a normal life as soon as possible. The aim of hospital treatment is first of all to save and to prolong human life, to control the symptoms of his disease, and to ameliorate his pain. This is of course a fundamental benefit for the patient and immense relief for his family. Yet, the hospital is not designed to help people die. Hospital death is always a sad event; it is lonely, technological, emotionally sterile, very often anonymous, and almost always public. In a way, it is always a tragic symbol of medical failure. Though some modern hospitals have, indeed, special units to care for terminally ill patients (extended care units - ECU), terminal care is not the main goal of hospital treatment. The function of the hospital is, above all, to cure the living and then, if it is possible, to care for the dying.

2.2. Hospice Death[2]

People go to a hospice because they know they are dying and they want to die well - with comfort and dignity. As the primary goal of the hospital is to cure for life, the principal goal of the hospice is to care unto death. The hospice staff are usually specially trained nurses, and volunteers who will stay and help terminally ill patients to say goodbye to life. The doctor plays a subsidiary role mostly as an advisor in choosing and administering appropriate palliative remedies or signing the death certificate. The hospice as a rule does not offer any life saving treatments or any aggressive medication. It is to provide total care in order to enable patients to have an enjoyable and full life until death occurs. If the main impediment in enjoying the last hours of one's life is pain, the hospice staff can offer a whole array of sophisticated techniques for controlling pain and a broad range of distressing symptoms like vomiting, breathlessness, diarrhoea or constipation. It seems that there is no drama in the hospice setting, because everybody including the patient himself has no illusions - death is near and the patient should be prepared for its arrival. Yet, even hospice death, however smooth and untroubled, has its own dramatic dimension. Here is a brief outline of death in a hospice:

2.2.1. The Stage

Preferably a ground floor structure, away from noisy traffic, in a green surrounding with easy access for weak and frail people and wheelchairs. There is no rush and no agonizing tension inside. People seem to be quiet and relaxed. The staff are rather invisible, patients usually doze in their beds, chat with their visiting relatives, or simply contemplate a sunset or a weeping willow tree through the window. Those, who can still walk can have some social life in the common room, or cook their favoured dishes in the patient's kitchen. There is a chapel, bar, and guest rooms for visiting members of the family.

2.2.2. Dramatis Personae

The main hero of the hospice drama is the patient and his death. The nurses are to care for the quality of his life and control pain. There are also some relatives, clergy, and other visitors.

2.2.3. The Drama

According to the philosophy of the hospice movement the patient has a right to die peacefully and with dignity. He is an active participant of the dying process, who as far as he is competent, controls and is fully responsible for his own death. The range of that control is, indeed, quite extensive. He can demand and receive increasing doses of pain-killers, he can demand to be left totally alone, he can wish to be moved home and his wish will be granted. But he can never demand to be assisted in hastening his death. Assisted suicide and euthanasia are absolutely incompatible with hospice care. Thus, if a patient whose bodily pains are fully controlled, is, nonetheless, so tired of living that he pleads for a final shot of morphine, his request will never be granted. The hospice nurse is to assist in death but she ought never to delay, or to hasten death.

2.2.4. Conclusion

The time the patient spends under hospice care is usually short - several weeks, rarely more than a few months. He lives and dies in comfort as an object of permanent and total care. However, there is something in hospice death which makes some people uneasy. For many people, particularly those living outside of the world of highly medicalized western culture, the very idea of dying in a special house of death is abhorrent. They are so terrified by death that they cannot simply accept the fact that they may be dying and that there are people specially trained to facilitate dying. In hospital care there is always a faint hope of a miracle. There is no such a hope in the hospice. Those who enter the hospice gate should indeed abandon all hopes. And that can be very, very depressing. There is another reason to be slightly sceptical about hospice death. The hidden assumption of the hospice movement is the belief that the process of dying is extremely painful. Terminal care becomes then identical with managing terminal pain. But how and where should we care about those patients who do not die from advanced cancer, multiple sclerosis, AIDS, or other painful conditions? Is there any place in the modern hospice for severely demented patients? The painless death, and that is the essence of the hospice death, can still be a very tragic and sad event. But if someone is dying painlessly, with alert mind, and even with the total acceptance of his death, does it follow that he is dying with dignity?

2.3. The Nursing Home Death[3]

Although nursing home residents comprise a large and rapidly growing section of the society, it is amazing how very little is known about nursing home death. It is very well known what the average age of the nursing home resident is, what the sex and race distribution is, and what are the typical causes of death in nursing homes. Yet, our idea of dying in the nursing home is rather vague, blurry, and quite disturbing. I have not been able to find a reliable description of death in the nursing home. So what follows is rather a collage of various personal memories and readings.

2.3.1. The Stage

There is no typical nursing home. There are nursing homes for very rich, and nursing homes for very poor people. There are countries where nursing home care is absolutely free and accessible for every citizen, and there are countries where so-called for-profit nursing homes are a growing success. For those of you who have had a chance to watch a charming BBC serial 'Waiting for God', the nursing home seems to be a pleasant and comfortable place where the old people enjoy themselves by making practical jokes to each other. But to those who have had a chance to visit some nursing homes in different parts of the world, very often they appear as incredibly gloomy and depressing places. These are the places where the very old, chronically ill, and totally abandoned people die.

2.3.2. Dramatis Personae

A typical resident (they are called - residents, not patients) of a nursing home is an old man or woman, who is not able to care for himself or herself because of old age or illness. There are many degenerative diseases of old age but the most distinctive and the most frightening among them is dementia (*dementia senilis*). As cancer is a paradigmatic disease of the hospice, dementia is a paradigmatic disease of the nursing home. So a typical hero of a nursing home is a more or less demented patient and their nurse and physician. The other parties, like relatives, lawyers, and even ministers of religion, are rarely seen there. It is a small, closed community which is far beyond the mainstream of normal life.

2.3.3. The Drama

People go to hospital because they want to live. People go to a hospice because they want to die in comfort and with dignity. Very few people voluntarily go to a nursing home. They are usually referred there either by their GP or community services, or (what is more frequent) by the family who are not able or cannot afford to provide everyday care. Whoever makes the decision, the status of the nursing home residents is ambiguous. On the one hand, if they suddenly get ill, they are treated exactly like hospital patients. So if a resident contracted cancer, got severe diarrhoea, or caught a serious upper respiratory tract illness or pneumonia, he will certainly be treated, because that is exactly how we should care for sick people whatever their age. On the other hand, however, very often pneumonia is regarded as 'the old man's best friend'. That characteristic chasm between cure and care can be easily explained by the time factor. In the hospital or hospice a typical length of stay is measured in days, weeks, but rarely in months. These are by definition short-term care institutions. The nursing home is a long-term care institution and some people spend almost their whole lives there. An average length of stay is related of course to disease, but in the case, for example, of Alzheimer's disease it is typically about three years of life from admission. So if it is not a place to be cured from the primary condition (we do not treat the old age), if it is not a place to die in comfort and with dignity (because the nursing staff are not usually trained in terminal care and the resident can live there sometimes for years), what is the proper role of the doctor in caring for terminally ill residents of nursing homes?

The main moral problem of long-term care is to treat or not to treat. And if not to treat, what is the proper model of palliative care for the old residents of nursing homes. Medical intervention may or may not improve the quality of a patient's life and if it prolongs their life, its effectiveness is measured in weeks rather than in months. Hence the typical moral problems of nursing home doctors: should we provide intensive care for a patient with advanced Parkinson's or Huntington's disease? How about cardiopulmonary resuscitation for elderly cancer patients? What shall we do if an old demented patient with Alzheimer's disease has suddenly fallen ill with pneumonia? It is quite easy to treat pneumonia with antibiotics, but is the doctor morally obligated to administer the life saving drugs? What about withdrawing and withholding treatment in some cases?

2.3.4. Conclusion

It is impossible to predict precisely how much of life is left for a nursing home resident. If it is a 56 year old, mildly demented patient, perhaps we should treat his pneumonia. He can still enjoy another ten years of peaceful life. But if someone is 76, or 86 and severely demented, should he not be allowed to die peacefully? As I said, an average length of stay for demented patients in the nursing home is about 3 years since admission. It is impossible to be a terminal patient for three years, especially if you are really not afraid of death and do not essentially suffer. So what can you do as a highly qualified medical professional? In the hospital, death was seen as a medical failure, in the hospice - good, peaceful death was regarded as a success; the mission has been accomplished. The main feature of the nursing home death seems to be relief - relief for the family, and relief for the staff. What is missing is the caring respect for the human person who is, in a way, suspended somewhere between life and death. Sometimes she is closer to life, but frequently she approaches death. We seem not to have a sufficiently strong moral motivation to do everything to prolong that person's life, and at the same time we lack enough courage to delay or to hasten her death. This is why it is so difficult to envisage the nursing home death.

2.4. Death at Home[4]

People do not die normally at home nowadays. Yet, for ages it was the most natural place of dying. Of course, people were dying in battlefields and wars, in fatal accidents, or from catastrophic illnesses, but generally death was regarded as a perfectly natural, although sometimes violent or quite unexpected, end of life. It was rapid growth of medicine and medical technology, dramatic change in the structure of the family and conditions of life, an increase in average length of life that, together with many other factors, haves initiated the process of the institutionalisation of death. According to recent data only 23% of British patients die at home, and 71% die in institutions; 54% of deaths occur in hospitals, 13% in nursing or residential homes, and only 4% in hospices (Thorpe, 1993). Even if a patient wants to die at home, very often there is nobody around to care for him. It seems, however, that the trend is changing. For example, more and more terminally ill AIDS patients die at home (Kelly, *et al.*, 1993), and we slowly learn that it is even possible to care for advanced cancer patients at home.

An instructive guide on this topic was published by the American College of Physicians early 1998 on the Internet as *Home Care Guide for Advanced Cancer*[5]. What follows here is not a description of the typical home death; it is rather an ideal picture of the traditional way of dying but in conditions of the modern civilisation.

2.4.1. The Stage

Your own home filled up with artifacts of your life. Your favourite armchair, a lamp, your books, your paintings, your Hi-Fi set, and collection of CD's with really good music. You feel safe. This is your home. This is your private space which you have used to live in and to meet your most beloved people: your life partner, your children, your friends. This is your own world with its own order and everyday routine - a place full of feelings and memories.

2.4.2. Dramatis Personae

The patient, the family, and the palliative care personnel. By 'family' I mean everyone who has a strong emotional affiliation to and with the patient, either by marriage, blood, or a long friendship. By 'palliative care personnel' I mean all those people who are involved in specialist palliative care and who are ready to advise and assist the members of the family in terminal care for the patient.

2.4.3. The Drama

If there is any drama in death at home, it is a drama of leaving one's dearest people for ever. But it is another natural fact of life. It is possible to be together for life but not for eternity. Death at home requires, however, full acceptance and the utmost dedication of the family. It is a burden to care for the dying person all around the clock. It may be a very frustrating and, indeed, a depressing experience. But it is the real test of love and friendship which requires the family to learn and properly apply the subtle and difficult art of caring for a terminally ill relative.

2.4.4. Conclusion

Yet, death at home is the most natural and probably the most desirable way of ending one's life. It is the death where both the patient and his family take full responsibility for the dying. It would be impossible without love, mutual trust, and utter dedication. If the family decides to care at home for their dying member, it is like suddenly deciding to run their own private hospice, without any earlier experience and knowledge in palliative care. One must admire such courage and dedication but one cannot compel anyone to love and compassion. Love on order is simply logically impossible. Saigyo, a 12th century, Japanese poet and priest wrote in one of his poems: "How I long to die in spring under cherry blossoms". It is still a customary wish of Japanese people (Suzuki, 1997). But death at home, however natural, and saturated with caring love, can bring a terrible fatigue and strain on the relatives of a dying person. At the same time, like the birth of a new member of the family, it is a memorable and important event in the life of a family and it can dramatically shorten the period of the grieving process.

Graham Thorpe (1993), a British specialist in palliative care, and committed advocate of dying at home lists in his paper the following conditions of death at home:

- adequate nursing care
- a night sitting service
- good symptom control
- confident and committed general practitioners
- access to specialist palliative care
- effective coordination of care
- financial support
- terminal care education.

I do not think there are many places in the world where these conditions can be successfully met. In many countries, especially in Central Europe and the third World countries, there is hardly any system of palliative care for the dying. Nevertheless, Thorpe makes a good case for a convincing alternative for dying in health care institutions and perhaps this is the ideal we should aim at.

I have presented four different places where people usually die. I, for one, would like to die at home if, of course, my family is ready to share with me the burden of my passing away. But what is particularly staggering in those four different kinds of death is a peculiar distribution of responsibility for one's death. In the first two settings (hospital and hospice) we shift our responsibility for the process of our dying on to the medical professionals. Even if they cannot effectively treat the patient, they can provide at least competent, professional palliative care. The nursing home death is a sad example of the situation where nobody in fact feels responsible for the patient's death. It is only death at home where all the main protagonists of the dying drama feel full solidarity and take total responsibility for the end stage of human life. I have no doubts, that we are not ready yet to return to the traditional way of dying at home. Our flats or houses are, as a rule, too small to accommodate the terminally ill person and the caring team. Our families have lost the traditional three generation structure and are often dispersed around the world. And, what is more important, we have the wrong idea of death as something thoroughly alien, frightening, obscene, or evil. This is, I think, the main reason why we try to avoid responsibility for our own death and the death of our neighbours.

3. HOW SHOULD WE DIE?

The simplest answer is - we should die well and to die well is merely to have a good death. The concept of good death, although very attractive and philosophically respectable, is yet quite difficult to define not to mention its practical implementation. Therefore, I suggest a different strategy in approaching our problem. The principal question should be not 'How should we die?' but 'What is good for the dying person?' If we can roughly explain the meaning of that question, perhaps we shall be able to say with some confidence what conditions are to be met to accomplish a good death.

There seem to be four main categories of good relevant to the dying person. These are: the life, the good of the mind, the good of the body, and the good of the communal life.

The dying patient is still alive and probably it is the most valuable time of his life. If he is aware of his death, he knows that it is an irreversible event and that he will never have a chance to live again. But being still alive means being still able to enjoy the last moments of one's life. Life is then the necessary condition of all experiences and we cannot suffer, nor enjoy life unless we are alive. If I am suffering or if I am in pain, it means I am still alive.

Sometimes philosophers distinguish between biological life and biographical life (Rachels, 1986). It is an important distinction. If I happened to be in a permanent vegetative state, if only some rudimentary physiological systems of my body have been operating without any significant higher brain activity, would it still be good for me to be kept alive? I dare say not. Life is a good for the dying patient only if he is still aware and capable of experiencing the world. The irreversible permanent vegetative state or irreversible coma are not indicative of the good of the dying patient. If there is no personal life, there is no justification in keeping a living human corpse alive.

The situation is however dramatically different if I am still able to experience the world, if I am still able to remember my youth, to worry about my family, to listen to music or even to feel an excruciating pain. I am alive, I can still have some wishes and desires, I still care about something, I can still decide, and I can get something from my life. This is the proper place for the old Stoic and Epicurean moral psychotherapy ("death is indeed nothing to be afraid of") and that is the proper time to begin palliative care. Because I cannot enjoy the last hours of my life if my mind is troubled by anxiety and my body is in agonizing pain. The word *ataraxia* means in Epicurean philosophy freedom from trouble and anxiety. You could not reach happiness *(eudaimonia)*, Epicurus taught, unless your mind is peaceful and the body is not tormented by any pains or discomforts. So the two great goods of the dying person are tranquillity of the mind (*ataraxia*) and peace of the body.

That is the state in which you can die peacefully and even with a faint smile on your face.

It is not our ideal of good death to die in solitude or abandonment. We are by nature social animals. We need our fellow beings to share our emotions, be it ecstatic happiness or deep sorrow. Normally we want to share with others our intensive joy of having a newborn son or daughter, and we want to share with others our profound sadness and anguish of our ultimate departure from life. Being together, feeling a friendly touch of someone's hand, or listening to someone's friendly, soothing whisper has a profound meaning for the dying person. He is not dying alone, he has a friend to see him off. This is a very important good.

So far I have tackled with two extreme situations: when the dying person has practically no biographical life (comatose and PVS patients), and when the dying person is more or less conscious and a competent participant of the dying process. The most difficult cases are, however, between the two extremes when the dying person is still able to experience some human emotions but his competence and autonomy are significantly diminished or almost non-existent. I cannot tell confidently what the good of a dying demented patient is. If it is a mild form of dementia, perhaps we should err on the side of life. The patient can still enjoy some sensual pleasures. So although we cannot cure him from his Alzheimer's disease we can and we perhaps should treat his pneumonia. What shall we do, however, if it is an old and profoundly demented patient at the end stage of his life? Would it be morally right to treat his pneumonia if he happens to fall ill with it? Is it not better for him to die? If this is the case, should we not confine our medical interventions to palliative care only? But if it is the right conclusion, it means that, paradoxically, death may also be a good for the dying person. Indeed, we have no doubts that in some cases it is better to die than to live. So, we are back to the fundamental question: how do we know what is the good for the dying person?

At the beginning of this chapter it was argued that all human life is a process and that it has its inevitable end. We are all mortal beings and we cannot escape death. I have said also that it is a good thing to die well and that there are some goods of the dying person, including death itself. I could have said also that we all have a right to die with dignity but I did not do it simply because it is a slogan which is used and abused by all the parties involved in the debate on terminal care. Thus, instead of making yet another contribution to the debate on death with dignity, I will venture to explain why all involved parties (professional health carers, the hospice movement, right to die and right to life activists) cannot reach a minimal consensus in the debate on terminal care.

The quintessence of that debate may be boiled down to the following practical syllogism:

Every terminally ill patient has a right to terminal care
X is a terminal patient

X has a right to terminal care

I do not expect that one may want to challenge the validity of the major premise of that argument. That is what we mean when we are talking about our right to die with dignity. We may of course have a considerable problem with defining the concepts of palliative and terminal care (Ashby and Stoffell, 1991) and argue for hours, whether, for example, feeding and hydration are forms of medical intervention or simply a palliative care (Craig, 1994, 1996; Ashby and Stoffell, 1995), or what is the difference between ordinary and extraordinary means of treatment and if that distinction is relevant in terminal *care?* I think, however, that this is not the conceptual problem surrounding the adequate definition of terminal care that is the main cause of the whole controversy. The central problem is: who is a *terminal* patient? What does the word 'terminal' mean? If someone is not a terminal patient, it means that he has still a chance of recovery, a chance of a sensible life, therefore there is no need to instigate the whole ritual of terminal care whatever its definition. But having decided 'This is a terminal patient' in a way we offer justification for terminal care only. It would be totally absurd to offer terminal care for someone who is not a terminal patient. So the principal question is: how do we know who is and who is not a terminal patient? To ask the same question in more general philosophical terms: how do we know when we are becoming former persons?

If individual human life is a continuous process there is always a critical point in that process when our life starts to decline. We enter into that line of shadow slowly without noticing it but there is always a point when we enter the class of former persons only. Is it possible to define precisely all the necessary and sufficient conditions of becoming a former person? What would constitute the criteria for being a former person? I can certainly say that a terminal patient is someone who is becoming a former person because all mortal beings (as we are) are becoming former persons. But the question is: when exactly does the process of dying begin and, more important, when does it become absolutely irreversible?

I am afraid that there is no good answer to these questions because the problem does not consist in empirical research or *discovering* a set of relevant criteria to ascertain the beginning (or end) of the irreversible process of dying. The real problem lies in *deciding* what is that particular point on the continuum of our life which is so morally relevant that we can begin the count down to our death. It is, of course, a moral decision. If a doctor orders "Palliative care only" it means usually that he has just declared the terminal status of the patient. It means "I have done everything that was possible. I cannot prolong his life any more. Let him die now in peace". Of course, he had to have his reasons to say so and it is very likely that some of his reasons will be good and others may be bad. But in each case what is at stake is the good of the dying patient.

There are two possible strategies in making such decisions. I shall call them *affirmation of life* and *affirmation of death* strategy. In the first case the most general assumptions are that it is always better for a patient to live than to die; human life has an intrinsic value, and the doctor should never do anything to harm his patient. Causing the death of the patient is absolutely forbidden. Very few people however, hold such a radical view. Normally, affirmation of life means: although human life has an intrinsic value, and a physician should never do anything to destroy it, he has nevertheless no moral obligation to apply any heroic, extraordinary, or futile means to sustain it. It is sometimes evident that death is approaching and inevitable; however, that is no moral reason to actively hasten it, even if such is the wish of the patient or his family.

The affirmation of death approach does not deny that human life has a value but it has a value only if it has a certain level of quality. So it is natural that sometimes it is better for the patient to die than to live particularly if he himself finds his life totally unbearable and wants to die. Consequently, there is nothing morally wrong with suicide, assisted suicide, euthanasia, withdrawing or withholding treatment, withdrawing or withholding feeding and hydration if they can bring about a good death. Because even if human life has a great value we are not obliged to live or to save human lives at absolutely any cost.

Both sides of the controversy allow that our life is a continuum and there is a critical threshold on that continuum when we slowly begin to die and that there is such a relevant point in that threshold at which the process of dying becomes irreversible. However, the pro-life side does not admit that death may be good for the dying patient and consequently they are always inclined to err on the side of life and, what is more important, they refuse to take any responsibility over the patient's death.

On the other hand, pro-choicers believe that there is such a critical point in the process of our declining that we must courageously decide if there is any relevant moral reason to sustain the withering life. The biological life *per se* is only a necessary condition of experiencing a meaningful, personal life. If the dying patient (or in the case of his incompetency, his family or legal proxy) decides that there is no reason for him to be kept alive with no chance of worthwhile and significant personal life, then there is nothing wrong, and perhaps it is even our moral obligation, to actively help the patient die.

The gist of the controversy is, then, the issue of possibility of rationally deciding who is and who is not a terminal patient. Those who will take an agnostic and pro-life position (we shall never discover any set of rational criteria for declaring the terminal status of the patient) will advise a cautious and traditional policy of keeping the patient alive as long as it is possible, whatever the moral and economic cost of that policy. Those who are ready to take on responsibility for their own and other people's death will have no moral problem in declaring the terminal status of the patient. But whatever model of terminal care we shall choose, there is always a price to pay. The first model, based on the concept of respect for human life, ignores the value of human dignity. The second one assumes that we should respect, not the human biological life only, but the whole human person, and we cannot respect the whole person if we do not respect her freedom of choice, and her right to self-respect. Care for the artificially sustained but absolutely personless human life, is not a proper terminal care but is rather post-terminal care, and as such requires other, special justification.

I am afraid it is impossible to reconcile these diametrically opposite moral positions. We have all been brought up in the tradition of respect for human life. We feel also strongly that we are morally responsible for our own life. The concept of being responsible for one's own death seems something new and upsetting. But if death is simply the main concluding event of our life, being responsible for one's own life implies also being responsible for one's own death. This is the only realistic way of de-medicalizing and de-institutionalizing death by treating it as a completely natural event in human life.

NOTES

1. See: Hanson and Danis (1991); Henley (1986); Lynn (1989); Talone (1996); Walton (1983).
2. For Example: Kübler-Ross (1978); McLean (1993); Munley (1983); Saunders and Baines (1989); Saunders *et al.* (1981).
3. See: Agich (1993); Applebaum, *et al.* (1990); Binstock, *et al.* (1992); Corey, *et al.* (1992); Kane and Kaplan (1990); Kemper and Murtauagh (1991); Muder, *et al.* (1996); Zweig (1997).
4. See, for example: Bowling (1983); Copperman (1983); Doyle (1994); Hinton (1994); Kelly, *et al.* (1993).
5. The American College of Physicians, *Home care guide for advanced cancer.* INTERNET edition. Available from: http://www.acponline.org./public/homecare/

REFERENCES

Agich, G.J.: 1993, *Autonomy and long term care.* Oxford University Press, Oxford.

Applebaum, G., King, J.E. and Finucane, T.E.: 1990, The outcome of CPR initiated in nursing homes. *Journal of American Geriatrics Society* 38: 197-200.

Ashby, M. and Stoffell, B.: 1991, Therapeutic ratio and defined phases: proposal of ethical framework for palliative care. *British Medical Journal* 302: 1322-1324.

Ashby, M. and Stoffell, B.: 1995, Artificial hydration and alimentation at the end of life: a reply to Craig. *Journal of Medical Ethics* 21(3): 135-140.

Binstock, R.H., Post, S.G. and Whitehouse, P.J. (eds.): 1992, *Dementia and aging: Ethics, values, and policy choices*. John Hopkins University Press, Baltimore.

Bowling, A.: 1983, The hospitalisation of death: should more people die at home? *Journal of Medical Ethics* 9: 158-161.

Copperman, H.: 1983, *Dying at home*. Wiley, Chichester.

Corey, T.S., Weakley Jones, B., Nichols, G.R. and Theuer, H.H.: 1992, Unnatural deaths in nursing home patients. *Journal of Forensic Sciences* 37(1): 222-227.

Craig, G.M.: 1994, On withholding nutrition and hydration in the terminally ill: has palliative medicine gone too far? *Journal of Medical Ethics* 20(3): 139-143.

Craig, G.M.: 1996, On withholding artificial hydration and nutrition from terminally ill sedated patients. The debate continues. *Journal of Medical Ethics* 22(3): 147-153.

Doyle, D.: 1994, *Caring for a dying relative: a guide for the layman*. Oxford University Press, Oxford.

Hanson, L.C., Danis, M.: 1991, Use of life-sustaining care for the elderly. *Journal of the American Geriatrics Society* 39(8): 772-777.

Henley, A.: 1986, *Good practice in hospital care for the dying*. King Edward's Hospital Fund for London, London.

Hinton, J.: 1994, Can home care maintain an acceptable quality of life for patients with terminal cancer and their relatives. *Palliative Medicine* 8(3): 183-196.

Kane, R.A. and Caplan, A.L. (eds.): 1990, *Everyday ethics: Resolving dilemmas in nursing home life*. Springer, New York.

Kelly, J.J., Chu, S.Y. and Buehler, J.W.: 1993, Aids death shift from hospital to home. *American Journal of Public Health* 83(10): 1433-1437.

Kemper, P. and Murtaugh, C.M.: 1991, Lifetime use of nursing home care. *New Engl J Med* 324: 595-600.

Kübler-Ross, E.: 1978, *To live until we say good-bye*. Prentice-Hall, Englewood Cliffs, NJ.

Lynn, J. (ed.): 1989, *By no extraordinary means: The choice to forgo life-sustaining food and water*. Indiana University Press, Bloomington (Ind.).

McLean, G.L.: 1993, *Facing death: conversations with the cancer patients*. Churchill Livingstone, Edinburgh.

Muder, R.R., Brennen, C., Swenson, D.l. and Wagener, M.: 1996, Pneumonia in a long-term care facility: a prospective study of outcome. *Archives of Internal Medicine* 156(20): 2365-2370.

Munley, A.: 1983, *The hospice alternative. A new context for death and dying*. Basic Books, New York.

Rachels, J.: 1986, *The end of life. Euthanasia and morality*. Oxford University Press, Oxford.

Saunders, C., *et al.* (eds.): 1981, *Hospice: the living idea*. Edward Arnold, London.

Saunders, C. and Baines, M., 1989, *Living with dying. The management of terminal disease* (2nd ed.). Oxford University Press, Oxford.

Suzuki, S.: 1997, Terminal care at home of aged. *Japanese Journal of Geriatrics* 34: 17-22.

Talone, P.A.: 1996, *Feeding the dying: Religion and end-of-life decisions*. Lang, New York.

Thorpe, G.: 1993, Enabling more dying people to remain at home. *British Medical Journal* 307: 915-918.

Walton, D.N.: 1983, *Ethics of withdrawal of life-support systems. Case studies on decision making in intensive care*. Praeger, New York.

Zweig, S.C.: 1997, Cardiopulmonary resuscitation and do-not-resuscitate orders in the nursing home. *Archives of Family Medicine* 6: 424-429.

FRANZ JOSEF ILLHARDT

WITHHOLDING AND WITHDRAWING TREATMENT

1. INTRODUCTION

First of all, I would like to outline the use of the terms 'withholding treatment' and 'withdrawing treatment'. They belong to the conceptual category of 'medical decisions at the end of life', including euthanasia. In contrast to killing, euthanasia only takes place when death is imminent - then and only then. To withhold treatment implies not initiating the treatment. To withdraw life support systems and procedures implies discontinuing a method of treatment, which was initiated to support life, but in the meantime has proven its uselessness and now serves only to prolong life, or rather: to prolong the process of dying.

The justification for the moral distinction between passive and active ways of terminating the life of a dying patient traditionally is based on the following distinction: Some acts hasten or cause the death of a patient by killing him/her (e.g., by poison). Other acts do nothing to hasten or cause death, but either do nothing against it, or intend something good for the patient (e.g., pain relief), although it leads to his/her death.

Recently, there has been an intensive discussion about the reasons why active termination of a human life is morally less justifiable than a passive termination, and why omitting is better than committing, though both are actions and lead eventually to the same effect (Birnbacher, 1995). Solving this issue is the never-ending task of moral philosophy, and we cannot support our arguments by referring to a consensus within the medical caregivers' community. We can only provide some action-guides to give orientation regarding this problem. Withdrawing or withholding treatment is consonant with the general pattern of medical treatment because it is based on a decision which respects the will, or at least the presumed will of the patient, the therapeutic relationship as well as the challenge to prevent suffering.

Knowledge of the most important problems of their patients seems to be the most significant criterion to determine whether caregivers are competent in caring for their patients. Baines presents a list of the eleven most frequent

H.A.M.J. ten Have and B. Gordijn (eds.), Bioethics in a European Perspective, 453-467.

problems of the dying (1992). Comparing the admittance criteria for terminally ill patients to regular hospitals, palliative wards or hospices, the following list resulted:

1. Weakness
2. Anorexia
3. Pain
4. Dyspnoea
5. Nausea/vomiting
6. Constipation
7. Cough
8. Insomnia
9. Confusion
10. Bedsores
11. Catheter/incontinence

The problem is that we are not fully aware of the situation of the dying. The French historian Ariès has stressed the ambivalence of respect and aversion when we cope with dying persons, e.g. with one of the above mentioned problems.

Perhaps, in the past, human beings could deal more adequately with the phenomenon of dying. Ariès called it the period of the 'tamed death'. Over the centuries this picture changed to the period of the 'wild death' (see chapter 12). Recently, we have experienced death as a catastrophe. What are the reasons for this shift? Some reasons refer to the psychologic situation of the individual caregiver, some refer to the specific culture to which the caregiver or the patient belongs. Let us illustrate these theoretical considerations by using the following case:

An internal ward having difficulties dealing with a patient called upon a geriatrist and an ethicist for consultation. Rarely do health care providers take sufficient time to deliberate.

Mr. B., 81 years of age, had been admitted to hospital two weeks before with a massive haemorrhage in the brain. He had already suffered repeated attacks. Recovery in such cases is, in principle, possible; however, after consulting the medical staff the recovery was found to be most unlikely. Even if the patient did recover, his disabilities would be severe. The ward staff was to decide whether the patient's therapy, care and nutrition were to be continued or withdrawn. Mr. B. showed no response to his relatives nor to the staff on the ward. All were in favour of reducing therapeutic treatment and of not interfering with the imminence of Mr. B.'s death.

The fact that Mr. B. was constantly pulling out his intravenous tubes and the staff putting them back was a recurring theme in the discussions. Although the staff realized that this behaviour was an unconscious reflex, they could not be certain that the patient might very well have been unconsciously protesting, even if only reflexively, against the way he was being treated. It became unbearable for the medical staff to perform their therapeutic duties in such a hopeless situation. The nursing staff felt frustrated, particularly because the patient's family continually assured them that neither they nor their father, had he been conscious, would be able to continue the treatment for an extended period of time.

The family was invited for another consultation, in which the medical staff on the ward also took part, in order to find out how they felt, and why they were in favour of discontinuing their father's treatment. The ethics consultant explored three options:

1. administering parenteral liquids and feeding should be maintained whatever may come;

2. discontinue feeding, continue administering liquids in order not to prolong the process of dying, but not actively contribute to an acceleration of the process and thus avoid problems of bad faith and guilt;

3. discontinue each form of prolonging life, while maintaining the necessary nursing procedures taking the positive effects of the patient's dehydration into consideration.

The parties agreed to meet the following day to go over the pros and cons of the third option as it was considered the most honest of the three. A final decision was to be made. Several hours prior to the appointment the patient died.

2. THE THEORETICAL BACKGROUND

Before working up the important dimensions of this case we must discuss two theoretical issues:

2.1. Discovering the Patient's Situation

The major dimensions of the patient's situation are the medical status on the one hand and his values, preferences, and moral resources on the other hand. A decision to withdraw or withhold treatment must be oriented by both the medical and the ethical dimension.

2.1.1. The Medical Dimension

The patient's medical status can only be illustrated on a case-to-case basis. For example, an apoplexy may be reversible or permanent, caused by panencephalitis, anoxia, or trauma. An apallic syndrome may occur, and perhaps the coma vigile is either a transitory stage or proves to be irreversible. Organ failure may be compensated for after a crisis or may result in a complete breakdown of the body system. Clinical experience has documented this ambivalence. There are no rules to cope with it except by pursuing the following procedure. First of all, a complete record of the medical data must be made. Furthermore, one must have all the available data concerning the biography and the patient's will. Finally, all the medical options available for the patient must be explored. Health care providers must resist making a medical decision based solely on a quality-of-life judgement.

Although the rule of futility is apparently well-defined, recently some criticism has brought forth evidence to the contrary. Whether medical treatment is futile can never be regarded as absolutely certain (Truog, Brett and Fader, 1992). It is worth bearing in mind that medical judgements are the result of an inductive procedure and a hypothesis that subsequently lead to a viable thesis on the nature of the patient's disease. Uncertainty occurs if the case is not compatible to either the derived hypothesis or the thesis. The concept of certainty (Seale and Pattison, 1994) also plays a significant role in moral judgements on the futility of starting or continuing treatment. Health care professionals only get an idea of certainty only when they have sufficient time to make an appropriate decision. But certainty is not a function of time, it is intrinsic of and dependent on the process of developing a judgement rather than on its speed.

A patient's medical status can only be definitively ascertained by consulting many professionals. In the case reported above we have experienced the need to have the assessment of the patient's intern, the neurologist, the nurses, and the relatives; furthermore, we needed to have the records. Not a judgement made in isolation, but our consensus made us certain of our shared decision. It would be sarcastic to reproach modern health professionals for their individual judgements and thus discourage them from making decisions; yet a decision made in 'splendid isolation' is to misunderstand the problem. This problem can be well illustrated by way of metaphor: about 15 years ago, a television series entitled 'Behind us only the Lord' (*Hinter uns steht nur der Herrgott*) was broadcasted in Germany. Each dramatic episode, set shortly after World War II, portrayed another chapter in the biography of a surgeon, whose isolated decisions demonstrated a pattern suitable for a cowboy in the Wild West, but by no means adequate for an appreciation of the medical complexity and its inductively obtained conclusions. Nevertheless we are looking for a clear-cut rule. If the patient's last will is available or can be reconstructed, and the medical staff agrees on the patient's medical status, problems do not generally arise. Most dilemmas, however, involve borderline cases.

Thus, once all relevant medical and biographical data have been gathered, the following rules for a medical ethical discourse should be observed:

1. Ascertain the degree of medical certainty
2. Feel responsible for guaranteeing a mutually shared setting for medical discourse (consilium)
3. State the irreversibility and futility of a treatment plan

4. Check the available means and organize their fair allocation
5. Do not harm any other patient's care by extraordinary means

2.1.2. The Ethical Dimension

First of all, we must state that determining the patient's values is not generally considered to be part of clinical routine.

It has become trite to say *voluntas aegroti suprema lex* (the patient's will is the highest rule) as long as there is uncertainty and unwillingness with regard to the assessment of the patient's will. The best way of dealing objectively with this uncertainty is to use the patient's advance directive, in which he declares his will while still in a condition of good health. Some doubts arise whether the advance directive is an appropriate procedure. I think it is because it neither exempts the physician of his duty to uphold the principle of beneficence, nor of his duty to interpret the tenor of the written last will. Above all geriatrics invoke the exigence of the advance directive (Dallas, 1987). It is unreasonable to let caregivers play the role of a fortune teller (MacKay, 1991) for what a patient would have said if ...

A similar objective way for patients is to appoint a close and trusted friend to represent them and declare their will for them. It is presumed that at the moment immediately prior to death no one is able to state what he wants. It is more appropriate to medicine to know who is the representative entitled by the patient than to play a fortune teller (Clarke, 1994).

The most subjective way of reconstructing the patient's will is based on information obtained from relatives and others close to the patient. Although friends and relatives cannot automatically serve as the patient's substitute or attorney and are sometimes entangled in a conflict of interests, they may be very useful in the reconstruction of the patient's will. Physicians must determine and understand the relevant values as well as reconstruct their significant features. Unfortunately, conventional medical training has failed in meeting such a requirement.

To summarize these considerations: If the patient is incompetent or incapacitated, decision-makers must

1. look for an appropriate substitute for the patient's will,
2. be 'free from a conflict of interests', and
3. be familiar with the patient's values and preferences (Pellegrino, 1989).

2.2. Two Types of Arguments

In the contemporary discussion we meet two positions: Sanctity of Life (SoL) and Quality of Life (QoL) (Keenan, 1996)

SoL Each idea what life and the protection of life can be, stems from religious traditions on the one hand or from natural law traditions on the other hand. The religious tradition says: The value of life is defined by the creator God.

> The secularized theory (e.g. the *droit de l'homme*, constituted after the French revolution of 1789, and most modern constitutions) says: The value of the human being is founded on the intrinsic nature from which every human being is a part.

Hence there cannot be an exception (except only the action with double effect).

> QoL Life is life as it stands in a social context. It is life, the more this life realized its social qualities, such as skills of communication and interaction or personal values of well-being, and having future. If these qualities fail there is no reason why to protect this life.

This theory is open for all experiences. When prolonging the life of a terminally ill human being the physician must evaluate its qualities.

Currently the argumentation in bioethics is shifting from SoL to QoL. We must not misunderstand this shift. It does not imply the total breakdown of the old ethos of protection of life. Whatever the caregiver's position may be, he or she continues

- to protect human life (both positions set limits for killing),
- to prevent the patient from the dominance of others, and
- to avoid that the patient may enter a prolonged vegetative life.

3. DIMENSIONS OF THE CASE

3.1. Experiences from Ethics Consultation

It would be disastrous if medical ethics were to question the caregivers' courage to act by raising doubts, theoretical problems, and uncertainty. Our fundamental experience is not accustomed to tolerate ambivalent feelings and cope with uncertainty. A symptom of this lack of skills is the implicit rule for medical decisions regarding terminally ill patients: pursue a course of action, e.g., prolonging life, as long as you do not feel right deciding not to do so.

First, we must emphasize the common problem of ambivalence. In the case described earlier the decision is made to discontinue treatment for the patient (e.g., hydrating and feeding). This act is good (due to our humane and merciful intentions) as well as bad (due to the difficult and troubling consequences). If decisions could escape this ambiguous situation, neither ethics consultation (Fletcher, Quist and Jonsen, 1989) nor medical ethics would be required at all. Perhaps someone will succeed in distinguishing good from bad, - theoretically. But in practice, according to the title by the book of the psychiatrist Watzlawick (1986), you must respect the bad in the good. Thus tolerating ambivalence, originally a psychoanalytical term, means tolerating an action that is good and bad at the same time. It means to be ready to act without being protected by some absolute good, i.e. to act in the moral twilight.

The ward staff requested ethics consultation because they felt that treating someone who must die is a cruel act. Before dealing with this issue we must explore the question who has the problem: the dying patient or the caregivers? We must be precise. Do we stop treating a patient because we feel uncomfortable doing this? However, the patient is our major concern. We cannot pass judgement on withdrawal or withholding treatment from the professionals' perspective alone.

During case consultation all concerned had to be ready to communicate with the others involved and explain their reasons for preferring one of the options, their hopes for a specific outcome and their anxiety about failing. A net of communication must be brought into the clinical setting. Some psychiatrists referred to this net of communication as the "sharing of uncertainties" (Gutheil, Bursztajn and Brodsky, 1984). This understanding of sharing must embrace the patients or their relatives as well as the caregivers and their consultants. It presupposes that every participant shows a deep rooted commitment, i.e. the ability to cope with this situation through the sharing of uncertain emotions.

Underlying this idea of shared decision-making is the concept of the 'moral community' (Duff, 1988; Illhardt, 1996). This idea should replace the image of the paternalistic helper in splendid isolation and emphasize the mutual aspects in a decision-making process. An ethical decision cannot be found in a drug store system where 'here is the good and there is the wrong'. The basis of an ethical decision is not the pursuit of the right principle but rather the best possible understanding of all factors influencing action. The 'moral community' in this sense consists of a group of persons on the patient's side and on the caregivers' side who attempt to explore these critical moments. Thus, they are bound to each other in finding the patient's good and not in finding what they believe is the patient's good.

During this consultation process we experienced ourselves that death, instead of us, has made a tough decision. It was a mixed feeling: Inability in finding the morally right way is for many caregivers in conflict with their concept of themselves as professionals: A professionalised helper feels obliged to be able (1) to do everything, (2) at every moment, and (3) immediately. But having problems does not appear to be adequate for that concept of self for a health care professional. It is necessary to correct this hypermorality.

3.2. Patient Autonomy

Do we have any methods to study the patient's will (Klaschik and Sandgathe-Husebö, 1998)? Cervantes gave Sancho Pansa, the servant of Don Quixote, a very important proverb: "Man's will is his heaven." If we do not respect the patient's will and right to self-determination, how can we justify turning the patient's life into a hell (see the pleadings of Küng, 1995, and Jens, 1995)?

Let us return to the case. Though patient autonomy must be based on a good doctor-patient-relationship and must be discussed, it is significant that the patient in the case actually showed his autonomy only by pulling out the tubes. The relatives reported that Mr. B. had some feelings against a life sustained only by medical means. They failed to make an assessment of his values and preferences which could be of help in a situation like this. Some ethicists support an ethical assessment because it is an instrument to guarantee the patient's autonomy.

Another way to warrant patient autonomy is the advance directive (Brenahan, 1994; Yellen, Elpern and Burton, 1994). It should not be handled like a driving-licence, it should be understood as the challenge to start an interaction between the patient who takes responsibility and a caregiver who respects autonomy. If Mr. B. would have had an advance directive, no one would have substituted his or her medical or bioethical role with the role of a fortune teller or detective, trying to find out what a patient would have said, if ...

3.3. Difficulty to Assess the Chance of Regaining Health

K.-H. Wehkamp (1998) describes how caregivers have several difficulties when assessing the outcome of the patient's illness:

1. Withholding or withdrawing treatment is a 'non-subject' in medical training;
2. Not providing medical treatment cannot be equivalent with being responsible;
3. Caregivers, particularly doctors, are not trained to determine and emotionally accept the turning point of the disease from curable to incurable.

As long as these difficulties are not removed, the situation of withholding/withdrawing treatment cannot be changed. It is not important to stress the moral point that a physician cannot be obliged to treat a patient when treatment is futile, or to treat a patient who requires the caregivers to accept a principle that they do not understand and support. It is much more important to know the possible sources medical doctors and nurses use for getting advice and consultation. A recent German survey (Schöne-Seifert and Eickhoff, 1996) provides a list of these sources:

Table 2.

OPTIONS	DOCTORS	NURSES
No answer	14%	10%
experience	6%	5%
codes of conduct	11%	24%
no assistance from outside	12%	2%
pastoral service	12%	24%
policies of the hospital	14%	0%
decision of the team	24%	17%
penal law	26%	45%
hospital ethics committee	28%	50%
conscience	42%	33%
medical ethics	62%	71%

Many give priority to the sources of advice and consultation, that are intrinsic to health care professionals like 'conscience', 'medical ethics', 'decision of the team'. They seem to overrule the sources coming from outside. But are the caregivers adequately trained enough to 'consult' the intrinsic sources?

3.4. Change of Perspective

The question of whether the patient shall live or not, cannot be answered by any other person than the patient him- or herself. If the patient has not chosen a representative or an attorney we are not allowed to end his/her life - e.g., by withdrawing nutrition or hydration; the physician can only withdraw medical treatment if he considers the treatment futile.

If we take the example of the case again: Were we right to withdraw nutrition? Whose interests had been touched upon? Those of the patient? Those of his family? Or those of the caregivers? Is the judgement 'futile treatment' an objective medical judgement, or a subjective judgement of caregivers who could not cope any longer with this man and could not see him having a meaningful life?

3.5. Confusion on Hydrating and Feeding

Many physicians and ethicists consider hydrating and feeding vital symbols. They represent basic care of the body. Some experts are convinced that beyond neuronal cognition the body perceives certain kinds of suffering. Thus, discontinuing hydration and feeding becomes comparable with torture. But we must differentiate here. There comes a point in the course of some terminal diseases where patients experience great weight loss, and their forced feeding and hydrating usually involves intense feelings of inconvenience and pain. Not feeding and hydrating patients will disturb some brain functions and produce a feeling of euphoria. Generally, the dehydration seen from the nursing perspective reduces secretion in the throat and bronchi, makes edemas disappear, enhances the analgesia without drugs, and brings the uneasiness of the terminally ill patient under control.

During a 1992 symposium in Basel, the following criteria for discontinuing hydration of the dying person were discussed (Stähelin, 1993; Printz, 1988; Hill, 1994):

1. rapid deterioration,
2. no curative treatment possible,
3. no palliative treatment possible,
4. death occurs any moment from up to several days to a few weeks,
5. presumable consent of the dying patient.

It seems much more convenient to the dying patient to be close to someone who cares about him, i.e. makes his mouth moist, instead of having a balanced osmotic state.

3.6. Realistic Self-esteem of Medicine

We must give up the modern illusion that technique can solve all problems. Also the medical system must accept that it cannot solve all problems. Let me give an example. F. Bacon (1623, republished 1958) illustrated the modern dilemma by calling active and some kinds of passive euthanasia the 'euthanasia exterior'. He suspected the physician who stopped treatment to be stupid because he seemed to be helpless. Did Bacon refuse to accept that medicine sometimes reaches its limits? If we remember the case and ask ourselves the following question: Why did the caregivers have this uncertainty whether they must respect the futile medical process of Mr. B.? Fears for the lawyers cannot explain this uncertainty. Medical professionals know what they are able to do medically. If they are in doubt they ask for counselling. But they do not know why they should do that, it and what the good of it will be.

3.7. The Duty to Care for the Doctor-Patient-Relationship

The psychiatrist and philosopher Karl Jaspers (1986) was convinced that doctor and patient can meet only when they accept that each is able to make rational decisions and choices. Regularly caregivers consider their patients as helpless beings not able to make rational decisions and choices. It is unclear how much caregivers invest in a good relationship. Often it seems to be clear that a physician does not need a good relationship with his/her patient, because medicine appears to work without relationships (Thomasma, 1992). But if this were right, the physician would not have any criterion to decide whether or not to stop treatment. This question has a non-medical impact, and the caregiver must have non-medical competence, too.

In the case discussed, how could the staff interpret the patient's pulling out the intravenous tubes? Was it only an unconscious reflex or a symptom of his suffering? But this problem has been discussed outside the social context, rather as being merely a problem of neurology. On this level it remains an unsettled question. Why have the caregivers not attempted during the previous days to reconstruct the patient's preferences, choices, as well as the meaning of being dependent?

3.8. Consciousness of Finitude

Keeping in mind that all things in life will come to an end, must be the basic idea of all caregivers who are confronted with the final stages of life. This basis can be worked out as an 'ethics of finitude' (Pellegrino, 1993). Its major consequence is that all medical efforts to cure must be limited, because their aims and healing concepts are limited.

In the case discussed, it is significant that the staff put the question forward whether the life of the patient should be prolonged or not. Finitude of life was worth questioning. The caregivers did not regard themselves as ruthless fighters against death and disease. In this context the medical task must be realistic.

3.9. Respect of the Caregivers' Moral Resources

Increasingly confronted with end-of-life decisions health professionals suffer from burn-out syndrome and emotional distress. Emotional distress is one cause of burn-out, and health professionals afflicted by burn-out spread emotional distress among their colleagues. There is a gap between ideal and reality, and to take moral resources into account may provide the means to build a bridge between reality and ideal (Illhardt, 1994) and may reduce the many sources for burn-out, suicidal tendencies among caregivers, their destroyed partnerships and other less dramatic forms of disappointment and dissatisfaction. These points can be illustrated with J. Steinbeck's *East of Eden* (1952). This novel proceeds from the philosophical claim that all people have to work out the implicit concept of themselves as well as of their fellow beings. They find their identity within this concept. The shaping of the physician's self-concept follows the same pattern. Whatever he or she does, it is not done because he or she 'must' do it, but because he or she has chosen to go this way. If the doctor prefers a particular way of action, he or she prefers a certain pattern of being, a specific self-concept.

One episode in Steinbeck's novel illustrates this aspect. Samuel Hamilton, one of the main characters in this novel, is the open-hearted neighbour of the Trask family. He witnessed the cruelty between the brothers Cal and Aaron Trask. Deeply concerned, he visits a wise Chinese man in his village. They discuss the problem of guilt and failure, what they are, and how and why history is affected by them. The wise Chinese man refers to the story of guilt and failure told in the Bible and explains:

> The American Standard translation *orders* men to triumph over sin, and you can call sin ignorance. The King James translation makes a promise in "Thou shalt," meaning that men will surely triumph over sin. But the Hebrew word ... "Thou mayest" ... gives a choice. It might be the most important word in the world. That says the way is open ... That makes a man great, that gives him stature with the gods, for in his weakness and his filth and his murder of his brother he has still the great choice. He can choose his course and fight it through and win ... (Steinbeck, 1952, p. 398)

Evil is not the disobedience to a normative principle, but betraying the self-concept. If somebody (as in the novel) kills his brother, or, as in the case, hurts the integrity of the other, he not only breaks a moral rule, but he is no more the man he wants to become; his own self-concept of humaneness has become obliterated. Killing a patient can never be regular medical practice, not because such acts are forbidden, morally wrong or whatever, but because they darken the self-concept of the professional. Those who want to kill can be influenced neither by judicial norms, nor by considerations of professional identity. That means: Before asking what the physician ought to do, we must ask what are his or her moral resources, what are important elements of his or her self-concept, and how can both be mediated.

4. CONCLUSION

Withholding and withdrawing treatment are not a soft version of active euthanasia. Some people think that the decision to terminate treatment for the dying is morally wrong. But there is nothing to argue against terminating treatment when the decision is according to the structure of therapeutic deliberations. Caregivers who deny their duty to follow the (intrinsic) order of life pay a high price.

1. They deny the medical competence for questions concerning life. Physicians, nurses and other health care professionals give up this competence, if they prolong the processes of life without serving them.
2. They ignore that medical actions are frequently based on probabilistic judgements; absolute certainty in medicine is impossible. They must learn strategies to make themselves certain, e.g., by consultation, (re-)constructing the patient's will, and by identifying the turning point from being ill to dying.
3. They fall behind the principle that no health care professional can neglect the order of life as well as the respect for the autonomy of living persons. They make the same error by active euthanasia as well by prolonging futile treatment, because they disregard the physical and personal order of life.
4. They disregard the proportionality between the different needs for care, because nobody has a right to care when there is no chance of restoring health. Therefore they harm the state of the art, when doing more. No physician can be obliged to do futile medical activities.

Withholding or withdrawing treatment is not a soft version of active euthanasia or a borderline action. Medicine is obliged to withhold and withdraw any treatment, which is no longer covered by the patient's aims and medicine's possibilities. Withholding and withdrawing treatment belong to the essence of medicine if medical practice follows three requirements:

a. It accepts all the above mentioned parameters (patient's autonomy, change of perspectives, realistic selfesteem, doctor-patient-relationship, the ethics of finitude, respect of the own moral resources).

b. It does not decide to stop treatment due to QoL- or SoL-ratio, but for medical reasons. The term of medical reference is 'futility'. Though futility is a vague term it requires the physician to make a responsible and a multidimensional consideration and 'shared decision'.

c. It takes the moral resources of the caregivers into account. They negate their professional - and perhaps - personal identity, if they commit acts of killing and not of treating. Cure and caregiving, but not 'mercy giving' is the job of medical professionals.

ACKNOWLEDGEMENT

Thanks for translation assistance and substantial discussions to E. Herych, M.A., M.A., and J.P. Tossic, M.A.

REFERENCES

Ariès, Ph.: 1975, *Essais sur l'histoire de la mort en Occident du moyen-âge à nos jours*. Edition du Seuil, Paris.

Bacon, F.: 1958, *De dignitate et augmentis scientiarum* [1623]. Spedding, J.D., Ellis, L.R. and Heath, D.D. (eds.): *The works of F. Bacon*. Longman, London. Vol. 1, p. 421-837.

Baines, M.J.: 1992, Symptom management and palliative care. In: Evans, J.G. and Williams, T.F. (eds.): *Oxford Textbook of Geriatric Medicine*. Oxford University Press, Oxford/New York/Tokyo, p. 685-696.

Birnbacher, D.: 1995, *Tun und Unterlassen*. Reclam, Stuttgart.

Brenahan, J.F.: 1994, Advance directive in the 1990s: Medical care of the dying and the myth of Sisiphus. In: Monagle, J.F. and Thomasma, D.C. (eds.): *Health Care Ethics. Critical Issues*. Aspen Publishers, Gaithersburg, p. 213-216.

Clarke, D.B.: 1994, The patient self-determination act. In: Monagle, J.F. and Thomasma, D.C. (eds.): *Health Care Ethics. Critical Issues*. Aspen Publishers, Gaithersburg, p. 93-108.

Dallas, M.H.: 1987, Planning for decisional incapacity. A neglected area in ethics and aging. *Journal of the American Geriatrics Society* 35: 814-20.

Duff, R.S.: 1988, Unshared and shared decisionmaking. In: King, N.M.P., Churchill, L.R. and Cross, A.W. (eds.): *The physician as captain of the ship: a critical reappraisal*. Reidel, Dordrecht/Boston, p. 191-221.

Fletcher, J.C., Quist, N. and Jonsen, A.R. (eds.): 1989, *Ethics consultation in health care*. Health Adminstration Press, Ann Arbor.

Gutheil, T.G., Bursztajn, H. and Brodsky, A.: 1984. Malpractice prevention through the sharing of uncertainty. Informed consent and the therapeutic alliance. *New England Journal of Medicine* 311: 49-51.

Hill, T.P.: 1994, Ethical issues in the use of fluids and nutrition: When can they be withdrawn? In: Monagle, J.F. and Thomasma, D.C. (eds.): *Health Care Ethics. Critical Issues*. Aspen Publishers, Gaithersburg, p. 217-225.

Illhardt, F.J.: 1996, Wem gehört der Mensch? Erwägungen über Besitzrechte am eigenen Körper. *Berliner Medizinethische Schriften* Heft 3, Humanitas Verlag, Dortmund.

Illhardt, F.J.: 1994, Das Selbstbild des Arztes: 'Missing Link' zwischen Ideal und Praxis? In: Leven, K.-H. and Prüll, C.-R. (eds.): *Selbstbilder des Arztes im 20. Jahrhundert. Medizinhistorische und medizinethische Aspekte*. (Freiburger Forschungen zur Medizingeschichte. Est. Aschoff, L. Ed. by Seidler, E. and Kuhlo, B. Neue Folge Bd. 16) HF Schulz, Freiburg, p. 1-37.

Jaspers, K.: 1986, Die Idee des Arztes. In: Jaspers, K. (ed.): *Der Arzt im technischen Zeitalter*. Piper, Mainz/München, p. 7-18.

Jens, W.: 1995, Si vis vitam para mortem. Die Literatur über Würde und Würdelosigkeit des Sterbens. In: Jens, W. and Küng, H.: 1995, *Menschenwürdig sterben: ein Plädoyer für Selbstverantwortung*. Piper, München, p. 87-129.

Keenan, J.F.: 1996, The concept of sanctity of life and its use in contemporary bioethical discussion. In: Bayert, K. (ed.): *Sanctity of life and human dignity*. Kluwer Academic Publishers, Dordrecht/Boston/London, p. 1-18.

Klaschik, E. and Sandgathe-Husebö, B.: 1998, Das Leben des Sterbenden - Palliativmedizin. In: Illhardt, F.J., Heiss H.W. and Dornberg, M. (eds.): *Sterbehilfe - Handeln oder unterlassen?* Schattauer, Stuttgart, p. 105-111.

Küng, H.: 1995, Menschwürdig Sterben. In: Jens, W. and Küng, H.: 1995, *Menschenwürdig sterben: ein Plädoyer für Selbstverantwortung*. Piper, München, p. 13-85.

MacKay, C.R.: 1991, The physician as fortune teller: A commentary on 'the ethical justification for minimal paternalism'. *Journal of Clinical Ethics* 2: 228-238.

Pellegrino, E.D.: 1989, Withholding and withdrawing treatments: Ethics at the bedside. *Clinical Neurology* 35: 164-184.

Pellegrino, E.D.: 1993, A philosophy of finitude: Ethics and the humanities in the allocation of resources. In: Winslow, G.R. and Walters, J.W. (eds.): *Facing limits*. Westview, Boulder, p. 31-53.

Printz, L.A.: 1988, Is withholding hydration a valid comfort measure in the terminally ill? *Geriatrics* 43: 84-87.

Schöne-Seifert, B. and Eickhoff, C.: 1996, Behandlungsverzicht bei Schwerstkranken: Wie würden Ärzte und Pflegekräfte entscheiden? Eine Vignettenstudie. *Zeitschrift für Ethik in der Medizin* 8: 183-216.

Seale, C. and Pattison S. (eds.): 1994, *Medical knowledge: doubt and certainty* (Completely rev. version). Open University Press, Buckingham.

Stähelin, H.B.: 1993, Flüssigkeitszufuhr bei terminalen geriatrischen Patienten. *Schweizerische Rundschau für Medizin (Praxis)* 82: 792-94.

Steinbeck, J.: 1952, *East of Eden*. Viking Press, New York.

Thomasma, D.C.: 1992, Models of the doctor-patient relationship and the ethics committee: Part one. *Cambridge Quarterly of Health Care* 1: 11-31.

Truog, R.D., Brett, A.S. and Frader, J.: 1992, The problem with futility. *New England Journal of Medicine* 326: 1560-64.

Watzlawick, P.: 1986, *Vom Schlechten des Guten. Oder Hekates Lösungen*. Piper, München.

Wehkamp, K.-H.: 1998, Therapieverzicht: Dilemmata ärztlichen Entscheidens. In: Illhardt, F.J., Heiss, H.W. and Dornberg, M. (eds.): *Sterbehilfe - Handeln oder unterlassen?* Schattauer, Stuttgart, p. 25-33.

Yellen, S.B., Elpern, E.H. and Burton, L.A.: 1994, Patient concerns about advance directives. In: Monagle, J.F. and Thomasma, D.C. (eds.): *Health Care Ethics. Critical Issues*. Aspen Publishers, Gaithersburg, p. 109-115.

HENK A.M.J. TEN HAVE

EUTHANASIA

1. HISTORICAL BACKGROUND

In 1969, the Dutch psychiatrist J.H. van den Berg published a booklet *Medical Power and Medical Ethics* (van den Berg, 1969). The author was professor of psychiatry at Leiden University. Immediately, this early critique of the increasing power of modern medicine became a best-seller, with 25 editions and translations in many languages. Van den Berg argues that the new technological capabilities of medicine had been counterproductive in many cases because they were guided by the traditional medical-ethical prescript to maintain, restore and protect human life. The book contains photographs, for example, of children with phocomelia and hydrocephalus and of a patient whose entire body below the lungs and stomach had been removed. These photographs convey the message that modern medicine applied every technique and intervention possible, without considering how the patients were benefitting from applications of medical technology. As long as medicine is dogmatically based on the moral principle of sanctity of life, medical interventions will only create harmful and cruel conditions for patients. Van den Berg argues that modern medicine should reorient itself, from preserving biological human life to sustaining meaningful personal life. As a phenomenological psychiatrist and disciple of one of the protagonists of anthropological medicine, van den Berg intended to criticize the current image of man in scientific medicine.

Van den Berg's book soon was followed by other critiques, such as Ivan Illich's *Medical Nemesis* (1975), illustrating with a wealth of empirical data that the medical establishment has become a major threat to health. Illich showed how death had become a clinically supervised event with people losing their right to preside over their act of dying.

While less an attack on the medical establishment, the contemporary work of Elisabeth Kübler-Ross (1975) also called for more emotional support and counselling in the final phases of human life rather than technical interventions. The dying person should be allowed to die his or her own

H.A.M.J. ten Have and B. Gordijn (eds.), Bioethics in a European Perspective, 469-486.

death; but the ethos of medical interventionism precludes the patience and tolerance required by the respect for the process of dying one's own death.

A review of the literature in the 1960s and 1970s on medicine's understanding of and approach towards death shows how contemporary medicine was criticized for aggressive treatments associated with unnecessary suffering and lack of compassionate care for the terminally ill (Clark and Seymour, 1999). Better understanding of this social context can help to explain why the euthanasia movement has arisen in the Netherlands, and many other Western countries during the last decades. This movement started as a protest against medicine's growing tendency to alienate individuals from their own death and dying. It therefore emphasized respect for the patient's right to self-determination as the ultimate moral principle in matters of life and death. Provided it is the expression of the autonomous wish of the patient, the attending physician should respect the patient's decision to discontinue medical treatment. Similarly, if a patient's request for euthanasia is autonomous, physicians should comply with such a request. In sum, the moral principle of respect for individual autonomy was the cornerstone of the original ideological theory justifying the medical practice of euthanasia. Patients wanted to regain control over their own life, including its final stages.

What is difficult to explain, however, is why this social context that is similar to that of many Western countries has given rise to such different responses. Somewhat schematically, it can be argued that within the same period three different movements have originated: (1) in the U.S.A., death and dying issues primarily focused on the topic of withholding and withdrawing medical treatment (with the Quinlan case as a first landmark in 1976), (2) in the U.K., the focus was first of all on the hospice movement (with Cicely Saunders' foundation of St Christopher's Hospice in 1967), (3) in the Netherlands, death and dying issues came to be centred around the euthanasia topic (with the Postma case in 1973 as the legal paradigm). All movements emerged from analogous critical settings; they also had the same preoccupations with good death, self-determination, and limiting medical interventions.

Euthanasia now is a major topic of medical-ethical and legal debate in many countries. The debate in the Netherlands, however, is exceptional, not only because it started earlier than in other countries, but also because it is a rather open and public debate in which all parties in society participate. The legal context contributes significantly to these peculiarities of the debate since the legal system tolerates practising euthanasia, although it is at the same time prohibited by law. Since the first court case in 1973, public debate on euthanasia in the Netherlands has become more intense (de Wachter, 1992). However, the past decades have also shown a shift in the Dutch euthanasia debate from the level of critical medical-ethical arguments, justifying or opposing euthanasia within the physician-patient relationship, to the socio-ethical and political problems of whether and how to regulate the actual practice of euthanasia given newly accumulated empirical data. Medical-ethical viewpoints regarding euthanasia in clinical practice have been moved to the background. Emphasis has now been put more on regulation and effective control. Recently, more attention has been paid to possible preventive strategies, such as efficient palliative care (Janssens, Zylics and ten Have, 1999).

2. THE MEDICAL PRACTICE OF EUTHANASIA

Although active euthanasia has always been a criminal offence, in the past two decades medical specialists and general practitioners have been quite open about their euthanasia practice, publishing case reports in influential Dutch medical journals (for example, Meyboom-de Jong, 1983; Weisz, 1994). This professional candour has coincided with (and probably was fostered by) a considerable judicial lenience towards physicians practising euthanasia under strict conditions. Among a variety of conditions, three have been referred to in court decisions and bills more frequently: (1) the patient's voluntary and persistent request, (2) the hopeless situation of the patient and (3) consultation of a colleague (Welie, 1992).

Yet in spite of this professional openness and legal lenience, many physicians who perform euthanasia were not prepared to face the risk of the legal consequences of their practice and completed death certificates incorrectly. Consequently, the overall incidence of active euthanasia in medical practice remained unknown for a long time; estimates varied from 2,000 to 20,000 per year. In January 1990, the 'Remmelink Committee', comprising three lawyers and three physicians, was established to obtain an empirical understanding of the frequency and nature of euthanasia in medical practice. A random population of some 400 physicians were retrospectively interviewed about their own experience with end-of-life decisions; furthermore, the same physicians were asked to prospectively provide (anonymously) information about the true cause of death of each of their dying patients in the next sixth months; and finally, an attempt was made to verify the cause of death of a random sample of some 8,500 recent deaths. In September 1991, the Committee issued its report (Commissie Onderzoek Medische Praktijk inzake Euthanasie, 1991). The results of the empirical part of the Remmelink Report have been published elsewhere in the English language (both the complete report and summary articles; van der Maas, *et al.*, 1991, 1992).

According to research findings, physicians made decisions about their patients' deaths in some 49,000 of approximately 130,000 patients who died in the year 1990. These decisions included whether to discontinue life support, provide increasing doses of pain medication, withhold treatment, assist in suicide, or commit euthanasia. Assisting in suicide was found to occur in some 400 cases a year. Euthanasia, 'intentionally ending life', is practised some 2,300 times, or in 5% of those 49,000 cases. Unexpectedly, it was also found that there were 1000 cases of termination of life without explicit request. At the same time, a separate descriptive, retrospective study of morbidity, age and sex of patients whose family practitioner helped them to die, was carried out by van der Wal (1992). He found that in euthanasia or assisted suicide cases, 85% of the patients suffered from malignant neoplasm (euthanasia mortality rates among patients with cancer were twice the normal cancer mortality rate); euthanasia or assisted suicide were also relatively frequent among patients with AIDS or multiple sclerosis (more than ten times the normal AIDS mortality rate). In approximately 20%, a secondary, usually chronic disease had been diagnosed. Van der Wal also sent questionnaires to a random sample of family practitioners asking them to rate 24 aspects of the suffering of the last patient they had euthanized as well as to assess the patient's life expectancy. According to the respondents, 90% of these patients showed severe physical suffering and 71% severe emotional suffering. 'General weakness or tiredness', 'dependence or being in need of help', 'loss of dignity' and 'pain' were the most frequently identified aspects of suffering. In 63% of the cases, life expectancy at the moment of execution of the request, was estimated as less than 2 weeks; in 39% it was less than 1 week; in 3% less than a day; in 10% it was more than 3 months (van der Wal, 1992).

The survey of the Remmelink Committee was replicated five years later. In 1996, van der Maas and van der Wal published the findings of this second extensive study of end-of-life decisions (van der Wal and van der Maas, 1996; van der Maas, *et al.*, 1996; van der Wal, *et al.*, 1996). The quantitative data of both surveys are reproduced in the table.

Table 3. End-of-life decisions in the Netherlands, 1990-1995

	1990	1994
Deaths in the Netherlands	128,800	135,700
Requests for euthanasia		
- later in the disease	25,100	34,500
- explicit	8,900	9,700
Euthanasia	2,300	3,200
Assisted suicide	400	400
Life-termination without explicit request	1,000	900
Withholding/withdrawing treatment	22,500	27,300
- at the explicit request of the patient	5,800	5,200
- without the explicit request of the patient and explicitly intended to shorten life	2,670	14,200
Intensification of pain and symptom management	22,500	20,000
- explicitly intended to shorten life	1,350	2,000

3. LEGAL SITUATION

In 1993, both Houses of the Dutch Parliament agreed with new legislation concerning euthanasia. On June 1, 1994 the new law came into force. The Netherlands is therefore the first country with explicit legislation on this topic. However, it should be emphasized that the new law does not legalize euthanasia. The law takes as starting points that: (1) euthanasia is to be defined as an active medical intervention to intentionally terminate life at the explicit request of the patient; withholding or withdrawing treatment that is medically futile or refused by the patient (improperly characterized as passive 'euthanasia') is generally accepted and legally sanctioned (since a 1973 court case), (2) euthanasia must remain open to legal audit (which is possible only if it remains a criminal offence).

Consequently, the Criminal Code has not been changed and the practice of euthanasia remains illegal. The new law emphasizes that a case where death has resulted from medical actions accelerating death may not be reported by the attending physician as a 'natural death.' The attending physician should report the case to the local coroner who will inform the prosecutor. The latter will then decide on a case-by-case basis whether prosecution should follow. The legal foundation for this reporting procedure should enable the public prosecutor to assess each case of euthanasia. The physician involved in a case of euthanasia must submit a written report to the coroner that includes the following information:

1. the patient's medical history;

2. the request to end life, as well as to whom this request was voiced; here, two situations could occur:
a. there was an expressed, voluntary and carefully considered request of the patient himself;
b. there was no expressed request of the patient; in this case, the physician must explain the reason for the absent request;
3. the consultation of a colleague;
4. the means used to end life.
On the basis of this report, the public prosecutor will decide whether or not to prosecute.

The main objective of the legislation is to allow better public control of the practice of euthanasia. The law is in fact an amendment of the Burial Act. When a patient dies, a physician has to fill in a death certificate. Here, he has two options: natural death or unnatural death. In the latter case, the police and prosecution office will start an investigation into the cause of death. Given these two options, most doctors performing euthanasia in the past did notify 'natural death', avoiding a police investigation. The law in fact created a 'third' category: doctors performing euthanasia should indicate this category, fill in a questionnaire, notify the coroner (a physician), who will check whether the jurisprudential criteria have been met; the coroner will then consult the prosecution office, which will not prosecute in cases where the criteria have been satisfactorily fulfilled. Therefore, the law introduces a better public control by the legal authorities of the medical practice of euthanasia. The law, however, does not define that or in which cases euthanasia is legal. One criticism of the law is that it does not make any distinction between termination of life at the request of the patient and without explicit request. Both conditions have to be reported. Another criticism is the paradoxical nature of the legislation. Since termination of life is formally a crime, physicians who break the criminal law are obliged by the law to report unlawful activities, and provide the evidence on the basis of which they might be prosecuted.

However, the law apparently does not realize its objective. The 1995 survey shows that the majority of physicians do not report life-terminating acts. In case of euthanasia (on request), 59% of all euthanizing physicians continues to complete a natural-death certificate after having committed euthanasia (although this was a higher percentage than found in the 1990 survey). Of the 900 cases in which the life of a patient was terminated without his or her explicit request, a total of three cases was reported.

Recently, Dutch parliament accepted new special legislation. This law accepts that each case will be reviewed retrospectively by a special ethics committee. If the case is compliant with the jurisprudential criteria, the prosecutor will no longer be involved. The new law came into force at the close of 2001.

4. MORAL EVALUATIONS

It is difficult to conclude from the empirical findings that euthanasia in the Netherlands apparently is not as frequent as assumed by both protagonists and antagonists. One problem is that this conclusion fails to take into account that many physicians do not interpret and classify their actions as euthanasia cases, even when those actions strictly fall under the range of the definition employed in the Netherlands, that is, an active medical intervention to intentionally terminate life at the explicit request of the patient (Gunning, 1991; ten Have and Welie, 1992). For example, the empirical data from both surveys reveal that hastening death was the explicit intention of the administration of high doses of 'pain' medication in at least some of the total number of cases in which such analgesics with a possible lethal effect were administered (in 1990 in 1% of all deaths; in 1995 in 1.5% of all deaths). And in an additional 5.2% (1990), respectively 2.1% (1995), death was at least partly intended. Thus, there is no longer a case for indirect effect, since death is the intended, direct effect. Also in cases of withholding and withdrawing life-sustaining treatment (including tube-feeding) it was the explicit intention to shorten life, even without the explicit request of the patient, in 2.1% of all deaths in 1990, and in 10.5% of all deaths in 1995.

Van Delden and colleagues (1993a) have argued that the formulations of the intentions (i.e., hastening death) seem to be the same in the above mentioned cases as they are in euthanasia proper, but that the 'sameness' of the intentions can be questioned. They claim that intentions ultimately are private and, therefore, beyond public evaluation. It may be agreed that in many a case it will be very difficult to *prove* the intentions of the physician who hastens death; but this is primarily a lawyers' problem. Moreover, the intention of the actor has been made an essential element in the official definition of euthanasia, and even in the legal definition, in spite of the foreseeable difficulties to prove the physician's intentions.

The point is that if the intention to terminate the life of the patient is definitive for euthanasia, the number of euthanasia cases in the Netherlands is considerably higher than the 'official' Dutch definition of euthanasia suggests, and with that the research data under this category.

Another difficulty with the interpretation of the medical practice of euthanasia is whether or not there is a slippery slope. From the surveys it can be concluded that within a short five-years period the number of requests for euthanasia increased as well as the number of requests granted. It is also obvious that for an increasing number of patients requesting euthanasia has become normal behaviour; the number of requests for euthanasia in due course, as a kind of guarantee early in the disease process against suffering, has increased with 37%. Furthermore, the majority of Dutch physicians has been personally involved in life-terminating acts (53% of all physicians; 63%

of all general practitioners). Nonetheless, the 1995 survey also showed that only 41% of cases were reported. Also, consultation of a colleague (one of the formal conditions for permissible euthanasia) did not occur in 21% of cases. These facts, combined with the substantial number of life-terminating actions without explicit request, have led some authors to the conclusion that the Dutch practice is sliding down the slippery slope (Hendin, 1997; Jochemsen and Keown, 1999). Other commentators argue that it is not (Angell, 1996). Whether or not there is a worsening of the situation, interpreted as a slippery slope, Gillon's conclusion seems correct:

> What is shown by the empirical findings is that restrictions on euthanasia that legal controls in the Netherlands were supposed to have implemented are being extensively ignored and from that point of view it is surely justifiable to conclude ... that the practice of euthanasia in the Netherlands is in poor control (Gillon, 1999, p. 4).

An unexpected finding, given the inclusion of the explicit request in the 'official' definition of euthanasia, was the number of cases without explicit request of the patient. The surveys show that there are about a 1,000, resp. 900 patients whose death has been caused or hastened by physicians without an explicit request. This number pertains to patients who no longer were competent to make decisions, yet apparently suffered severely. Not included are cases where medically futile treatments were withheld or withdrawn, since such treatments always have to be withdrawn. Although it is not clear how many of these cases are *in*voluntary (i.e., the patients would have expressed the wish *not* to be euthanized, had they been able to speak about it), the absence of an expressed request precludes qualifying these cases as euthanasia cases proper. Nonetheless, the Remmelink Committee felt in its 1991 report that these 1,000 cases of *non*voluntary termination of life should not be of concern either; moreover, they should be thought of as 'providing assistance to the dying'; nonvoluntary termination of life was justified because the suffering of those patients had become 'unbearable' and standard medical practice prescribed considering their life as 'given up'. Death would have occurred quickly anyway (usually within a week), if the physician had not acted. Elsewhere, the Committee adds that actively ending life when "the vital functions have started failing", is "indisputably normal medical practice" (Commissie Onderzoek Medische Praktijk inzake Euthanasie, 1991, p. 15 en p. 32).

Although about a quarter of these 1,000 patients had previously expressed the wish to die, interestingly this was not always the leading argument for the physicians euthanizing them. Only 17% of the physicians involved in these 1,000 cases mentioned 'previously uttered request of the patient' as their reason to terminate their patients' life. The researchers explained this discrepancy by arguing that physicians more often are guided by their own 'empathy' with the patient's unspoken but probable wishes, than by explicit oral or written patient requests (*O.c.*, p. 51).

This explanation indicates a significant shift in moral justification. Respect for autonomy had always been the prime argument in favour of active euthanasia, publicly defended by the physicians, lawyers and courts. But now that quite a number of cases turned out to happen without an explicit patient request, other arguments are brought forward to defend this practice. Thus, a paradox emerges between this line of reasoning, and the very opposite reasoning by the original advocates of voluntary euthanasia that suffering is a purely subjective phenomenon, that, consequently, only the patient can decide whether his or her suffering has become unbearable, and that, finally, termination of life is only justifiable when the patient so requests. It seems that some advocates of euthanasia use the latter strategy when defending the right of the competent patient to autonomously opt for euthanasia, and the former strategy when defending the practice of euthanasia on the mentally incompetent patient. A similar ambiguity is shown in Van der Wal's study (van der Wal, 1992). His conclusion that the majority of euthanized patients *has* severe physical and emotional suffering, does not follow. It merely can be concluded that the physicians in retrospect *think* this about their patients.

What the medical practice of euthanasia in the Netherlands reveals is that the ethical justification has been shifting from *respect for autonomy* to *relief of suffering*. But this has created a tension within the justificatory strategies regarding euthanasia. The two arguments are mutually exclusive. It only makes sense to talk about respect for autonomy if a physician *refrains* from making judgements about the patient's benefits. It simply is logically impossible to base a euthanasia decision on *both* autonomy and beneficence, as suggested (van Delden, *et al.*, 1993b). Moreover, the primacy of the bioethical principle of respect for patient autonomy has always been grounded in the presumed inability - or virtual inability - of physicians (or any other third persons) to make reliable judgements on the patient's well-being or suffering. If, on the other hand, physicians are now considered to be very well able to make such judgements, the decisive factor is no longer the patient's own explicit request for euthanasia but the physician's judgement concurring with the patient's assessment of the suffering being unbearable. The physician will only comply with the 'autonomous' request of the patient if (s)he agrees that the patient's suffering is, indeed, unbearable or the quality of the patient's life is so low, that the patient is better off dead. In fact, the patient's request will only be regarded as an *autonomous* request if it is rational from a *medical* point of view. In this line of reasoning, nothing changes when the same medical rationality indicates that euthanasia is appropriate, but the patient is no longer able to express an autonomous wish.

When the patient is incompetent or his views are unavailable, the physician is still capable of making the assessment.

Both empirical research and political debate reveal that in daily practice two moral considerations compete with each other: respect for autonomy and relief from suffering. From the physician's point of view, the latter consideration appears to be the most important; it is the prime motive to perform euthanasia in cases of incompetent patients who, in the judgement of the physicians, suffer unbearably. It is also a strong motive in cases of competent patients since less than a third of all requests are fulfilled.

The emphasis on suffering as predominant in the moral justification could have been expected, considering the history of euthanasia. The term derives from the Greek for 'good' or 'merciful' death. Definitions of euthanasia often refer to suffering from incurable diseases as the fundamental condition. It is also argued that the crucial difference between euthanasia and murder is the motive; murder would be killing for reasons other than kindness (Thomasma and Graber, 1990). The history of euthanasia primarily is the history of 'mercy killing'. It is the argument of compassion, not the argument of respect for autonomy which has been the most basic moral justification for euthanasia (Meerman, 1991).

This observation implies that the outcome of the euthanasia debate is paradoxical. Physicians now seem to have ultimate control over the moral justification of active euthanasia. If the life of a patient is terminated because the physician fell morally justified to do so on the basis of the unbearable suffering of the patient, it is difficult to distinguish the compassionate involvement of the doctor from physician paternalism. The doctor is not wicked or criminal; he has the best possible motives and offers the most compassionate care available, but yet, this is paternalistic behaviour. The doctor knows best when it is your time to die.

The best thing to do is also to trust your physician. The moral sensibility and judgement of the medical profession apparently is the only safeguard against abuses. It is, nonetheless, unclear why and how medical professionals are competent to make judgements about their patients' suffering, and whether they have the right to do so - a question raised but often unanswered (van der Wal, 1993). Intervention-driven medical technology has led to the euthanasia movement in the first place, yet there is no sign that this characteristic of medicine has significantly changed. What was the initial cause of the problem, is now considered the prime solution to it (ten Have, 1998).

5. RECENT DEVELOPMENTS

Since the law has been changed, public debate on euthanasia has in fact intensified. Several developments are significant. First, there seems to be a growing tendency to regard active termination of life as normal medical practice: it is the rational response in hopeless cases; it is also a professional activity of a compassionate physician. Such at least have been the messages of the documentary 'Death on request' broadcasted on Dutch television in October 1994, and since then shown in various countries. A major impact on public debate was also made by a ruling of the Supreme Court in June 1994 in the so-called Chabot case. The Court accepted a case of a severely depressed woman who was assisted in suicide by the psychiatrist dr. Chabot. The Court agreed that there can be cases of non-physical suffering in which death is a better option. This ruling caused tremendous upheaval, because many doctors now felt that they could be requested to end the lives of patients with existential and psychic problems, where in fact medicine is considered as an instrument to end the life of a patient who does not want to live anymore, for whatever reasons. Psychiatrists in particular argued that it is wrong and dangerous to consider death wishes of depressed patients as an indication of a euthanasia request. The Court ruling in fact eliminates a basic criterion prominent in classical euthanasia cases, *viz.* the terminal phase of a physical illness. Euthanasia in the classical case used to be regarded as an ultimate refugium within a process of caring for dying patients and counselling the patient in the terminal phase of his or her life. Now the Court ruling emphasized that a state of suffering is more significant than the existence of somatic illness or the terminal phase of an illness. And what should be regarded as suffering is in fact within the competency of the patient himself or herself.

A second development has undermined another major criterion of the classical case: the voluntary and explicit request. The current definition of euthanasia stresses the request of the patient. The government surveys already disclosed that there were a considerable number of annual cases in which the life of a patient was ended without explicit request of the patient. Public debate now concentrates on the question whether it can be justified to medically end the life of an incompetent patient. The survey data caused little outcry in the Netherlands. In fact, the number of advocates of the possibility of nonvoluntary termination of the life of an incompetent patient only seems to have increased. Professional committees (of the Royal Dutch Medical Association, the Dutch Society of Pediatricians, and the Dutch Society of Psychiatry) have argued that actively terminating the life of certain comatose patients and disabled newborns can be morally justifiable. And with regard to patients suffering from a psychiatric illness, it is argued that their disorder does not necessarily reduce their voluntary decision-making competency;

hence, the acceptability of euthanasia cannot be ruled out a priori (Nederlandse Vereniging voor Kindergeneeskunde, 1992; Nederlandse Vereniging voor Psychiatrie, 1992; Commissie Aanvaardbaarheid Levensbeëindigend Handelen KNMG, 1997). In April 1995, the Court of Alkmaar was the first to consider a case of active termination of the life of a severely handicapped newborn, the so-called Prins case (named after the gynaecologist who gave the baby a lethal injection). The physician was exempted from punishment because the court ruled that in this case the suffering of the neonate was indeed so severe that it provided a moral justification for ending her life; another justification was found in the parents' wish that she would die. This case was followed by another one, the Kadijk case, in which a lethal injection was given to a baby with trisomy 12, a severe genetic abnormality. In both cases, the physicians were finally acquitted by the Court of Appeal because of futility of continued treatment, the acceptability under these circumstances of active termination of life, and the applicability of *force majeure*: the physicians had faced an unavoidable conflict of duties between the duty to prolong life and the duty to alleviate unbearable suffering (Klotzko, 1997).

Such events indicate a significant change in attitude, not only towards death and dying, but towards suffering, growing old, and being dependent on long-term care. Public opinion polls show that among the population agreement with life-termination without request is even more frequent than agreement with euthanasia (69% versus 56%; Sociaal en Cultureel Planbureau, 1997). The Dutch Society for Voluntary Euthanasia advocated in 1992 that everyone who fears to be admitted to a nursing home should have a declaration ready requesting euthanasia in case of impending admittance to a nursing home (Jannink-Kappelle, 1992). Nursing home physicians reacted against the Society's proposals, arguing that they only enhance popular prejudices against nursing homes without taking into account the positive effects that treatment and care can accomplish (Hertogh, *et al.*, 1992). Nonetheless, in 1999 for the first time a doctor was legally acquitted for terminating the life of a patient in the early phase of dementia.

Social developments towards nonvoluntary termination of life, despite the ideological framework emphasizing patient's autonomy, are difficult to deny. One of the underlying reasons for this development is the ambivalence of ethical justifications, discussed above. A second reason is the blurred distinction between active and passive euthanasia. Authoritative Dutch ethicists, such as Kuitert, a protestant theologian, have argued that the distinction between active and passive euthanasia is no longer relevant: the physician is thought to bear responsibility for the patient's death, not only when actively terminating the patient's life, but also when withdrawing life-saving treatment (Kuitert, 1993). Ignoring the moral significance of the intentions of the physician has far-reaching consequences. Not only does it nullify the moral distinctions between withholding/withdrawing medical treatment (resulting in the patient's death) and active termination of life (causing the patient's death); it also implies that if a physician has withdrawn life-sustaining treatment, and the patient does not die, the physician must be prepared to actively end the patient's life. Moral consistency would require this second step - which is what Kuitert defends.

But this line of arguments clearly misrepresents actual medical practice. Undeniably, the empirical surveys have shown that cases do occur where treatment is discontinued primarily to hasten the patient's death. But there is no doubt that death is not always the intended outcome of a physician's decision to discontinue treatment. Treatment in many cases is withdrawn because it is deemed medically futile and, therefore, no longer beneficial to the patient. Conversely, if available medical remedies were not judged medically futile, treatment would have been continued and the patient would not have died. From the fact that *some* cases of letting die are justified, it does not follow that all such cases are justified. Some cases of letting die are clearly *unjustified*, for example, when a patient is left to die when beneficial treatment is available. Here the treatment is not given because someone simply wants the patient to die. In such cases, there is, indeed, no morally relevant difference between killing and letting die, but this is not true for all cases.

There is yet another reason to uphold the moral significance of the distinction between discontinuing medically futile treatment and withdrawing beneficial treatment in an attempt to hasten death (ten Have, 1996). Categorizing all forms of withholding and withdrawing medical treatment in the category of euthanasia is possible only on the assumption that such withdrawals are *acts* of ending human life. But that assumption is mistaken. When medically futile treatment is discontinued, the life of a patient ends; but the agent ending the patient's life is not the physician. It is the disease, the underlying pathology, or the patient's general condition that is the cause of death. In such instances, the physician is not morally responsible for the patient's subsequent death. Any other interpretation leads to hypertrophy of physicians responsibilities: they would be morally responsible for the entire course of regretful events, even if they are unable to give that course of events a positive turn. This hypertrophy is counter-productive since it fails to recognize that medicine has its limits: some conditions are not manipulable by medical interventions. When a patient is suffering and no longer benefitting from medical care, *treatment* should be terminated, not the patient's life. Because of its aggressive interventionism and its inability to acquiesce, medicine could be instrumental in creating conditions that bring patients to request active euthanasia.

The third issue that continuously provoked public debate is the question of public control. How is review and evaluation of the euthanizing physician's justifications to be carried out? As noted above, the 1994 legalisation enacting new notification procedures, is not very effective: the majority of euthanasia cases continue to go unreported. The main reason for physicians not to report these cases is their dislike of the complex judicial procedure that the notification of euthanasia involves (55% cited this as one reason among others; see, van der Wal and van der Maas, 1996). In only 18% of the cases in which a physician did not report euthanasia is another colleague was consulted. In cases of life termination without the explicit request of the patient it is 3%.

For several reasons it is doubtful whether the purpose of better public control can be accomplished. First, society's possibilities for controlling the practice of euthanasia are entirely dependent upon the cooperation of those who are involved in this illegal practice. Second, the new law itself is unusual. Maintaining that euthanasia is illegal under the Criminal Code, it is paradoxical to amend another law requiring the trespassers to disclose their illegal deed. Normally, criminals cannot be required to assist in their own conviction; actually, they have all kinds of rights hindering their conviction. It is quite paradoxical to require physicians to assist in their own arrests by disclosing their actions, even more so when those actions legally constitute one of the most serious crimes (i.e., murder). The law is also unusual in that it prescribes how to commit a serious crime; it even provides physicians with official documents to be used when disclosing this crime. This paradox aptly illustrates the ambiguous stance of the government (and society more at large) towards euthanasia. On the one hand it considers killing human beings, though seriously ill and at the verge of death, an extremely problematic practice requiring legal restrictions and judicial audit. On the other hand, the government seems unwilling to rigorously scrutinize the medical profession. It sides with the profession's view that deviations from medical practice by individual physicians within the intimate relationship with their patient are to be thought of as indications of respect for personal wishes and conscientious decisions. Obviously, with this mutual trust and respect between physician and patient, the privacy of their relationship cannot be open to public scrutiny. If euthanasia is the prerogative of physicians, their conscientiousness cannot at the same time be doubted. Remarkably, this line of arguments is widely accepted. Despite decades of sharp criticism of the power of the medical profession, when it comes to decisions about life and death, Dutch physicians seem to only have gained in unconditional trust of society.

Recently, the Dutch government expanded the notification procedure with a retrospective review procedure. Early in 1999, five regional committees were established - each consisting of a lawyer, an ethicist, and a physician. Each committee will judge, retrospectively, whether the euthanizing physician had acted in compliance with the procedure. The committees will subsequently forward their findings to the public prosecutor, who cannot disregard the advice of these committees. For life termination without the request of the patient a central committee will be established. The expectation is that these measures, by diminishing the judicial character of the present procedure, will improve the willingness of physicians to report cases of euthanasia.

The effects of this new review mechanism will be visible in future. However, the workload of the committees will be considerable. If all cases will be reported, each committee will have to review an average number of 720 cases annually (2 cases a day). The central committee, judging the cases of life termination without the patient's request, will have to discuss on average 3 cases every day if, indeed, all cases are reported. Also, the effectiveness of these committees remains uncertain. Will they simply check whether the procedures are satisfied or will they thoroughly judge each case on the basis of medical, ethical, and judicial standards? In the first scenario, the willingness to report cases will certainly increase, whereas this is less likely in the second scenario. If the focus is primarily on increasing reporting behaviour, there is a risk of bureaucratisation of euthanasia. Furthermore, criticism is likely to continue, regarding the moral implications of this review mechanism. If the committees will focus on procedural issues, the participation of an ethicist is clearly misleading. After all, reviewing procedures is rather different from ethical analysis. Finally, it is argued that because of the focus on controlling the euthanasia practice, relatively little attention has been paid so far to preventability of euthanasia. In order to prevent euthanasia, prospective consultation with physicians considering committing euthanasia would have to receive more attention. If palliative care is truly to advance in the Netherlands, professional and institutionalized consultation services will have to be established that precede the decisions of the physicians involved to provide, or to not provide, euthanasia. Creating optimal quality of care for the dying requires easily accessible, prospective palliative care consultation, rather than retrospective case review. Optimal use of expertise in palliative care can make treatment alternatives more visible, and, in all likelihood, may decrease the number of euthanasia cases (ten Have and Janssens, 1997).

6. CONCLUSIONS

The Dutch experience with euthanasia shows that it is imperative to make distinctions between three ethically different conditions:

6.1. Withholding or Withdrawing Life-sustaining Treatment

In these cases, the treatment is ended and the patient can die. It is not the doctor who is the cause of death, but the underlying disease or condition of the patient. Setting limits to medical interventions can help to prevent situations where patients can only die by asking the doctor to end their lives or to assist in causing death. In many countries, bioethical debate is now very much focused on non-treatment decisions. Many hospitals, for example, are developing policies for not-resuscitating patients in certain conditions. Better palliative care and pain management, as other alternatives for euthanasia are now also receiving much attention in many European countries.

6.2. Active Termination of Life at the Request of the Patient

Debates over the last three decades have been focused almost exclusively on this situation. The basic moral question is whether individual persons are allowed to end their own life. If this basic question is answered positively, the next question is whether another human being can assist in ending this life. In the discussion, it is generally assumed that only a physician can do so. But it is unclear what the implications are for medical practice in general. If a doctor can end a human life, medicine will have completely new goals: it will be executing the wishes of autonomous persons (being a kind of service on request), or it will be fighting suffering even if it means to kill the sufferer (being a kind of ultimate care), or both. A more fundamental philosophical concern has to do with the ideal of total self-control which shows itself in practices of managing mortality. The idea of a human being as *causa sui*, producing his or her individual wellbeing and being in control of his or her life and existence, negates the finiteness of human beings. Moral debates on euthanasia apparently revive a very old ideal, for example defended by Pelagius (fifth century AD), that the powers of a human being himself or herself suffice for achieving his or her own perfection, and also that since perfection is possible for human beings, it is obligatory. However, in Western culture these perfectionist ideals have always been criticized and relativized (from Augustine onwards) as a denial of human frailty and a refusal to accept that human life in principle is uncontrollable and beyond personal autonomy.

6.3. Active Termination of Life without the Request of the Patient

In this situation, relief of suffering is regarded as the primary goal of medicine. It is evident, however, that what is regarded as suffering implies a subjective judgement. Physicians are not better equipped than other persons to accept or disqualify various conditions of suffering; their judgements about which kind of suffering is unbearable, may therefore vary widely. In this situation, one may also fear a diminishing acceptability of various conditions (for example in the case of handicapped newborns or demented elderly) within society at large.

REFERENCES

Angell, M.: 1999, Euthanasia in the Netherlands - good news or bad? *New England Journal of Medicine* 335(22): 1676-1678.

Berg, J.H. van den: 1969, *Medische macht en medische ethiek*. Callenbach, Nijkerk.

Clark, D. and Seymour, J.: 1999, *Reflections on palliative care*. Open University Press, Buckingham.

Commissie Aanvaardbaarheid Levensbeëindigend Handelen KNMG: 1997, *Medisch handelen rond het levenseinde bij wilsonbekwame patiënten*. Bohn Stafleu Van Loghum, Houten/Diegem.

Commissie Onderzoek Medische Praktijk inzake Euthanasie: 1991, *Medische beslissingen rond het levenseinde*. Staatsdrukkerij, 's-Gravenhage.

Delden, J.J.M. van, Pijnenborg, L. and Maas, P.J. van der: 1993a, Dances with data. *Bioethics* 7(4): 323-329.

Delden, J.J.M. van, Pijnenborg, L. and Maas, P.J. van der: 1993b, The Remmelink study. Two years later. *Hastings Center Report* 23(6): 24-27.

Gillon, R.: 1999, Euthanasia in the Netherlands - down the slippery slope? *Journal of Medical Ethics* 25: 3-4.

Gunning, K.F.: 1991, Euthanasia. *Lancet* 338: 1010.

Have, H.A.M.J. ten and Welie, J.V.M.: 1992, Euthanasia: Normal medical practice? *Hastings Center Report* 22(2): 34-38.

Have, H. ten: 1996, Een eigen-zinnige visie op mens en zorg. In: Pijnenburg, M. and Vosman, F. (eds.): *Tegendraadse levensvisies. Visies in de gezondheidszorg op leven en lijden onder kritiek*. Van Gorcum, Assen, p. 52-62.

Have, H.A.M.J. ten and Janssens, R.M.J.P.A.: 1997, Regulating euthanasia in the Netherlands. Ethics committees for review of euthanasia? *HEC Forum* 9(4): 393-399.

Have, H. ten: 1998, Euthanasia and the power of medicine. In: Thomasma, D.C., Kimbrough-Kushner, T., Kimsma, G.K. and Ciesielski-Carlucci, C. (eds.): *Asking to die: Inside the Dutch debate about euthanasia*. Kluwer Academic Publishers, Dordrecht, p. 205-220.

Hendin, H.: 1997, *Seduced by death: doctors, patients and the Dutch cure*. W.W. Norton, New York.

Hertogh, C.M.P.M., Ribbe, M.W., Konings, J.W.P.M. and Muller, M.T.: 1992, Het nieuwe levenstestament van de Nederlandse Vereniging voor Vrijwillige Euthanasie. *Medisch Contact* 47: 1532-1534.

Illich, I.: 1975, *Medical Nemesis: The expropriation of health*. Calder and Boyars, London.

Jannink-Kappelle, W.W.: 1992, Euthanasieverklaring/Weigering behandeling; Nieuwe tekst Nederlandse Vereniging voor Vrijwillige Euthanasie. *Medisch Contact* 47: 845-847.

Janssens, M.J.P.A., Zylics, Z. and Have, H.A.M.J. ten: 1999, Hospice and euthanasia in the Netherlands. An ethical point of view. *Journal of Medical Ethics* (in press).

Jochemsen, H. and Keown, J.: 1999, Voluntary euthanasia under control? Further empirical evidence from the Netherlands. *Journal of Medical Ethics* 25: 16-21.

Klotzko, A.J.: 1997, What kind of life? What kind of death? An interview with dr. Henk Prins. *Bioethics* 11(1): 24-42.

Kübler-Ross, E.: 1975, *Death: The final stage of growth*. Prentice-Hall, Englewood Cliffs, NJ.

Kuitert, H.M.: 1993, *Mag er een einde komen aan het bittere einde?* Ten Have, Baarn.

Maas, P.J. van der, Delden, J.J.M. van, Pijnenborg, L. and Looman, C.W.N.: 1991, Euthanasia and other medical decisions concerning the end of life. *Lancet* 338: 669-674.

Maas, P.J. van der, Delden, J.J.M. van and Pijnenborg, L.: 1992, *Euthanasia and other medical decisions concerning the end of life*. Elsevier, Amsterdam.

Maas, P.J. van der, Wal, G. van der, Haverkate, I., *et al.*: 1996, Euthanasia, physician-assisted suicide, and other medical practices involving the end of life in the Netherlands, 1990-1995. *New England Journal of Medicine* 335(22): 1699-1705.

Meerman, D.: 1991, *Goed doen door dood te maken. Een analyse van de morele argumentatie in vijf maatschappelijke debatten over euthanasie tussen 1870 en 1940 in Engeland en Duitsland.* Uitgeversmaatschappij J.H. Kok, Kampen.

Meyboom-de Jong, B.: 1983, Klinische lessen: Actieve euthanasie. *Nederlands Tijdschrift voor Geneeskunde* 127(22): 946-950.

Nederlandse Vereniging voor Kindergeneeskunde: 1992, *Doen of laten: Grenzen van het medisch handelen in de neonatologie*. Ned.Ver.Kindergeneeskunde, Utrecht.

Nederlandse Vereniging voor Psychiatrie: 1992, Mededelingen Bestuur. *Nieuws en Mededelingen* 86(2): 2-3.

Sociaal en Cultureel Planbureau: 1997, *Sociale en culturele verkenningen 1997*. SCP, Rijswijk, p. 162-163.

Thomasma, D.C. and Graber, G.C.: 1990, *Euthanasia. Toward an ethical social policy*. Continuum, New York.

Wachter, M.A.M. de: 1992, Euthanasia in the Netherlands. *Hastings Center Report* 22(2): 23-30.

Wal, G. van der: 1992, *Euthanasie en hulp bij zelfdoding door huisartsen*. WYT Uitgeefgroep, Rotterdam.

Wal, G. van der: 1993, Unrequested termination of life: Is it persmissible? *Bioethics* 7(4): 330-339.

Wal, G. van der and Maas, P.J. van der: 1996, *Euthanasie en andere medische beslissingen rond het levenseinde*. SDU uitgevers, Den Haag.

Wal, G. van der, Maas, P.J. van der, Bosma, J.M., *et al.*: 1996, Evaluation of the notification procedure for physician-assisted death in the Netherlands. *New England Journal of Medicine* 335(22): 1706-1711.

Weisz, F.H.: 1994, Hulp bij zelfdoding. Verslag van een huisarts. *Medisch Contact* 49(21): 700-702.

Welie, J.V.M.: 1992, The medical exception: Physicians, euthanasia and the Dutch criminal law. *Journal of Medicine and Philosophy* 17: 419-437.

MARCEL F. VERWEIJ

RESUSCITATION POLICIES

1. INTRODUCTION

It was one of those meetings *post factum*. It had happened over the weekend but the team was not happy about the way things had gone, and wanted to evaluate the case in a team discussion presided by the hospital's clinical ethicist.

2. CASE

David, a 17-year old youngster, had been admitted to the hospital in the middle of the week before, because of severe coughing and shortness of breath. A first round of diagnostic exams and tests revealed that he was suffering from a relapse of his malignant tumour (located close to the lungs). The young man was informed of the fact that his malignancy was incurable and, understandably, he did not take the news lightly. It took him the rest of the week to slowly recover from this blow and the nursing staff of the paediatric ward was really pleased to see him play the pin ball machines again on Friday afternoon.

However, during the Friday evening team meeting, the attending physician voiced the opinion that if, unexpectedly, David would suffer a cardiopulmonary arrest, he should not be resuscitated. The physician was also of the opinion that David should not be informed about this, at any rate, not before the weekend. David was only now recovering from that first major blow, and during the weekend the care providers who had helped him cope with that blow would not be on call. Besides, the chance of such an arrest was really small.

The team agreed that it would be better not to tell him now, but to wait until after the weekend. But what to do next? If the DNR decision was official, it should be written into the patient's file and he should be told. Nevertheless, it was decided not to tell him and not to make such a formal entry into the patient's file. But what if the unexpected event would happen and he would suffer an arrest? Lacking formal instructions, the nurses on call would have to start resuscitation, then call up the attending physician, who would make a

H.A.M.J. ten Have and B. Gordijn (eds.), Bioethics in a European Perspective, 487-493.

presumably ad-hoc decision that further resuscitation was futile and, hence, should be discontinued.

Fortunately, the weekend went by without such calamities and everybody was relieved. But they also worried about this rather paradoxical protocol. Was it at all an ethically warranted protocol?

3. DISCUSSION

3.1. A First Reaction

The nurses and physicians of the paediatric ward decided to discuss, retrospectively, the ethical dimensions of their Friday-evening decision. Should they have decided differently from a moral point of view? Apparently, the decision sounded suspicious to some of the staff members. Indeed the case of David provokes many moral as well as factual questions. Its most salient feature, perhaps, is that it remains rather unclear what the staff had decided at all. It seems as if all agreed that David should not be resuscitated in case of an arrest, but that, simultaneously, all agreed that a 'full' Do-Not-Resuscitate order would lack a moral basis. If this interpretation of their decision is correct, one could characterise their protocol as a Do-Not-Resuscitate order in disguise. One can imagine that, especially the nurses will feel uncomfortable with this decision. After all, in case of an arrest they should start cardiopulmonary resuscitation (CPR) while knowing that the attending physician would discontinue their attempts on arrival. Their attempts would be more like a theatrical performance rather than a genuine attempt to keep David alive. The protocol reminds us of the so-called 'slow-code': a special way of calling for the hospital resuscitation team, meaning that the team should come, but not run as fast as possible. In the 1970s, when DNR-orders were not yet fully accepted and it was considered appropriate to resuscitate any patient who suffered from an acute cardiopulmonary arrest, such codes were quite common in some hospitals. Physicians and nurses used the slow-code in order to 'save' severely ill patients from resuscitation attempts that would have had doubtful effects. Nowadays, many agree that, if there are doubts about the benefits of resuscitation of a patient, care-givers and patient should discuss the possibility of a DNR-order. Regarding any patient, it should be fully clear to care-givers whether or not they should attempt to resuscitate in case of an arrest. And if they should, they should do it as well as possible. If nurses start resuscitating the patient with the intention and expectation to stop as soon as the physician arrives, they deceive patients and family and perhaps rather harm than help their patient.

3.2. Analyzing the Case

Apparently, it is rather easy to focus only on the team's actual decision concerning David and argue that it cannot survive moral criticism. However, this first critical reaction does not do justice to the moral problem in which the nursing staff and attending physician were trapped on Friday evening. DNR-decisions in disguise and 'slow-codes' may be criticised if they are chosen as an alternative to an open discussion with the patient about the (im)possibilities of life-saving treatment. In David's case, however, care-givers, are certainly willing to talk to David about how they think they should act in case of a cardiopulmonary arrest. Yet they are convinced that the topic should not be discussed right now, this weekend. They assume that David should not simply be 'left alone' during the weekend with more bad news and painful medical decisions. Possibly, information disclosure would be required in order to respect David's autonomy, yet if it is not possible to support him coping with the new information, disclosure could be harmful or even merciless as well.

In analyzing the case, it is important to assess the facts of the case on Friday evening, during the team meeting. The problem arises as a result of the attending physician's voiced opinion that David should not be resuscitated if he were to suffer from a cardiopulmonary arrest. Although it appears that the team agrees with this statement, the reasons for withholding resuscitation are not articulated. Why would it be good or at least acceptable to forgo attempts to save David's life in case of a cardiopulmonary arrest? The answer to that question is indispensable for assessing (1) whether or not a DNR-order would be justified and (2) whether or not David should be consulted or at least informed about such an order.

There are three lines of argument that may support the moral statement that no attempts should be made to keep David alive in case of an arrest. First, if David would have explicitly refused resuscitation, the staff should respect his right to refuse treatment. However, given the case description, it seems unlikely that such an explicit refusal has occurred. Second, if, considering David's condition, resuscitation should be judged medically futile, this would be a good reason for the physician to forego CPR. This seems to be indeed the physician's assumption, as the physician and nursing staff agree that in case of an arrest, the former could make a 'presumably ad hoc decision' that further attempts would be futile. Apparently, in case physicians consider an intervention to be futile, they are allowed to decide unilaterally that it may be foregone. Many physicians, courts and ethicists accept this assumption. However, it is not yet clear whether in David's case, the attending physician's specific assessment of medical facts ('CPR of David would be futile') is warranted. This will be discussed in the next section.

Finally, a third line of argument would be to say that, in the context of David's goals and values, and considering his present medical condition, resuscitation would cease to be a meaningful, beneficial intervention for him, i.e. it would not increase his possibilities to attain certain goals in his life. The patient's own moral perspective and view of life are central in this third line of argument. Therefore, the argument only applies if the patient himself, in casu David, considers treatment to be meaningless. In the Friday evening meeting, David's own view did not play a major part in the debate about resuscitation. However, in discussions among care-givers it is often difficult to distinguish statements like "CPR would be medically futile" from "David would not really benefit from CPR."

3.3. Various Criteria of Futility

Is the physician right when he considers all attempts to resuscitate David, in case of an arrest, to be medically futile? That would depend on David's medical condition, one would say. Unfortunately, the case description contains hardly any accurate medical information. In the past David has been treated for a malignant tumour 'close to the lungs' and now such a relapse has occurred that the physicians deem David's illness incurable. Probably they have (or had already) detected metastases. Nothing is said about David's life expectancy. Does the attending physician have enough factual information to decide that CPR would be futile? The answer to that question partly depends on the definition and criterion of futility that are assumed. An intervention may be called futile if it has no effect or if its effects are of no use. The crux of such a definition is in the phrase 'of no use'. Which standards of usefulness (or success) does the physician assume if he deems an intervention futile? This assumption needs to be clear if physicians are granted unilateral decision-making power to stop life-prolonging treatment they consider to be medically futile. Definitions and criteria of futility have been a major theme in medical-ethical debate during the last decade (Brody and Halevy, 1995). Several types of standards of medically useful/futile treatment can be distinguished:

1. Life-prolonging treatment is medically futile if the patient's demise is imminent and will unavoidably occur within the short term. This approach will be quite acceptable to many people if 'the short term' is limited to several days. However, it seems unlikely that David's life-expectancy is *that* short. It seems more likely that David's physicians would estimate that he could live for some weeks or months, maybe even more. In that case, it is far from self-evident that the physicians should be granted the right to decide that further life-prolonging treatment (i.c. CPR) would be futile. After all, these last months could be highly valuable to David, and it is possible that he would accept aggressive treatment in order to increase the chance that he will

live just a little bit longer. It would only be natural to have David judged himself whether or not CPR would make sense. Hence, this should not be considered a unilateral *medical* judgement.

2. Life-prolonging treatment is medically futile if it will result in an extremely low quality of life of the patient, or if it cannot improve such a low quality of life. Many people may consider all interventions futile if they would be in an extremely bad and hopeless condition, for example a persistent vegetative state. Yet others believe that even such a life deserves respect and should be sustained. Therefore, to say that in such a case physicians may withhold treatment is rather controversial. Anyway, unless the attending physician is almost certain that a cardiopulmonary arrest and subsequent CPR would leave David with severe neurological damage, it will be difficult to maintain that CPR is futile in this second sense. There may of course be good reasons to talk to David about CPR in order to see whether it would be a meaningful intervention to him. Yet this is not the same as deciding that CPR is *medically* futile.

3. Life-prolonging treatment is medically futile if it cannot produce the physiological effects intended by the physician. This approach may be the best acceptable one, as it restricts the physician's value-judgements to the direct effects of the intervention; this approach leaves no room for physicians to stop treatment if they consider the remaining length or quality of life of the patient too poor (Waisel and Truog, 1995).

In evaluating the effects of CPR, often two specific standards are mentioned. Many empirical studies focus on the effectiveness of CPR either in terms of restoration of cardiopulmonary functions, or in terms of survival to hospital discharge. These standards could fit within the 'physiological effect' approach and the 'imminent demise' approach to futility, respectively. Not surprisingly, both standards of measurement result in rather different evaluations of effectiveness of CPR. For example, in a meta-analysis of 30 studies, J. van Delden calculated the average chances of success of CPR in a large group of hospitalised patients: in 38% CPR resulted in restoration of cardiopulmonary functions, while only 14 % of all resuscitated patients left the hospital alive (Van Delden, 1993, 52). How now should the effectiveness of CPR in David's case be estimated? Empirical studies show that only few cancer patients survive after CPR. According to Ebell (1992) only 5.8% of all cancer patients survive after CPR; regarding patients with a metastasized tumour this percentage has dropped to 0%. Apparently, the chance that David would survive after CPR must be considered very small. Yet even if this chance is almost nil, some people may find it unacceptable that physicians would issue a DNR-order without consulting the patient. After all, they could argue that the benefit to be gained is very great: it is a matter of life and death.

3.4. The Importance of Talking to David

Suppose the attending physician would be right in his medical assessment that, from a medical perspective, CPR of David would be futile, and that he and his colleagues agree to forgo resuscitation in case of a cardiopulmonary arrest. In that case, there is still the problem about informing David. Two questions should be distinguished: (1) is the attending physician obliged to inform David about the DNR-decision? (2) Should a DNR-order be postponed as long as David has not been informed?

Physicians sometimes argue that there is no moral basis to inform patients about decisions to forego treatment if those decisions are (legitimately) made by physicians, on medical grounds (Verweij and Kortmann, 1997). Their argument is twofold. On the one hand *informed consent* is not necessary as the patient's wish cannot influence the medical decision. Therefore, information would lack a rationale. On the other hand, disclosure of more 'bad news' could be harmful to the patient who may get depressed and loose all hope. In combination, these arguments may support the belief that it is cruel to disclose information about a medically indicated DNR-order. Openness can sometimes be virtuous, but physicians who are dealing with vulnerable patients should not disclose potentially harmful information that is useless to those patients. If this line of argument can be supported, then the attending physician would be justified in issuing a DNR-order Friday-evening.

Yet is the argument correct? Indeed, disclosure of all information to patients can be cruel if the information is painful as well as useless. However, many patients may consider it important to be informed about a medical decision to forego treatment in certain life-threatening situations. Moreover, probably many patients will expect that, in case of a cardiopulmonary arrest, attempts will be made to keep them alive. For them it could be important to be informed that their expectation is unjustified. Such information may certainly be a painful experience, and therefore physicians will sometimes refrain from discussing the topic. Yet it is not true that such information would lack any rationale: disclosure of information is an important feature of the professional-patient-relationship (Beauchamp and Childress, 1994, p. 396-397). Furthermore, a pragmatic reason may be given for disclosing information about a medically indicated DNR-order: it may be difficult for care-givers to prevent the situation that the patient to his surprise discovers that he will not be resuscitated in case of an arrest. In the Netherlands, patients have a legal right to read their medical files. The professional-patient-relationship may be seriously harmed if the patient, reading his file, discovers that the care-givers agreed to a DNR-order and also agreed not to tell him anything about it. If there is no such a legal right to read one's dossier, then secrecy about DNR remains possible. The question whether or not secrecy is morally acceptable, however, cannot be answered by referring to legal requirements.

Finally, should the DNR-order be postponed if, during the weekend, it is impossible for the care-givers to talk to David and support him in a compassionate way? This question is only relevant, evidently, if one has concluded that information disclosure *is* mandatory. As in many cases, to decide about what is prudent to do, the precise circumstances of the case are highly important. In David's case, the very small chance that, during the weekend, an arrest would occur, is a relevant fact. After all, if this chance is really small, it does not seem necessary to agree on a DNR-order and issue it immediately on the awkward time of Friday-evening.

REFERENCES

Beauchamp, T.L. and Childress, J.F.: 1994, *Principles of biomedical ethics* (4th edition). Oxford UP, New York.

Brody, B.A. and Halevy, A.: 1995, Is futility a futile concept? *J Med Philos* 20: 123-144.

Delden, J.J.M. van: 1993, *Beslissen om niet te reanimeren*. Van Gorcum, Assen.

Ebell, M.H.: 1992, Prearrest predictors of survival following in-hospital cardiopulmonary resuscitation: a meta-analysis. *J Fam Pract* 34: 551-8.

Verweij, M.F. and Kortmann, F.A.M.: 1997, Controversen bij ziekenhuisrichtlijnen voor niet-reanimeerbesluiten. *Ned Tijdschr Geneeskd* 141: 575-577.

Waisel, D.B. and Truog, R.D.: 1995, The cardiopulmonary resuscitation-not-indicated order: futility revisited. *Ann Intern Med* 122: 304-8.

RIEN J.P.A. JANSSENS

PAIN MANAGEMENT

1. INTRODUCTION

For a long time, pain used to be regarded as a consequence of an underlying physical cause. If you take away the cause, the pain will disappear. Well-known is the picture of Descartes which shows a boy who holds his foot near the fire. We see a line, running from the foot to the brain which represents the nerve transmitting the pain stimulus to the brain. When the stimulus has reached the brain certain animal spirits (*esprits animaux*) are sent back to the muscles of the foot which as a consequence contract and, subsequently, make the foot withdraw from the fire. Medical knowledge has developed since Descartes and his description of the pain stimuli has been modified. But the underlying paradigm still pervades medical practice. Medicine still approaches pain as an unpleasant symptom with an underlying physical cause which should, through medical treatment, be taken away as soon as possible. Eventually, so the paradigm presumes, medicine will disclose the mechanisms of the body. The body is seen as an object, separated from the soul, or, in modern terms, separated from the person. The *body* defines the domain of medicine and medicine approaches the body scientifically, that is, with methods derived from the natural sciences. The *person* seems to be irrelevant if one adheres to this paradigm.

This paradigm has been under criticism. And, not surprisingly, it was moral discomfort with the way medicine dealt with pain that first gave rise to criticism. Pain confronted medicine with its boundaries as it often seemed to occur without an assignable physical cause. Chronic pain confronted the doctor with the imperfectibility of the biomedical paradigm. In the midst of the 20th century German phenomenologists such as Scheler, von Weizsäcker and Plessner, French phenomenologists such as Leriche and Merleau-Ponty and Dutch phenomenologists such as Buytendijk and Metz demonstrated the pitfalls of the biomedical paradigm. Within the biomedical paradigm, pain is only approached under the perspective of combatting it by technological means. But according to the phenomenologists this is only one possible perspective, perhaps suitable for the treatment of acute pain, but other

H.A.M.J. ten Have and B. Gordijn (eds.), Bioethics in a European Perspective, 495-507.

perspectives are equally legitimate. Phenomenology wanted to approach pain more directly as a phenomenon that reveals itself in the experience of the person suffering from it. Pain is seen as a way of being in the world. Within this perspective, questions concerning the meaning of pain arise. Severe, lasting pain makes the person become more conscious of his own bodiliness. Physical pain seems all that is real for the suffering person. At the same time the person is forced to relate to it, to give meaning to that which is seemingly meaningless (Buytendijk, 1943). The person who suffers from pain has lost his relationship with the world. (S)he is in complete isolation. And if that is the case, pain treatment should be focussed, not so much on medical interventions, but rather on communication, revealing the life histories of the patients (Metz, 1964). One can of course question the topicality of these insights. Be that as it may, for the western European continent, phenomenology has created room for broader perspectives on pain and pain management, which not only explain medical treatment but also the psychological and the social context in which the pain experience occurs.

In the United Kingdom, with philosophical traditions different from the European continent, it was not so much the philosophical debate which demonstrated the failure of the biomedical paradigm. One important domain where moral discomforts with regard to the biomedical paradigm arose was the hospice movement. Within the biomedical paradigm, the death of a patient meant the defeat of the doctor. Every death occurring from incurable disease confronted the doctor with the imperfectibility of his knowledge and methods. When it appeared that a patient's disease was incurable, the doctor moved away. The hospice movement wanted to form a counterbalance against medicine's lack of attention for the dying. Started in 1879, when the Sisters of Charity founded Our Lady's Hospice for the Dying in Dublin, the hospices showed that care for the dying can be highly rewarding. In the light of new developments in medicine in the area of pain and symptom management and in the light of publications describing the distress of the dying (Hinton, 1963), hospices started to professionalize and in 1967 Cicely Saunders thought the time was ripe for the establishment of the first professional hospice which was not only occupied with patient care but also with educational and research activities. Eventually, the hospices wanted to integrate their model of care into mainstream medicine. As Cicely Saunders herself stated she had to move out of mainstream medicine so that new attitudes and new knowledge could move back in (Saunders, Summers and Teller, 1981). In 1967, in a famous publication of Saunders, she used the concept of total pain, indicating that pain consists of physical, psychological, social and spiritual aspects (Saunders, 1967).

At that time pain and symptom treatment were still not given the attention they deserve. Since the beginning of the 20th century, medicine has developed drastically but with a strong focus on cure. Treatment was aimed at removing the underlying cause of pain and symptoms, not at pain and symptoms themselves. Because of the worldwide spread of the *concept* of hospice (the term does not in the first place denote an institution) in the following years, more attention was given to the treatment of pain, not only to its medical treatment, but also to psycho-social and spiritual aspects of pain treatment. By now, the literature in the area of palliative care is enormous and it is hard to find literature on pain management in palliative care which does not address psycho-social and spiritual aspects and acknowledge the total pain concept. When at the beginning of the 1980s, palliative care was introduced on the European continent, the concept of total pain seemed to go well together with the intentions of the phenomenogical views on pain treatment.

In the light of these developments, there would be reason for optimism but some critical remarks have to be made. First, it is not obvious at all that the practical implications of the total pain concept have already been acknowledged by and integrated in 'mainstream medical practice'. We suggested above that the biomedical paradigm is still pervasively present in medicine and the medical literature seems to support this view (e.g., Crul, 1997; Leclercq and Jongemans-Liedekerken, 1997; ten Have, 1997). According to this literature, pain and symptom treatment are underdeveloped areas of medicine. Not only is too little attention given to the medical treatment of pain and to the development of expertise, also the psycho-social and spiritual dimensions of pain are underestimated in medical practice. Secondly, the total pain concept itself has come under criticism. For example, it is said that the division of pain in four distinct areas (physical, psychological, social and spiritual) still implies a duality between mind and body and should therefore be seen as an exemplification of the Cartesian model. If that is the case, even palliative care, emerging from a moral discomfort with the biomedical paradigm, has failed to leave this paradigm aside and replace it with another (Lanceley, 1995).

It seems hard to support or reject the latter view, even though anecdotal evidence seems to contradict it. But what can be concluded from the literature is that medicine, to a more or lesser extent, is suffering from a dichotomy between theory and practice. Even though in theory the wider scope of pain and pain management is acknowledged, in practice the biomedical paradigm still prevails. This dichotomy between theory and practice in the area of pain management has been called *'the pain of medicine'* (Vrancken, 1989). The 'management' of this 'pain' seems to be one of the greatest challenges for medicine in the nearby future. And, supposedly, the concept of palliative care, has the potentials to play an important role in this respect.

2. CASE

A woman of 58 years old suffered from a small-cell lung carcinoma. When the patient was informed of her diagnosis, the first thing she said to the oncologist was that, when suffering would become unbearable for her, she wanted to have euthanasia, which in the Netherlands is a tolerated practice.

Sixth months after the diagnosis, she contacted a hospice physician. She asked for information about hospice care in order to make adequate arrangements for the future. She was told that euthanasia was not performed in the hospice but that the carers would never abandon her. If she persisted in her euthanasia request, in spite of all efforts of the interdisciplinary team, she could be discharged to the hospital where euthanasia would be provided.

The woman used to be a lawyer and during her entire life she had always been extremely busy. Her marriage had been a failure and she was divorced. One of her two daughters lived with her ex-husband and one lived with her. However, her relationship with both her children was very detached. The children themselves had not seen each other for years.

Thirteenth months after the diagnosis, she was admitted to a university hospital with pain in her back. Metastases in the brain and spinal column were diagnosed. It was decided to start radiotherapy and chemotherapy aimed at life prolongation and palliation of symptoms.

Three months later she decided not to undergo another chemotherapy and, at her request, she was admitted to the hospice. At the time of her admittance, she used opioids and was suffering from dyspnoea, constipation and nausea. It appeared that the dose of opioids was far too high. When the dose was decreased her symptoms alleviated. Corticosteroids were prescribed to alleviate headache and nausea.

Because she was paraplegic, she was forced to let others take care of her, which she did not like at all. She felt guilty towards the caring team. Towards her children she was very demanding and unreasonable. All carers stated time after time that they enjoyed looking after her and that there was no reason for feeling guilty. In this period she never talked about her oncoming death.

One morning the hospice physician entered her room and like a bolt from the blue she started crying. She was afraid to die, she felt guilty towards her children as she had never been able to take care of them. She was afraid they would never see each other again after her funeral. From that time on, many things changed. She talked about her oncoming death to her children as well as to the pastoral worker. The children stopped avoiding each other. Together with them she arranged her funeral service. She started to enjoy the high standard of care she received and stopped quarrelling on the phone with the lawyer's office she had always worked at. This led to an improvement in the contacts with the people surrounding her.

Nineteenth months after her diagnosis, she met her sister who lived in India. Now she felt everything was said and done. Apart from fatigue she did not suffer from any symptoms. She had enjoyed her months in the hospice and had accepted her fate. During one of the last discussions she had with the hospice physician she requested euthanasia again, an idea which was disliked by the daughters. She wanted to die. She was afraid that the dying process would be endless and that she would deteriorate slowly. As an alternative for euthanasia in the hospital, the hospice physician discussed the option with her and her daughters to discontinue the corticosteroids and instead administer sedatives which would relieve her mental anguish. She said she needed one day to consider this and phoned a schoolfriend, a rheumatologist, who confirmed the rationality of this option. She agreed and felt relieved. She wondered why other people so seldom choose this way of dying. Four days after the withdrawal of the corticosteroids, she died peacefully in the presence of her daughters.

3. CASE DISCUSSION

In this section we will indicate that the case of the woman suffering from small-cell lung carcinoma can be seen as an example of total pain management. The practical implications of the concept of 'total pain' were acknowledged and the interdisciplinary team acted accordingly. The medical, psychological, social and spiritual needs and wishes of this patient were addressed which led, in retrospect, to a meaningful death of the patient and to satisfaction on the side of the daughters. In the case discussion we will draw attention to the decision of the patient and the physician to discontinue the corticosteroids, administer the sedatives and refrain from transferral to the hospital to undergo euthanasia. Also, the importance of psycho-social and spiritual pain 'management' will be underlined.

3.1. Autonomy of the Patient

The patient was in a physically excellent condition, given the circumstances, when she uttered the wish to die. She had looked forward to the visit of her sister who had come and had returned to India. Also, she sensed that her daughters would manage rather well without her. She told the hospice physician that she wanted to be transferred to the hospital to have euthanasia. Although she suffered from brain tumour, it was beyond question that her request was well-considered and stable. She had to be respected as an autonomous person. There is no doubt that, if she were to be transferred, many doctors in the Netherlands would not be reluctant to meet her request for euthanasia. The hospice physician understood the wish of the patient, coming forth out of fear for an endless dying process which might even lead to severe brain damage due to tumour growth. As she had always been a highly intelligent, independent and ambitious woman, the thought of dying 'insane' (as she herself put it) was unbearable for her and also her daughters did not want their mother to die this way. Moreover, life had been just enough for her. Her dying process had already lasted much longer than initially expected. She wanted to die. There seemed no way out of her mental distress. Therefore, the hospice physician confronted her with an option. The corticosteroids she received could be replaced by sedatives. Discontinuing the corticosteroids and starting with sedatives would certainly shorten her prognosis drastically and it could be assumed that death would occur within a week. After a discussion with a friend she decided to withdraw the corticosteroids and have the sedatives.

There is a consensus that a physician has to inform the patient adequately about the medical treatment he provides. Also, it is widely acknowledged that a physician cannot administer medications against the autonomous will of the patient. But even though this argument contains truth and is certainly backed by an international consensus, it does not suffice to clarify important dimensions of this case. It pays too little attention to the context of the decision-making process. After all, the patient utters an autonomous request for euthanasia. And it is only then that the physician informs the patient of the possibility to start with the sedatives. He wants to provide the patient with an alternative for euthanasia, a practice which he himself criticizes. Thus, the information he gives is not neutral. One can suppose that he very much hopes the patient will respond positively to his proposal. For him and for the caring team, it would be unsatisfying if the patient, after having cared for her during so many weeks, would have to be transferred to a hospital in order to die there. One can even suppose that, perhaps implicitly and between the lines, the physician's discomfort with the patient's request for euthanasia is very well realized by the patient. And, furthermore, there may even have been a feeling of embarrassment on the side of the patient. Perhaps she thought that, if she would opt for euthanasia, she would be ungrateful to the team that cared so well for her. Perhaps more importantly, she may well have been afraid that she would have disappointed her daughters, who disagreed with their mother's request for euthanasia. Her consent to the physician's proposal may have come forth out of a feeling of moral obligation to the team and to her daughters. So was her choice autonomous after all? Or, in other words, did the physician respect her autonomy? Or was his attitude rather paternalistic, implicitly limiting the patient's freedom to be transferred to a hospital?

We are caught in an ethical discourse which is inappropriate for this case. The supposed incompatibility of an old paternalistic ethics and a modern, liberal ethics of autonomy has to be transcended. In order to account for compassionate care, it is imperative to acknowledge the patient's freedom to make choices that fit best to him-/herself but it is also imperative to acknowledge the patient's vulnerability, contingency and situatedness in a community of people. Solely focusing on autonomy may well lead to the abandonment of the patient, leaving him/her alone with a range of options (s)he has to choose from. Therefore, as an alternative for the principle of autonomy, authenticity has been proposed as a guiding principle for medical practice (Welie, 1994; Welie, 1998; Arnason, 1994). More than autonomous choices, authentic choices would relate to the person of the patient, acknowledging his/her life history, current situation and hopes for the future. Authentic choices do not only relate to the present but also to the past and future. It is said that the fostering of a patient's authenticity would be foundational for compassionate medical care. Also within the newly developed 'ethics of care' these dimensions are acknowledged but, instead of replacing the principle of autonomy by another principle, these dimensions have been included in the concept of autonomy.

So, in light of the above, we have to ask once more: was the patient's choice autonomous, in the sense that it fitted with her personality? And did the physician, together with the children, leave enough room for such an autonomous choice? In retrospect, the feeling of relief may indicate the autonomy of her choice. The patient had spent a period of two months in the hospice. During that time, the caring team had started to admire this woman for the way she coped with the disease and the patient had started to admire the team for their compassion and sympathy. A bond had been established between them. And also her relationship with both her daughters, initially detached, ameliorated drastically. Her daughters had also started to admire their mother the and mother realized, for the first time since long, that she loved both her children. Supposedly, it was not just the team and the children that felt responsible for the patient's well-being, it was also the patient who felt a moral commitment to her daughters and to the carers she had come to know and respect. In this regard, it can be suggested that the relief of the patient after having made her decision can be understood in its social context. She was left a choice, she could have chosen for transferral, but eventually, she wanted to die in the presence of the people she had spent her last weeks with, not in a hospital where an unknown physician would provide her with euthanasia. It can also be assumed that she did not want to disappoint her daughters by opting for transferral.

3.2. Euthanasia

But now another question urges itself upon us. For if it was so important for her to die in the hospice, in the presence of her carers, one could make a strong argument for providing her with euthanasia in the hospice. Then, one could say, her wish to die would have been met in an even better way because that is what she asked for in the first place.

Clearly, this patient wanted to die and clearly she wanted to die fast. But it can be seriously questioned if she wanted to be killed by the people who had cared for her. That would have been incongruent with the high quality of care she had received for more than two months. The people who had cared for her all this time would then suddenly become the ones who were to end her life. Assumably, because of the authentic psycho-social and spiritual care she had been given, she would not have wanted those people to put an end to her life.

But even if that was not to be the case, even if she had wanted those people to end her life, then still objections would have to be made. Because the principle of autonomy should not be seen as completely individualistic, let alone that it would imply a positive right, but instead has to be situated in a larger social context, for patients cannot make their carers do things they do not support. We will not go into the range of arguments that have been put forward by the hospice movement against euthanasia. Let it therefore suffice to say that autonomy, conceived of as an individualistic principle, fails in palliative care practice (Saunders, 1994). Autonomy takes at least two. It is this self-evident fact only that reveals the failure of an individualistic notion of autonomy. After all, we all agree that a physician who refuses euthanasia on good grounds, cannot be forced to provide it.

3.3. Slow Euthanasia or Total Pain Management?

Some ethicists would speak in this case of 'slow euthanasia' which is a term recently used in a discussion on a similar case in the *Journal of Palliative Care* (e.g., Billings and Block, 1996, criticized by Mount, 1996). Clearly, they would state, it is the intention of the hospice physician to shorten the life of this patient. And clearly, the whole medical decision-making process is not aimed at improving the quality of life, but at the occurrence of death, albeit in a slower manner than in the case of strict voluntary euthanasia.

Above, we have argued that it would not suffice to parry this reproach simply through pointing towards the principle of autonomy (i.e. the autonomous decision of the patient not to take the corticosteroids any longer and start with sedatives). After all, if this case actually was an example of euthanasia, be it slow or not, patient autonomy would not be the decisive moral factor. What is more, the physician would not even have presented the patient with the option in the first place. Even if one supposes that the patient would have asked him or herself for sedation as an alternative means for corticosteroids, his/her wish would simply not have been met, at least not in the context of hospice.

Parrying the reproach that this case is an example of slow euthanasia requires thus a more substantial argument. Crucial in this respect is the intention of the physician. Provided that one is critical of euthanasia, the principle of double effect is paramount in the moral justification of one's decisions (Mount, 1996). The intention of the physician was to relieve the patient's mental suffering. The patient wanted to die. Her dying process had lasted long enough for her, and the physician acknowledged the pain this process entailed for her. The mental pain this patient experienced was for him a reason to discuss the option of discontinuing the corticosteroids and instead start with sedatives to relieve her mental distress and to treat her headache and nausea. He did not *want* to shorten her life. If he had wanted that, starting with sedatives would, from a moral point of view, make no significant difference with euthanasia. Instead, he wanted to treat her (total) pain. He acknowledged that total pain is foundational for any treatment decision in palliative care. This has far-reaching consequences. Mental, social or spiritual pain can, under certain circumstances, be an indication for sedation. When this patient's mental condition changed, from high satisfaction to distress, the physician saw, in starting with sedatives, the possibility to relieve this discomfort.

There are however two crucial presumptions one has to make in order to convincingly argue that this case is not an example of euthanasia. First, the presumption has to be that other options to relieve her discomfort (e.g., seeing a psychologist) were no longer available. If other options to relieve her mental distress had still been available, administering sedatives would have been disproportional. It would be impossible to justify the decision because the 'evil' unintended but foreseen side-effect - the shortening of life - could have been prevented. The physician would in that case be rightly suspected of intending the shortening of life of the patient. In retrospect, one can discuss whether there were other options with less 'evil' side-effects present. Note that the patient was seeing the psychologist daily. Psychiatric treatment would probably not have taken away her mental distress, and even if it had, she would certainly have refused that option and probably she would then have asked for transferral right away. Certainly, according to the physician, there were no other options available. Sedation seemed the only way out of her mental distress.

The second presumption one has to make is that the sedatives were not given in a disproportionally high dose. If one states that the primary intention is the relief of mental distress, one should administer the necessary dose to accomplish that. Disproportionally sedating the patient, in order to shorten her life, is from a moral point of view the same as euthanasia. It is beyond question that this second presumption was met in this case.

Through a discussion of the double effect principle we are now able to conclude that the decision to withdraw the corticosteroids and to administer sedatives, essentially is total pain management. The management of total pain of patients makes out the heart of palliative care. Medical, sociological, psychological and spiritual dimensions are intermingled and it takes interdisciplinarity (a term that denotes the need for co-operation and teamwork, more than multidisciplinarity) to deal adequately with these dimensions. Let us, at the end of this case discussion, look somewhat closer to the role of the psychological, social and spiritual dimensions of this case.

3.4. Psycho-social and Spiritual Dimensions of Care

During the first weeks of the patient's stay in the hospice, her attitude was defensive. She was hardly able to accept her dependency on the care of others. She did not want to see the team's psychologist and pastoral worker. She behaved rudely towards her daughters, blaming them for not visiting every day (they both lived at the other side of the country) and, during their presence, it took tolerance not to start rows in reaction to their mother's criticism.

Randall and Downie have adequately argued that unwanted psycho-social and spiritual care should not be inflicted on the patient (Randall and Downie, 1996). Patient autonomy is for the domain of psycho-social or spiritual care as important as it is for the domain of medical care. It is good and vital for palliative care that psychologists, social workers and pastoral workers are available but before they start doing their work they have to assess whether the patient wishes to have psycho-social or spiritual care.

Clearly, in the beginning of her stay in the hospice the patient did not want to receive this kind of care and, principally, nothing is wrong with that. Even though the carers hoped that her defensive attitude would after a while make way for more openness, they did not want to force themselves upon the patient. It was only after the woman told the physician, in tears, everything about her agony, that she left her defensiveness behind and opened up to the team. Guilt transformed into enjoyment and gratitude. Her daughters enjoyed visiting her, taking her for walks in the forest and shopping into town. Her openness created the possibility for her to tell them the history of her life, and, to some extent, to settle things that had happened in the past. Not only the psychologist and the pastoral worker, but also the hospice physician, the physiotherapist, the voluntary workers and even the undertaker provided her with occasions to express her hopes, anguish, and worries. Randall and Downie state that it is especially non-professional care that is paramount in this domain, rather than professional care. They underline the limitations of professional expertise which would call for modesty. They state that the belief that carers can be trained to empathize is a dangerous illusion: "The possibility of understanding a patient is therefore remote" (Randall and Downie, 1996, p. 19). Their sober-minded view is important because it underlines the necessity to acknowledge the limits of good palliative care. But we would argue that in palliative care, empathic understanding of the patient is very well possible. It is true that palliative care cannot and should not aim at relieving all suffering. Grief, anger, and despair are emotions that belong to any dying process. Total pain management does not mean that these emotions should be 'treated' in order to remove them. Rather, dealing with total pain implies that room is created to utter these emotions. And if the patient feels the need to discuss these emotions, (s)he should not feel that (s)he is talking to a *black box*. Of course it is impossible to fully understand a patient, as it is impossible to fully understand any human being. One of the characteristics of being human is that some mystery always remains. We can be completely surprised at times by people we have known for years. Complete understanding of a human being is impossible and aiming for it would eventually come down to objectivisation. But what can and should be done in the context of palliative care is thoughtfully listening to what a patient has to say. Often, when people are confronted with the end of their lives, their *soul wounds* from the past emerge. Dealing with these soul wounds requires communication with friends and relatives, but it may also require professional care. Feelings of guilt may be eased through prayer with a priest or minister. Also, they may be uttered in the presence of a psychologist or social worker, less involved in the past of the patient than friends or relatives are and trained in communication techniques. Understanding the message of the patient and responding adequately to it, requires an empathic attitude. In other words, it requires from the professional carers that they not only logically understand what the patient is saying but that they also to a certain extent live the situation the patient is in. There has to be a certain correspondence between the emotions of the patient and the emotions of the carers. The carers will have to be able to imagine for themselves the mental state the patient is in, in order to respond adequately to the message of the patient.

Empathy, like Randall and Downie state, cannot be taught at university. It requires a willingness to let the patient be the guide. Furthermore, it requires practical experience. But the communicative means which at the same time presuppose and foster empathy can very well be taught. In order to provide professional psycho-social and spiritual care, these communication techniques, together with an empathic attitude, seem indispensable. It is imperative that people who are confronted with their oncoming death are given the possibility to communicate with people who are trained in communication techniques and who are willing to let the patient be the guide. The importance of professional psycho-social and spiritual care should therefore not be underestimated in palliative care. At the same time, its limitations should be acknowledged.

4. CONCLUSION

We have argued that the decisions taken in this case were essentially examples of total pain management. The practical implications of the total pain concept are important. Total pain implies that the medical domain is intrinsically connected with the psycho-social and spiritual domains. Medicine reaches beyond the limits of the body, it addresses the person of the patient. In order to make adequate medical decisions it is imperative to address the psycho-social and spiritual state the patient is in. Addressing the total pain of the patient can therefore not be achieved without interdisciplinary care. Decisions in palliative care practice mostly entail more than one dimension of care.

In this case, the decision of the patient to have the corticosteroids withdrawn and instead opt for sedatives cannot be understood without rendering an account of her mental situation (her wish to die) and her social situation (the role of her daughters, the friend who came to visit her, the physician). Perhaps, if one assumes an individualistic notion of autonomy, she would have opted for euthanasia, although even that is hypothetical. But people do not live in solitarity, we are essentially *fellow* human beings. We depend on others and bear responsibilities for others, whether we are healthy or almost dying. Other people constantly shape our lives and influence the choices we make, perhaps more often than we would like to admit. With the help of this range of communities we are members of, we have to try to keep and foster our authenticity. And this is exactly what total pain management is all about. It is not about removing all pain. Rather, it is, as Cicely Saunders put it, about helping the dying *live* until they die. The quality of their lives, from a physical, mental, social and spiritual point of view, is in this respect the only moral criterion for the decisions that are made.

ACKNOWLEDGEMENT

I am grateful to the Commission of the European Communities for funding the Pallium project, of which this paper forms a part.

REFERENCES

Àrnason, V.: 1994, Towards authentic conversations. Authenticity in the patient-professional relationship. *Theoretical Medicine* 15: 227-242.

Billings, A. and Block, S.: 1996, Slow euthanasia. *Journal of Palliative Care* 12: 21-30.

Buytendijk, F.J.J.: 1943, *Over de pijn*. Het Spectrum, Utrecht.

Crul, B.J.P.: 1997, Pijnbestrijding in de laatste levensfase. *Nederlands Tijdschrift voor Geneeskunde* 141: 2097-2100.

Have, H.A.M.J. ten: 1997, Pijn als medisch probleem? Een kritische reflectie. In: Uerz, I. (ed.), *Pijn*. Nijmegen University Press, Nijmegen.

Hinton, J.M.: 1963, The physical and mental distress of the dying. *Quarterly Journal of Medicine* 32: 1-21.

Lanceley, A.: 1995, Wider issues in pain management. *European Journal of Cancer Care* 4: 153-157.

Leclercq, R.M.F.M. and Jongemans-Liedekerken, A.W.: 1997, Pleuramesothelioom in de huisartsenpraktijk: gecompliceerde pijnproblemen. *Nederlands Tijdschrift voor Geneeskunde* 141: 1081-1085.

Metz, W.: 1964, *Het verschijnsel pijn. Methode en mensbeeld van de geneeskunde*. De Erven F. Bohn N.V., Haarlem.

Mount, B.: 1996, Morphine drips, terminal sedation, and slow euthanasia. Definitions and facts, not anecdotes. *Journal of Palliative Care* 12: 31-37.

Randall, F. and Downie, R.S.: 1996, *Palliative care ethics. A good companion*. Oxford University Press, Oxford.

Saunders, C.: 1967, *The management of terminal illness*. Hospital Medicine Publications, London.

Saunders, C., Summers, D. and Teller, N.: 1981, *Hospice. The living idea*. Edward Arnold, Leeds.

Saunders, C.: 1994, The dying patient. In: Gillon, R. (ed.), *Principles of health care ethics*. John Wiley and Sons, Chichester.

Vrancken, A.M.E.: 1989, *Chronische pijn. Het kruis van de geneeskunde*. Haveka, Alblasserdam.

Welie, J.V.M.: 1994, Authenticity as a foundational principle of medical ethics. *Theoretical Medicine* 15: 211-225.

Welie, J.V.M.: 1998, *In the face of suffering. The philosophical-anthropological foundations of clinical ethics*. Creighton University Press, Omaha, Nebraska.

ROBERTO MORDACCI

COMATOSE PATIENTS

1. INTRODUCTION

The ethical and philosophical issues in regard to comatose patients are extremely hard and complex. A number of clinical cases have become widely known and the legal decisions concerning them have marked turning points in the history of bioethical debate. The case presented here has been selected because it is one of the most recent, because a good deal of bioethical literature has already been devoted to it and because of its European origin, since it took place in the United Kingdom.

2. THE BLAND CASE

Tony Bland was a seventeen year-old boy who supported the Liverpool soccer team; on April 15th, 1989, he attended the match Liverpool vs. Nottingham Forrest at the Hillsborough Stadium in Sheffield, when the crowd, out of control, started to push him, together with several other persons, against the fence protecting the game field from invasions. The pressure of the crowd caused the death of ninety-five persons and severe harms to many others. Pressed against the fence, Tony Bland could not breathe for a number of minutes; his brain suffered severe damage due to prolonged anoxia. At the Airedale General Hospital, where he was admitted, the physicians discovered that the cortical area was destroyed and only the brain stem was functioning.

Tony Bland's condition never improved. He was diagnosed to be in a persistent vegetative state (PVS) and therefore was given artificial nutrition and hydration (ANH); antibiotics and medications were given when local infections occurred; he showed no signs of consciousness, his limbs were shrunk and his face contracted in a grin. The attending physician reported to the coroner that he was going to stop ANH, but the coroner advised him that withdrawing treatment could lead to incrimination for homicide. The hospital solicited the opinion of the High Court concerning the interruption of treatment; the opinion was favourable, but the official solicitor, who was in charge of protecting Tony

H.A.M.J. ten Have and B. Gordijn (eds.), Bioethics in a European Perspective, 509-514.

Bland's interests, appealed first to the Court of Appeal and then to the House of Lords. Tony Bland had no advance directives and there was no witness of any declaration of him concerning his wishes in such a situation. The opinions given by the members of the High Court, the Court of Appeal and the House of Lords were all centred on the idea that the treatments offered (ANH, medication, antibiotics) were 'of no benefit for the patient' and that therefore they could legitimately be discontinued. This was done and Tony Bland died in 1993.

3. THE DIFFICULT DIAGNOSIS OF PVS

The Persistent Vegetative State is a condition characterized by

1. no evidence of awareness of self or environment and an inability to interact with others;
2. no evidence of sustained, reproducible, purposeful, or voluntary behavioral responses to visual, auditory, tactile, or noxious stimuli;
3. no evidence of language comprehension or expression;
4. intermittent wakefulness manifested by the presence of sleep-wake cycles;
5. sufficient preserved hypothalamic and brain-stem autonomic functions to permit survival with medical and nursing care;
6. bowel and bladder incontinence; and
7. variably preserved nerve reflexes (pupillary, oculo-cephalic, corneal, vestibulo-ocular, and gag) and spinal reflexes (Multi-Society Task Force on PVS, 1994).

It is one of the possible developments of *coma*, a state of unconsciousness with eyes closed, and it must be distinguished from the *locked-in syndrome*, in which consciousness and cognition are still present but the patient is unable to communicate with others (Celesia, 1997). There are various difficulties in diagnosing a PVS with enough certainty; various medical and interdisciplinary groups have defined different periods of observation, varying from one month to one year or more, before formulating a diagnosis, also depending on the etiology of the state (Celesia, 1997). Patients in this state, if cared for with minimal life support, can survive for a long time.

4. FOUR APPROACHES TO THE CASE

The case of Tony Bland (and similar cases) can be analyzed by at least four different theoretical approaches.

1. A first approach would be to emphasize the role of the wishes of the patient, trying to find out what was his or her opinion concerning the treatment. This route was followed in another famous case concerning a PVS

patient, that of Nancy Cruzan (Angell, 1990). In this case, the Supreme Court of the United States ruled that stopping ANH would be permissible if there was a 'clear and convincing evidence' that the patient had declared that she did not want to be kept alive in such conditions. In this perspective, it is important to have signs of the wishes of the patient in order to decide on the continuation of treatment, through any kind of records of his or her opinions on the hypothesis of a similar situation, expressed before he or she lost consciousness. Ideally, such records, in particular written ones, constitute an 'advance directive' which the caregivers should take into account in the process of decision-making. Thus, the heart of the argumentation lies not in a judgement concerning the life of the patient (its quality or its sacredness) but on his or her autonomy, as expressed in the declarations given while conscious and competent. The advance directives have raised a complex debate: first, their reliability as *actual* wishes of the patient could be questioned because they were signed by the patient in a completely different condition from the present one; many patients change their opinion when faced with the illness they refused so strongly to endure before. Second, it is hardly possible to define clearly the legal status and weight of these documents in order to justify the decision to withhold or withdraw treatment from a particular patient; the medical team has in any case the responsibility of the decision taken, particularly concerning the diagnosis and prognosis of the patient, which is of course of paramount importance to determine the applicability of the advance directive. These documents can nonetheless offer an orientation concerning the overall good of the patient, which is made of medico-biological, as well as psychological-relational and personal dimensions which cannot be totally disregarded. Third, such an approach is of no use when, as in the case of Tony Bland, we have no evidence of the patient's wishes in such a situation and it is unrealistic to think that anybody should be obliged to sign an advance directive in order to be able to decide in the eventuality of such a condition. Fourth, these kind of documents do not seem to have equal meaning in different cultural contexts, because they presuppose an individualistic background that contrasts with the emphasis on interpersonal bonds that are perceived in some countries, for example in southern Europe (Sanchez-Gonzalez, 1997).

2. A second approach is based on the idea of 'sanctity of life': even if the condition in which Tony Bland was offered him no possibility of recovery and was extremely poor in terms of communication, perception and consciousness, nonetheless his life could be recognized as a diminished but still human one; since the taking of human life is in principle forbidden, stopping the minimal treatment required to sustain him would mean to kill him indirectly. The residual life of Tony Bland was still that of a person, in the sense that a human individual is a person even if he is not able to exercise the typical functions of autonomous agents any more; the *nature* of

person of that individual persists as long as his body is alive. Such an approach does not imply that intensive care must always be given: the poor condition and the lack of any hope of recovery allow, in fact, a low level of effort to sustain this human life; thus, intercurrent infections or pathologic conditions requiring intensive treatment can be omitted in this perspective, while ANH should always be given.

3. The third approach seems to be the one followed by the Courts and the House of Lords in the Bland case: the decision to stop treatment, including ANH, is based on a quality of life judgement that regards the PVS as a condition incompatible with the minimal requirements of an acceptable life. The interests of the patient, interpreted in this perspective, incline more towards putting an end to a life totally deprived of consciousness than towards prolonging an artificially sustained life with no hope of recovery. The important philosophical point is that quality of life judgement impose an interpretation of life based on its appreciability by the subject, usually in the sense of the ability of experiencing pain and pleasure; this is why this approach is favoured by utilitarian thinkers (e.g., Singer, 1994, ch. 4).

4. Quality of life judgements seem to be implied, but in a different framework, also in the fourth possible approach, based on the notion of *proportionate treatment* (Cattorini and Reichlin, 1997): the diminished condition of patients in PVS deserves respect and care as long as it is the residual form of life of a human person. This enables one to accept a presumption to treat as a general rule with PVS patients, meaning that it is usually proportionate to offer nutrition and hydration and nursing care for such patients, since a personal life deprived of the possibility to express its most typical traits justifies a low level of intervention. When a patient in PVS develops intercurrent infections or worsens, higher levels of treatment are less justified because they cannot restore the health of the patient, but may be seen as prolonging the process of death. The notion of proportionality is significantly different from the traditional distinction between 'ordinary' and 'extraordinary' means: the former allows to take into account not only a list of means, but the relation of each technical solution with the individual condition and the subjectivity of the patient. A treatment is proportionate if it can contribute a meaningful good to the patient. 'Meaningful good' can be considered the good of the patient in all its dimensions: biomedical, psychological-relational, personal (relative to personal values) (Pellegrino and Thomasma, 1988). In the case of PVS patients, the only dimension left is the biochemical one, thus a very poor content of the notion of good of the person, so that minimal treatments can be justified but the more complicated they become the less justified they are; in some cases, even ANH can become onerous and burdensome because of repeated local infections or the need for repositioning or modifying the ways of administration; in these conditions, they may be discontinued. A way to take into account the other dimensions

of the good of the patient might be the advance directives: they give an idea of the possible *interpretation* that the patient would give to the treatment if he was conscious; the same treatment could in fact be interpreted by different patients as an aggression, and therefore in principle rejected, or as a minimal sign of care meaning that he has not been abandoned by the family and the health care team, and therefore in principle desired. In the presence of such declarations, the decision-makers have the opportunity of construing a richer notion of the good of the patient. In this approach, the notion of quality of life is not only connected with the capacity to experience pain or pleasure, but with the overall value of human life as a rich texture made of lived corporeality and meaning.

5. CONCLUSION

The issue of ANH in PVS patients also raises a question concerning the meaning of care and the caring professions. PVS is a condition of extreme deprivation of the capacity of living a fully human life: it creates a situation of complete dependence of the patient on the compassionate care of others. The fact that any human person is in need of some help for all his life is exacerbated in this condition; the mutual help which sustains the life of everyone has in these circumstances a particularly intense symbolic meaning (Callahan, 1983); nourishing and hydrating a patient in this state displays a commitment to the mutual bond of care which ties together all human beings, while stopping the treatment at an early stage or when the patient is in a stable condition may look as an abandonment. Should such an attitude become systematic in a certain community it is likely that the common perception of the mutual bonds of care would slowly fade away.

REFERENCES

Angell, M.: 1990, Prisoners of technology: The case of Nancy Cruzan. *New England Journal of Medicine* 322: 1226-1228.

Callahan, D.: 1983, On feeding the dying. *Hastings Center Report* 13: 22.

Cattorini, P. and Reichlin, M.: 1997, Persistent vegetative state: A presumption to treat. *Theoretical Medicine* 18: 263-281.

Celesia, G.C.: 1997, Persistent vegetative state: clinical and ethical issues. *Theoretical Medicine* 18: 221-236.

Multi-Society Task Force (MSTF) on PVS: 1994, Medical aspects of the persistent vegetative state. *New England Journal of Medicine* 330: 1572-1579.

Pellegrino, E.D. and Thomasma D.C.: 1988, *For the Patient's Good.* Oxford University Press, New York.

Sanchez-Gonzalez, M.A.: 1997, Advance directives outside the USA: Are they the best solution everywhere? *Theoretical Medicine* 18: 283-301.

Singer, P.: 1994, *Rethinking life and death*. Text Publishing Co, Melbourne.

PART 6

PRACTICAL INFORMATION

HENK A.M.J. TEN HAVE

THE DISCIPLINE OF BIOETHICS: PRACTICAL INFORMATION

1. INTRODUCTION

Practising bioethics as a scientific discipline requires the existence of a certain infrastructure as well as the availability of basic tools and services, as well as professional co-operation. Like any other science, bioethics is an organized social activity. While the collective goal of the work is the advancement of knowledge, individual professionals can specialize in particular methods, theories, topics and skills. In the early days, bioethics was very much an individual affair. In most countries, some individual personalities developed into specialists of bioethics, sometimes monopolizing public debate. Later, when more scholars entered the profession of bioethics, the institutional basis of bioethics expanded. At the moment, we can witness a growing number of university departments, centres, institutes or units, specifically focused on ethics (sometimes in combination with other disciplines, such as medical philosophy, history of medicine, medical humanities, social medicine). This institutionalisation not only shows that job opportunities for bioethicists are multiplying, but also that the discipline itself is recognized as contributor to medical education, clinical training and practical health care.

Furthermore, we can notice that the discipline of bioethics is in a mature stage because of the growing number of scientific journals as well as the increasing quantity of published materials. Communication among bioethicists is finally developing through a rising number of learned societies and associations as well as different ways of Internet exchanges.

In this chapter, useful instruments for the practice of bioethics will be presented: handbooks, journals, and associations.

2. HANDBOOKS

The series of textbooks and handbooks in bioethics is immense. Many countries now have introductions into medical ethics and synopses of the

H.A.M.J. ten Have and B. Gordijn (eds.), Bioethics in a European Perspective, 517-532.

major issues in the national languages. However, for any novice who wishes to be introduced into the state of the art in bioethics, the following materials are particularly advisable. First, standard textbooks in the English language will be discussed shortly. Next, some brief references will be presented to textbooks in other European languages.

2.1. Standard English Textbooks

T.L. Beauchamp and J.F. Childress: *Principles of Biomedical Ethics*. Oxford University Press, New York/Oxford.

This is one of the most important and most frequently used textbooks for medical ethics. It is required reading for anyone interested in the subject. The first edition of 1979 has helped to establish the developing field of bioethics. It is a powerful and well-composed text because it presents a clear theoretical structure (the well-known four principles) as well as case histories, exemplifying applications of the theory. Starting with chapters on morality and types of ethical theory, subsequent chapters discuss the principles of respect for autonomy, nonmaleficence, beneficence, and justice. Another chapter focuses on the professional-patient relationship. The closing chapter examines virtues and moral ideals. In the Appendix, 10 cases are analyzed. In revised editions, the book has been gradually expanded and enriched. The fifth edition has been published in February 2001.

R. Gillon (ed.): *Principles of Health Care Ethics*. John Wiley & Sons, Chichester, 1994.

This voluminous book (1118 pages) presents an overview of current thinking in bioethics. Over 100 authors from various national, professional, cultural and religious backgrounds discuss most of the ethical dilemmas of modern health care. The theoretical framework of the book is the same set of four moral principles, but the analysis is much more applied to issues, problems and topics, arising in medical practice. The first part presents approaches to applied health care ethics. The second part addresses issues arising in professional relationships, such as paternalism, confidentiality, and consent. Moral problems in particular health care contexts are analyzed in the third part. Here, we find a range of standard topics as abortion, fertility treatment, neonatological problems, psychiatric ethics, and care for the elderly. Part four focuses on health care ethics and society (for example, management, economics, health promotion, research). The final part presents ethical problems of scientific advance, such as arising in genetics, transplantation and animal experimentation.

W.T. Reich (ed.): *Encyclopedia of Bioethics*. Simon and Schuster Macmillan, New York.

This standard work, first published in 1978, was the first encyclopedia in the field of bioethics. It has been a major force in the establishment of this discipline. It organized the body of knowledge, it presented definitions and demarcations, and gave a comprehensive set of entries in basic topics and practical issues, selecting appropriate terminology. Reflecting the extensive changes in the field, a completely revised edition was published in 1995. This edition offers 464 articles by 437 contributors in 5 volumes (2950 pages). All articles, from Abortion until Zoos, are original contributions; very few were carried over from the first edition. In the section on history of medical ethics, several contributions discuss developments in Europe in historical and geographical order.

R. Chadwick (ed.): *Encyclopedia of Applied Ethics*. Academic Press, San Diego, 1998.

A recent effort to present a comprehensive overview of all major issues in applied ethics has been published in 4 volumes. This work addresses all subject areas in applied ethics: theories of ethics, ethical concepts, medical ethics, scientific ethics, environmental ethics, legal ethics, ethics in education, ethics and politics, business and economic ethics, media ethics, ethics and social services, and finally, social ethics. The whole range of applied ethics is covered in 281 separate full-length articles. The index contains more than 5000 entries. The 4 volumes cover a total of 3101 pages.

H.T. Engelhardt and S.F. Spicker (eds): *Philosophy and Medicine*. Kluwer Academic Publishers, Dordrecht/Boston/London.

This series of books started in 1975 and continues until today with the publication of volume 61 in 1999. The series was also a major force in the establishment of philosophy of medicine as a scientific discipline, and in particular of bioethics. The focus is on philosophical reflection on present-day problems in medicine and health care. Many volumes however have concentrated on particular topics in bioethics, such as justice (1981), abortion (1983), euthanasia and the newborn (1987), death (1988), suicide and euthanasia (1989), and infertility (1997). Most volumes are edited collections of reviewed scientific papers centred on a specific theme. Representing the philosophical intentions of the series, many volumes focus on fundamental notions, such as clinical judgement (1979), responsibility (1982), explanations (1984), medical knowledge (1990), competency (1991), the human body (1998). Occasionally, monographs have been included, for example, L. Nordenfelt: *On the nature of health* (1987; 1995), and S.K. Toombs: *The meaning of illness* (1992).

2.2. Textbooks in Other European Languages

2.2.1. Danish:

D. Andersen, C.E. Mabeck and P. Riis (eds.): *Medicinsk etik.* Fayl's forlag, Copenhagen, 1987.

K. Kappel: *Medicinsk etik. En filosofisk diskussion af etiske grundprincipper.* Gyldendal, Copenhagen, 1996.

H.R. Wulff: *Den samaritanske pligt. Det etiske grundlag for det danske sundhedsvaesen.* Munksgaard, Copenhagen, 1995.

2.2.2. Dutch:

I.D. de Beaufort and H.M. Dupuis (eds.): *Handbook Gezondheidsethiek.* Van Gorcum, Assen/Maastricht, 1988.

H.A.M.J. ten Have, R. ter Meulen and E. van Leeuwen: *Medische ethiek.* Bohn Stafleu van Loghum, Houten, 1998.

2.2.3. French:

G. Durand: *Introduction generale a la bioethique: Histoire, concepts et outils.* Fides/Cerf, Montreal, 1999.

D. Folscheid, B. Feuillet-le-Mintier, J.F. Mattei: *Philosophie, ethique et droit de la medecine.* Presses Universitaires de France, Paris, 1997.

G. Hottois: *Les mots de la bioethique.* De Boeck, Bruxelles, 1993 (revised edition in 2001).

2.2.4. German:

E. Loewy: *Ethische Fragen in der Medizin.* Springer Verlag, Wien, 1995.

A. Bondolfi & H. Muller: *Medizinische Ethik im arztlichen Alltag.* EMH Schweizerischer Arzteverlag AG, Zurich, 1999.

G. Poeltner: *Grundkurs Medizinethik.* Uni-TB, Stuttgart, 2001.

2.2.5. Hungarian:

B. Blasszauer: *Orvosi Etika.* Medicina Konyvkiado Rt, Budapest, 1999, 2nd ed.

J. Kovacs: *A modern orvosi etika alapjai - Bevezetes a bioetikaba.* Medicina Konyvkiado Rt, Budapest, 1997.

2.2.6. Italian:

E. Sgreccia: *Manuale di bioetica.* Vol. 1: *Fondamenti di etica biomedica.* Vita e Pensiero, Milano, 1999.

C. Viafora: *Fondamenti di bioetica.* Ambrosiana, Milano, 1989.

M. Mori (ed.): *Questioni di bioetica.* Editori Riuniti, Roma, 1988.

2.2.7. Polish:

T. Slipko: *Granice zycia. Dylematy wspolczesnej bioetyki.* Akademia Teologii Katolickiej, Warszawa, 1988.

Z. Szawarski (ed.): *W kregu zycia I smierci. Moralne problemy medycyny wspolczesnej.* KiW, Warszawa, 1987.

2.2.8. Russian:

B. Iudin and P. Tishchenko (eds.): *Introduction to Bioethics.* Progress-Tradition, Moscow, 1998.

A. Orlov (ed.): *Foundations of Bioethics.* Ofset, Krasnoiarsk, 1997.

A. Bartko and E. Mikhailovska-Karlova: *Biomedical ethics: theory, principles and problems.* MMSI, Moscow, 1999 2nd ed.; 1995 1st ed.

2.2.9. Slovakian:

L. Soltes, *et al.*: *Vybrane kapitoly z medicinskej etiky.* Comenius University, Bratislava, 1994.

J. Glasa and L. Soltes (eds.): *Osetrovatelska etika 1.* Martin, Osveta, 1998.

2.2.10. Slovenian:

M. Dolenc: *Medical Ethics and Deontology II.* Mihelac, Ljubljana, 1997.

2.2.11. Swedish:

H. Fagerberg, E. Bischofberger, L. Jaconsson and G. Lindmark: *Medicinsk etik och manniskosyn.* Liber, Stockholm, 1984.

T. Tannsjo: *Vardetik.* Raben & Sjogren, Stockholm, 1990.

2.2.12. Turkish:

E.M. Atabek: *Tibbi Deontoloji Konulari.* Yenilik Basimevi, Istanbul, 1983 (new edition with assistance of M.Deger, published in 2000).

A. Erdemir, Y. Oguz, O. Elcioglu and H. Dogan (eds.): *Klinik Etik.* Nobel Tip Kitabevi, Istanbul, 2000.

3. SCIENTIFIC JOURNALS

The number of scientific journals in bioethics has increased rapidly. In many countries, one or more journals in the national language are published. In this section, we will include what may be regarded the most prestigious journals, using two related criteria:

1. international journals published in the English language. These journals are not only widely distributed and available throughout the world, but they also allow professional communication and international exchange of arguments and opinions.

2. peer-reviewed journals. The editorial policy of these journals is in agreement with the stringent requirements of the best scientific journals in other disciplines. Submitted manuscripts are anonymously evaluated by at least two scholars. Depending on the outcome of peer-review evaluation, decisions are made about publication.

Bioethics journals that comply with both criteria are subdivided in two categories. The first category are so-called indexed journals. If a journal exists for some time, and if it publishes interesting and useful materials, it will have an impact on the scientific community. Colleagues will refer to the articles in the journal, and cite its publications. Citations (reflected in the Impact Factor of journals) are more and more important since medical schools use citation analysis as an indicator of the quantitative value of research. Several bioethics journals are indexed in the Science Citation Index or Social Sciences Citation Index.

The second category includes journals that, although they are international and peer-reviewed, do not belong to the most cited journals in the field. This is due to fact that they either have come into existence recently (citations are impossible) or address a rather small readership.

Finally, we will include the most important journal in several national languages, in order to provide a starting-point for those wanting to orientate themselves within the area of bioethics.

3.1. International Journals with Impact Factor

Bioethics

- aims and scope: The journal is the official journal of the International Association of Bioethics. It seeks to publish rigorously argued articles discussing ethical issues raised by medicine and the biological sciences.
- 4 issues
- first volume: 1987
- editors: Helga Kuhse and Peter Singer (Clayton, Australia)
- publisher: Basil Blackwell Ltd, 108 Cowley Road, Oxford OX4 1JF, United Kingdom.

Cambridge Quarterly of Healthcare Ethics

- aims and scope: This journal explores both broad issues in health care and society and organisational concerns that arise in institutions where ethics committees work. To respond to the diverse needs of ethics committee members, the journal publishes articles devoted to medicine, law, philosophy, economics, research, theology, education, and behavioural and social sciences, with a focus on practical application in committee settings.

- 4 issues
- first volume: 1992
- editors: David Thomasma (Chicago), Thomasine Kushner (Berkeley) and Steve Heilig (San Francisco)
- publisher: Cambridge University Press, The Edinburgh Building, Shaftesbury Road, Cambridge CB2 2RU, United Kingdom.

Hastings Center Report

- aims and scope: As one of the first periodicals in bioethics, the journal is published by the Hastings Center, a nonprofit, nonpartisan organisation that carries out educational and research programmes on ethical issues in medicine, health care, technology, and the environment. Its contents are varied: scientific contributions, case studies, thematic supplements, reviews.
- 6 issues
- first volume: 1971
- editor: Gregory E. Kaebnick (Garrison, NY)
- publisher: The Hastings Center, Garrison, NY 10524-555, U.S.A.

Health Care Analysis

- aims and scope: The journal is explicitly focused on the interactions of health care, philosophy and policy. It promotes debate about the fundamental rationale of all aspects of health systems and health care provision, including: public policy and health; health-related education; health services organisation and decision-making; health care professional practice.
- 4 issues
- first volume: 1993
- editor: Alan Cribb (London)
- publisher: Kluwer Academic Publishers, P.O. Box 322, 3300 AH Dordrecht, The Netherlands.

Journal of Clinical Ethics

- aims and scope: The journal has been focused on the clinical context. Rather than theoretical expositions, it wants to provide practical recommendations and assistance, and to give voice to moral concerns and case analyses of practitioners. Each issue usually has a special section focused on a particular topic, besides usual rubrics as cases, personal perspectives, research, practice, law.
- 4 issues
- first volume: 1990
- editor: Edmund G. Howe (Bethesda)

- publisher: The Journal of Clinical Ethics, Inc., 107 East Church Street, Frederick, Md., 21701 U.S.A.

Journal of Medical Ethics
- aims and scope: The journal has been established to promote the study of contemporary medico-moral problems. It includes papers on all aspects of health care ethics, analyses ethical concepts and theories and features case conferences and comments on clinical practice.
- 6 issues
- first volume: 1975
- editor: Julian Savulescu (London)
- publisher: BMJ Publishing Group, Tavistock Square, London WC1H 9JR, United Kingdom.

Journal of Medicine and Philosophy
- aims and scope: The journal explores the shared themes and concerns of philosophy, the health care sciences and professions. It aims to provide an ongoing forum for the discussion of these themes and issues. Almost all issues are thematic issues.
- 6 issues:
- first volume: 1976
- editor: H. Tristram Engelhardt (Houston)
- publisher: Swets & Zeitlinger, P.O. Box 825, 2160 SZ Lisse, The Netherlands.

Kennedy Institute of Ethics Journal
- aims and scope: An interdisciplinary journal for the members of the Kennedy Institute of Ethics, founded in 1971 as a research and teaching centre of Georgetown University, U.S.A.. It publishes opinion and analysis dealing with social, ethical, and public policy aspects of bioethics and related areas of applied ethics.
- 4 issues
- first volume: 1991
- editors: Carol Mason Spicer and Robert M. Veatch (Washington, D.C.)
- publisher: The Johns Hopkins University Press, 2715 North Charles Street, Baltimore, Maryland 21218-4319, U.S.A.

Theoretical Medicine and Bioethics
- aims and scope: The journal is a forum for interdisciplinary studies in the philosophy and methodology of medical practice and research. It mainly publishes issues focused on specific themes.
- 6 issues
- first volume: 1981

- editor: David C. Thomasma (Chicago)
- publisher: Kluwer Academic Publishers, P.O. Box 322, 3300 AH Dordrecht, The Netherlands.

3.2. International Refereed Journals

Ethical Perspectives

- aims and scope: This quarterly review is the journal of the European Ethics Network. The purpose of the Network is the promotion of cooperation and integration between university centres, departments, organisations and networks of professional ethics. It is not exclusively focused on medicine and health care, but also addresses ethical problems in business, law, politics, civil service, biotechnology and the media.
- 4 issues
- first volume: 1994
- editor: Bart Pattyn (Leuven)
- publisher: Peeters, Journals Department, Bondgenotenlaan 153, B-3000 Leuven, Belgium.

HEC Forum

- aims and scope: Health care Ethics Committee Forum is an interprofessional publication featuring original contributions of interest to practising physicians, nurses, social workers, risk managers, attorneys, ethicists, and other HEC members. The journal is specifically addressed to committee members and lay readers, with a variety of sections (documents, cases, critical annotations, health law).
- 6 issues
- first volume: 1989
- editors: Stuart F. Spicker (Boston) and Judith W. Ross (Orange)
- publisher: Kluwer Academic Publishers, P.O. Box 322, 3300 AH Dordrecht, The Netherlands.

Medicine, Health Care and Philosophy

- aims and scope: This is the official journal of the European Society for Philosophy of Medicine and Health Care. It is a forum for international exchange of research data, theories, reports and opinions in the discipline of bioethics, and the philosophy of medicine and health care in general. Particular attention is paid to contributions from all European countries, and to making accessible scientific work and reports from the practice of health care ethics, from all nations, cultures and language areas in Europe.
- 3 issues
- first volume: 1998

- editor: Henk ten Have (Nijmegen)
- publisher: Kluwer Academic Publishers, P.O. Box 322, 3300 AH Dordrecht, The Netherlands.

3.3. National Journals (in various European languages)

3.3.1. Danish:
although there is not a special bioethics journal, articles on medical ethics are frequently published in the journal of the Danish Medical Association: *Bibliotek for Laeger* (since 1809; 8 issues per year).

3.3.2. Dutch:
Tijdschrift voor Geneeskunde en Ethiek (since 1991, quarterly)

3.3.3. French:
Les Cahiers du Comite Consultatif National d'Ethique pour les sciences de la vie et de la sante (since October 1994; 4 issues per year;
Email: editass@club-internet.fr)
Ethica Clinica (Belgium) (since March 1996; quarterly;
Email: fih-w@openweb.be)
Journal International de Bioethique (since 1990; quarterly)
La lettre de l'Espace ethique (since 1996; quarterly;
Email: espace.ethique@sls.ap-hop-paris.fr)
Laennec: medicine-sante-ethique (since 1954; quarterly;
Email: Centre.Laennec@wanadoo.fr)

3.3.4. German:
Zeitschrift fur medizinische Ethik (since 1954; quarterly)
Ethik in der Medizin (since 1989; quarterly)

3.3.5. Hungarian:
Magyar Bioetikai Szemle (since 1994, quarterly)

3.3.6. Italian:
Bioetica (since 1993; 3 issues annually
Medicina e Morale (since 1950; 6 issues per year)
Bioetica e Culture (since 1992; 2 issues per year)
L'Arco di Giano (since 1993; 3 issues per year)

3.3.7. Polish:
Etyka (since 1966, yearbook)

3.3.8. Slovakian:
Medicinska etika & Bioetika (since 1994, quarterly; http://www.elis.sk/etika/etika.htm)
Acta medica christiana slovaca (since 1994; quarterly; http://www.elis.sk/christ/amcs.htm)

3.3.9. Swedish:
There is not a special bioethics journal, but medical ethics contributions appear in *Filosofisk Tidskrift*.

3.3.10. Turkish:
Tip Etigi (since 1993; initially 3 issues a year; since 1999 2 issues per year; Email: saksoy@harran.edu.tr)

4. BIOETHICS ONLINE

4.1. Bioethicsline

The Kennedy Institute of Ethics at Georgetown University produces an online database for bioethics literature which is now available free on the web at: http://guweb.georgetown.edu/nrcbl/.
Bioethicsline is part of the National Library of Medicine's MEDLARS system of searchable databases, and has over 60,000 references to English language bioethics literature, covering the health sciences, law, religion, philosophy, and the social sciences. Searches can be conducted by subject, title, or author.

4.2. Euroethics

Four European countries (Germany, France, Netherlands and Sweden) are cooperating for the establishment of a European Data Base Network in the field of ethics in medicine, health care and health professions. This standardized European information and Data Base Network aims at providing means to facilitate (1) an exchange of information across European borders, (2) comparison and analysis of different ethical viewpoints and standards between European countries, (3) the realisation of a European consensus. In the cooperating countries, national databases have been established, using the same thesaurus and format; these databases are joined together into the unified database 'Euroethics'. Search forms are available in 5 languages (English, Dutch, French, German and Swedish).
Online access via Internet: http://www.gwdg.de/~uelsner/euroeth.htm

5. SCIENTIFIC ASSOCIATIONS

Scientists need to communicate. First of all, journal publications aim at making available the results of research and the state of the art in particular areas or concerning specific topics. A second major instrument of communication is through conferences, seminars, website discussions, organized by associations and societies. Over the years, in bioethics a growing number of associations and societies have been established. The purposes are often quite similar: regular meetings and conferences to share the results of research and exchange views, sometimes in combination with publication of a journal or newsletter. The membership may be different, according to geographical orientation, professional background or areas of common interests. Here, we will mention societies and associations with an international membership.

Akademie für Ethik in der Medizin (AEM):
German-speaking philosophers, theologians and physicians founded this association in 1986. It is the largest German organisation of its kind in the field of ethics in medicine, with more than 300 members. The AEM is an institute at the University of Göttingen; it works in close co-operation with the faculty of medicine and other faculties and institutions of the university. Membership is awarded after selection of individuals following proposals by colleagues. The Academy has annual conferences focused on specific themes. It also publishes the quarterly scientific journal *Ethik in der Medizin*. Special committees, nominated by the board, prepare reports on particular topics (e.g., ethics teaching in the German medical curriculum). In 1992, the AEM set up an Information and Documentation Centre for Ethics in Medicine in Göttingen. This includes among others the literature data base 'Ethmed' (included in Euroethics) as a supplement to *Bioethicsline*.
Executive Office: AEM Secretariat, Dr. Alfred Simon, Institut für Geschichte der Medizin der Georg-August-Universität, Humboldtallee 36, D-37073 Göttingen, Germany. Email: simon@ethik.med.uni-goettingen.de

American Society for Bioethics and Humanities (ASBH)
This Society was established in 1998 as the successor of three separate North American societies, the Society for Health and Human Values (SHHV), the Society for Bioethics Consultation (SBC) and the American Association of Bioethics (AAB)
Address: ASBH, 4700 West Lake Avenue, Glenview, Illinois 60025-1485, USA. Email: info@asbh.org

Central and East European Association of Bioethics (CEEAB)
This association was founded in 1999, during a meeting of scholars from Croatia, Hungary, Lithuania, Slovakia, and Russia in the Hungarian city of Pecs. Membership is primarily from the ex-socialist countries. CEEAB, as a 'non-for-profit' and non-political public organisation, is based on a strong dedication of its members to universal values of human rights and respect for human dignity. The basis for specific CEEAB goals and tasks is found in the common economic, political and ideological problems of the transitional period in ex-socialist countries. The association aims to consolidate the efforts of scholars, medical professionals and lay people from ex-socialist and other countries in areas of reformation of health care systems, teaching of medical ethics and the philosophy of medicine, improvement of health care policies and legislation, protection of patients' rights, and the development of bioethical research.
Executive Office: CEEAB Secretariat, Maria u. 9, H-7621 Pecs, Hungary; Email: fact@mail.matav.hu

European Association of Centres of Medical Ethics (EACME)
This association was established in 1985. With almost 60 institutional members, it unites many centres, departments and institutes of bioethics throughout Europe. Full membership is only open to institutions, departments, units showing evidence of actual experience in the field of medical ethics, collaborative work, contributions to research and publications. Individuals can join as associate members. The official languages are French and English. The general aims of EACME are: creating an international research and communication network; providing members with information concerning ethical and health problems, developing research tools, contributing to the European debate on bioethics. Since its first conference in London (1987), the association organises annual scientific meetings focused on specific topics (e.g., ethics and mental health in 1998; genetics and ethics in 1999). Each member receives the monthly journal *The Bulletin of Medical Ethics*, which is the source of the most current information on what is happening in medical ethics and bioethics in Europe (editor: Richard Nicholson, London).
Executive Office: EACME Secretariat, Prof.dr. Guy Widdershoven, Department of Health Ethics, University of Maastricht, P.O. Box 616, 6200 MD Maastricht, The Netherlands.

European Society for Philosophy of Medicine and Health Care (ESPMH)
This society was founded in Maastricht, the Netherlands, in 1987. It aims at stimulating and promoting the development and methodology in the field of philosophy of medicine and health care in a wide sense; it also aims at facilitating contacts for European scholars in this field and to promote

international contacts between members in the various countries of Europe. The society pursues its goals through organizing annual conferences, publishing reports and materials, and liaising with other organisations. Since its inaugural meeting in 1987, annual conferences have been held in various European countries; each conference is focused on a particular theme (e.g., Medicine and culture, Greece, 1995; Research in health care, 1997, Italy; Epistemology and medicine, 1999, Sweden). Cooperating with a coalition of other organisations, ESPMH initiated the first world congress 'Medicine and Philosophy' in Paris (1994). The second world conference has taken place in Cracow, Poland in 2000.
Membership is open for individual persons and institutions. Over 400 physicians, philosophers, lawyers, nurses from all European countries have joined. Each member receives the journal *Medicine, Health Care and Philosophy*, published in 3 issues annually.
Executive Office: ESPMH Secretariat, Prof.dr. Henk ten Have, Department of Ethics, Philosophy and History of Medicine, University Medical Centre Nijmegen, P.O. Box 9101, 6500 HB Nijmegen, The Netherlands.

Hans Jonas Gesellschaft. Verein zur Förderung der Ethik in der Medizin
Established in Vienna, Austria, in 1993, this society specifically aims at fostering the interdisciplinary and inter-cultural dialogue between medicine, biology and ethics. The Society finds inspiration in the ideas of the philosopher Hans Jonas (1903-1993). Born in Germany, and disciple of Heidegger and Bultmann, Jonas taught in Israel, Canada and the U.S.A. In his work, while increasingly focusing on bioethics, he developed an ethics of responsibility, taken into account the advancements of technology and science. The society, commemorating his name, multi-professionality and multi-culturality, organized its first conference in Vienna on the subject of conscience (*Das Gewissen der Medizin*).
Executive Office: Hans Jonas Gesellschaft, Erdberger Lände 20/16, A-1030 Vienna, Austria.

International Association of Bioethics (IAB):
In 1990, the Australian bioethicists Singer and Kuhse took the initiative to establish this association. The IAB aims to be international, linking all those working in bioethics and related fields, facilitating mutual contact, and encouraging the discussion of cross-cultural aspects in bioethics.
The association organizes biennial world conferences (1996, San Francisco; 1998, Tokyo; 2000, London). The official journal of the association is *Bioethics*.
Executive Office: IAB Secretariat, Dr. Hans van Delden, Centre for Bioethics and Health Law, Utrecht University, P.O. Box 80105, 3508 TC Utrecht, The Netherlands. Email: IAB@ggl.ruu.nl

Societas Ethica. European society for research in ethics:
This is one of the oldest associations of ethicists in Europe. Founded in 1964, the membership of approximately 250 unites philosophers and theologians from more than 20 European countries, the majority Nordic and German-speaking. The languages of the society are German and English. The aim is to bring together university teachers and researchers to discuss current issues in ethics. The discussion should combine fundamental problems of philosophical and theological ethics with specific issues of applied ethics. Annual conferences since 1964 have been focused on particular topics (1997: Poland: Solidarity and community; 1998, Finland: Ethics and legislation; 1999, Italy: Ethics and emotions). Proceedings are published separately as *Jahresbericht*.
Executive Office: Societas Ethica, Address: Prof.Dr.theol. Svend Andersen, Dept. of Systematic Theology, Bygning 410, Univ. Aarhus, Hovedbygniden, Ringgade 1, DK-8000 Aarhus C, Denemarken.
Email: andersen@teologi.aau.dk

6. RESEARCH FACILITIES

In Europe, the major funding agency of transnational research in the area of bioethics today is the European Commission. The research policy of the European Community began in the 1970s; it has been organised in a multi-annual pattern (with a planning instrument called Framework Programme) since 1984. The European Parliament initiated ethics studies in the Second Framework Programme (1987-1991). As an ad-hoc measure, focusing on the Human Genome programme in 1990, 18 studies were selected which were granted limited funding for one year. In the Third Framework Programme (1991-1994) research in bioethics was extended. A new sub-area on Medical Ethics was included in the specific programme in the area of Biomedicine. On various topics 15 concerted actions involving 135 teams from different member states were implemented. Also the specific programme in the area of Biotechnology included support for a series of studies aiming at assessment of socio-economic impacts of biotechnologies; 26 projects were selected.

In the Fourth Framework Programme (1994-1998) bioethics research was further extended to the three specific programmes in the field of Life Sciences and Technologies. These programmes have an ELSA sub-area. ELSA stands for Ethical, Legal and Social aspects. About 30 million ECU in this period was spent as the EC participation on research activities on the societal issues concerning regulation and desirability of the life sciences and technologies.

In the Fifth Framework Programme, launched in 1999, and bringing EC research into the new millennium, both research into bioethics and ethical review of research are major features.

A catalogue of all projects funded by the European Commission has recently been published: European Commission, 1998: *Ethical, Legal and Social Aspects of the Life Sciences and Technologies Programmes of Framework Programme IV (EUR 18309).* Office for Official Publications of the European Communities, Luxembourg.

Information: ELSA implementing Unit, DG XII/E.5, European Commission, rue de la Loi 200 (SDME 8/3), B-1049 Brussels, Belgium.

INDEX

INDEX OF NAMES

INDEX OF SUBJECTS

International Library of Ethics, Law, and the New Medicine

1. L. Nordenfelt: *Action, Ability and Health.* Essays in the Philosophy of Action and Welfare. 2000 ISBN 0-7923-6206-3
2. J. Bergsma and D.C. Thomasma: *Autonomy and Clinical Medicine.* Renewing the Health Professional Relation with the Patient. 2000 ISBN 0-7923-6207-1
3. S. Rinken: *The AIDS Crisis and the Modern Self.* Biographical Self-Construction in the Awareness of Finitude. 2000 ISBN 0-7923-6371-X
4. M. Verweij: *Preventive Medicine Between Obligation and Aspiration.* 2000 ISBN 0-7923-6691-3
5. F. Svenaeus: *The Hermeneutics of Medicine and the Phenomenology of Health.* Steps Towards a Philosophy of Medical Practice. 2001 ISBN 0-7923-6757-X
6. D.M. Vukadinovich and S.L. Krinsky: *Ethics and Law in Modern Medicine.* Hypothetical Case Studies. 2001 ISBN 1-4020-0088-X
7. D.C. Thomasma, D.N. Weisstub and C. Hervé (eds.): *Personhood and Healt Care.* 2001 ISBN 1-4020-0098-7
8. H. ten Have and B. Gordijn (eds.): *Bioethics in a European Perspective.* 2001 ISBN 1-4020-0126-6

KLUWER ACADEMIC PUBLISHERS – DORDRECHT / BOSTON / LONDON

International Library of Ethics, Law, and the New Medicine

KLUWER ACADEMIC PUBLISHERS – DORDRECHT / BOSTON / LONDON

MIX
Papier aus verantwortungsvollen Quellen
Paper from responsible sources
FSC® C105338

If you have any concerns about our products,
you can contact us on
ProductSafety@springernature.com

In case Publisher is established outside the EU,
the EU authorized representative is:
Springer Nature Customer Service Center GmbH
Europaplatz 3, 69115 Heidelberg, Germany

Printed by Libri Plureos GmbH
in Hamburg, Germany